Handbook of
TEXTILE FIBRES

## II. MAN-MADE FIBRES

# Handbook of
# TEXTILE FIBRES

### By
## J. GORDON COOK
**BSc, PhD, CChem, FRSC**

## II. MAN-MADE FIBRES

**WOODHEAD PUBLISHING LIMITED**

Cambridge     New Delhi

Published by Woodhead Publishing Limited, Abington Hall, Granta Park,
Great Abington, Cambridge CB21 6AH, England
www.woodheadpublishing.com

Woodhead Publishing India Pvt Ltd, G-2, Vardaan House, 7/28 Ansari Road,
Daryaganj, New Delhi – 110002, India

Formerly published by Merrow Publishing Co Ltd
First published 1959
Second Edition 1960
Third Edition 1964
Fourth Edition 1968
Reprinted 1974
Fifth Edition 1984
Reprinted 1993, 2001, 2002, 2009

British Library Cataloguing in Publication Data
A catalogue record for this book is available from the British Library.

ISBN 978-1-85573-485-2

# FOREWORD

The manufacture of textiles is one of the oldest and most important industries of all. Its raw materials are fibres, and the study of textiles therefore begins with an understanding of the fibres from which modern textiles are made.

In this book, an outline is given of the history, production and fundamental properties of important textile fibres in use today. The behaviour of each fibre as it affects the nature of its fabric is discussed.

The book is in two volumes. Volume I deals with the natural fibres on which we depended for our textiles until comparatively recent times. Volume II is concerned with man-made fibres, including rayons and other natural polymer fibres, and the true synthetic fibres which have made such rapid progress in modern times.

The book has been written for all concerned with the textile trade who require a background of information on fibres to help them in their work. Every effort has been made to ensure that the text is accurate and up-to-date. The information on man-made fibres is based on facts supplied by the manufacturers of the fibres themselves.

In writing this book I have been given much encouragement and help by many individuals and organizations. The manufacturers of the man-made fibres mentioned in the text have gone to great trouble on my behalf in providing information and in checking the text before publication. I would like to acknowledge their help, with grateful thanks, and also that given to me by the following individuals and organizations:

D. A. Derrett-Smith, Esq., B.Sc., F.R.I.C., Linen Industry Research Association.
Dr C. H. Fisher, U.S. Dept. of Agriculture.
The Cotton Board, Manchester.
Stanley B. Hunt, Textile Economics Bureau.
Dr R. J. W. Reynolds, I.C.I. Dyestuffs Division.
Mr H. Sagar, I.C.I. Dyestuffs Division.
H. L. Parsons, Esq., B.Sc., F.R.I.C., Low and Bonar Ltd.
L. G. Noon, Esq., Wigglesworth and Co. Ltd.

Silk and Rayon Users Association.
J. C. Dickinson, Esq., International Wool Secretariat.
W. R. Beath, Esq., and his colleagues; Courtaulds Ltd.
E. Lord, Esq., B.Sc., Cotton Silk and Man-Made Fibres Research Association.
K. J. Brookfield, Esq., Fibreglass Ltd.
F. H. Clayton, Esq., Wm. Frost Ltd.
Burlington Industries Inc.                                    J.G.C.

# NOTE ON THE FIFTH EDITION

The man-made fibre industry has expanded greatly since the fourth edition of *Handbook of Textile Fibres* was published. Many new fibres have come into production in countries throughout the world, but the emphasis has been largely on development and modification of established fibre classes, rather than upon the introduction of fibres of new chemical types.

Within almost every chemical class there is now a family of fibres displaying a range of properties and applications limited only by the fundamental chemical structure of the fibre class. To include detailed information about every fibre in production would have meant producing a book of unmanageable and uneconomic size. In this volume, therefore, I have provided background information about each chemical class of fibre, based usually upon a fibre in current production which exemplifies its chemical class. More specific information about individual·fibres will be found in a supplementary volume.

Since the fourth edition was published, production of some classes of fibre has been suspended. I have, however, retained information about these fibre classes; they are of technical and historical interest, and there is always the possibility that production of these fibres may restart to meet changing economic and technical circumstances.

As in previous editions, I have been given much valued assistance by fibre manufacturers and textile organisations throughout the world. Many individuals have gone to great trouble on my behalf by providing information and checking the text before publication. I would like to acknowledge their help with grateful thanks.

                                                           J.G.C.

# CONTENTS

CONTENTS

# INTRODUCTION

## FUNDAMENTALS OF FIBRE STRUCTURE

During the last half-century, all the familiar materials that the world has been using for thousands of years have come under the microscope. Science has opened up a great era of exploration which is probing into the nature of material things. We want to know why different forms of matter behave as they do: and to find our answers we have had to study the atoms and molecules from which materials are made.

In this respect, natural fibres have proved to be one of the most interesting fields of modern scientific research. As raw materials of one of the world's greatest industries, and as peculiar forms of matter in their own right, fibres have long excited the curiosity of scientists. Now, research into the chemistry and physics of fibres has provided a satisfying explanation of the unusual and invaluable properties that they possess.

### Thread-like Molecules

All fibres have been found to share one thing in common; the fundamental particles, the molecules, are always long and thread-like. That is to say, the molecules of fibrous matter are in the form of hundreds or even thousands of individual atoms strung together one after the other. The molecule of cellulose, for example, is built up by the plant from hundreds or more of small glucose molecules, each of which in turn contains six carbon atoms. The cellulose molecule, therefore, is in the form of a long thin chain of atoms.

The molecules of a fibre are thus in shape very similar to the fibre itself. And just as the fibre bestows its characteristics on the yarn of which it forms a single strand, so does the fibre derive its properties from the thread-like molecules of the substance from which it is made.

One of the most outstanding properties of a fibre is its strength. Relative to its cross-sectional area, the strength of a silk fibre, for example, is extraordinarily high. A single strand, so fine as to be almost invisible to the naked eye, will support a weight of several grams. Yet, at the same time, this filament is flexible and resilient.

In silk, as in other natural fibres, the thread-like molecules tend to lie along the direction of the fibre itself; they are aligned in one direction like sticks in a bundle of faggots. It is almost as though the silkworm, in extruding its silk, the sheep in growing its wool fibres, and the plant in producing its cotton and flax could align the long thin molecules as the fibres are formed.

## Orientation

This orientation of the fibre molecules is not a precise geometrical arrangement. Rather is it a tendency for the majority of the molecules to lie in one direction. The effect on the fibre is analogous to the effect of the individual strands twisted into a rope – each one plays its part in taking up the strain on the rope as a whole.

Long molecules of this sort are a characteristic of the peculiar forms of matter we call plastics and rubbers, as well as fibres. Bu. it is only very special types of long molecule that are able to form fibres. They must, for example, be fairly regular in shape with a 'repeating' pattern of atoms in the molecule. They must not have large pendant groups of atoms sticking out from the sides, or the long molecules are unable to pack together.

When the long molecules are able to pack closely together, they can exert strong forces of attraction between each other. In a fibre the molecules are able to develop these forces, and it is this that is responsible for many of the fibre's characteristic properties.

Inside the fibre the long molecules lying alongside each other pack tightly here and there into their little bundles. But the molecules are so long that they can each be involved in many different close-packed bundles. In between these orderly regions, the fibre molecules run through regions in which the molecules are aligned to some degree along the fibre, but are not aligned with the precision that allows them to pack together into well-ordered bundles.

The effect of this wandering of molecules in and out of regions of tight-packing is that each individual molecule is tied firmly to its neighbours at intervals along its length. In between each 'tied' region there is a sector of freedom and disorder. It is this peculiar molecular arrangement that gives a fibre its combination of strength and flexibility.

Though nature herself produces many different sorts of long molecules, she has used few for making fibres. The differences between the natural fibres are the result of different characteristics in the constituent molecules.

# CRYSTALLINITY

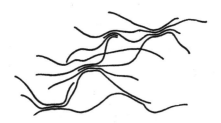

The long molecules of a typical fibre-forming material are able to pack together closely alongside one another, like sticks in a bundle of faggots. The regularity of structure brought about by this arrangement results in regions of crystallinity in the fibre. These are regions in which a number of molecules are aligned in such a way that strong forces of attraction hold the molecules together. The bonds developed in this way are not chemical bonds in the familiar sense, but they are stronger than the normal forces of attraction exerted between individual molecules.

The degree of order introduced by these regions of crystallinity is an important factor in determining the usefulness of a potential fibre. Individual molecules forming part of a region of crystallinity may wander through a tangled mass of molecules in random arrangement, and then form part of another region of crystallinity. In this way, the molecules forming the fibre are arranged into a structure consisting of regions with a high degree of alignment where the molecules hold tightly to each other, and regions of random arrangement where the molecules are not holding tightly to each other. The crystalline regions provide strength and rigidity, and the amorphous regions provide flexibility and reactivity.

The ratio of crystalline to amorphous material has an important influence on the properties of any fibre. In the case of natural fibres, this is an inherent property of the fibre which is fixed by nature. In the case of a man-made fibre, the crystalline-amorphous ratio may be controlled to a large degree by the conditions under which the fibre is produced.

## Cellulose

In the vegetable fibres, cellulose is the material that provides the thread-like molecules. This molecule, built up by nature from smaller glucose molecules, is regular in arrangement, but it is fairly rigid in structure. The cellulose fibres, in consequence, are strong and tough.

At regular intervals along the cellulose molecule there are groups of atoms which tend to attract water. When cotton, for example, is steeped in water, these groups encourage the relatively small water molecules to penetrate between the long thin cellulose molecules. As a result, the fibre structure is loosened up and softened. In a humid atmosphere, therefore, cotton fibres are not so inclined to break – a factor that may have helped the growth of the cotton trade in Lancashire.

This softening effect of water also explains how mercerization can give its special properties to a cotton fabric. The effect of the caustic treatment is to force water molecules into the fibre, making it soft and plastic. The cellulose of the cotton is thus able to 'flow' as it is stretched.

Animal fibres are made from proteins, the class of substances used in the animal world for so many building jobs. Protein molecules are, once again, long, thread-like chains of atoms.

In the plant world, cellulose holds a monopoly in fibre production. Whether it is in cotton or the trunk of a tree, in flax or in the fruit or leaves, the cellulose has the same chemical structure. But the proteins used in the animal world differ widely one from another.

## Proteins

The long molecules of a protein are built up from some twenty or so different types of small amino acid molecule. The proportion and arrangement of these different units determine the structure of the protein molecule and the nature of the protein itself.

By comparison with proteins, the regular cellulose molecule built from its glucose units is simple and straightforward. Protein molecules, with infinite possibilities for the arrangement of their many constituent amino acid units, are exceedingly complex. But chemists have been probing steadily into the mysteries of protein structure during recent years, and we are beginning to understand the intricacies of these complex molecules.

The protein molecules in wool are now regarded as being folded molecules. The long, thread-like chains of atoms do not lie straight alongside each other; they bend backwards and forwards like a meandering stream.

In the ordinary unstretched wool fibre, these folded molecules are arranged alongside each other and lie in the general direction of the fibre itself. When the fibre is stretched, the folds in the molecules are partly straightened out until they cannot unfold any further. The wool fibre is then at the limit of its elasticity.*

*Cross-links*

There is another fundamental difference between the cellulose molecules in cotton and flax, and the protein molecules of wool. Whereas the close-packed cellulose molecules are held together solely by electrostatic forces of attraction, the close-packed wool molecules are actually joined together here and there by chemical links. These 'cross-links' act as extremely strong ties between the molecules. They ensure that when the molecules are stretched out of their normal folded shape, they return to that shape when the stretching force is removed.

One of wool's most important characteristics is its thirst for water. As in the case of cotton, the small water molecules can penetrate between the long wool molecules. Under suitable conditions, wool can absorb half its own weight of water.

Unlike cellulose, however, the protein of wool is attacked fairly readily by water, which causes profound changes in the wool molecule. If wool is stretched, for example, and then heated in boiling water and allowed to cool whilst still stretched, it will remain in its stretched form.

The reason for this is found in the cross-links which join the wool molecules together. Hot water or steam can destroy these links so that the molecules are free to stay in the new positions they reached when the fibre was stretched. Moreover, prolonged heating will actually cause new links to form which anchor the molecules firmly in their new positions.

This is what happens when hair is given a permanent wave. The hair is bent and twisted into its curly shape, which stretches

* Stretching of wool fibres only partly straightens out the folds. Complete transition from α-keratin to β-keratin only takes place on destroying the cross-links.

and distorts the fibres. It is then steamed whilst held in this shape:
the links between the distorted molecules break down and then
rebuild themselves in their new positions. Once this has happened,
the hair fibres are fixed firmly in their new shape.

## Silk Protein

Silk, like wool, is an animal fibre, and it is once again a protein.
But the chemical structure of the silk protein is different from that
of wool. This difference is reflected in the difference between the
two fibres.

Where wool molecules are folded and capable of being stretched
out straight, the silk molecules are in the extended position to
start with. That is why silk possesses little 'returnable' elasticity
after a substantial degree of stretching. When it is stretched with
sufficient force, the molecules have to slide over each other and
do not return to their original positions when the stretching
force is released. The molecules of silk are not joined together
by chemical cross-links as are the molecules of wool.

Research has shown that in silk the protein molecules are
highly orientated – they lie in the direction of the fibre and can
pack tightly together. The forces of attraction between the molecules are thus able to come into play and give the molecular
bundles very great strength.

The effect of heat on silk is similar to its effect on wool. At a
high temperature, silk will burn. But silk is simpler than wool in
its molecular structure. There are no cross-links between the
molecules to break down or rebuild. Silk will thus stand higher
temperatures than wool without taking any harm.

Sericin, the gum that holds the twin strands of silk together,
is a protein similar to that of silk. But the molecules of sericin
protein are not aligned, and the material is thus not fibrous.

This difference between silk fibre and sericin gum is an example
of the requirements for fibre formation. It is not enough that the
molecules of the material should be long and thread-like. To
make a useful fibre, they must be of such a shape and structure
that they can be aligned and packed together alongside each other.

### Effect of Orientation

Differences in the properties of natural fibres of similar chemical
constitution can be explained in part by variations in the state of

alignment of the molecules. Flax and cotton, for example, are chemically almost identical; they are both cellulose fibres. But flax has tensile properties quite different from those of cotton; it has a tenacity of up to 55.6 cN/tex (6.3 g.p.d.), compared with 26.5-44 cN/tex (3-5 g.p.d.) of cotton.

These differences in the tensile properties of cotton and flax are caused by differences in the fine structure of the two fibres. Most important of all are the differences in the degree of alignment of the cellulose molecules themselves. In flax, the molecules are highly orientated along the fibre; they lie alongside one another in the same direction as the fibre itself. In cotton, this degree of alignment is not so high; there are more of the molecules lying 'out of true'.

When the flax fibre is subjected to a stretching force, the aligned molecules combine to resist the force. But in cotton, many more of the molecules are lying at an angle to the long fibre axis and contribute little to the strength of the fibre. In natural fibres, the 'degree of orientation' of the long molecules is controlled by nature, and there is little we can do about it. Nature gives flax and ramie their unusually high tensile strengths by aligning their molecules to a high degree. We have no way of increasing the degree of orientation of the cellulose molecules in cotton in order to make it compare in strength with, flax or ramie.*

In natural fibres, also, the way in which the molecules are aligned is complicated by various factors. The cellulose molecules in flax or cotton are organized in unit bundles or fibrils, which are in turn built up into larger units in the fibre. These molecular bundles are not laid down by the plant in a simple fore-and-aft fashion along the fibre; they are in spiral form.

As would be expected, the nature of the spiral in cellulose fibres has a great influence on the tensile properties of the fibre. The greater the angle of the spiral, the more the fibre can stretch before the aligned molecules have to take up the full strain of the tensile force. Cotton, with an angle of spirality of about 31 degrees, has a much greater elongation at break than flax, with its spiral angle of 5 degrees.

The effect of orientation and spirality can thus be seen quite clearly in the case of natural cellulosic fibres. Cellulose is a reasonably uniform straightforward molecule which consists of

* The degree of orientation of cellulose molecules in cotton can be increased to some extent by mercerization.

## COPOLYMERS

**(A)** RANDOM COPOLYMER  —X—X—Y—X—Y—Y—Y—X—

**(B)** BLOCK COPOLYMER  —X—X—X—X—X—Y—Y—Y—Y—

**(C)** ALTERNATING COPOLYMER —X—Y—X—Y—X—Y—X—Y—

**(D)** GRAFT COPOLYMER  —X—X—Y—Z—Z—Z—Z—X—Y—

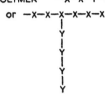

or  —X—X—X—X—X—X

Linear molecules may be produced by polymerization of a mixture of monomers, forming 'copolymers' in which the linear molecule contains two or more types of monomer unit. A number of different types of copolymer may be produced in this way, as shown above.

In a *random copolymer*, the monomer units are linked together in random fashion. In a *block copolymer*, one or more of the components may be polymerized to form sections of molecule containing only one type of monomer unit. These 'blocks' are linked together to form the linear molecule. In an *alternating copolymer*, the monomer units alternate in sequence along the linear molecule. In a *graft copolymer*, a block of third component may be grafted on to the linear molecule, forming part of the molecular chain itself, or forming a side chain.

---

many repeating units of smaller glucose sections. But the proteins from which the animal fibres are made are much more complex in their detailed chemical arrangement. The orientation of protein molecules such as those of wool, for example, is complicated by the existence of chemical bridges between the molecules, and by the folded state in which the molecules lie.

### Control of Orientation

We can do little to modify the degree of orientation of the molecules in natural fibres. But in making semi-synthetic or synthetic fibres the alignment of long molecules is an essential step in manufacture. We are able to control the alignment in such a way as to exert a major influence on the properties of the fibre itself.

The first step in making a synthetic or semi-synthetic fibre is to obtain a substance with the requisite long thread-like molecules. In the case of a synthetic fibre, this substance is built up from simpler chemicals; in the case of a semi-synthetic fibre, such as a rayon, the substance has been made by nature.

## MAN-MADE FIBRE YARNS

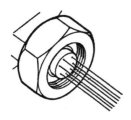

Man-made fibres are made by extrusion of fibre-forming substances in liquid form (molten or in solution) through fine holes in a spinneret. The jets of liquid are hardened in one of several ways to form solid filaments. These are drawn or stretched and may be twisted slightly together to form yarns of virtually any desired length, which are known as *continuous filament yarns*. The filaments may also be collected together into a thick rope or tow and then cut into short lengths to form staple fibre; this may be combed, attenuated and spun into yarns by techniques similar to those used for natural staple fibres such as cotton or wool, forming *staple or spun yarns*.

*Continuous Filament Yarns.* These consist of unbroken filaments which are held together into a yarn by a slight twist. They are smooth and generally compact and are used for satins, poults, taffetas, failles and similar fabrics.

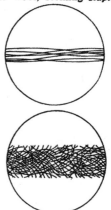

*Spun or Staple Yarns.* These consist of short fibres held together by the twist given to an attenuated strand of fibres. They are generally much fuller in handle than continuous filament yarns. The short fibres lie at various angles with respect to the long axis of the yarn, the degree of uniformity depending upon the combing and other treatments given to the fibre strands before being twisted together. The surface of a spun yarn is rougher to the touch, owing to the fibre-ends protruding from it, and spun yarns are in general fuller and warmer than continuous filament yarns. They are used for sports shirts, suitings, sheets, blankets, furnishing and other fabrics.

In its 'raw' state, a fibre-forming substance may be little more than an amorphous material such as, for example, the powdered casein from milk, the cellulose of wood pulp or the ribbon of tough horny plastic extruded from the nylon-manufacturing plants.

In this bulk-material, the long thread-like molecules are mixed up one with another in more or less random fashion, like the mass of fibres in a bundle of cotton wool. In order to turn the material into a textile fibre, we have to (a) shape it into the usual fibre form, i.e. a long, uniform rod of extremely fine cross-section, and (b) ensure that the long molecules of the material are aligned so that they tend to lie alongside one another in the same direction as the long axis of the fibre itself.

## Spinning

The first stage of fibre-production is carried out by rendering the mass of fibre-forming material into a liquid or semi-liquid state. This can be done either by dissolving the material in a solvent, or by heating it until it melts. In either case, the long molecules are freed from close entanglement with each other, and can move independently.

The liquid containing the fibre-forming material is then extruded through very small holes so that it emerges as fine jets of liquid. These jets are hardened, forming a solid rod which possesses all the superficial characteristics of a long filament such as silk.

In the production of man-made fibres, the extrusion of liquid fibre-forming material, followed by hardening of the fine jets to form filaments, is described as 'spinning'. It is similar to the 'spinning' process used by the silkworm or the spider, resulting as it does in the production of continuous filaments.

The hardening of the jets from the spinneret may be carried out in one of several ways:

(1) *Wet Spinning*. The solution of fibre-forming material may be extruded into an aqueous coagulating bath in which the jets are hardened. as a result of chemical or physical change.

Viscose, for example, is wet spun, the solution of cellulose xanthate being extruded into an aqueous solution of acids and salts. Cellulose is regenerated, and this is insoluble in water, forming solid filaments.

## SPINNING

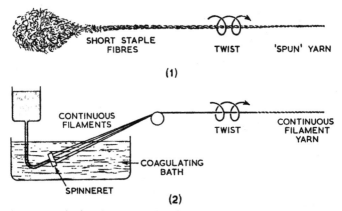

SHORT STAPLE
FIBRES

TWIST

'SPUN' YARN

**(1)**

CONTINUOUS
FILAMENTS

TWIST

CONTINUOUS
FILAMENT
YARN

COAGULATING
BATH

SPINNERET

**(2)**

The term 'spinning' is used to describe any process used in producing continuous yarns or threads. Confusion is often caused by the fact that it refers to two different techniques, each of which achieves the above result.

1. In its original textile sense, spinning is the process which has been used for thousands of years for converting a mass of short 'staple' fibre, such as cotton or wool, into continuous yarns. The mass of fibre is drawn out into strands, which may be subjected to combing and other processes, and the attenuated strands are then twisted so that the fibres grip each other to form a 'spun' yarn.

2. In making man-made fibres a liquid fibre-forming material is extruded from holes in a spinneret. The jets of liquid are hardened as they emerge, forming continuous filaments which may be virtually any length. A number of them twisted together form a yarn. This process, too, is called spinning.

Continuous filament yarns 'spun' in this way may be cut into short lengths of staple fibre, and this may then be 'spun' into yarns in the same way as cotton or other staple fibres.

---

(2) *Dry Spinning*. The fibre-forming substance may be dissolved in a solvent. As the jets of solution emerge from the spinneret, the solvent is evaporated, e.g. by a stream of hot air, leaving solid filaments. Acetate, for example, is dry spun by extruding acetone solutions of cellulose acetate into hot air.

(3) *Melt Spinning*. The fibre-forming material may be rendered liquid by heating it until it melts. The molten material is extruded

## STRETCHING

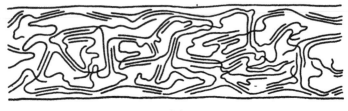

(A)

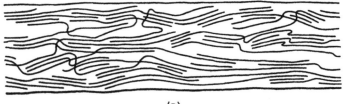

(B)

*Stretching (drawing) and Alignment.* The extrusion of fibre-forming material brings about some slight degree of orientation of the linear molecules in the direction of the fibre axis. This is most pronounced near the outer surface of the filament (the 'skin effect') (A).

The subsequent stretching or drawing of the filament continues the alignment of the molecules throughout the bulk of the filament material. The crystalline regions are orientated in the direction of the fibre long axis, and the molecules in the amorphous region are brought into greater alignment, increasing the degree of crystallinity of the material (B). The properties of the fibre are greatly influenced by the amount of stretch to which the filaments are subjected.

through spinnerets, and the jets harden as they cool on emerging from the spinneret. Nylon and 'Terylene', for example, are melt spun.

### Skin Effect

The extrusion process brings about some orientation of the long molecules inside the filament. This is especially pronounced on the outer surface of the filament, where the molecules have been influenced by the edges of the spinneret hole.

It is now established that the surface of an extruded filament is usually more highly orientated than the material inside the filament. This surface alignment is known as the *skin effect*. It has an important influence on the properties of the fibre.

## Stretching

Orientation of the long molecules is completed by stretching the filament. This has the effect of pulling the long molecules into alignment along the longitudinal axis of the fibre, so that they are able to lie alongside one another and develop their cohesive forces.

The degree of orientation depends upon the amount of stretch to which the filament is subjected, and by controlling the stretching (or 'drawing') it is possible to control the tensile properties of the filament to a high degree.

In the production of a synthetic fibre, we have control over the chemical nature of the fibre-forming substance, and hence can produce a fibre with well-defined chemical properties and behaviour. This control over the chemical structure of the fibre also enables us to control the shape and the physical behaviour of the long thread-like molecules that we make.

It is reasonable to expect, for example, that slender, uniform molecules will be able to pack alongside one another much more efficiently than irregular molecules with awkward knobs and angles destroying their uniformity. A bundle of bamboo canes, for example, will pack together more tightly than a bundle of twigs.

In making a synthetic fibre, therefore, we tend to design our long-chain molecules in such a way that they have an opportunity of packing together with reasonable efficiency. Large groups of atoms attached to the sides of the long molecules are generally undesirable, for example, as they prevent the close-packing which contributes so greatly to fibre strength.

## Crystalline and Amorphous Regions

Wherever the thread-like molecules are able to pack closely together in a fibre, there is a tendency towards an ordered arrangement of the atoms with respect to one another. These tight-packed bundles of thread-molecules are, in effect, regions of crystallinity; they possess the regular and precise arrangement of atoms that is characteristic of any crystal such as salt or copper sulphate.

In between these regions of crystallinity are regions in which the molecules have not been able to line themselves up with such precision. These are the amorphous regions of the fibre.

In this modern conception of fibre-structure we regard the long thread-like molecules as passing through regions of ordered crystalline arrangement which are embedded in amorphous material. The molecules in the amorphous regions are aligned to some degree, but have not been lined up with the precision that enables them to pack together in a well-defined crystalline form.

(A)

(B)

(C)

*Cross Linking and Chain Branching.* The production of long molecules during polymerization of a monomer X may take place in such a way as to form a linear molecule (A). It may, however, form branched molecules (B), and these may eventually link together to form network structures (C).

The formation of branches tends to reduce the ability of the linear molecules to pack together in such a way as to form regions of crystallinity, and branched molecules do not as a rule result in good fibre properties.

The formation of a network structure, in which the linear molecules are linked together, prevents movement of the chains of atoms relative to one another. Close-packing of the chains is not possible, and crystallization does not normally take place.

Network structures may be created after the linear molecules have been formed and aligned into fibres. This has the result of binding the molecules firmly together, and may improve certain fibre properties. Swelling may be reduced, for example, as solvents (e.g. water) cannot penetrate so readily between the long molecules. Cellulose molecules are cross-linked in modified rayons such as 'Topel' and 'Corval'.

During the stretching operation in synthetic fibre manufacture, the long molecules slide over one another as they are pulled into alignment in the direction of the fibre's longitudinal axis. As drawing continues, more and more of the molecules are brought to a state where they can pack alongside one another into crystalline regions; in these regions, the molecules are able to hold tightly together as a result of their cohesive forces. They will then resist further movement with respect to one another.

When nylon is drawn in this way after spinning, a filament may stretch to as much as five times its original length. Then, quite abruptly, the drawn filament will resist further stretching. Its molecules have aligned themselves as effectively as possible into crystalline regions and are holding tightly together. The filament will now withstand much greater force without stretching, and if the load increases it will eventually rupture as the molecules are dragged apart.

*Effect on Properties*

The degree of alignment of fibre molecules affects the properties of a fibre in several ways. The more closely the molecules pack together, the greater is the tenacity of the fibre. This increase in tenacity is accompanied by a decrease in the elongation at break; the molecules are not able to slide over one another as they could before alignment took place.

A high degree of orientation also tends to increase the stiffness or rigidity of the fibre. The molecules no longer have the freedom of movement that they had before alignment.

Water is unable to penetrate between the molecules in a crystalline region of the fibre as readily as it does in the amorphous regions. Increased alignment therefore tends to lower the moisture absorption of the fibre. This resistance to water-penetration affects the dyeing properties in a highly orientated fibre; molecules of dyestuff cannot migrate from the dyebath into the spaces between the fibre molecules.

This resistance to penetration by foreign molecules affects the general chemical stability of a fibre; highly orientated fibres are more resistant to chemical attack.

There is a marked change in the appearance of fibres as they are drawn. In the undrawn state, nylon is usually dull and opaque; as the filaments are drawn, and orientation increases, the filaments acquire a transparency and lustre which are characteristic of drawn nylon.

## MAN-MADE FIBRES

Though nature has used long thread-like molecules for many purposes, it is only in a relatively few cases that she has fulfilled the requirements for a textile fibre. In wool and cotton, flax and silk, nature has carried out the entire job of fibre production. All we have to do is to avail ourselves of nature's bounty.

In other cases, nature has associated her fibre-forming substance such as cellulose with extraneous materials that make it useless as a fibre. In wood, for example, the cellulose fibres are bound together by lignin and other gummy substances.

Yet again, nature may produce the necessary fibre-forming molecules, but omit to align them in the necessary way. Casein, for example, the protein of milk, will form a fibre if the molecules are arranged alongside each other.

## POLYMERIZATION

$$CH_2=CHCl + CH_2=CHCl + CH_2=CHCl$$

VINYL CHLORIDE

$$-CH_2-CHCl-CH_2-CHCl-CH_2-CHCl-$$

POLYVINYL CHLORIDE

(A)

$$NH_2(CH_2)_6 NH[H + HO]OC(CH_2)_4 COOH$$

DIAMINE      DIBASIC ACID

$$NH_2(CH_2)_6 NH.OC(CH_2)_4 COOH + H_2O$$

$$H[HN(CH_2)_6 NH.OC(CH_2)_4 CO]_n OH + _n H_2O$$

POLYAMIDE

(B)

Polymerization may take place in one or other of two ways:

(1) *Addition Polymerization.* This is a process in which monomer molecules link together without the elimination of atoms to form by-product molecules. The monomer molecules literally add together. Polyethylene, polypropylene, and polyvinyl chloride (A) are examples of fibre-forming polymers made by addition polymerization. The monomer molecules link together via the double bond in the molecule.

(2) *Condensation Polymerization.* This is a process in which the linkage of monomer molecules takes place by chemical action which results in the elimination of a by-product molecule, commonly water. Polyamides are produced by condensation polymerization (B).

The great modern rayon industries have developed from these natural long-chain molecules which nature has neglected to turn out in the form of ready-made textile fibres. Cellulose from wood is a raw material for rayon; it is separated from its undesirable gums and then re-made into fibres suitable for textiles. Casein from milk, and other proteins, are manipulated until their long molecules are lying side-by-side in fibrous form.

Finally, we have now learned to remain entirely independent of nature for our fibre production. We can start from scratch and actually make the long fibre-forming molecules themselves from simpler chemicals. This is what we have done in making nylon and the other synthetic fibres. As a result, we have opened up a great new field of scientific industry which can provide us with fibres unlike any that we have been able to derive from nature's limited selection of ready-made long-chain molecules.

### Classification of Man-Made Fibres

Man-made fibres fall naturally into two broad groups, depending on the origin of the fibre-forming materials from which they are produced (see page xxx ).

Man-made fibres are considered under two main headings:

A. NATURAL POLYMER FIBRES (in which the fibre-forming material is of natural origin).
B. SYNTHETIC FIBRES (in which the fibre-forming material is made from simpler substances).

These main sections are sub-divided as follows:

### Natural Polymer Fibres

The fibres in this group may be classified into the following sub-groups:
(1) Cellulose Fibres; Rayons (in which the fibre is wholly or mainly cellulose).
(2) Cellulose Ester Fibres.
(3) Protein Fibres.
(4) Miscellaneous Natural Polymer Fibres.

### Synthetic Fibres

Synthetic fibres may be classified with reference to their

chemical structure. The following synthetic materials have become the' basis of commercially-important fibres:

(1) Polyamides.
(2) Polyesters.
(3) Polyvinyl Derivatives.
     (*a*) Polyacrylonitrile.
     (*b*) Polyvinyl chloride.
     (*c*) Polyvinylidene chloride.
     (*d*) Polyvinyl alcohol.
     (*e*) Polytetrafluoroethylene.
     (*f*) Polyvinylidene dinitrile.
     (*g*) Polystyrene.
     (*h*) Miscellaneous Polyvinyl Derivatives.
(4) Polyolefins.
     (*a*) Polyethylene.
     (*b*) Polypropylene.
(5) Polyurethanes.
(6) Miscellaneous Synthetic Fibres.

This is not by any means the only effective way in which man-made fibres may be classified, but it is a simple and straightforward method of considering fibres on the basis of their chemical constitution. It is the classification which has been followed in the remaining section of the Handbook.

It should be remembered that modern synthetic fibres are often copolymers or modifications of polymers, and they may on that account be considered as belonging to two or more chemical sub-groups. For the purposes of this book, fibres are included in the sub-group represented by the major constituent of the polymer.

### Federal Trade Commission Fibre Identification Act 1958

In recent years, the number of synthetic fibres appearing on the market has given rise to considerable confusion regarding the true nature of textile products. In order to protect producers and consumers from misbranding and false advertising, the U.S. Federal Trade Commission established Rules and Regulations for Fibre Identification which came into force on 3 March 1960. After that date, the following generic names were obligatory for man-made textile fibres:

*Acetate (and Triacetate).* A manufactured fibre in which the fibre-forming substance is cellulose acetate. Where not less than

92 per cent of the hydroxyl groups are acetylated, the term 'triacetate' may be used as a generic description of the fibre.

*Acrylic.* A manufactured fibre in which the fibre-forming substance is any long chain synthetic polymer composed of at least 85 per cent by weight of acrylonitrile units ($-CH_2-CH(CN)-$).

*Anidex.* A manufactured fibre in which the fibre-forming substance is any long chain synthetic polymer composed of at least 50 per cent by weight of one or more esters of a monohydric alcohol and acrylic acid ($CH_2 = CH-COOH$).

*Aramid.* A manufactured fibre in which the fibre-forming substance is a long chain synthetic polyamide in which at least 85 per cent of the amide linkages ($-CO-NH-$) are attached directly to two aromatic rings.

*Azlon.* A manufactured fibre in which the fibre-forming substance is composed of any regenerated naturally occurring proteins.

*Glass.* A manufactured fibre in which the fibre-forming substance is glass.

*Metallic.* A manufactured fibre composed of metal, plastic-coated metal, metal-coated plastic, or a core completely covered by metal.

*Modacrylic.* A manufactured fibre in which the fibre-forming substance is any long chain synthetic polymer composed of less than 85 per cent but at least 35 per cent by weight of acrylonitrile units ($-CH_2-CH(CN)-$), except fibres qualifying under subparagraph (2) of paragraph (j) (rubber) of this section and fibres qualifying under paragraph (q) (glass) of this section.

*Novoloid.* A manufactured fibre containing at least 85 per cent by weight of a cross-linked novolac.

*Nylon.* A manufactured fibre in which the fibre-forming substance is a long-chain synthetic polyamide in which less than 85 per cent of the amide ($-CO-NH-$) linkages are attached directly to two aromatic rings.

*Nytril.* A manufactured fibre containing at least 85 per cent of a long chain polymer of vinylidene dinitrile ($-CH_2-C(CN)_2-$) where the vinylidene dinitrile content is no less than every other unit in the polymer chain.

*Olefin.* A manufactured fibre in which the fibre-forming substance is any long chain synthetic polymer composed of at least 85 per cent by weight of ethylene, propylene or other olefin units except amorphous (non-crystalline) polyolefins qualifying under category (1) of paragraph (j) (rubber) of Rule 7.

*Polyester.* A manufactured fibre in which the fibre-forming substance is any long chain synthetic polymer composed of at least 85 per cent by weight of an ester of a substituted aromatic carboxylic acid, including but not restricted to substituted terephthalate units $p(-R-O-CO-C_6H_4-CO-O-)$ and para-substituted hydroxybenzoate units $p(-R-O-C_6H_4-CO-O-)$.

*Rayon.* A manufactured fibre composed of regenerated cellulose, as well as manufactured fibres composed of regenerated cellulose in which substituents have replaced not more than 15 per cent of the hydrogens of the hydroxyl groups.

*Rubber.* A manufactured fibre in which the fibre-forming substance is comprised of natural or synthetic rubber, including the following categories:

1. A manufactured fibre in which the fibre-forming substance is a hydrocarbon such as natural rubber, polyisoprene, polybutadiene, copolymers of dienes and hydrocarbons, or amorphous (non-crystalline) polyolefins.

2. A manufactured fibre in which the fibre-forming substance is a copolymer of acrylonitrile and a diene (such as butadiene) composed of not more than 50 per cent but at least 10 per cent by weight of acrylonitrile units ($-CH_2-CH(CN)-$).
The term 'lastrile' may be used as a generic description for fibres falling within this category.

3. A manufactured fibre in which the fibre-forming substance is a polychloroprene or a copolymer of chloroprene in which at least 35 per cent by weight of the fibre-forming substance is composed of chloroprene units ($-CH_2-C.Cl=CH-CH_2-$).

*Saran.* A manufactured fibre in which the fibre-forming substance is any long chain synthetic polymer composed of at least 80 per cent by weight of vinylidene chloride units ($-CH_2-CCl_2-$).

*Spandex.* A manufactured fibre in which the fibre-forming substance is a long chain synthetic polymer comprised of at least 85 per cent of a segmented polyurethane.

*Vinal.* A manufactured fibre in which the fibre-forming substance is any long chain synthetic polymer composed of at least 50 per cent by weight of vinyl alcohol units ($-CH_2-CH(OH)-$) and in which the total of the vinyl alcohol units and any one or more of the various acetal units is at least 85 per cent by weight of the fibre.

*Vinyon.* A manufactured fibre in which the fiber-forming substance is any long chain synthetic polymer composed of at least 85 per cent by weight of vinyl chloride units ($-CH_2-CHCl-$).

# FIBRE CLASSIFICATION CHART

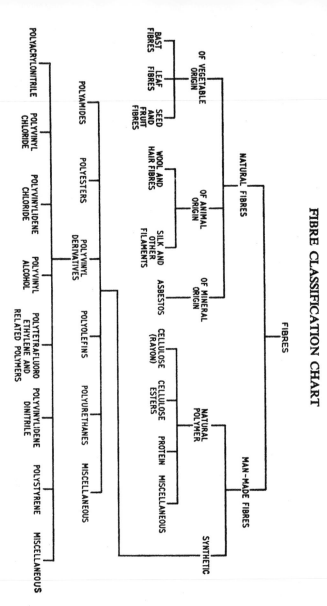

# *MAN-MADE FIBRES*

**A:** NATURAL POLYMER FIBRES

**B:** SYNTHETIC FIBRES

# A : NATURAL POLYMER FIBRES

1. CELLULOSE FIBRES; RAYONS

2. CELLULOSE ESTER FIBRES

3. PROTEIN FIBRES

4. MISCELLANEOUS NATURAL POLYMER FIBRES

## INTRODUCTION

In 1664, the famous English scientist Robert Hooke published a book called *Micrográphia*. Amongst the many subjects Hooke discussed was the possibility of imitating the silkworm to make an artificial fibre. Here was an insect that made the finest known textile fibre simply by forcing a liquid through a tiny hole in its head. Why could not we do the same thing mechanically, and make an artificial silk?

It was nearly two hundred years before Hooke's suggestion was successfully tried out. Only the silkworm knew how to make the liquid that hardened into silk after it had been squirted into the air. Nobody could suggest anything else to do the job.

In 1842, an English weaver, Louis Schwabe, devised a machine for making artificial filaments by forcing liquid through very fine holes. The material he used was glass, which was sufficiently plastic when molten to be forced through the holes, and yet would cool to a solid once it came into contact with the air.

This was the nearest thing yet to an artificial fibre. But glass fibre was not suitable for textiles, and Schwabe's entreaties to the scientists to provide something better were of no avail.

At that time, science had hardly begun to interest itself in the nature of fibrous materials. The existence of long thread-molecules, such as are needed for fibre-formation, had not even been suspected. But it was realized that in natural cellulose there was a potential raw material for making fibres. Nature herself made cellulose fibres in cotton and flax. Why should not man make use of the vast stores of vegetable cellulose for making additional supplies of textile fibres?

Unfortunately, the cellulose from wood and straw and similar sources was associated with gummy materials such as lignin. And though in some cases it was possible to separate the fibrous cellulose in useful form – for example, flax – most of it was useless for textile purposes. What was needed was some way of purifying the cellulose and obtaining it as a satisfactory fibre.

During the latter part of the nineteenth century, many attempts were made to use crude cellulose as raw material for a textile fibre. The problem resolved itself into finding a way of dissolving cellulose, separating the liquid from the impurities and then squirting it through tiny holes and hardening it into a fibre. Cellulose, however, would not dissolve in any suitable liquid.

## Nitrocellulose

In 1846 a scientist Friedrich Schönbein discovered that cellulose could be turned into another substance, nitrocellulose, when it was treated with nitric acid. This nitrocellulose was a highly flammable material; it was, indeed, explosive. Its discovery marked the beginning of the modern explosives industry; and, mixed with camphor, it gave us the first man-made plastic, celluloid.

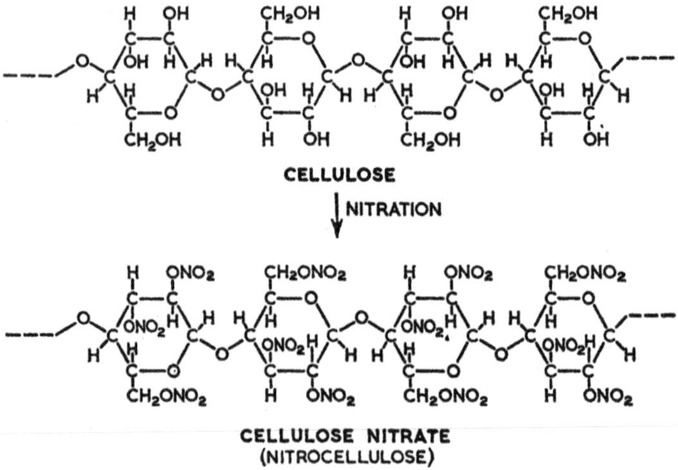

**CELLULOSE**

NITRATION

**CELLULOSE NITRATE**
(NITROCELLULOSE)

4

Nitrocellulose, unlike its parent material cellulose, dissolved readily; for example, in a mixture of ether and alcohol. In 1855, George Audemars discovered that if he dipped a needle into a solution of nitrocellulose and then drew it away, a filament was formed which dried and hardened in the air and could be wound up into a reel. The modern rayon industry had begun.

When cellulose is turned into nitrocellulose, the molecules remain in their long thread-like shape; small groups of atoms have been attached to their sides, making the new substance more soluble. When Audemars touched the nitrocellulose solution and then drew his needle away, the alcohol and ether evaporated in the air, leaving the solid nitrocellulose behind. And the long molecules were able to hold the nitrocellulose together in its new fibrous shape.

These nitrocellulose fibres were a great advance towards the production of a commercially useful fibre. They were soft, strong and flexible. But their flammability prevented any great use of the fibres for making textiles – nobody wanted to wear clothes made from guncotton.

So almost thirty years passed until, in 1883, Sir Joseph Swan began looking for some way of making filaments for his electric light bulbs. He wanted something that would give him an extremely fine filament of carbon, and he used nitrocellulose. Swan patented a process for squirting nitrocellulose solution through holes to form filaments, followed by a chemical treatment of the filaments which changed the dangerous nitrocellulose back into harmless cellulose.

In 1885, Swan exhibited textiles made from his 'artificial silk', But he was mainly interested in his filaments as a way of making fine carbon filaments for lamps, and he failed to follow up the textile possibilities.

### Chardonnet Silk

Meanwhile, in 1878, Count Hilaire de Chardonnet began experimenting in France. Chardonnet was a student at the École Polytechnique under Louis Pasteur at the time of the pébrine investigation in the silk industry. He became interested in the silkworm's ability to spin fibres, and determined to emulate it artificially.

In 1884, Chardonnet made his first artificial fibres from nitrocellulose solution which was squirted through tiny holes, hardened

5

in warm air and then treated chemically to convert it back to cellulose. Materials made from this artificial silk were exhibited at the Paris Exposition in 1889 and Chardonnet secured financial backing for the industrial development of his fibre. A factory was built in 1890 at Besançon and began producing 'Chardonnet silk'.

This was the first artificial fibre to be produced commercially, and it marked the beginning of our modern man-made fibre industry. But nitrocellulose is a highly flammable material, and the manufacturing process proved difficult and dangerous. Large scale production of Chardonnet silk was never realized, although the fibre was manufactured sporadically until 1949. In that year, the only remaining Chardonnet silk factory, in Brazil, was burned down.

The Chardonnet process is no longer used commercially. It had the advantages of simplicity, a stable spinning solution and a minimum of waste during manufacture. But it was slow in operation, potentially dangerous and expensive.

## Cuprammonium Fibre; Cupro

In 1890, a new process for making artificial fibres was invented, which made use of the discovery that cellulose could be dissolved in cuprammonium liquor. The solution was extruded through small holes into a coagulating bath, where the cellulose was regenerated to form continuous filaments.

The cuprammonium process was developed into a commercially important process, and it continues in operation today. Cuprammonium fibre has never achieved really large scale production, but the fibre has special qualities which have enabled it to establish important outlets in the textile trade.

## Viscose Fibre

In 1892, another method of making regenerated cellulose fibre was devised, in which the cellulose is converted to cellulose xanthate and dissolved to form the solution known as *viscose*. When fine jets of viscose are extruded into an acid coagulating bath, cellulose is regenerated in the form of filaments.

In the years prior to World War I, the viscose fibre industry developed rapidly, and viscose is now the most important natural polymer fibre of all.

6

## Acetate Fibre

It was not until after World War I that another type of artificial fibre came into successful production. This was the fibre we now know as acetate.

Once again, the raw material is cellulose, which is rendered soluble by being converted to a derivative, cellulose acetate, which dissolves in acetone and other solvents. In this respect, the production of acetate resembles that of Chardonnet silk, which was made by converting cellulose into cellulose nitrate and dissolving it in solvent.

Cellulose acetate solution is extruded through fine holes, as in the case of the other regenerated fibres. But instead of entering a coagulating bath, the fine jets emerge into a stream of warm air. The solvent evaporates, leaving filaments of solid cellulose acetate.

The filaments made in this way differ fundamentally from those made by the cuprammonium or viscose processes, in that they are not reconverted to cellulose. They remain cellulose acetate, and the properties of the fibre are thus different from those of cellulose fibres.

## Protein Fibres

Cellulose is by far the most important source of natural polymer fibres made today. But nature provides other materials which are capable of forming fibres, and some of these have been made into fibres of commercial value.

Proteins, for example, are used by nature in making natural fibres such as wool, hair and silk. But there are many other proteins which are not in fibrous form, and many of these can be manipulated to convert them into fibres. As in the case of cellulose, it is necessary to dissolve the protein and extrude the solution in the form of fine jets which can be hardened into filaments.

Casein (from milk), zein (from maize) and arachin (from groundnuts) have been made into useful fibres, but none has yet achieved major success in the textile field.

## Alginate Fibres

Alginic acid extracted from seaweed is a chemical relative of cellulose, and it has become a raw material from which fibres are spun. Alginate fibres have useful specialized applications, but

7

are of relatively minor importance by comparison with other natural polymer fibres.

## Nomenclature

In the early days of natural polymer fibre production, the fibres became known as 'artificial silk'. The reason for this lay in the fact that the fibres were produced in the form of continuous filaments; in this respect, they resembled silk rather than the short staple fibres of cotton or wool. Also, they often had a silk-like sheen.

Fundamentally, however, the fibres made by the cuprammonium and viscose processes are related to cotton, in that they are all cellulose fibres. Silk, on the other hand, is a protein. The name 'artificial silk' was obviously a misnomer, therefore, especially as it became apparent that the new fibres were establishing themselves in their own right in the textile trade. They were not merely substitutes for silk, but had unique properties which made them unlike any other fibres. In due course, the term 'artificial silk' was abandoned, and the fibres became known as *rayon*.

At first, this term included natural polymer fibres of all types, but it is now restricted to those fibres consisting wholly or substantially of regenerated cellulose. In practice, this means that it refers to fibres made by the cuprammonium and viscose processes. Fibres made from cellulose acetate are called *acetate* (or *triacetate*), and those from proteins are *azlon* fibres. (See U.S. Federal Trade Commission definitions on page xxvi).

## Natural Polymer Fibres Today

Viscose, cuprammonium and acetate make up the bulk of the natural polymer fibres produced today. The cellulose that serves as raw material is available in virtually unlimited quantities, and the manufacturing processes have been developed and improved until the production of natural polymer fibres is now one of the most efficient and important industries in the world.

The rapid rise of the natural polymer fibre industry is reflected in the manufacturing statistics. At the end of World War I, the world output of natural polymer fibres amounted to only 9000 tonnes (less than that of silk). By 1957 output had soared to some 2¼ million tonnes and in 1977 it was 3½ million tonnes.

## 1. CELLULOSE FIBRES; RAYONS

*Federal Trade Commission Definition*

The generic term *rayon* was adopted by the U.S. Federal Trade Commission for fibres of the regenerated cellulose type, the official definition being as follows:

*Rayon.* A manufactured fibre composed of regenerated cellulose, as well as manufactured fibres composed of regenerated cellulose in which substituents have replaced not more than 15 per cent of the hydrogens of the hydroxyl groups.

This definition includes three types of regenerated cellulose fibre in production today, i.e. *Viscose Rayon, Cuprammonium Rayon* (Cupro), and *Saponified Cellulose Acetate.*

## VISCOSE RAYON

### INTRODUCTION

The large-scale development of rayon was made possible by C. F. Cross, E. J. Bevan, and C. Beadle in England in 1892, who found that they could dissolve cellulose without first making it into nitrocellulose. The cellulose was treated with caustic soda, then with carbon disulphide, and the product dissolved in dilute caustic soda. This viscous liquid they called 'viscose'.

A method of producing textile filaments from the viscose was discovered by C. H. Stearn and C. F. Topham, the latter of whom invented the 'ageing' of viscose (to its correct condition for spinning), the multiple hole spinning jet and the famous Topham spinning box.

In 1904, the British rights of the viscose process were purchased by Courtaulds Ltd., who developed it into the most successful method of rayon manufacture in the world.

The viscose process is comparatively lengthy and some 300 accurately controlled steps are involved. The raw materials, however, are cheap. Viscose rayon can generally be produced cheaper than other rayons, and viscose is now manufactured in greater quantity than either cuprammonium rayon or acetate.

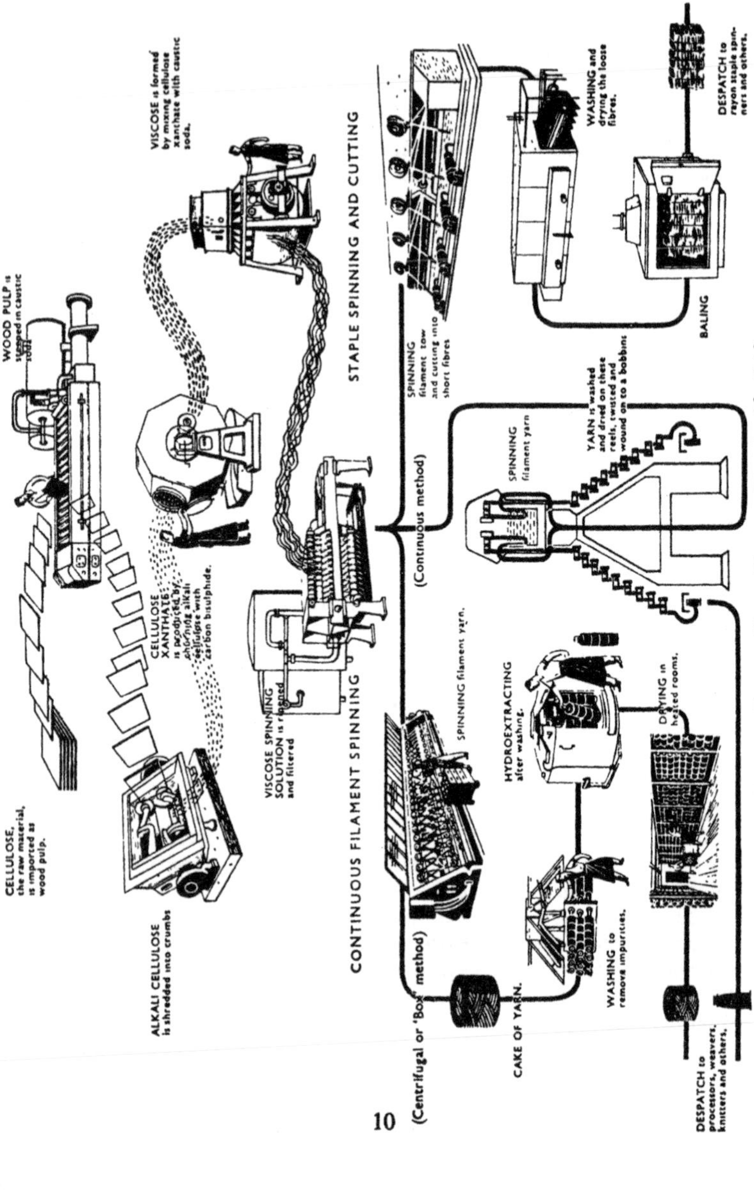

WOOD PULP is lapped in caustic soda.

VISCOSE is formed by mixing cellulose xanthate with caustic soda.

CELLULOSE, the raw material, is imported as wood pulp.

ALKALI CELLULOSE is shredded into crumbs

CELLULOSE XANTHATE is produced by chemical alkali cellulose with carbon bisulphide.

VISCOSE SPINNING SOLUTION is ripened and filtered

STAPLE SPINNING AND CUTTING

SPINNING filament tow and cutting into short fibres

WASHING and drying the loose fibres.

BALING

DESPATCH to rayon staple spinners and others.

(Continuous method)

SPINNING filament yarn

YARN is washed and dried on these reels, twisted and wound on to a bobbin.

CONTINUOUS FILAMENT SPINNING

SPINNING filament yarn.

HYDRO-EXTRACTING after washing.

DRYING in heated rooms.

(Centrifugal or 'Box' method)

WASHING to remove impurities.

CAKE OF YARN.

DESPATCH to processors, weavers, knitters and others.

*Viscose Rayon Flow Chart – Courtaulds Ltd.*

10

Water is needed in great quantity and many chemicals are used in viscose rayon manufacture. A kilogram of rayon fibre entails the use of more than 1,600 kg of water, nearly 2 kg of sulphuric acid, 1½ kg of caustic soda, 1¼ kg of wood pulp or cotton linters, ⅓ kg of carbon disulphide and smaller amounts of other chemicals.

### Continuous Filament and Staple

Until 1914, viscose rayon was produced almost entirely in the form of continuous filament yarn. During World War I, German and Italian firms began producing staple rayon fibre by chopping the filaments after extrusion.

The production of rayon staple made rapid progress during the 1930s, and by 1940 there was as much staple being used as continuous filament. After World War II, filament production exceeded that of staple until 1954, when staple once again took the lead. In 1961, some 60 per cent of the world production consisted of staple fibre.

In the 1960s and 1970s production of continuous filament diminished but staple production increased.

## TYPES OF VISCOSE RAYON

As control and understanding of the viscose process has increased, it has become possible to modify the properties of the fibre in a variety of ways. A range of viscose rayons is now available which includes fibres of widely differing characteristics.

Physical modifications of the viscose fibre range from changes in the form of the filament, e.g. hollow, shaped and surface-modified filament (see page 20) to changes in the fine structure as in the high tenacity rayons (see page 39) and high wet modulus (including polynosic) rayons (see page 47).

Chemical modification, likewise, has resulted in many types of modified viscose fibre, such as cross-linked, basified and grafted rayons (see page 38).

## PRODUCTION

### Raw Materials

The raw materials for making viscose rayon are either cotton linters (the short, useless fibres in the cotton boll) or wood pulp derived from such timber as northern spruce, western hemlock, eucalyptus or southern slash pine. These pulps contain about 94 per cent cellulose, and are most suitable for fibre manufacture.

Wood pulp is purified by boiling with caustic soda or sodium bisulphite solution. It is bleached and washed, and reaches the rayon factory in the form of sheets like thick blotting paper. The cellulose pulp is stored under controlled conditions of humidity and temperature until the moisture is distributed uniformly; this 'conditioning' may take several weeks.

## Formation of Alkali Cellulose (Soda Cellulose)

The first step in viscose rayon manufacture is the production of alkali cellulose. The cellulose pulp sheets are steeped in warm caustic soda for an hour, and then pressed to remove excess solution. The treated cellulose is broken up in a shredder to form powdery crumbs.

The crumbs are aged for up to a day, during which time the caustic soda reacts with cellulose to form alkali cellulose (soda

*Viscose Rayon. Stages in production*

cellulose) (see page 12). During the ageing process, the long cellulose molecules are attacked by oxygen from the air and broken up to some extent into shorter molecules.

## Sodium Cellulose Xanthate Production

The aged crumbs of alkali cellulose are mixed with carbon disulphide in a revolving drum. The almost white crumbs turn gradually yellow and then orange as sodium cellulose xanthate is formed (see page 12). The batch is tipped into a dilute solution of caustic soda, forming a thick orange-brown solution. There is a loose association at this stage between the sodium cellulose xanthate and the sodium hydroxide.

The lustre of the rayon is controlled at this stage. If rayon is produced from the sodium cellulose xanthate solution without adding anything to it, the rayon will have a silk-like sheen. Often however, a duller appearance is preferred, and this is achieved by adding a fine white pigment, usually titanium dioxide, to the spinning solution at this point in manufacture.

## Ripening

The sodium cellulose xanthate solution (viscose solution) is allowed to stand and ripen for up to a day at a carefully controlled temperature, during which time it is filtered repeatedly. Some breakdown of the long cellulose molecules into molecules of lower molecular weight takes place, and the viscosity of the solution falls initially.

On further standing, the viscosity of the solution begins to rise again as cellulose is regenerated by the breakdown of some of the sodium cellulose xanthate. If ripening is allowed to continue for a long time, cellulose is deposited from solution. In practice, however, ripening is allowed to continue until the solution has reached a state suitable for spinning. It is then subjected to vacuum to remove bubbles of air or other gases which would interfere with the smooth flow of the solution during spinning.

### Spinning*

The ripened viscose spinning solution is passed through a final

* In the manufacture of rayon and other man-made fibres, the term 'spinning' has come to be applied to the process of forcing liquid through tiny holes to form the fibre. The same term is used in the case of both natural and man-made fibres for the twisting together of short fibres into yarn.

filtering stage, and then forced through tiny holes bored in a cap of metal forming the spinneret. Spinnerets are made from gold, platinum, palladium, tantalum and other corrosion-resisting metals; platinum alloys are commonly used.

The holes in the spinneret are usually between 0.005 and 0.0125 mm diameter, and each spinneret will be pierced by up to 20,000 of them.

As it emerges from the hole in the spinneret, the jet of viscose enters a coagulating bath containing a mixture of acids and salts, typically of the following composition:

| | |
|---|---|
| Sulphuric acid | 4–12 parts by weight |
| Sodium sulphate | 10–22 parts by weight |
| Zinc sulphate | 1–5 parts by weight |

In the coagulating bath, the sodium cellulose xanthate is converted back into cellulose. This is insoluble in the liquid of the bath, so that the fine jet of viscose solution is changed into a solid filament of cellulose.

The action of the spinning bath is complex. The sodium sulphate brings about the coagulation of the sodium cellulose xanthate to form a filament. This is then converted to cellulose by one or other of two routes:

(a) the sodium cellulose xanthate is converted into cellulose xanthic acid, which decomposes into cellulose.

(b) the sodium cellulose xanthate is converted first into zinc cellulose xanthate, which is then converted into cellulose xanthic acid and finally into cellulose.

The conversion of zinc cellulose xanthate into cellulose xanthic acid takes place more slowly than the conversion of sodium cellulose xanthate into cellulose xanthic acid, and route (b) is slower than route (a).

In the coagulation of viscose, using a bath as outlined above, the zinc sulphate is in low concentration, and it penetrates only a short distance into the filament in the time that the acid penetrates into the centre of the filament. The bulk of the filament, including, the core, is thus regenerated via the direct route (a). Only the outer layer is regenerated via the slower route (b).

The slower regeneration taking place in the outer layer of the filament results in a more uniform deposition of cellulose, and creates the skin effect that is typical of a regular viscose fibre.

14

As the core shrinks, the skin becomes wrinkled and the filament acquires its lobed cross-section.

In the production of high-tenacity fibres of all-skin construction (see page 43), regeneration retardants are added to the coagulating bath, slowing up the regeneration of cellulose by the acid, and so allowing the slower route (b) to bring about regeneration throughout the fibre. This effect is intensified by using a higher concentration of zinc salts in the spinning bath. The slower regeneration obtained in this way allows time for stretching and orientation to be carried out more effectively.

There are three ways in which the filaments are treated after leaving the coagulating bath; the processes are known as pot or box spinning, bobbin spinning and continuous spinning, respectively.

*Pot Spinning; Box Spinning*

In pot spinning the bunch of filaments from each spinneret is led out of the bath and around a wheel. This wheel – called the godet wheel – pulls the filament from the jet at a controlled speed. It is this speed, together with the rate of extrusion, which determines the diameter of the rayon fibre. The faster it is pulled as it leaves the jet, the thinner the fibre will be.

On leaving the godet wheel, the fibres pass around a second wheel which is moving faster than the first. The fibres are therefore stretched between the two wheels – a process which has a profound effect on the final fibre. This stretching of the still-plastic rayon tends to orientate the molecules of cellulose along the direction of the fibre. The long molecules are packed more tightly together so that their mutual attraction comes into play. They hold strongly to each other, giving a stronger fibre.

The more the rayon is stretched while it is still plastic, the stronger is the fibre. But at the same time, this tight packing of the molecules reduces the 'stretchability' of the fibre, so that excessive stretching will achieve high strength usually at the expense of other desirable properties. The treatment is therefore regulated to suit the conditions the fibre will have to withstand.

After stretching, the fibre passes into a Topham box. This is a hollow container about 18 cm (7 in) in diameter which whirls like a spinning top. The filament is led through a hole in the top of the box and is flung against the side by centrifugal force. In this way it is pulled continuously through the hole and builds up into a cake of filament inside the box. A mechanical device ensures that the cake is built up evenly from top to bottom, and the spin of the box gives a twist to the fibres, usually about 1.2 turns to the cm (3 turns to the inch).

The Topham box rotates some 10,000 times per minute. The sides are perforated, so that most of the liquid is flung off from the wet fibre cake. Up to 63 m (70 yd) a minute are fed into the box; it takes several hours to build up a complete 'cake', which is then washed and may be treated with sodium sulphide solution to remove residual sulphur compounds. The fibre is bleached, usually with sodium or calcium hypochlorite or peroxide, rinsed in dilute acid, washed and dried.

*Bobbin Spinning*

In bobbin spinning, the filaments of rayon emerging from the coagulating bath may be wound without twist on to bobbins. The bobbins, which have perforated barrels, are purified and bleached under pressure. The yarn is then dried and oiled, and after twisting is wound up again ready for winding into skeins or cones.

*Continuous Spinning*

In the pot and bobbin spinning processes, packages of viscose filament are collected and then subjected to desulphurizing, bleaching, washing and drying before the rayon is ready for use. The process is thus an intermittent one, in which batches of filament are handled separately as they become available.

Batch processes of this type are inevitably costly in labour and operating charges, and the intermittent operating tends to introduce variations in the quality of the product from batch to batch. In modern industry there is a tendency to favour processes in which the product moves from one stage to the next in a continuous stream. Continuous processes of this sort are usually cheaper to operate, and can be controlled to produce a highly uniform product.

It has long been realized that the production of viscose rayon could be adapted to operate on a continuous basis. But many practical problems had to be solved before continuous spinning became a reality. The main difficulty lay in the time required for purification, bleaching, washing and drying of the filament after leaving the coagulating bath. In the production of rayon by pot spinning, for example, these stages might take 30 minutes or more to complete. And in this time, more than a mile of filament might be spun. If continuously-produced filament took the same time to process, the purification, washing and drying equipment would have to accommodate a comparable length of filament.

The successful development of continuous spinning of viscose was made possible to a large extent by the design of mechanical devices which could hold immense lengths of filament in continuous movement as they passed through the processing train.

### Industrial Rayon Corporation Process

During the 1930s, many firms experimented with continuous viscose spinning techniques. One of the first to achieve commercial success was a process developed by Industrial Rayon Corporation, U.S.A., which came into operation at Painesville, Ohio, in 1938.

The problem of handling great lengths of filament during processing was solved by using thread advancing reels of ingenious design. Each reel consisted of a pair of rollers with axes set on the skew, i.e. not parallel to each other. When filament is fed to one end of a pair of moving rollers of this sort, and passed round the rollers as though round a pair of pulleys, the filament tends to form a spiral which moves along the pair of rollers until it reaches the other end. The direction of movement of the spiral, the distance between the coils, and the length of filament carried, depends upon the angle of skew between the axes of the two rollers.

A pair of skew-set rollers can thus be used for carrying great lengths of filament in a very small space, without physical contact between individual coils of filament taking place. Individual strands of filament can be subjected to processing liquids and environments in a most direct way, by contrast with filaments which are wound together into a cake or other package. This, in itself, reduces the processing time needed for the treatment of continuously-produced filament, as compared with the processing of filament in package form.

In the Industrial Rayon Corporation continuous spinning process, the thread advancing reels consist of pairs of skew-set hollow rollers which rotate one inside the other. A succession of these reels carries the filament from the coagulating bath and stretching equipment, through desulphurizing, bleaching, washing, oiling and drying stages, until eventually a clean, dry filament is delivered ready for shipment to the textile manufacturer. The technique has now been developed and refined, and Industrial Rayon Corporation continuous spinning machines are in widespread use throughout the world. World rights to the process, excluding U.S.A. and South America, are held by Courtaulds Ltd., U.K.

A modern continuous spinning machine of this type is 6 m (20 ft) high, and has three operating levels. On the top are the coagulating bath and the stretching mechanism, from which the filament moves downward to pass through a train of ten processing stages. Each stage consists of a thread advancing reel, and during its passage through the reel the filament is subjected to the appropriate processing liquids or environments. Finally, the filament passes through a drying reel enclosed in a heated chamber. The dry filament emerges from this reel and is twisted and wound on to bobbins which carry up to 4.54 kg (10 lb).

The doffing of the bobbins is automatic, and there is no interruption to the operation of the machine. The entire process is continuous, the filament being wound on to the bobbin little more than 5 minutes after being produced in the coagulating bath.

### Nelson Process

A continuous spinning process was devised in the U.K. by S. W. Barber and J. Nelson during the early 1930s. By 1934, the process was in operation. It has since been developed by Lustrafil Ltd., and has become known as the Nelson Process.

In the Nelson Process, a combination of two techniques is used to overcome the problem of carrying great lengths of filament during the processing stages. Firstly, skew-set rollers are used to carry the filament, as in the Industrial Rayon Process; secondly, the stages in processing are reduced, desulphurizing and bleaching being omitted.

The thread advancing device, in the Nelson Process, is similar in principle to that used in the Industrial Rayon Process, but differs in the details of its operation. The two rollers, instead of rotating inside one another, are arranged one above the other like the rollers in a wringer, with their axes set on the skew. The rollers, about 1 m (3.3 ft) long, carry more than 100 coils of the filament spiral as it moves from one end to the other.

As the filaments emerge from the coagulating bath, they are carried upwards to pass over the upper roller and then downwards to pass under the lower roller, and so on. The first coils of the spiral are sprayed with acid, and coagulation is completed during this first stage. The filament is then washed by water sprays as it moves along the rollers, passing finally over the end sections of the rollers, which are heated. Dry filaments leave the rollers, having spent some 3 minutes traversing from one end to the other, moving through more than 100 coils of the spiral on the way. The dry filaments are twisted and collected on to bobbins, usually by a cap spinning mechanism.

Despite the omission of the desulphurizing stage, filaments produced by the Nelson Process contain only 0.1 – 0.3 per cent sulphur. If, as is usual, the yarn subsequently passes through a wet processing treatment, such as scouring or dyeing, this small proportion of sulphur is removed. If the subsequent handling of the yarn does not include a wet processing operation, the trace of sulphur may be removed easily by washing the fabric.

The use of high quality wood pulp has rendered the bleaching stage unnecessary for most applications, but bleaching can also be carried out if necessary at fabric stage.

## American Viscose Corporation Process

This is a high-speed continuous spinning process in which the filaments leave the coagulating bath via a jet of coagulating liquid, in which the filaments move for a distance of about 15 cm (6 in). Filaments then pass round the reels of a thread-advancing mechanism at such a speed that excess liquid is thrown off by centrifugal force. Coagulation of the cellulose continues as the filaments travel along their spiral path, and stretching is carried out when less that 70 per cent regeneration has taken place.

## Kuljian Process

In this process, filaments are carried from the coagulating bath by godet wheels, and pass on to a system of rollers which can be controlled to apply a desired degree of stretch. The filaments are treated as they travel through the roller system, and then

19

dried by hot air. The dry filaments are wound on to bobbins by a ring spinning mechanism.

## Kohorn 'Okomatic' Process

A process of continuous spinning devised by Von Kohorn International Corporation, known as the 'Okomatic' Process, carries the yarn forward by means of a system of skew-set glass rollers.

The production of yarns by these continuous spinning processes has now become established practice in the rayon industry. The quality of the yarns is fully equal to that of yarns produced by batch techniques, and the uniformity of the filaments is high. This increased uniformity is reflected in the reduction of breakages during weaving, and consequent improvement in the quality of the fabrics.

### MODIFICATION OF FILAMENT

By manipulation and modification of the spinning process, the physical structure and form of the rayon filament can be changed in many ways.

## Cross-Section

The cross-sectional shape of the filament may be varied by extruding through spinneret holes of suitable shape. Modification of filament cross-section is becoming of increasing importance today, as it can cause profound changes in the characteristics of yarns and fabrics. Circular cross-section filaments, for example, are poorer in covering power than lobed cross-sections typical of the normal viscose filament. Many synthetic fibres are now being produced in non-circular forms, such as dog-bone and trilobal cross-sections.

Viscose rayon has been made experimentally in a variety of cross-sectional shapes, and some have become of commercial importance. Straw filaments, for example, are produced by some manufacturers. Flat filaments are made by extrusion of viscose through slit orifices instead of circular ones; these filaments have improved covering power, but tend to be of harsh handle.

The diameter of a filament may be varied continuously between thick and thin, providing rayon filaments which make up into special effect fabrics.

## Bubble-filled Filaments

The covering power of viscose filaments may be increased by spinning in such a way that bubbles of air or other gas are trapped inside the filament. This may be done by spinning a viscose solution which has been agitated to produce a form in which air bubbles are entrapped.

In 1976 Courtaulds Ltd marketed a hollow viscose fibre 'Viloft' which is made by generating carbon dioxide inside the filament. The fibre has greatly increased bulk and high moisture absorption. In blends with polyester fibres, it provides increased covering power.

Blends of hollow viscose with cotton are used in shirtings and dress fabrics and for terry towel pile. Hollow viscose fibre is widely used in non-wovens, particularly in fields such as surgical and medical fabrics where high moisture absorption and moisture holding properties are important.

During World War II, a bubble-filled viscose filament called 'Bubblefil' was produced in U.S.A. by du Pont, using a technique by which air was injected into the filament as it was extruded. This produced a continuous filament containing discrete bubbles 3–6 mm ($\frac{1}{8}$ –¼ in) long, which was used as a substitute for kapok in life jackets, pontoons, insulated clothing etc.

## Spun-dyed Filament and Staple

Control of the spinning process in rayon production enables the manufacturer to mix finely-dispersed pigments with the viscose solution before spinning. The pigments are locked inside the filaments after spinning, providing 'spun-dyed' filaments which are unusually fast to light and to washing. White titanium dioxide is used in this way for dulling the natural sheen of rayon.

## Crimp

The spinning qualities of staple fibre are usually enhanced if the fibre has a waviness or crimp (cf. wool), and filaments which are to be made into staple are commonly treated to provide a crimp. This may be done mechanically, for example by passing the filament between gear-like rollers, or chemically by controlling the coagulation of the filament in such a way as to create a fibre of asymmetrical cross-section.

Chemical crimp has resulted very largely from experimental work carried out in Japan, and much of the viscose staple produced in Japan is now crimped in this way. The crimp is introduced by spinning viscose into a coagulating bath containing less acid and more salt than is usual, followed by carefully controlled stretching. The filament is then cut into staple and dried.

Filaments produced in this way have an asymmetrical cross-section, one side being thick-skinned and almost smooth, the other side being thin-skinned and highly serrated. When the fibres are wet, they swell much more on the thin-skinned side than on the thick-skinned side, so that there is a tendency to curl.

A similar effect may be introduced into rayon by using the 'bicomponent' technique which has been developed successfully in the production of some synthetic fibres (see 'Orlon' Bicomponent Fibre). This consists in the extrusion of twin filaments through orifices set side by side, in such a way that the two filaments join as they coagulate. The composite filament is made from viscose solutions of different characteristics, and the two portions of the filament have different swelling properties. In water, the filament tends to curl as one side swells more than the other.

The amount of crimp that is put into a fibre depends upon the decitex. Fibres of 1.7 dtex (1.5 den) may have 5 crimps per cm and 3.3 dtex (3 den) fibres 3 crimps per cm. If fibres are given too much crimp, neps may be caused during processing; if they are given too little crimp, the cohesion during processing may be too low.

*Crimped Viscose Rayon.* By suitable control of spinning conditions, viscose filaments may be spun in which the skin is thicker on one side of the filament than on the other. The swelling and other characteristics of the two sides of the fibre are different, and the wet filament contracts more on one side than on the other. This produces a crimp (cf. the relationship between the twin-core structure of the wool fibre cortex and the crimp of wool) – *After Courtaulds Ltd.*

## Surface-Modified Fibre

The nature of the fibre surface influences the processing properties of a fibre, and affects its behaviour in use. The striated surface of a regular viscose fibre, with its typical lobed cross-section, influences the spinnability of viscose staple, and also affects the appearance and handle of viscose yarns. On the other hand, indentations of this sort tend to cling to particles of dirt, and fibres of this type are often more difficult to clean than similar fibres with a non-serrated surface.

The unique properties of wool are due in some measure to the scaly surface of the fibre and many attempts have been made to create a surface of this type on rayon and other man-made fibres. Such fibres would be expected to provide improved blends with wool.

Surface modified rayon fibres have been produced by means of finishing treatments, and by using vibrating spinnerets.

### High Tenacity Rayon

During the extrusion and stretching which form part of the process of producing rayon filaments, the molecules of cellulose are aligned and orientated to some degree. Where molecules are able to pack together in orderly fashion, they form regions of crystallinity, or crystallites, which are separated from one another by regions of amorphous cellulose.

In this respect, the rayon filament resembles the cotton fibre, which also consists of cellulose molecules partly in crystalline and partly in amorphous form. But rayon differs from cotton in a number of significant ways.

During viscose manufacture, the cellulose molecules undergo some degradation, and in a normal viscose fibre the molecules will contain perhaps 200–700 glucose units. In cotton, the cellulose molecules are much longer, and may contain 2,000–10,000 glucose units.

The proportion of crystalline material in a normal viscose fibre is commonly in the region of 25–30 per cent, whereas in cotton it is as high as 70–75 per cent. The crystallites in rayon are smaller than in cotton; in rayon, they are, on average, about 300 angstroms long and 40 angstroms wide; in cotton they are some 600 angstroms long and 60 angstroms wide. Moreover the orientation of the crystallites along the fibre axis is greater in cotton than in rayon.

These factors all contribute to the differences in physical properties between rayon and cotton. In particular, they explain the
relative weakness of rayon, especially when wet. And they have
pointed the way towards the development of viscose rayons of
increased strength and dimensional stability.

The success achieved in this field is demonstrated by the introduction of high-tenacity viscose fibres (see page 39). Since World
War II, the use of high-strength rayon in tyres, conveyor belting,
transmission belting, hose pipes, etc., has become commonplace.
These yarns are produced by applying a high degree of stretch
during manufacture, when the individual filaments are in a
pseudo-plastic state. The stretching is effected by suitable choice
and control of the chemicals in the spinning solution and in the
spinning bath.

The acid bath coagulates the cellulose xanthate solution and
permits stretch to be applied in one or more stages. The process
used in the production of 'Tenasco' yarns by Messrs. Courtaulds
Ltd. is of this type, providing yarns which are three times as
strong as normal viscose rayon.

The additional stretch given to viscose filaments in producing
high tenacity rayons increases the degree of alignment of the
cellulose molecules. This has the effect of increasing the proportion of crystalline material, and orientates the molecules and
crystallites more highly in the direction of the fibre axis. The
physical structure of the fibre changes, resulting in increased
strength and diminished extensibility.

*Skin Effect*

When viscose solution is extruded into the coagulating bath, a
skin of cellulose forms on the outside of the filament. As coagulation continues, the core of the filament hardens and shrinks,
causing a wrinkling of the outer skin of the filament. The result
of this can be seen in the serrated cross-section of the normal
viscose filament. The skin itself can be distinguished by examining
a dyed and leached fibre. The core dyes more readily than the
skin, and it likewise loses its dye more readily on leaching, leaving
a darker shade which can be seen quite easily through the microscope.

The skin and core are both cellulose, but they differ in the
nature and orientation of the crystallites. In the skin, the
crystallites are smaller than in the core, and the molecules in

24

the amorphous regions of the skin are less tangled and haphazard in arrangement than in the core. The skin is thus of more uniform structure, and the orientation of the crystallites is higher in the skin than in the core.

In the production of high-tenacity rayons, the coagulation and stretching of the fibre are controlled in such a way as to influence the internal structure of the filament. This is accompanied by an increase in the proportion of skin, and a decrease in the proportion of core, to the point at which the core disappears completely. The fibre is coagulated in a more uniform way, and the cross-section becomes less serrated as the core-shrinkage effect is diminished. In the case of a 'whole-skin' fibre such as 'Tenasco Super 105', the cross section is almost circular.

The increased uniformity and orientation of the molecules in an all-skin fibre of this type results in an increase in tensile strength. If the molecules in a filament are not aligned and have a poor degree of orientation, or if the orientation and crystallinity vary greatly throughout the fibre, the resistance to a tensile stress will be taken by only a small proportion of the available molecules at a time. As the fibre is stretched, these molecules will break, and others will take up the strain. These too will break, and so on.

If the degree of orientation of the molecules and crystallites is high, and the structure of the fibre is uniform, a greater proportion of molecules will cooperate in taking the strain when the fibre is pulled. The fibre will have a greater tensile strength.

## High Wet Modulus Rayons; Polynosic Fibres (see page 47)

Many new types of viscose have emerged during recent years, as textile scientists have increased their understanding and control of spinning and processing techniques. Among the most important are the high wet modulus modal and polynosic fibres. These have been developed in a number of countries; in Britain, a modal fibre is manufactured by Courtaulds Ltd. under the name 'Vincel'; in the U.S., modal fibres are available as 'Avril' (Avtex Fibers Inc.) and 'Moynel' (Courtaulds North America Inc.).

Modal and polynosic fibres are high-tenacity, high wet modulus rayons in which modification of the molecular structure has resulted in a fibre with many of the attractive characteristics of cotton. The high strength, especially when wet, results in good dimensional stability and firmness.

### Staple

The manufacture of viscose staple fibre has assumed increasing importance since the end of World War II. Staple is made by chopping filaments, which may have been crimped mechanically or chemically, into short uniform lengths, commonly 38—200mm (1½—8 in), after they emerge from the coagulating bath. The staple fibre is then washed and dried, and packed in bales.

In producing staple, it is not necessary to control the uniformity of the fibre to such a fine degree as in the case of continuous filament production. (Uniformity is nevertheless very good, and intrinsically better than that obtained with natural fibres.) Also, the filaments can be spun from spinnerets which provide a thick rope or tow consisting of thousands of filaments. These two factors tend to lower the cost of producing staple as compared with continuous filament yarns.

Viscose staple may be blended with wool, cotton or other fibres, and spun into yarn by the various systems used for staple fibres. Yarns made from viscose staple are naturally fuller and rougher in handle than those made from filament yarns (cf. wool and silk).

### *Tow to Top Conversion*

The conversion of staple fibre to yarn involves the realignment of the mass of short fibres which have resulted from the cutting of the filaments, bringing them back, in effect, towards the state of alignment that they had when they were in the form of uncut tow. In order to avoid an apparently unnecessary disorganization and realignment, tow to top conversion techniques have been developed in which the filaments in the tow are cut or broken into staple and drafted into sliver as a continuous process.

'*Tenasco*'. Opposite: High tenacity viscose rayons of the 'Tenasco' type are produced by extrusion of the viscose into an acid bath containing zinc and sodium sulphates, followed by stretching of the newly-formed filaments in hot aqueous acid. This results in increased molecular orientation, and an increase in the proportion of 'skin' to 'core'.

Yarns made in this way are stronger than normal viscose, and this extra strength is achieved with only slight reduction in extension at break. This means that the fibres have a high work of rupture.—*After Courtaulds Ltd.*

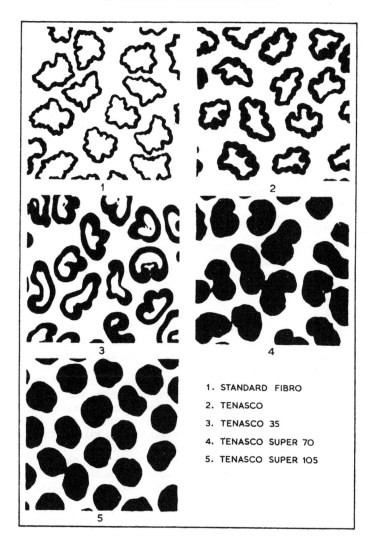

1. STANDARD FIBRO
2. TENASCO
3. TENASCO 35
4. TENASCO SUPER 70
5. TENASCO SUPER 105

## PROCESSING

### Desizing

Viscose filament yarns are commonly sized with water-soluble sizes which are removed by scouring. Staple yarns sized with starch may be desized by treatment with enzymes.

### Scouring

As in all wet processing of viscose, great care must be taken to avoid causing distortion of the rayon goods when wet. Scouring may be carried out using the usual techniques for cellulosic fibres, such as soap and sodium carbonate; soap, surface active agent and trisodium phosphate; soap, surface active agent and tetra-sodium pyrophosphate. Scouring should be followed by thorough rinsing.

### Bleaching

Strong oxidizing agents should be avoided. Bleaching may be carried out with hydrogen peroxide at temperatures up to 50°C., neutral sodium hypochlorite, perborate, and potassium permanganate followed by sodium bisulphite.

### Dyeing

Viscose rayon is a regenerated cellulose fibre, and as such is chemically almost identical with cotton. It can be dyed with all cotton dyestuffs, but techniques of dyeing are influenced by special factors.

Rayon has a greater affinity for dyes than has cotton. Affinity may vary according to manufacturing conditions, and will not necessarily be identical in rayons produced from the same plant.

Rayon swells to a greater extent than cotton when it is immersed in water. The fibres are weakened, and yarns or fabrics must be handled with great care in the dyebath.

Viscose is dyed in the form of hanks, staple fibre, piece goods and 'cake' (i.e. as collected on the spinning machine). The high degree of swelling can cause difficulties when rayon is dyed in packaged form. Cakes for dyeing should have an open wind.

REACTIVE DYESTUFFS which chemically bond to the cellulose molecules are often used.

DIRECT COTTON DYESTUFFS are used more widely than any other class for dyeing viscose rayon. They provide the dyer with a good range of colour and are reasonably fast.

BASIC DYESTUFFS often provide bright shades which are unobtainable with other dyestuffs but are fugitive to light.

·SULPHUR DYESTUFFS are used when excellent washing fastness is essential. Sulphur dyes also give a high degree of fastness to milling and to cross-dyeing.

With the exception of sulphur black, these dyes are not often used on continuous filament viscose rayon. Sulphur black is applied to continuous filament in cake or hank form.

Staple fibre is dyed with a range of sulphur dyestuffs.

AZOIC DYESTUFFS. Viscose rayon has a high degree of affinity for azoic dyestuffs (higher than mercerized cotton). The colours are intense, and have good fastness to boiling, cross-dyeing, rubbing and sunlight.

VAT DYESTUFFS are the fastest of all the dyestuffs used on viscose rayon. They resist the effects of light and washing just as they do on cotton.

The application of vat dyes to viscose rayon is complicated by the sensitivity of the fibre. Rayon swells in the alkaline solutions used in dyeing with these dyestuffs, and the leuco compounds are absorbed more rapidly than in the case of cotton. This tends to cause poor levelling, and only dyes with first-rate levelling characteristics should be used.

### Spun-dyed Rayon

Much rayon is now produced in the spun-dyed form, in which coloured pigments are incorporated during spinning. This process is of great importance, providing a large range of colours generally of very high quality and fastness.

### Drying

Wet viscose must be dried in such a way as to avoid placing any unnecessary strains on the fabric. Hydroextraction of the cloth at full width may be followed by drying on a stenter. Loop or festoon drying may also be used, followed by adjustment of width on a stenter equipped with steam box.

Cylinder drying should be avoided, as this may cause a harsh handle.

**Finishing**

The finishing of viscose fabrics is concerned largely with minimizing the shortcomings inherent in viscose, notably its sensitivity to water. Resin finishes of many types are now used effectively for this purpose, providing increased dimensional stability during washing, improved wrinkle resistance and crease resistance. These finishes should be used with care, as they may cause loss of abrasion resistance and tear strength, and produce a harsh, boardy handle.

Viscose fabrics may be calendered carefully to increase the fullness of handle. Decatising is used to produce a wool-like finish.

*Mercerizing*

Rayon is a lustrous fibre, and there is rarely any call for lustre to be increased, e.g. by mercerizing. It may be desirable, however, for blends of cotton and rayon to be mercerized to improve the lustre of the cotton, and bring its dyeing qualities more into line with those of the rayon.

Viscose fibre will withstand the caustic soda solution used in mercerizing, but disintegrates during subsequent washing. Special techniques are used to overcome this problem, including the use of caustic potash, or mixtures of caustic potash and caustic soda, in place of caustic soda.

STRUCTURE AND PROPERTIES (Regular Viscose)

**Fine Structure and Appearance**

The filament of viscose rayon is smooth and straight. It may be crimped ('Sarille') but there are no convolutions as in cotton. The surface is however, marked by longitudinal channels which are caused by contraction in volume of the filament during coagulation. These channels or striations give the cross-section of viscose rayon a characteristic outline, which is deeply serrated.

When rayon has been dulled with titanium dioxide, or 'spun-dyed' during manufacture, the particles of pigment are seen as dark specks embedded in the filament.

As rayon is a manufactured material, the diameter of the filament can be varied through wide limits. Viscose is commonly made in a range of dtex. Typical staple fibre dtex are 1.7, 3.3, 5.0, 9.0, 17, 40, 56 (1½, 3, 4½, 8, 15, 18, 44, 50 den); staple lengths 32–200 mm (1¼–8 in).

## Tensile Strength

Ordinary viscose rayon has a tenacity of 18–23 cN/tex (2.0–2.6 g/den) dry; 9.0–13.2 cN/tex (1.0–1.5 g/den) wet. Tensile strength of normal viscose rayon is 2109–3234 kg/cm$^2$ (30,000–46,000 lb/in$^2$).

## Elongation

Normal viscose will stretch by about 17–25 per cent of its original length before breaking, and 23–32 per cent when wet.

## Elastic Recovery

Cotton and other natural cellulose fibres have little inherent elasticity. Viscose rayon, however, has even less. It has a small elastic stretch of about 2 per cent from which it will recover when relaxed. But more persistent stretching will tend to cause permanent deformation as the long cellulose molecules slide over one another.

Elastic recovery (60 per cent r.h.):

                 1 per cent extension:  67 per cent
                 2 per cent extension:  60 per cent
                 3 per cent extension:  38 per cent
                 5 per cent extension:  32 per cent
                10 per cent extension:  23 per cent

## Average Stiffness

98 cN/tex (11.1 g/den).

## Initial Modulus

477 cN/tex (54 g/den).

## Work Factor

0.62.

## Specific Gravity

1.50 to 1.52.

**Effect of Moisture**

In natural cellulose fibres such as cotton, the cellulose molecules are packed together in orderly fashion wherever alignment of the molecules makes this possible. These ordered, crystalline regions confer strength and rigidity on the fibre; the amorphous regions, on the other hand, where cellulose molecules are arranged in random fashion are responsible for the flexibility, 'stretchability' and swelling properties of the fibre.

When natural cellulose fibres are dissolved during viscose manufacture, the molecules are set free from one another, and are able to move around more or less independently in the liquid. The extrusion of the liquid, followed by coagulation and stretching, tends to restore the alignment of the cellulose molecules and encourages the formation of crystalline regions again. In general, however, the molecular line-up is not restored to such a high degree as in the original natural state. Although the filament of viscose rayon consists of cellulose, it differs in this respect from cotton. It behaves in many ways like a cotton in which the cellulose molecules have been shortened (i.e. by chemical action during ripening and ageing) and aligned with rather less precision than in cotton. The actual degree of alignment and crystallinity depends upon the amount of stretch that is given to the filament during manufacture.

The reduced crystallinity of the cellulose in viscose rayon renders the fibre more responsive to water-penetration. The molecules of water can force their way between the loosely organized cellulose molecules in the amorphous regions of the rayon. Viscose rayon will absorb twice as much water naturally from the air as cotton does. Viscose has a moisture regain of 13 per cent under standard conditions. (Water imbibition: 100–110 per cent.) When soaked in water, viscose rayon will increase in length by 3–5 per cent and swell to double its original volume.

This increased water penetration is reflected in the change in tensile strength when rayon is wetted. Viscose loses as much as half its strength when wet, and is more easily stretched. The strength returns on drying, increasing as the rayon becomes bone-dry.

**Thermal Properties**

*Effect of High Temperature*

Rayon is not thermoplastic, and does not melt or become tacky on heating. It begins to lose strength at 150ºC. after prolonged heating, and begins to decompose at 185–205ºC. (depending on time factor).

### Flammability

Rayon burns readily with a characteristic odour of burnt paper.

### Effect of Age

So slight as to be almost nil.

### Effect of Sunlight

Viscose rayon withstands exposure to sunlight without discoloration; prolonged exposure causes a gradual loss of tensile strength. This is more severe if the fibre contains titanium oxide.

### Chemical Properties

### Acids

Similar to cotton. Viscose rayon is attacked by hot dilute or cold concentrated mineral acids, which weaken and disintegrate the fibre.

### Alkalis

Like cotton, viscose rayon has a high degree of resistance to dilute alkalis. Strong solutions of alkali cause swelling, with loss of tensile strength.

### General

The cellulose of viscose rayon undergoes some depolymerization during the manufacturing process. The fibre reacts to chemicals in a manner similar to cotton, but is generally more sensitive. It is attacked by oxidizing agents such as high-strength hydrogen peroxide, but will withstand normal hypochlorite or peroxide bleaches.

### Effect of Organic Solvents

Viscose rayon is insoluble in most organic solvents; it dissolves in a few complex solutions, such as cuprammonium. Dry cleaning solvents do not have any deleterious effect.

**Insects**

Viscose is resistant to insect attack but is attacked by silver-fish.

**Micro-organisms**

Mildews do not readily attack the cellulose of the fibre itself, but will feed on the size that is left on the fibres after processing. Mildews will cause discoloration, and weaken the fibre if the attack is severe.

**Electrical Properties**

The high moisture absorption of rayon tends to detract from its value for insulation purposes. The dielectric strength of dry fabrics is fair. Under ordinary conditions, viscose rayon does not develop static charges but antistatic agents are usually added if the relative humidity is less than about 30 per cent.

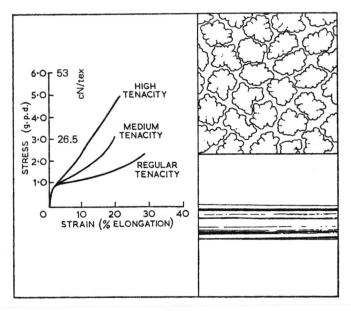

*Viscose* ·

## VISCOSE RAYON IN USE

In the man-made fibre field, rayon plays a role similar to that of cotton in the field of natural fibres. It is produced in greater quantity than any other man-made fibre; it is relatively cheap, and has a wide range of applications.

Although viscose rayon is similar to cotton in its cellulosic structure, it provides a range of yarns and fabrics with their own characteristic properties. The cellulose of rayon has been modified to some degree during manufacture, and the alignment of the molecules is not identical with that of natural fibres such as cotton. Also, the fact that rayon is a manufactured material enables us to control the physical characteristics of the final product. We can make the rayon coarse or fine, alter its strength and elasticity, modify its lustre and colour. Moreover, as a manufactured material, viscose rayon is not subject to changing economic and climatic circumstances such as those that affect the properties and price of a natural fibre.

Viscose rayon conducts heat more readily than silk does, and the rayon has a cooler feel against the skin. Viscose is also highly absorbent, and this enhances its value as a clothing material.

The loss of strength which rayon undergoes when wet is probably its most serious shortcoming, but modern resin finishes have done much to overcome this problem. Properly finished rayon garments have high dimensional stability when wet.

The introduction of rayon staple has enabled manufacturers to blend rayon with other natural and synthetic staple fibres, and rayon staple is used very largely in this way. Rayon contributes its moisture absorption and other 'cellulosic' characteristics to blends of stronger and less absorbent fibres, including most of the synthetics. Blends with polyester staple are of particular importance.

Blends of rayon with other fibres may be processed by any of the familiar techniques. Staple lengths are provided to suit particular blends and systems; a 50 mm (2 in) staple, for example, can be handled on cotton machinery and a 100–150 mm (4-6 in) staple may be used with wool.

### Washing

In general, viscose rayon fabrics wash like cotton; they are cellulosic fibres. But viscose goods are much less strong than cotton, especially when wet, and this must always be remembered when rayon garments are laundered. The use of resin finishes has done much to increase the dimensional stability of viscose fabrics when wet, but it is always necessary to take great care when wet rayon fabrics are being handled.

Rayon fibre itself does not shrink appreciably, but a woven fabric may undergo progressive shrinkage, even when it has been treated with a resin finish to provide dimensional stability. Much depends upon the way the cloth has been constructed.

Rayon fabrics will usually withstand temperatures up to boiling, but it is recommended that most garments should be washed in hot water (60°C., 140°F.). In the case of some knitted and lightweight garments, hand-hot (48°C., 118°F.) or even warm (40°C., 104°F.) water should be used.

In general, the washing temperatures for other fibres (except cotton) are more restricted than for rayons, and when rayon is blended with other fibres the washing instructions for these fibres should be followed.

Washing and bleaching agents may be used as for cotton. Soap does not affect the fibre under normal washing conditions, and rayon will withstand hypochlorite bleaches.

### Drying

Rayon is an absorbent fibre, and it dries slowly. Heavy garments must be supported carefully when they are hung up to dry, or they may stretch and lose their shape. After the fabric has dried, the rayon retains its original strength.

Spin driers and tumbler driers may be used with rayon garments, but spinning for too long may increase the amount of ironing required. Hard wringing must be avoided.

### Ironing

Ironing presents no special difficulties. The fibre is not unduly sensitive to heat, and ironing temperatures for other fibres (except cotton) are more restricted than those for rayons. When ironing blends, the instructions for the other fibres in the blend should be followed.

Rayon fabrics iron well with a medium-hot iron (HLCC setting 3) when slightly damp. If damping is necessary, it is better to roll the garment in a damp towel than to sprinkle it.

**Dry Cleaning**

Viscose is not affected by the usual dry cleaning solvents, and viscose fabrics may be dry cleaned as effectively as cotton.

**End Uses**

A great variety of fabrics can be made from viscose rayon, and it is now possible to use rayon in making almost any of the traditional patterns and weaves that have long been made from natural fibres.

To meet specific needs, viscose rayon is produced in a wide variety of types and sizes, and it is possible to ring the changes on fibre properties by suitable choice of rayon type. The softness of handle of a fabric is increased, for example, by using finer filaments.

Rayon in its many forms is astonishingly and uniquely versatile. It is used in every branch of the textile industry; men's women's and children's outerwear and underwear; furnishings and carpets; household textiles and medical fabrics.

*Crimped Rayons*

These are finding particularly important outlets in tufted carpets and rugs, tufted chenilles, curtains, upholstery, and non-woven fabrics for surgical use.

*Spun-dyed Rayon*

Curtains and car upholstery are applications in which spun-dyed rayon is of special interest, providing exceptional light stability. In knitted goods, blends with acrylic fibres are popular.

Jersey knit fabrics have long been a preserve of cotton, which excels in low shrinkage, good handle and cover. The introduction of foam-backed fabrics and laminates, however, has largely cancelled out the advantages enjoyed by cotton in these respects, and spun-dyed rayon has made good headway in this field.

*Flat Filament*

Flat filament viscose is used where increased lustre and a firmer handle are required.

An unusual but potentially important outlet is in the blending of very short flat-filament staple with wood pulp, in the production of paper.

37

## POLYMER-MODIFIED VISCOSE RAYON

By mixing substances into viscose solution before it is spun, the manufacturer can alter the composition of the extruded filament. Spun-dyed fibre, for example, is made by adding finely-dispersed pigments to the viscose solution (see page 21); titanium dioxide is added to dull the lustre of the filaments.

This technique may be used to modify the character of viscose fibre in more subtle ways, giving it characteristics which are required for particular applications. Additives may increase the water resistance of the rayon, for example, or give it an affinity for dyes outside the usual range of viscose dyes.

The possibilities inherent in this technique are almost infinite, and thousands of modified rayons have been made. A few have become of commercial importance.

### Incorporated Rayon Staple

Viscose may be blended with non-cellulosic polymers, including polyacrylonitrile and polyvinyl alcohol, commonly to the extent of about 10–20 per cent. These so-called incorporated rayons may be made in a great variety of forms, depending on the nature and amount of added polymer. Features which are generally common to all include affinity for acid dyes, increased elasticity, bulkiness and wool-like handle.

### Cross-linked Rayon

Polymers added to viscose may bring about cross-linking of the cellulose molecules. This reduces the freedom of movement of the molecules with respect to each other, and has the effect of reducing water absorption and swelling in water and alkali, and increasing the wet initial modulus.

These improvements in resistance to the effects of water are reflected in increased dimensional stability and washability. Fabrics also have a better handle and increased covering power.

### Grafted Rayon Staple

Polyacrylonitrile, polystyrene and other polymers may be added to viscose solution in such a way as to produce graft polymers. The characteristics of these depend upon the nature and extent of the grafting, but the technique is used commonly to improve the dimensional stability of the viscose. Grafted viscose rayons have increased resistance to water and alkali, with reduced swelling. The elastic recovery of the fibre is increased, the handle is more wool-like, and cover is improved.

### Basified Viscose

The introduction of rayon staple provided a fibre that could be blended with other staple fibres, including wool. Viscose is a cellulosic fibre, however, and it does not have an affinity for acid dyes comparable with that of wool. The dyeing of blends of viscose and wool may therefore present difficulties.

By incorporating basic constituents in the viscose solution, it is possible to produce fibres which have improved dyeability with respect to acid and other dyestuffs used in dyeing wool.

The additives may take the form of synthetic resins, as used in 'Rayolanda', or casein as in 'Cisalpha' and 'Lacisana'.

## HIGH TENACITY VISCOSE RAYON

### INTRODUCTION

The stretching of filaments during the spinning operation is a feature of all modern viscose rayon production processes. It stems from the technique of using two godet wheels, one rotating faster than the other, perfected by L. P. Wilson of Courtaulds Ltd in 1914. This provided a simple and effective way of bringing about the continuous orientation of cellulose molecules in the newly-formed filament, and so increasing the strength of rayon to a satisfactory level.

Since that time, research on the spinning of rayon has been intensive and unceasing. Every aspect of viscose production and spinning has been studied, and increased understanding of the factors involved has made possible the production of rayons with characteristics that are suited to particular applications. Perhaps the most important of these are the high tenacity rayons.

The use of textile yarns as reinforcement in industrial applications has been increasing rapidly in importance during the last half century. High strength yarns are used in hose pipes, conveyor belts, tyres and other applications of this type. Most important of all is the part played by reinforcement yarns in tyres, and as the car and truck industry has grown, so has the demand for high strength yarns increased.

Until the 1930s, this demand was met almost entirely by the use of cotton yarns. By this time, the market for tyre yarns and associated industrial reinforcement yarns had grown to an enormous degree, and had become an attractive field for the sale of any fibre which could meet the requirements. It was of obvious interest to the manufacturers of viscose rayon.

The normal textile rayon in production at that time was not strong enough for use as a tyre yarn. Research was directed, therefore, to producing a viscose filament which would be of a tenacity high enough for this application, and would have the other properties which were required.

The search for high tenacity rayon resolved itself into attempts to increase the degree of stretching to which the filament could be subjected. In 1926, Lilienfeld had shown that it was possible to stretch the still-plastic filament of viscose to produce tenacities in the region of 44 cN/tex (5 g/den). This work led eventually to the production by Courtaulds Ltd. of a high tenacity yarn, 'Durafil'. The early 'Durafil' had a tenacity of 49 cN/tex (5.6 g/den) dry, 34 cN/tex (3.9 g/den) wet.

Unfortunately, the increased tenacity had been attained at the expense of the extensibility, which was 7 per cent dry, and 8 per cent wet. The fibre was brittle and easily broken during processing, a characteristic which detracted from the increased tenacity of the filament itself.

This decrease in extensibility that accompanied an increase in tenacity on stretching was a problem that seemed to be inherent in the production of high tenacity rayon. Research became directed, therefore, towards the discovery of methods of increasing tenacity without decreasing the extensibility to an unacceptable degree.

*'Tenasco'*

In 1935, a major step in this direction was taken with the development of a new process by Courtaulds Ltd., in which viscose was spun into a bath containing a high proportion of zinc sulphate, followed by continuous stretching in a bath of hot water or dilute acid. Under these conditions, the coagulation of the viscose is slowed down, enabling the stretching of the still plastic filament to be carried out under effective control.

This process became the basis for the production of the well-known 'Tenasco' yarns, providing rayon with tenacity of about

31 cN/tex (3.5 g/den), with extensibility slightly less than that of normal rayon (about 17 per cent). The yarn had improved resistance to the effect of water, retained a tenacity of 20 cN/tex (2.3 g/den) with extensibility of 23 per cent.

The development of 'Tenasco' yarns and of high tenacity yarns by other manufacturers allowed viscose rayon to enter the important field of tyre and industrial yarns in the late 1930s. These yarns were stronger and more uniform than the cotton yarns which were being used, and they had better resistance to heat and to fatigue. These advantages enabled them to compete effectively in the industrial fibre field.

During World War II, the demand for tyres increased, and reinforcement yarns were called upon to meet high performance requirements. Heavy duty tyres of all types were brought into use, and synthetic rubber took the place of natural rubber which could no longer be obtained in the quantities required. Under these conditions, the ability to retain tenacity and other characteristics at the elevated temperatures generated in synthetic rubber tyres became of great importance, and the high tenacity rayons proved superior to cotton in this respect. Rayon took over a large share of the tyre cord market during the war.

When the war ended, the trend towards higher and higher performance in tyres continued, and high tenacity rayons continued to make progress in this field. But in due course they, in their turn, began to lose ground in competition with the nylon tyre cords which had now become available. Great efforts were made by the rayon manufacturers to find new types of high tenacity rayon with even greater strength and improved performance. The problem resolved itself, to a large extent, into discovering how tenacity could be increased further without causing a significant decrease in extensibility.

The efforts of the rayon manufacturers were rewarded by the discovery of new techniques of coagulating and stretching the rayon filaments, notably by using special types of additive in the coagulating bath. These additives slow up the coagulation of the viscose, enabling the stretching to be carried out in such a way as to increase the uniformity and regularity of molecular structure in the strand. The effect of this is to increase strength and abrasion resistance, reduce the moisture absorption and the degree of swelling that takes place in water. Extensibility is reduced, but can be held at acceptable levels; this effect is more

41

than compensated for by the increased tenacity, especially when wet.

These fibres have become the basis of the so-called 'Super' high tenacity rayon yarns, such as 'Super Tenasco', 'Super Cordura' and 'Suprenka'. Manufacturers of the super high tenacity yarns cooperated in the marketing of their tyre cords through the Tyrex Inc. organization.

## Molecular Structure

The physical characteristics of rayon depend upon the molecular structure of the filaments. All rayons consist of regenerated cellulose, but the cellulose molecules may be of varying lengths, and they may be positioned in all manner of ways with respect to each other and to the filament itself. The uniformity of the filament may vary with respect to the nature and positioning of the cellulose molecules.

The length of the cellulose molecules is controlled largely by the conditions under which the viscose solution is made. These may be such as to bring about severe breakdown of the cellulose molecules into shorter ones. Or they may be such that breakdown is kept to a minimum, and little depolymerization takes place.

The positioning of the cellulose molecules in the filament is controlled primarily by the conditions under which regeneration and coagulation take place in the spinning bath. The molecules may be produced in such a way that the formation of crystallites is at a minimum, i.e. the degree of crystallization is low. Conditions of spinning and stretching will also influence the size of the crystallites, and the way in which they are orientated with respect to each other and to the long axis of the fibre. These conditions may also be used to control the uniformity of the filament structure; some filaments may be of the same structure throughout, whereas others may have a structure in the centre that is different from that of the outer layers.

These factors all have a major influence on the tenacity and other properties of filaments that are produced when rayon is spun. By understanding and controlling the technique of regenerating, coagulating and stretching rayon, therefore, it is possible to produce rayons to meet particular specifications. It is in this way that modern high tenacity rayons have been developed.

The manner in which control of filament structure influences the characteristics of a rayon may be seen in the range of 'Tenasco' rayons produced by Courtaulds Ltd. There is a progressive increase in tenacity from normal viscose rayon, through 'Tenasco', 'Tenasco 35', 'Tenasco Super 70' and 'Tenasco Super 105'. This change is accompanied by a change in the cross-sectional structure of the filament as shown in the illustration on page 27, notably in the outline of the filaments and in the uniformity of internal structure.

## PRODUCTION

In the production of high tenacity rayons, such as the 'Tenasco' yarns, the coagulation and stretching of the fibre are controlled in such a way as to increase the proportion of skin, and decrease the proportion of core in the filament. The fibre is coagulated in a more uniform fashion, the cross-section becoming less and less serrated as the core-shrinkage effect is diminished.

These changes in the structure of rayon filaments may be seen as a steady progression in the 'Tenasco' yarns, as we move from the ordinary viscose filament to the yarn of maximum tenacity, 'Tenasco Super 105'.

Ordinary viscose rayon has a thin skin, and is highly serrated. In 'Tenasco' high tenacity filament, the thickness of skin has increased, and the serrations have become less pronounced. This trend is maintained in 'Tenasco 35', in which the tenacity h .s increased. In 'Tenasco Super 70', the core has disappeared entirely, and the serrations have almost disappeared, leaving only a bean-shaped cross-section. Finally, in 'Tenasco Super 105', the cross-section is almost round.

There are now many techniques and additives which may be used in producing these super-high-tenacity rayon yarns without reducing extensibility and other desirable properties tu an unacceptable degree. Research in this field of rayon production is as active today as it has ever been, and it seems likely that we shall see many new types of rayon emerging in the future.

## STRUCTURE AND PROPERTIES (High Tenacity Rayons)

High tenacity rayons are chemically similar to regular viscose;

C*                                      43

they are regenerated cellulose. The differences between regular and high tenacity viscose rayons are to be found in the degree of degradation of the cellulose which has occurred during preparation of the viscose, the degree of crystallization, the size of the crystallites, the degree of orientation, and the fine structure and uniformity of the filament.

In those characteristics which derive from the fact that regular and high tenacity rayons are cellulose, the two types of fibre are similar. But the mechanical properties of regular and high tenacity viscose which are influenced so greatly by molecular structure, differ greatly.

### Fine Structure and Appearance

High tenacity rayons commonly have a lobed or serrated cross-section similar to that of regular viscose. There is a thicker skin, however, which may be seen if the filament is dyed.

Super high tenacity filaments are commonly of circular or near circular cross-section, with only a vestige of serration remaining in some cases, e.g. to give a bean-shaped cross-section. These filaments are usually of all-skin structure.

### Tensile Strength

Tenacity is in the region of 26–44 cN/tex (3–5 g/den) dry, 17–22 cN/tex (2.0–2.6 g/den) wet. The wet tenacity is thus about 60 per cent that of dry (cf. 45–55 per cent for regular viscose).

Tenacity of super high tenacity rayons is in the region of 35–45 cN/tex (4–5 g/den) dry, and 26–35 cN/tex (2.9–4 g/den) wet. The wet tenacity is thus about 80 per cent that of dry.

### Elongation

High tenacity rayon: 9.5–11.5 per cent, dry; 20–22 per cent, wet. Super high tenacity rayon: 11.0–12.0 per cent, dry; 24–25 per cent, wet.

### Elastic Recovery

The elastic recovery of high tenacity rayon from 2 per cent extension is typically in the region of 70–100 per cent (cf. regular viscose, about 60 per cent).

**Initial Modulus**

883–1104 cN/tex (100–125 g/den) (cf. regular viscose 485–662 cN/tex (55–75 g/den)).

**Average Stiffness**

230 cN/tex (26 g/den) (cf. regular viscose, 97 cN/tex (11 g/den)).

**Specific Gravity**

1.52–1.54.

**Effect of Moisture**

The high tenacity rayons show a decreased sensitivity to moisture, with lower regain and a higher ratio of wet to dry strength (see Tenacity).

Water imbibition is lower, and the high tenacity fibres do not swell as much as regular viscose.

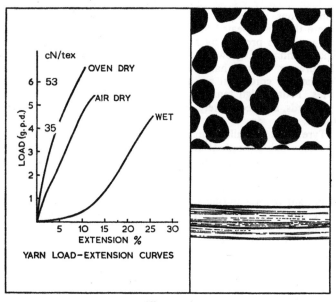

'*Tenasco*'

**Thermal Properties**

High tenacity rayons have an improved performance at elevated temperatures; tensile strength and other mechanical properties are affected less than in the case of regular viscose.

## HIGH TENACITY VISCOSE RAYON IN USE

The highly oriented molecules of cellulose in high tenacity viscose rayons present a barrier to water molecules and to molecules of dyestuffs in solution or suspension. High tenacity viscose rayons do not, as a rule, dye easily or effectively, and the majority of their applications are in fields where colour is of minor importance. They are predominantly industrial fibres, more often than not being buried out of sight in a mass of rubber or similar material.

High strength, and the ability to retain high strength under severe environmental conditions, are the most valuable features of high tenacity viscose rayons. They are used, for example, in applications where elevated temperatures are encountered, or where there is repeated flexing.

Tyre cords provide by far the largest outlet for these fibres. The phenomenal growth of the car and truck industry during the present century has created a huge market for tyres, and for the tyre cords that are used in reinforcing the rubber in the tyres. Today, high tenacity rayons supply a large part of these tyre cords, and seem likely to continue doing so in the forseeable future.

Tyre cords are called upon to provide great tensile strength, and to retain high strength at the considerable temperatures generated inside the tyre during use. They must withstand repeated flexing, and resist deformation. High tenacity viscose rayons have much to offer in these respects, and they have ousted cotton from this important market during recent years. The arrival of nylon has diminished the hold that high tenacity rayons had established, but rayon retains a large portion of the tyre cord market, and it is unlikely that nylon will change this situation.

Nylon competes with greatest effect in the reinforcement of heavy duty tyres for aircraft, earth-moving equipment and the like. In these applications, its phenomenal resistance to shock loads gives nylon the edge on high tenacity rayon, and it also has a higher strength/weight ratio. But in tyres for lighter purposes, including the mammoth car tyre market, high tenacity rayon retains its hold. The ability to sustain high tenacity and dimensional stability at the temperatures generated in car tyres, the excellent resistance to fatigue, and the price advantage of high tenacity viscose rayons have enabled these fibres to withstand competition from nylon and other synthetics in this field.

The combination of properties which serves high tenacity rayons in tyres has been equally effective in other important industrial applications. Flexible rubber belting is used for conveying all manner of materials, from coal and iron ore to parts and products moving down innumerable assembly lines. The rubber in these conveyor belts requires reinforcement, just as it does in a car tyre. And again, high tenacity viscose rayons have come into widespread use for this purpose. They provide the high strength, dimensional stability, fatigue resistance and flexibility that are needed, and at modest cost.

Power transmission belts form another important sector of this field, requiring the same combination of properties in the yarns that are used to reinforce them. Nylon's superior resistance to shock loading gives it a useful advantage over high tenacity rayons in some power belt applications, but high tenacity rayon remains competitive where this is not a vital requirement.

High tenacity viscose rayons have many other applications in industry, including the production of tarpaulins and protective fabrics, sewing threads and umbrella fabrics, the reinforcement of hoses and of plastics used for bearings and other heavy-duty purposes.

## HIGH WET MODULUS (POLYNOSIC) RAYONS

### INTRODUCTION

Viscose rayon has now been in production for more than half a century. When first produced, it was a filament yarn of high lustre which bore a superficial resemblance in these respects to natural silk. Its properties did not, however, bear comparison with those of silk, and it competed initially in the continuous filament field on the basis of relative cheapness and novelty value.

In due course, viscose rayon settled down and began to find its proper niche in the textile field. It was accepted as a cellulosic fibre, in this respect resembling cotton, which could be produced extremely cheaply from the cellulose available in wood. Its shortcomings were accepted, and it found its market in those applications where cheapness was of over-riding importance.

As experience of viscose production grew, and the process came under scientific investigation, improvements were made in fibre quality. This trend has continued to the present day, and modern viscose rayon has established for itself a wide range of applications and outlets. The production of viscose rayon exceeds that of any other man-made fibre, and this position seems likely to remain for a very long time to come.

Despite this remarkable progress made by viscose rayon, and the continuous improvement in the quality of the fibre, viscose retains unattractive characteristics which have been associated with it since the earliest times. These shortcomings have prevented viscose rayon from competing as effectively as it might with the natural fibre it most nearly resembles – cotton.

As a manufactured fibre, viscose rayon has certain advantages over cotton. It is produced as a continuous filament, of uniform linear density and composition. It is cut into staple of any length, or mixture of lengths. The production costs can be assessed and controlled more accurately than is possible with cotton, which is subject to all the fluctuations of price and production typical of a natural product. And rayon, produced from a cheap and abundant raw material, is the least expensive textile fibre now available.

These advantages have enabled rayon to sustain an important position in the textile field. But rayon manufacturers have long understood that the potential of viscose rayon is only partly being realized; if its shortcomings could be overcome, and its properties brought more nearly into line with those of cotton, viscose rayon could become the most important textile fibre of all.

The deficiencies of modern viscose rayons are the same as those that have been with it since the first filaments were produced at the end of the last century. Improvements have been made, but these have been largely a matter of degree.

Viscose rayon is sensitive to the effects of moisture. When rayon is wet, it absorbs water and swells, the diameter of the filament increasing by more than 25 per cent. At the same time, the tenacity falls by about 50 per cent, and the extensibility increases by some 20 per cent. The initial modulus of the rayon falls, and the filament will stretch in response to only a small tensile stress. Elastic recovery from such stretching is poor.

This deterioration in the mechanical properties of rayon when wet is reflected in the behaviour of yarns and fabrics. Rayon goods do not possess the wonderful wet-stability and washability of cotton. They tend to deform when handled without due care, and undergo progressive shrinkage.

Also, rayon does not have the crisp, firm handle that is so characteristic of cotton. Rayon fabrics tend to have a limp and floppy feel.

In recent years, much has been done to improve rayon in these respects. High tenacity rayons, for example, have enabled rayons to compete effectively in the important field of industrial textiles. Cross-linked and chemically modified rayons have increased the resistance to water (see page 38), and resin finishes have done much to provide dimensional stability and washability.

Despite these advances in rayon technology, however, viscose rayon is still no match for cotton in its behaviour with respect to water, or in the character and crispness of its handle.

## Structural Differences

In chemical composition, viscose rayon and cotton are alike; they are both cellulose. The differences between the fibres stem from differences in the physical structure of the filaments. It is reasonable to assume, therefore, that by modifying the structure of the viscose filament, it should be possible to produce a rayon that more nearly resembles cotton.

The micro-structures of cotton and viscose rayon have been studied extensively, and the differences between them are well understood. In cotton, the cellulose molecules consist of some 2,000 to 10,000 glucose units linked together (i.e. cotton has a degree of polymerization of 2,000 to 10,000). These long cellulose molecules are laid down in a wonderfully precise and ordered way (see Vol. 1), forming a highly orientated, uniform structure in which there is a proportion of crystalline material amounting to about 70–80 per cent.

The crystallites in cotton are orientated with respect to each other, forming fibre-like groups or micro-fibrils; the micro-fibrils, in turn, are arranged into fibrils, and the fibrils into filaments. The structure of the cotton fibre is, in fact, 'fibrous' all the way through.

If a cotton fibre is disintegrated, e.g. by chemical treatment, its micro-fibrillar structure is displayed as it breaks up into ever-finer filaments. With the help of the electron microscope, it is possible to follow the filamentous disintegration until eventually the cellulose molecule itself is reached; this is the finest fibrous element of all.

In this wonderfully organized micro-fibrillar structure of cotton we have the explanation of many of cotton's unique characteristics. The high degree of orientation and crystallization, and the uniformity of the structure, enable the cellulose molecules to cooperate effectively in resisting a tensile stress. Cotton has a high tenacity.

The crystalline regions of the cotton fibre are not readily penetrated by water molecules; the amorphous regions, into which water can find its way, form only a relatively small proportion of the whole. Swelling takes place as water enters the cotton fibre, but without affecting drastically the strength-providing crystalline structure; the ratio of wet to dry strength is high.

The highly-crystalline, highly-orientated micro-fibrillar structure of the cotton fibre enhances the rigidity and stiffness that is inherent in the cellulose molecule itself. Cotton is a stiff fibre, and this stiffness plays a part in giving cotton fabrics their characteristic crispness.

The possibility of reproducing this cotton structure in a viscose filament seems remote indeed. The cotton fibre grows slowly, and its architecture is established gradually and with great precision. Viscose rayon, on the other hand, is created rapidly by regeneration and coagulation of cellulose in the coagulating bath. Simultaneous stretching aligns the cellulose molecules to some degree, the extent of alignment depending on the conditions used. But even under the most favourable circumstances, the positioning of the cellulose molecules cannot be expected to match the precise organization that we find in the cotton fibre.

Despite the obvious difficulties that face the manufacturer in his attempt to model his rayon on the cotton plan, great progress has been made in this respect in recent years. The production of high tenacity rayons has taken us some way along the road; some of these rayons have micro-fibrillar structures that begin to look like that of cotton. Even more impressive progress in this direction has taken place with the development of the new types of viscose rayon which have become known as high wet modulus (HWM) modal and polynosic rayons.

## High Wet Modulus Modal and Polynosic Rayons

The main structural differences between viscose rayon and cotton can be summarized as (a) differences in the degree of polymerization of the cellulose molecules, and (b) differences in the arrangement of these molecules in the filament.

By the 1930s, understanding of the viscose rayon process was such that methods of improving rayon in both these respects were known. It was realized, for example, that breakdown of the cellulose molecules took place during the ageing of the alkali cellulose, and in the ripening of sodium cellulose xanthate solution during viscose production. By avoiding these stages, viscose solutions could be made in which the molecules of sodium cellulose xanthate were longer than in normal viscose.

The development of high tenacity rayons had demonstrated, also, that the structure of the filament could be influenced greatly by control of the spinning conditions. In general, the slower the regeneration and coagulation of the cellulose, the more effectively could stretching be used in orientating the cellulose molecules.

By the late 1930s, rayons were being made in which the degree of polymerization of the cellulose was increased by modification of the viscose production process, and the orientation of the cellulose was improved by slowing the regeneration and coagulation of the cellulose filaments (see High Tenacity Rayon, page 39).

During World War II, further progress was made along these lines, notably in Japan. In 1951, this work culminated in the application for a patent by S. Tachikawa, covering the production of viscose rayon by a technique which yielded fibres of novel type. In particular, the Tachikawa rayons were stronger than regular viscose, with reduced elongation, and they had a greatly improved ratio of wet to dry strength (76 per cent, compared with the 56 per cent of regular viscose). This increased resistance to the effect of water was reflected also in a high wet modulus, with lower water imbibition and reduced swelling.

The new type of rayon differed structurally from regular viscose rayon. The cellulose molecules were longer, with a degree of polymerization in the region of 500 (cf. ordinary rayon about 250). Also, disintegration of the filament displayed a microfibrillar structure with a resemblance to that of cotton.

Development of the Tachikawa process in Japan led to the production of high strength, high wet modulus rayons which were

## HWM MODAL POLYNOSIC FIBRES

| Country | Firm | Fibre Trade Mark |
|---------|------|------------------|
| Austria | Chemiefaser-Lenzing | Superfaser |
| Belgium | Fabelta | Z 54 (Zaryl) |
| England | Courtaulds Ltd. | Vincel |
| France | | Z 54 |
| Germany | | Polyflox, Super Polyflox |
| Italy | Snia Viscosa | Koplon |
| Switzerland | Viscose Suisse | Z 54 |
| Japan | Daiwa Spinning Co. | Polyno |
| | Fuji Spinning Co. | Junlon |
| | Mitsubishi Rayon Co. | Hipolan |
| | Teijin Ltd. | Polycot |
| | Toho Rayon Co. | M 63 (Tovis) |
| | Toyobo Co. Ltd. | Tufcel |
| U.S.A. | Avtex Fibers Inc. | Fiber 40, Avril. |
| | American Enka | Zantrel |
| | Courtaulds N. America Inc. | W 63 (Lirelle) |

marketed as 'Toramomen' and later 'Tufcel'. Similar types of rayon have been developed in other countries, and are now in production.

NOMENCLATURE

In the late 1950s, viscose rayons produced by the new techniques were being described in Europe as 'polynosique' rayons. This term was derived, presumably, from a combination of 'poly', to indicate a high degree of polymerization, and 'cellulosique'. The term was subsequently modified to 'polynosic'.

In its original sense, the term 'polynosic' was restricted to fibres of the high wet modulus type produced by techniques similar to that described in the Tachikawa patent. In the U.S.A., an official Federal Trade Commission definition was coined, using a high wet modulus as the criterion, i.e. extension at 0.5 g./den (4.4 cN/tex) being not more than 3.5% in water (see page 53). The term 'polynosic' has since been used with less precision, to describe higher-strength rayons of the increased wet-strength type. These do not necessarily meet the requirements of the F.T.C. definition.

In practice the term "high wet modulus" (HWM) is commonly used to describe a broad range of fibres of this type, the term "polynosic" being used for those with the highest wet modulus. "Modal" is widely used as a generic term for regenerated cellulose fibres obtained by processes giving a high tenacity and a high wet modulus.

## Textile Institute Definitions (UK)

*Polynosic Fibre* A regenerated cellulose fibre that is characterised by a high initial wet modulus of elasticity and a relatively low degree of swelling in sodium hydroxide solution.

*Modal Fibre* Generic name for regenerated cellulose fibres obtained by processes giving a high tenacity and a high wet modulus.

## Federal Trade Commission Definition (U.S.A.)

The term *polynosic fibre* has been defined by the U.S. Federal Trade Commission as follows:

*Polynosic Fibre.* A manufactured cellulosic fibre with a fine and stable micro-fibrillar structure which is resistant to the action of 8 per cent sodium hydroxide solution down to 0°C., which structure results in a minimum wet strength of 2.2 g/den (19.4 cN/tex) and a wet elongation of less than 3.5 per cent at a stress of 0.5 g/den (4.4 cN/tex).

## TYPES OF HWM MODAL FIBRE

HWM modal fibres all have the following properties in common:

(1) high wet modulus, i.e. resistance to extension when wet
(2) increased ratio of wet to dry breaking tenacity
(3) increased resistance to swelling by caustic alkalis
(4) high degree of polymerization of cellulose
(5) micro-fibrillar structure.

These characteristics are shared with cotton and other natural cellulosic fibres, and for this reason HWM modal fibres are sometimes called 'artificial cottons'.

### Three Types

Despite these characteristics which all HWM modal fibres have in common, the individual HWM modal fibres differ from one another

## HWM MODAL FIBRES – COMPARATIVE PROPERTIES

| | HWM Modals | | | Cotton Uppers | Rayon Staple |
|---|---|---|---|---|---|
| | High Strength | Standard | High Elong'n | | |
| | Super Poly-flox Junlon W 63 (Lirelle) | Z 54 Vincel Polyflox Koplon Polyno Hipolan Polycot | Superfaser Fiber 40 (Avril) | | |
| Tenacity (cN/tex) | | | | | |
| dry | 41–46 | 28–35 | 33–42 | 32 | 22 |
| wet | 30–35 | 18–27 | 21–30 | 35 | 12 |
| Extensibility (%) | | | | | |
| dry | 6–10 | 8–12 | 12–14 | 9 | 18 |
| wet | 8–14 | 9–16 | 16–20 | 10 | 22 |
| Wet Modulus (cN/tex) per 100% ext'n | | | | | |
| at 2% ext'n | 132–221 | 98–159 | 53–80 | 106 | 35 |
| at 5% ext'n | 221–353 | 124–247 | 109–115 | 159 | 44 |
| Water Imbibition (%) | 65–75 | 55–70 | 65–75 | 50 | 90–100 |

in properties over a wide range. Air-dry tenacities, for example, may be 28.3–47.7 cN/tex (3.2–5.4 g/den); wet elongations may range from 8 to 20 per cent. Within this wide range of properties, however, it is possible to classify the polynosic fibres into three main groups, as proposed by J. D. Griffiths of Courtaulds Ltd. (*Text. Inst. Industr.*, 1965, **3**, No. 3, p. 54):

(1) *High Strength HWM Fibres*. These are characterized by high tenacities, dry and wet, e.g. 40.6–45.9 cN/tex (4.6–5.2 g/den) dry; 30–35.3 cN/tex (3.4–4.0 g/den) wet.

(2) *Standard HWM Fibres*. This group includes the majority of polynosic fibres. Tenacities 28.3–35.3 cN/tex (3.2–4.0 g/den) dry; 17.7–26.5 cN/tex (2.0–3.0 g/den) wet. Elongations in the range 8–12 per cent dry; 9–16 per cent wet.

(3) *High Elongation HWM Fibres*. These are characterized by high elongations, dry and wet, e.g. in the range 12–14 per cent dry; 16–20 per cent wet.

Fibres belonging to each of these three groups, and the ranges of properties they display. are shown in the table on page 54. The table also includes corresponding properties of ordinary rayon staple and of a representative cotton (uppers).

## PRODUCTION

The principles followed in the production of high wet modulus rayon by the Tachikawa technique are (a) reduction in the amount of cellulose breakdown which takes place in the preparation of the viscose solution and (b) slowing down of the regeneration and coagulation of the filament, permitting stretching to be carried out gently and in stages.

The stages in production of a HWM modal rayon by the Tachikawa process are as follows:—

(1) Soda cellulose (alkali cellulose) is produced by steeping the cellulose in caustic soda, followed by pressing and shredding as in the production of regular viscose. The conditions are carefully controlled to ensure that the temperature does not rise above 20°C., and the process is completed within 2 hours.

(2) The theoretical quantity of carbon disulphide is used in xanthation (less than the theoretical quantity is used in producing regular viscose), and the addition is made over $2\frac{1}{2}$ hours. The temperature is held below 20°C., and then raised to 25°C. for 1 hour.

(3) Sodium cellulose xanthate is dissolved in water to provide a solution containing the equivalent of 6 per cent cellulose and 2.8 per cent sodium hydroxide. (In the regular viscose process, xanthate is dissolved in caustic soda solution.)

(4) The solution is spun by extrusion into a bath of very dilute (1 per cent) sulphuric acid at 25°C. The filaments are stretched in stages to three times their spun length. (Regular viscose is extruded into a bath containing 10 per cent sulphuric acid, 1 per cent zinc sulphate and about 18 per cent sodium sulphate maintained at 45–55°C. The degree of stretch depends upon the type of rayon being made.)

Under these conditions, the degradation of cellulose is held to a minimum by the omission of ageing and ripening stages, and by the milder conditions used in preparing the viscose solution. The regeneration and coagulation of the cellulose takes place slowly and gently in the dilute acid of the spinning bath, which contains little or no salt. This permits stretching to be carried out gradually, allowing the molecules to assume a high degree of orientation and crystallization. The filaments produced are of more uniform composition; the cross-section is round.

The degree of polymerization of HWM modal fibres produced in this way is about 500, i.e. about twice that of ordinary rayon.

The conditions described above are typical of those used in HWM modal fibre production, but they may be varied in a number of ways to provide fibres of the desired characteristics within the HWM modal range. The coagulation bath, for example, may contain sulphuric acid and sodium sulphate in varying proportions. Zinc salts may be used to slow the regeneration of cellulose by forming zinc cellulose xanthate (see page 14). Formaldehyde may be added to the viscose solution or to the spinning bath, forming an ester between the xanthate and the formaldehyde, which also serves to slow the regeneration process.

## PROCESSING

HWM modal fibres are essentially cotton-like in character, and the initial emphasis in staple production has been to provide staple lengths and linear densities suitable for use on cotton spinning machinery. The setting of cards, drawframes, etc., require adjustment to suit the particular fibre being processed, but there are no major difficulties in producing HWM modal yarns to a wide range of counts, from coarse to fine.

HWM modal fibres share with other man-made fibres the advantages of uniformity of staple and linear density, and yarns may be produced from them to a much higher standard of uniformity than is possible with cotton.

HWM modal fibres are also spun and cut to linear densities and staples suitable for processing on woollen worsted and flax equipment.

Weaving and knitting of HWM modal yarns is straightforward, the best results in weaving fine count yarns being obtained at relative humidities of 70% and above, as in the weaving of cotton.

### Bleaching

Hypochlorite, chlorite and hydrogen peroxide bleaches may be used safely on HWM modal fibres. The greater cleanliness of these fibres means, however, that less drastic conditions may be used to produce whites equivalent to those obtained with other fibres. As with all cellulosic fibres, overbleaching may cause degradation and should be avoided.

### Dyeing

All the usual types of dyestuffs for cellulosic fibres may be used on HWM modal fibres, including direct, vat, azoic, reactive etc. In general, the dyeing properties of HWM modals are nearer to cotton than to ordinary viscose rayon, but the affinities of individual HWM modal fibres varies considerably. This is particularly noticeable with direct dyes. Thus, a dyestuff may have an affinity for one HWM modal equivalent to that it shows for cotton, whereas the same dyestuff will have an affinity for another HWM modal which is closer to that it displays for ordinary viscose.

The range of available dyestuffs is so wide that almost every shade of required fastness to washing, light, perspiration, etc., can be produced on any HWM modal fibre.

### Finishing

HWM modal rayons are generally similar to cotton in their chemical and physical structure, and they respond to finishing in much the same way as cotton. HWM modals have a greater resistance to swelling in caustic alkalis than ordinary viscose, and they will withstand mercerizing conditions. The resistance of HWM modals is not as great as that of cotton however, and it is inadvisable to use the full mercerization process with 100% HWM modal fabrics. This is, in any case, unnecessary as HWM modal fabrics may be fully set, stabilized and given increased dye affinity by treatment with caustic soda not exceeding 6–7 per cent. This concentration causes no significant loss of properties, and it is useful as a pretreatment for various resin applications, and to increase colour yield in printing.

Resin treatments used with cotton may be applied to HWM modals, e.g. to provide increased stability and to bestow ease-of-care properties. HWM modal fabrics subjected to resin treatment suffer loss of tensile and tear strength to a lesser extent than cotton given the same treatment. This means that HWM modal fabrics of given crease-recovery and ease-of-care properties will be stronger than cotton or rayon fabrics of equivalent crease-recovery and ease-of-care properties. The amount of resin needed to attain a particular level of these properties with HWM modal fabrics is less than with cotton or rayon fabrics.

## STRUCTURE AND PROPERTIES

### Fine Structure and Appearance

HWM modal fibres are typically of round cross-section, and do not display any skin effect. The micro-structure is fibrillar, the filament breaking up into smaller and smaller fibrils when disintegrated for example by nitric acid. The fibrils are distributed uniformly throughout the filament cross-section, producing a homogeneous structure. The degree of polymerization is in the region of 500. The degree of crystallinity of HWM modal fibres is in the region of 55 per cent, compared with 40–45 per cent for ordinary rayon and 70–80 per cent for cotton. The crystallites in HWM modal fibres are larger than those in ordinary rayon.

The degree of orientation of the long cellulose molecules, in both the amorphous and the crystalline regions of the fibre, is higher in HWM modal fibres than in ordinary rayon.

### Tenacity

High Strength: 41–46 cN/tex (4.6–5.2 g/den) dry; 30–35 cN/tex (3.4–4.0 g/den) wet.

Standard: 28–35 cN/tex (3.2–4.0 g/den) dry; 18–26 cN/ tex (2–3 g/den) wet.

High Elongation: 34–42 cN/tex (3.8–4.8 g/den) dry; 21–30 cN/ tex (2.4–3.4 g/den) wet.

### Elongation

High Strength: 6–10 per cent dry; 8–14 per cent wet.
Standard: 8–12 per·cent dry; 9–16 per cent wet.
High Elongation: 12–14 per cent dry; 16–20 per cent wet.

### Elastic Recovery

Elastic recovery is higher for HWM modals than for cotton or rayon staple, especially in the wet state. The following diagram shows the work done in stretching a fibre when wet, and the permanent set (i.e. 100 per cent minus elastic recovery per cent) for the three classes of HWM modals, and for cotton and ordinary rayon staple. The superiority of HWM modals in this respect means that fabrics made from them are less subject to permanent deformation during wet treatments than either cotton or ordinary rayon staple fabrics.

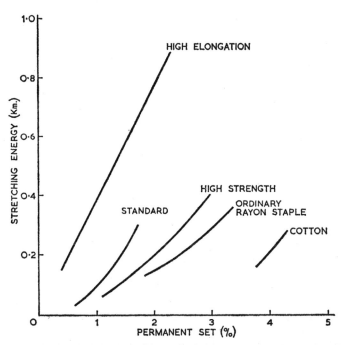

*HWM Modal Fibres.* The relationship between work done in stretching a fibre when wet and the permanent set. The superiority of HWM modal fibres with respect to ordinary rayon staple and cotton means that HWM modal fabrics are less subject to permanent deformation when given wet treatments than either cotton or ordinary rayon fabrics —

*Courtesy J.D. Griffiths.*

## Initial Modulus, Wet

Less than 3.5% elongation in water at 4.4 cN/tex (0.5 g/den).
High Strength HWM Modal:

    132–221 cN/tex (15–25 g/den) per 100% ext'n at 2% ext'n.
    221–353 cN/tex (25–40 g/den) per 100% ext'n at 5% ext'n.
Standard HWM Modal:

    88–159 cN/tex (10–18 g/den) per 100% ext'n at 2% ext'n.
    124–247 cN/tex (14–28 g/den) per 100% ext'n at 5% ext'n.
High Elongation HWM Modal:

    53–79 cN/tex (6–9 g/den) per 100% ext'n at 2% ext'n.
    88–115 cN/tex (10–13 g/den) per 100% ext'n at 5% ext'n.

### Effect of Moisture

See Tenacity and other tensile properties.
Water Imbibition:

    High Strength HWM Modals: 65–75 per cent.
    Standard HWM Modals: 55–70 per cent.
    High Elongation HWM Modals: 65–75 per cent.
Increase in diameter on wetting: 11.5-15 per cent. This is intermediate between cotton and ordinary rayon staple. HWM modal fibres swell less readily in aqueous solutions than ordinary rayon staple does.

### Effect of Alkalis

HWM modal fibres swell much less than ordinary rayon staple. They will withstand mercerizing conditions.

### Chemical and Biological Properties

Generally similar to other cellulosic fibres.

## HIGH WET MODULUS RAYON FIBRES IN USE

HWM modal fibres have brought to the textile industry a viscose rayon which approaches cotton in character, notably in its behaviour with respect to water. The high initial wet modulus of HWM modal fibres is reflected in fabrics which are highly resistant to deformation when wet; they will withstand the stresses imposed during laundering and wet processing generally without undergoing shrinkage to a significant extent, and without being pulled out of shape.

This dimensional stability under wet conditions is perhaps the most important characteristic of HWM modal fibres from the practical point of view. But HWM modal fibres may also be given a soft, silky handle that differs from cotton or viscose.

## Washing

Fabrics made from HWM modal fibres may be washed repeatedly without undergoing deformation or progressive shrinkage. Laundering characteristics are generally similar to those of cotton.

## Ironing

HWM modal fabrics iron like cotton.

## Dry Cleaning

HWM modal fibres are virtually pure cellulose, and are not affected by dry cleaning solvents. Fabrics made from HWM modal fibres may be dry cleaned as readily as cotton.

## End Uses

The dimensional stability of HWM modal fabrics when wet has given them the entrée into virtually every field of application in which cotton is used. HWM modal fabrics are strong, hard wearing, dimensionally stable, uniform and of good handle and appearance.

The poor stability of ordinary viscose when wet has denied it access to many important applications, including woven and knitted shirtings, blouses, knitted underwear and outerwear garments. But HWM modals have now enabled viscose to compete effectively with cotton in end uses of this type, especially those in which a soft, silky handle is advantageous. In printed dress fabrics, for example, HWM modals provide a combination of stability, subdued lustre and attractive handle.

Dimensional stability at varying humidities is an important factor in the field of curtain materials. Curtains made from ordinary viscose staple are liable to alter in length from season to season, and even from morning to evening, as a result of changes in humidity. But this phenomenon may be overcome by the use of HWM modal yarns in the warp of the fabric.

Cotton interlock fabrics tend to become hard and boardy, and

## STRESS-STRAIN DIAGRAMS – HWM MODAL FIBRES

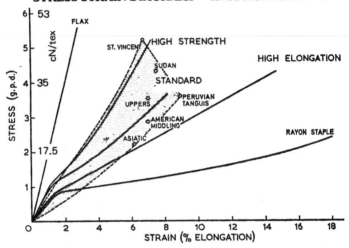

1. *Air-Dry*. This diagram, based on air-dry conditions, compares the three types of HWM modal fibre with cottons, flax and rayon staple. *Cotton*. The region covered by the curves for various cotton fibres is indicated by the shaded area. The upper margin of the area represents the stress-strain curve of a Sea Island cotton (St. Vincent); the lower margin represents the curves of typical Asiatic cottons (Oomras and Bengals). The breaking tenacities of various Egyptian, American and Peruvian cottons are indicated by points.
*HWM Modal Fibres*. Curves typical of the three groups of HWM modal fibres are shown. It will be seen that (a) the curve of the high strength HWM modal group closely resembles that of Sea Island cotton, (b) the curve of the standard HWM modal group resembles those of uppers, American Middling and Peruvian cottons, (c) the curve of the high elongation HWM modal group lies within the area of the cotton curves for about the first 5 per cent of extension, but beyond this point its high extension takes it well outside the area.
*Flax; Rayon Staple*. The curves for flax and rayon staple represent approximately the two extremes of the total range of cellulosic fibre curves.

See diagram opposite for fibres in the wet state.

the surface takes on a felted appearance, when subjected to repeated washing and tumble drying. HWM modal interlock suffers none of these defects, and retains its initial appearance and handle throughout most of its life.

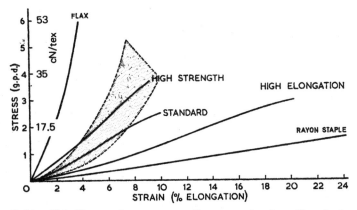

2. *Wet*. This diagram shows corresponding curves for these fibres in the wet state.

*Cotton*. The breaking tenacity of cotton fibres has increased slightly, but there has been a significant drop in initial modulus.

*HWM Modal Fibres*. These all show lower breaking tenacities and increased breaking extensions, which are evident from the lowering of the slopes of the curves.

High strength and standard HWM modal fibres remain largely within the region of the cotton curves. High elongation HWM modal fibres now lie between the cotton curves and that for rayon staple.

In tensile properties, the HWM modal fibres resemble cotton more closely than does any other type of regenerated cellulosic fibre.

*Courtesy J.D. Griffiths.*

## Blends

The end uses described above relate largely to yarns made from 100 per cent HWM modal fibres. But HWM modals are also making rapid headway as constituents of yarns spun from fibre blends.

### Cotton Blends

The resemblance of HWM modal fibres to cotton indicates that these two types of fibre would be compatible in blends. Much of the HWM modal fibre now produced is being used in blends with cotton, and many excellent fabrics are made in this way, including dress materials, sheetings, furnishings etc.

HWM modal fibres contribute increased regularity and uniformity to the blended yarn.

### Flax Blends

HWM modal fibres have higher bending and torsional rigidities than either cotton or rayon staple, and in heavier deniers they provide fabrics with a linen-like handle. This is especially true in the case of the high strength and standard HWM modal fibres. Blended with flax, HWM modal fibres of this type provide fabrics of full linen handle which is retained after repeated washings.

### Wool Blends

Blends of HWM modal fibres with wool provide fabrics which display reduced shrinkage, even with the HWM modal constituent in comparatively small amounts. The handle and appearance of the fabrics remain substantially unchanged, resembling the all-wool fabric.

### Synthetic Fibre Blends

HWM modals have replaced cotton to a considerable extent in blends with polyester fibres; the fabric retains all the desirable characteristics of the cotton blend, but is clearer and more regular. Also, such blends of HWM modal and polyester may be cheaper, as it is not necessary to comb the HWM modal fibre; cotton is commonly combed before blending.

The standard HWM modals have a load-extension curve which is very similar to that of polyesters for the first few percentage extension; this part of the curve is of greatest importance in fabric performance. The high elongation HWM modals have air-dry curves which resemble the polyester curves over almost the whole of their range. Blend yarns of different ratios show a more nearly linear relationship than do similar blends with other types of HWM modal fibre, and for this reason the high-elongation type of HWM modal is favoured for polyester blends.

Blends of HWM modal fibre with a relatively small proportion of low density fibre, e.g. polypropylene, provide fabrics with handle and shear properties very similar to those of cotton fabrics. The addition of the low density fibre increases the bulk of the HWM modal yarn, which tends to be less bulky and more compact than cotton, largely owing to the circular cross-section of the HWM modal fibre.

## CUPRO (CUPRAMMONIUM)

### INTRODUCTION

Cellulose will dissolve in a mixed solution of copper salts and ammonia, called cuprammonium liquor, and regenerated cellulose fibres are produced by extrusion of this solution into a coagulating bath. The yarn produced by the cuprammonium process consists of regenerated cellulose; it is now widely known by the name of *cupro*.

The cuprammonium process had its beginnings in the very early years of cellulosic fibre manufacture. In 1890, a French chemist, Louis Henry Despeissis, discovered that he could dissolve cellulose in cuprammonium liquor, and spin a fine filament of 'artificial silk' from the solution. Despeissis died, however, and for two years his invention was forgotten.

In 1892, Max Fremery and Johann Urban at Oberbruch in Germany made use of the cuprammonium process for making carbon filaments used in early electric light bulbs. In 1899, with the help of others, they began manufacturing cellulosic fibres for textile purposes; this was the beginning of the Vereinigte Glanzstoff-Fabriken A.G., which became one of the largest manmade fibre producing organizations in the world (Glanzstoff A.G.).

After about 10 years, the cuprammonium process was abandoned in favour of the viscose process, and it remained neglected until after World War I. In 1919, a technique of stretch spinning developed by J. P. Bemberg A.G. revived interest in the cuprammonium process, and since then the production of the fibre has continued. The cupro fibre has become widely known as 'Bemberg' yarn.

### NOMENCLATURE

#### *Federal Trade Commission Definition*

Fibre produced by the cuprammonium process is regenerated cellulose which falls within the class described as *rayon* under the U.S. Federal Trade Commission definitions, the official description being as follows:

*Rayon.* A manufactured fibre composed of regenerated cellulose, as well as manufactured fibres composed of regenerated cellulose in which substituents have replaced not more than 15 per cent of the hydrogens of the hydroxyl groups.

## Cupro

The term *cupro* has now come into widespread use throughout the world to denote any regenerated cellulose fibre produced by the cuprammonium process.

## Cuprammonium Rayon

Cupro is also still described as *cuprammonium rayon* to distinguish it from viscose rayon.

## PRODUCTION

In its essentials, the process for making cupro is similar to that used in making viscose. Cellulose is dissolved, and the solution is forced through holes in a spinneret. The jets of solution are coagulated, the cellulose being regenerated as a solid filament.

### Raw Material

Cotton linters and wood pulp are both used as raw material in making cupro. Cotton linters is a source of very pure cellulose, and for this reason was preferred initially as raw material. Latterly, however, wood pulp has been used on an increasing scale, largely because of its lower cost. For high quality productions, cotton linters cellulose is still used exclusively.

Cotton linters is purified by kier-boiling with dilute caustic soda at about 150°C., followed by bleaching with sodium hypochlorite.

Wood is selected and purified to yield a material of high alpha cellulose content (above 96 per cent).*

Cuprammonium liquor is prepared by dissolving basic copper sulphate in ammonia to form a solution of cupritetrammino hydroxide and cupritetrammino sulphate in the ratio 3:1, containing 3-4 per cent copper and 5-8 per cent ammonia.

* Alpha cellulose is that which does not dissolve in 17.5–18.0 per cent caustic soda solution after 30 minutes at 20°C. It consists of cellulose which has undergone a minimum of degradation, and it is the most satisfactory cellulose for use in fibre-manufacture.

### Preparation of Spinning Solution

Purified cotton linters or wood pulp is mixed into cuprammonium liquor at low temperature. Stabilizing agents and caustic soda are added, the latter in sufficient quantity to convert the cupritetrammino sulphate into hydroxide. The cellulose content of the solution is about 10 per cent.

The spinning solution is filtered by passing it through a succession of nickel filter screens. It is then deaerated and is ready for spinning. The solution is stable and may be stored for considerable periods without appreciable deterioration; in this respect, it contrasts strongly with viscose solution.

### Spinning

(a) *Batchwise Spinning (Reel or Pot Spinning)*

The filtered spinning solution is pumped to a nickel spinneret, and extruded through holes of 0.8 mm. diameter. The jets of solution emerging from the spinneret holes flow into a glass funnel, where they meet a stream of pure water which is flowing down through the funnel. The water dissolves most of the ammonia and about one third of the copper from the jets, bringing about coagulation of the cellulose to form plastic filaments. The filaments are carried along by the stream of water, and are stretched continuously to form filaments of usually about 1.4 dtex (1.3 den).

The loose thread of filaments emerging from the bottom of the funnel is carried round a guide rod, most of the water being flung off. The thread then passes round a roller which rotates in a trough of sulphuric acid; the remaining copper and ammonia are removed as copper sulphate and ammonium sulphate respectively.

The filaments are then wound either into skeins (Reel Spinning), or into cakes in a Topham box (Pot Spinning). The skeins or cakes are washed to remove acid and any remaining copper sulphate or ammonium sulphate, softened by adding lubricants, and dried. The yarn is commonly given a second wash in soap and oil emulsion, or (if it is to be twisted later) in a soaking bath. It is then dried again.

(b) *Continuous Spinning*

As in the production of viscose rayon, the production of cupro has been modified to operate on a continuous basis. A continuous spinning process was introduced first in Germany and in 1944 in the U.S.A. The following description refers essentially to the U.S. process.

Up to the point at which the filaments emerge from the funnel, the continuous process is virtually identical with the batchwise process. The thread of filaments from the funnel is passed through an enclosed bath of hot dilute acid called the pretreatment pan. This continues the coagulation of the cellulose, reducing the filaments to about one third of their original diameter. The oriented filaments of cellulose are sheathed in a film of unaligned cellulose, and this is washed away in the pretreatment pan. If left, the unaligned cellulose would act as a glue, holding the filaments together.

After leaving the pretreatment pan, the thread of filaments passes through an acid trough where remaining copper is removed as copper sulphate. The acid is washed away as the thread moves through a water trough, and lubricants, sizes etc. are added as required by passing the thread over a preparation roll.

The thread passes through a succession of driers and over a roll which applies coning oil before being wound on to flangeless spools. Untwisted threads may also be wound on to beams which are used directly in warp knitting, or combined to provide a weaver's beam.

Throughout the continuous process, the thread of filament is never handled, and imperfections are thus held at a minimum. The filaments are of highly uniform structure and dimensions, and the properties are excellent. After conditioning for a few days at controlled humidity, the cupro is ready for despatch.

Cupro filaments adhere to each other, and are separated only by a comparatively strong force. Unlike viscose yarns, they may be used for many purposes in an untwisted condition.

A wide variety of cupro yarns is produced, ranging from 17 to to 330 dtex (15–300 den) and more. Weaving and knitting yarns are commonly in the range 56 to 110 dtex (50–100 den).

*Novelty Yarns*

Cupro manufacturers have been particularly successful in their production of novelty yarns, such as slub and nubby yarns.

These may be made, for example, by extruding the spinning solution through spinnerets with two sets of orifices. The filaments from one set of orifices are allowed to collect on a flat surface to form bundles which adhere together; these are carried away at intervals to join the filaments extruded from the other set of orifices, forming a composite yarn with the bundles creating slubs at intervals.

## PROCESSING

### Scouring and Desizing

Water soluble sizes, such as the polyvinyl alcohol types often used, may be removed by soaking, followed by a neutral scour at the boil.

### Bleaching

Cupro is an unusually white fibre, and bleaching is not generally required. If it should prove necessary, the usual techniques for cellulosic fibres may be used, e.g. hypochlorite or hydrogen peroxide.

### Dyeing

Cupro is a cellulosic fibre of relatively low crystallinity (e.g. by comparison with cotton), and it is produced usually in the form of fine filaments. Water penetrates quickly into the fibre and dyeing takes place rapidly and effectively. The types of dyestuff used for cotton and other cellulosic fibres are used for cupro. The shades obtained with cupro are deeper than those obtained under comparable conditions with other cellulosic fibres.

## STRUCTURE AND PROPERTIES

### Fine Structure and Appearance

Cupro is the most 'silk-like' of all cellulosic yarns. It is smooth-surfaced and shows no markings or striations. In cross-section it is almost round.

The filaments are extremely fine, usually 1.4 dtex (1.3 den), and have been manufactured in 0.45 dtex (0.4 den).

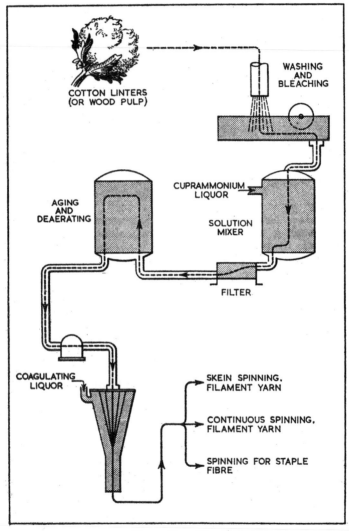

COTTON LINTERS
(OR WOOD PULP)

WASHING
AND
BLEACHING

CUPRAMMONIUM
LIQUOR

AGING
AND
DEAERATING

SOLUTION
MIXER

FILTER

COAGULATING
LIQUOR

SKEIN SPINNING,
FILAMENT YARN

CONTINUOUS SPINNING,
FILAMENT YARN

SPINNING FOR STAPLE
FIBRE

*Cupro Flow Chart*

## Tensile Strength

The tenacity of cupro is 15—20 cN/tex (1.7—2.3 g/den) dry; 9.7—11.9 cN/tex (1.1—1.35 g/den) wet.

The tensile strength of cupro is about 2100—3150 kg/cm² (30,000—40,000 lb/in²).

## Elongation

Cupro has an elongation of 10–17 per cent when dry and 17–33 per cent when wet.

## Elastic Properties

Cupro has an elastic recovery of 20–75 per cent at different elongations.

## Specific Gravity

1.54 conditioned at 11 per cent moisture.

## Effect of Moisture

Cupro swells in water and loses strength. The moisture regain is 12.5 per cent under standard conditions. The commercial standard is 11 per cent.

## Effect of Heat

Decomposition begins at about 250°C. without melting. Yarns and fabrics burn readily, leaving little ash.

## Effect of Age

Similar to viscose.

## Effect of Sunlight

Prolonged exposure causes some degradation and loss of strength.

## Chemical Properties

*Acids*

The fibres are disintegrated by hot dilute or cold concentrated acids.

*Alkalis*

Dilute solutions do not have any appreciable effect. Strong solutions cause swelling and degradation.

*General*

Cupro behaves generally like other cellulosic fibres. It is not affected by weak oxidizing agents or by bleaches such as hypochlorite or peroxide solutions. Strong oxidizing agents cause degradation.

**Effect of Organic Solvents**

Like other cellulosic fibres, generally insoluble.

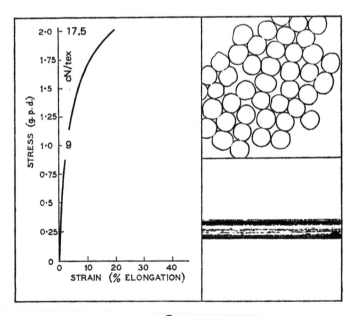

*Cupro*

**Insects**

Cupro is· moderately attacked by some insects.

**Micro-organisms**

Wet fibre is attacked by mildews.

**Electrical Properties**

Moderate dielectric strength when dry.

**Other Properties**

The fibre has a soft silk-like handle and a characteristic lustre.

## CUPRO IN USE

Cupro is in general more expensive than other man-made cellulosic yarns. Its extra fineness and strength, attractive handle,
subdued lustre and good draping properties enable it to carry
this extra cost in the manufacture of high quality goods.

**Washing, Ironing, Dry Cleaning**

Cupro is similar to other cellulosics in its general behaviour
towards laundering and dry cleaning processes. It should be
treated in the same way as viscose.

**End Uses**

Cupro is made into chiffons, satins, nets, ninons and all manner
of very sheer fabrics. Much of this yarn goes into underwear,
dress fabrics and linings.

Novelty yarns, such as slub yarns, are used in a great variety of
applications, especially as weft.·Slub yarns are used in dresswear,
sportswear and fine drapery fabrics.

A speciality end use lies in the production of yarn-dyed fabrics
for high quality silk-like linings, dress and upholstery fabrics.
Reel spun yarns are especially suited to these applications; they
are produced in skeins ready for yarn-dyeing in the untwisted
state. The dyed yarn is used untwisted for the weft and twisted
for the warp.

## SAPONIFIED CELLULOSE ESTER

### INTRODUCTION

During the early 1930s, every effort was being made to develop techniques for increasing the tenacity of viscose rayon filaments by modification of the spinning and stretching techniques. The heart of the problem lay in the difficulty of maintaining the extruded filament in a plastic condition after it entered the coagulating bath, and so providing an opportunity of stretching the filament to orientate the cellulose molecules (see page 40).

During this same period, attempts were also being made to produce high tenacity cellulose acetate yarns by stretching filaments of cellulose acetate. In this case, however, the project was simplified by the readiness with which cellulose acetate could be brought into a plastic condition after it had been made.

The hardening of a cellulose acetate filament is achieved by evaporating solvent from the jet of solution emerging from the spinneret. And the filament may be rendered plastic again by treatment with a solvent, which will first swell and then dissolve the cellulose acetate. Cellulose acetate is also a thermoplastic material, and may be softened by heating.

These differences in the behaviour of cellulose and cellulose acetate towards solvents and heat are a consequence of the differences in their molecular structure. The hydroxyl groups of cellulose do not encourage solution of the molecule in organic solvents, whereas the acetyl groups of the cellulose acetate molecule do. Also, cellulose molecules are able to pack closely together and develop powerful forces of attraction associated with high crystallinity; cellulose acetate molecules, with their large pendant groups, do not permit of the close packing that results in high crystallinity, and cellulose acetate is softened by heat.

This readiness with which cellulose acetate can be rendered plastic, either by solvent or by heat, provides an easy solution to the problem of stretching filaments to create a high degree of orientation. Filaments of cellulose acetate which have been softened by solvent or by heat may be stretched to many times their original length, the long molecules of cellulose acetate sliding readily over one another as they are drawn into alignment.

74

During the early 1930s, the stretching of solvent-plasticised cellulose acetate resulted in the production of yarns with tenacities in the region of 44–53 cN/tex (5–6 g/den). These yarns retained the essential characteristics of cellulose acetate, but they could be converted into cellulose by saponification with caustic soda solution, providing highly-oriented filaments of regenerated cellulose. The molecules in these saponified cellulose acetate filaments were in a more highly oriented and crystalline condition than could be obtained by stretching filaments produced during coagulation of viscose rayon.

This elegant technique of creating highly-oriented cellulose filaments forms the basis of the process used by the Courtaulds group of companies in producing high tenacity rayons under the trade name 'Fortisan'.

## NOMENCLATURE

### Rayon

Fibres produced by the saponification of cellulose acetate fibres are regenerated cellulose, and are properly described as *rayon* under the rules of the U.S. Federal Trade Commission (see page xxvi).

## PRODUCTION

Saponified cellulose acetate fibres are made by heating cellulose acetate filament yarns in steam at about 2.1 kg/cm$^2$ (30 lb/in$^2$), and stretching the softened yarn by 4 to 10 times its original length. The stretched yarn is wound on to perforated bobbins and saponified by treatment with caustic soda solution. The yarn is then washed, oiled, dried and rewound.

Very fine filaments of regenerated highly-oriented cellulose may be produced in this way.

## PROCESSING

### Dyeing

Saponified cellulose acetate yarns have dyeing properties similar to those of cotton or viscose rayon, the high degree of orientation rendering dyeing slower and less effective.

## STRUCTURE AND PROPERTIES

Saponified cellulose acetate yarns consist of regenerated cellulose, and their characteristics are essentially those of highly-oriented cellulosic fibres. They are similar to high tenacity yarns derived by the viscose and the cupro techniques (see page 43).

The mechanical properties of the yarns depend upon the degree to which the filaments have been stretched.

The properties described in the following section are based upon 'Fortisan'.

### Fine Structure and Appearance

Filaments are of somewhat lobed, almost round cross-section. The indentations are seen as striations when the filament is viewed lengthwise.

### Tensile Strength

Tenacity:  53-62 cN/tex (6–7 g/den) dry; 44–53 cN/tex (5–6 g/den) wet.
Tensile strength:  9520 kg/cm$^2$ (136,000 lb/in$^2$).
Loop strength:  about 50 per cent of standard.

### Elongation

6 per cent, dry or wet.

### Elastic Recovery

60–80 per cent at 2 per cent extension.

### Initial Modulus

1500–2207 cN/tex (170–250 g/den).

### Average Stiffness

1033–1192 cN/tex (117–135 g/den).

### Specific Gravity

1.5.

76

**Effect of Moisture**

High resistance to stretch is retained under both wet and dry conditions, giving high dimensional stability.
Regain: 10.7.

**Other Properties**

Similar to cotton and viscose.

## SAPONIFIED CELLULOSE ESTER FIBRES IN USE

Saponified cellulose acetate yarns are used for applications typical of high strength cellulosic rayons (see high strength viscose rayons, page 46). They are used where a high ratio of strength to volume and excellent dimensional stability are advantageous, e.g. in parachute ropes and fabrics, tyre cords, belting, hoses and balloon fabrics. The very fine filaments made by this technique have enabled saponified cellulose acetate yarns to replace natural silk in applications such as electrical insulation materials, e.g. in hearing aid equipment. Coated fabrics provide light, strong tarpaulins and protective fabrics.

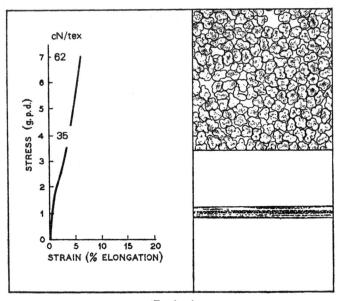

*'Fortisan'*

## 2. CELLULOSE ESTER FIBRES

## CELLULOSE ACETATE FIBRES (ACETATE)

### INTRODUCTION

Viscose and cuprammonium rayons are basically similar, in that the fibre produced at the end of the process consists of cellulose. In both processes, the raw material cellulose – wood pulp or cotton linters – is brought to a soluble form so that it can be extruded through fine holes to form filaments. The filament is created by regeneration of the solid cellulose as the liquid jet enters the coagulating bath.

Viscose and cuprammonium are therefore regenerated cellulose. In their chemical structure they resemble cotton or flax. But there is another important form of cellulosic man-made fibre, in which the wood or cotton cellulose is changed into a different substance to render it soluble and spinnable, and after being spun is left in its changed chemical form. This fibre is made from a chemical derivative of cellulose; cellulose acetate.

Like the nitrocellulose from which early rayons were spun, cellulose acetate has bulky groups of atoms attached to the long cellulose molecule at intervals throughout its length. These acetate groups tend to keep the molecules apart, preventing the alignment and close packing into regions of regularity which make for crystalline structure. The hydroxyl groups which exert so powerful an attraction on each other in the cellulose molecule have been reduced or eliminated, diminishing the grip exerted by each molecule on its neighbours. It is easier for the molecules of an organic solvent to penetrate between the molecules of cellulose acetate than between the molecules of cellulose itself, and cellulose acetate will dissolve in suitable solvents.

The nitro groups in nitrocellulose had the disadvantage of making the material highly flammable. But cellulose acetate is no more flammable than cotton.

Cellulose acetate was first prepared by Schutzenberger in 1865, by heating cotton with acetic anhydride at 130–140°C. in a closed vessel. In 1894, Cross and Bevan discovered a more practical process, in which the acetylation was carried out at atmospheric pressure, using sulphuric acid or zinc chloride as catalyst.

## Primary Acetate

The cellulose acetate produced in these early experiments was a completely acetylated cellulose, in which all three hydroxyl groups of each glucose unit in the cellulose molecule is acetylated. It was cellulose triacetate, which later became known as primary acetate (see page 82).

Cellulose triacetate was obtained as a tough, horny solid which was not readily washed free of acid, and it also contained sulphuric ester groups which rendered it unstable. It was soluble only in toxic and expensive solvents, such as chloroform and tetrachloroethane.

## Secondary Acetate

In 1906, Miles discovered that the triacetate could be partially hydrolysed to produce a cellulose acetate in which some of the acetate groups of the triacetate had been removed, and reconverted to hydroxyl groups as in the original cellulose. This new material was, in effect, a partially acetylated cellulose obtained by complete acetylation and subsequent partial hydrolysis. It became known as 'secondary acetate'.

The secondary acetate was soluble in a relatively cheap and non-toxic solvent, acetone. Also, it could be washed free of acid much more easily than the primary acetate.

During the period leading up to the outbreak of World War I, many workers experimented with the production of cellulose acetate filaments from solutions of the primary and secondary acetates. The secondary acetate, which dissolved in the cheaper and less toxic solvent, appeared to offer the greatest prospect of success, and it was upon this material that much of the work was carried out.

Among the most active workers in the cellulose acetate field at that time were the brothers Drs. Henry and Camille Dreyfus in Switzerland. When war broke out in 1914, the brothers Dreyfus were invited to Britain by the Government, who were interested in the use of cellulose acetate as a varnish for the fabric wings of aircraft. In the early days of the war, nitrocellulose had been used for this purpose, but its extreme flammability had caused heavy casualties. Cellulose acetate was much more satisfactory for the purpose.

The Dreyfus brothers established a cellulose acetate factory at Spondon, Derby, where secondary acetate solution or 'dope'

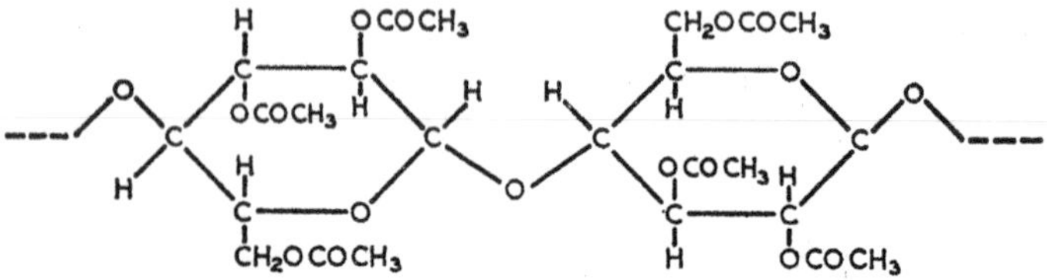

**CELLULOSE TRIACETATE**
(PRIMARY ACETATE ; FULLY ACETYLATED)

## CELLULOSE ACETATE

*Primary Cellulose Acetate (Cellulose Triacetate)*. Complete acetylation of cellulose yields cellulose triacetate (shown above) in which all three of the hydroxyl groups of each glucose unit are converted to acetate groups. This is called primary cellulose acetate.

Primary cellulose acetate is soluble in chloroform and methylene chloride. It is insoluble in acetone.

*Secondary Cellulose Acetate.* (Partial hydrolysis of cellulose triacetate converts some of the acetate groups back into hydroxyl groups, to form secondary cellulose acetate. This is sometimes called cellulose diacetate, implying that each glucose unit has two of its hydroxyl groups acetylated. This name is incorrect, as each glucose unit has, on average, 2½ of its hydroxyl groups acetylated in secondary cellulose acetate.

Secondary cellulose acetate is soluble in acetone.

*Acetyl Value*

The degree of acetylation of the cellulose molecule in cellulose acetate is commonly expressed as the *acetyl value*, which is the acetic acid content quoted as a percentage.

Cellulose triacetate has an acetyl value of 62.5 per cent. Commercial secondary cellulose acetate has an acetyl value of about 52–55.5%.

---

was ultimately produced in considerable quantity. When America entered the war, Camille Dreyfus went to the U.S. to establish production of cellulose acetate there. The war ended before the factory was completed and Dreyfus returned to England.

### Acetate Fibres

The end of the war found the Dreyfus brothers with a large cellulose acetate factory in Britain, and no longer any demand for aircraft dope. They set to work, therefore, to try and develop a cellulose acetate fibre. By 1921, most of the technical difficulties had been overcome, and a filament yarn was being made in Britain from secondary cellulose acetate, under the name 'Celanese'. Some six years later, Courtaulds Ltd. were also producing a secondary cellulose acetate yarn. In 1934, a million pounds of acetate yarn was exported to America, and by this year British dyestuffs chemists had solved the biggest problem that was holding up the progress of the new fibre – how to dye it.

Viscose and cuprammonium rayons, consisting as they do of regenerated cellulose, can be dyed with dyestuffs used for cotton and other natural cellulosic fibres. Cellulose acetate, however, differs chemically from cotton in that many of the hydroxyl groups have been replaced by acetate groups, and the affinity characteristics of the fibre with respect to dyestuffs have been changed. It was necessary to develop new types of dyes for acetate fibre.

Once this problem had been solved, cellulose acetate fibres made rapid headway. Filament yarns, staple and tow are now produced in many countries, and are available in a wide range of sizes and spun-dyed colours. They are now known simply as *acetate* (see below).

## NOMENCLATURE

### Acetate

Until comparatively recent times, fibres spun from secondary cellulose acetate (see page 82) were described as 'acetate rayon'. In modern terminology, 'rayon' is used only in describing fibres consisting of regenerated cellulose, including viscose rayon and cupro, and fibres spun from secondary cellulose acetate are now known simply as *acetate*. Fibres spun from primary cellulose acetate are called *triacetate*.

### Federal Trade Commission Definition

The generic terms *acetate* and *triacetate* were adopted by the U.S. Federal Trade Commission for fibres of the cellulose acetate type, the official definition being as follows:

*Acetate (and Triacetate).* A manufactured fibre in which the fibre-forming substance is cellulose acetate. Where not less than 92 per cent of the hydroxyl groups are acetylated, the term 'triacetate' may be used as a general description of the fibre.

## PRODUCTION

(A) ACETYLATION OF WOOD PULP OR COTTON LINTERS

### Raw Material

As in the production of viscose rayon and cupro, the cellulose used as raw material in acetate manufacture comes either from cotton linters or predominantly from purified wood pulp.

Cotton linters are purified by kier-boiling for several hours with a solution of sodium carbonate or caustic soda. The linters are then washed and bleached with sodium hypochlorite, washed again and dried.

### Steeping

The purified cotton linters or wood pulp are steeped in glacial acetic acid to swell the fibres and increase their chemical reactivity. Modern processes use vapour-phase activation based on acetic-water mixtures.

### Acetylation

The swollen cellulose is transferred to a closed reaction vessel containing a mixture of acetic acid and anhydride. The mixture has the following weight ratio:

> Purified linters 1 part
> Acetic anhydride 3 parts
> Glacial acetic acid 5 parts

The reactants are mixed, and a small quantity (0.1 part) of sulphuric acid dissolved in glacial acetic acid is added. The acetylation of the cellulose now proceeds. The reaction is exothermic, and the vessel is cooled to maintain a predetermined temperature profile. After a period the temperature is allowed to rise and maintained at a higher temperature for a further period.

During this time, complete acetylation of the cellulose takes place. The three hydroxyl groups on each glucose unit of the cellulose molecule are all acetylated, and the product is cellulose triacetate. This is known as primary acetate.

### Hydrolysis (Ripening)

The solution of primary acetate is transferred to another vessel, and mixed with dilute acetic acid. The residual acetic anhydride reacts with the water to form acetic acid and this, together with the

84

residual acetic acid from the acetylation mixture, forms a solution of acetic acid in water.

The cellulose triacetate is allowed to stand in this solution of acetic acid in water for up to 20 hours. During this time, partial hydrolysis of the cellulose triacetate takes place, some of the acetyl groups being removed and hydroxyl groups formed again. This hydrolysis stage is, in effect, a partial reversal of the acetylation process, the completely acetylated cellulose triacetate being converted into a partially-acetylated cellulose acetate.

The restoration of some of the hydroxyl groups on the cellulose acetate molecule changes the solubility characteristics of the acetate. Cellulose triacetate is soluble in chloroform, but insoluble in acetone; the partially-acetylated cellulose acetate is insoluble in chloroform, but soluble in acetone.

The partially-acetylated cellulose acetate made in this way is often called cellulose diacetate, signifying that one third of the total acetyl groups have been removed from the cellulose triacetate, so that each glucose unit now has two of its three hydroxyl groups acetylated. But, in fact, the secondary acetate used in spinning acetate fibre does not correspond precisely with cellulose diacetate. The hydrolysis of the triacetate is allowed to proceed until each glucose unit in the cellulose molecule has, on average, about $2\frac{1}{2}$ of its hydroxyl groups acetylated The secondary acetate structure lies part way between that of triacetate and diacetate.

During the hydrolysis process, tests are carried out at intervals to indicate how the de-acetylation is proceeding. When it has reached the desired point, the solution is poured into an excess of water, and the secondary acetate is precipitated as white flakes. These are washed thoroughly and may be ground into finer particles.

The acetic acid is recovered from the residual solution and used again.

### Dry Spinning

The spinning solution is made from a blend of secondary acetate containing material from a large number of batches, in order to secure as high degree of uniformity as possible. The blended acetate is dissolved in acetone containing a small proportion of water (up to 5 per cent on the weight of acetone), and pigments may be added at this stage. Finely-divided carbon black (2 per

85

cent on the weight of acetate) is used in making black fibre; titanium dioxide (1–2 per cent on the weight of acetate) is added in making dull fibre.

The spinning solution, now known as 'dope', is filtered carefully and deaerated, and is ready for spinning. It contains 20–30 per cent cellulose acetate. It is pumped to the spinnerot, undergoing a final filtration on the way. From the spinneret holes, jets of spinning solution, 25–75 $\mu$ in diameter, emerge into a spinning tube. This is an enclosed vessel through which hot air is flowing at a temperature of about 100°C. As the jets of spinning solution meet the hot air stream, acetone is evaporated to leave solid filaments of cellulose acetate. More than 90 per cent of the acetone in the jets is evaporated during the fraction of a second that the jet is moving downwards through the spinning tube.

The newly-formed filament of cellulose acetate is stretched slightly while still plastic, to align the long molecules and develop the strength of the filament.

After moving downwards through the spinning tube for a distance of several feet, the filaments are sprayed or wiped with lubricant. They pass round a guide roller and emerge from the spinning tube on to a constant speed feed roll. From this, they are led to a winding mechanism which may wind the untwisted filaments on to a cylindrical tube, or insert twist and then wind the twisted yarn on to bobbins. Ring mechanisms and cheese collection are commonly used.

Acetate filament yarns produced in this way are ready for immediate use, without any of the washing or purification treatments that are necessary with wet-spun fibres.

The technique of dry spinning is made possible by the fact that secondary cellulose acetate dissolves in a readily-evaporated solvent, acetone. The process is simpler and faster than the wet spinning processes used with viscose and cuprammonium rayons, and spinning may be carried out at very high speeds of 1000 metres per minute. There is no handling of the filament between extrusion and collection.

Acetate filaments are produced in a range of counts, the most popular filaments being of 3.3–4.4 dtex (3–4 den) with yarn counts 44–2000 dtex (40–1800 den). The count is controlled by three factors, (1) the rate at which spinning solution is pumped to the spinneret, (2) the size of the spinneret holes, and (3) the rate at which the filament is drawn away from the spinneret.

Staple fibre is produced by crimping the filaments mechanically and then cutting them into short lengths ranging from 38–150 mm (1½–6 in). The staple length is chosen to blend suitably with other fibres in making blended yarns.

Although the dry spinning process simplifies the production of acetate fibre, it makes use of a number of chemicals which are needed in considerable quantity. Every kg of acetate fibre produced needs about 0.65 kg of cellulose, 1.5 kg of acetic anhydride, 4.0 kg of acetic acid, 0.05 kg of sulphuric acid, 3.0 kg of acetone and 45 litres (10 gall) of water. The commercial production of acetate fibre is made possible only by the careful recovery of a high proportion of these raw materials, so that they can be recycled.

### Wet Spinning

The final stage in the production of secondary cellulose acetate consists in the hydrolysis of primary acetate by allowing it to stand in acetic acid and water. The secondary acetate is obtained as a solution, from which it is precipitated by dilution with water.

In the normal way, this precipitation is carried out in such a way as to yield flakes of secondary acetate which are subsequently dissolved in acetone for dry spinning. But there is an obvious short cut to this procedure; the solution of acetate resulting from hydrolysis could be extruded through a spinneret, and the cellulose acetate precipitated in the form of filaments in an aqueous coagulating bath.

Wet spinning techniques of this sort have been developed for spinning cellulose acetate.

### Melt Spinning

Cellulose acetate is a thermoplastic fibre; it melts when heated to temperatures in the region of 230°C., and molten acetate is sufficiently stable to undergo melt spinning.

With the development of melt spinning techniques for synthetic fibres, the melt spinning of cellulose acetate has become a practical possibility, and many experimental fibres have been spun. The filaments differ in some ways from filaments produced by dry spinning, notably in their reaction to boiling water. Dry spun filaments tend to delustre in boiling water, but melt spun filaments do not. The tenacity of melt-spun filaments is increased.

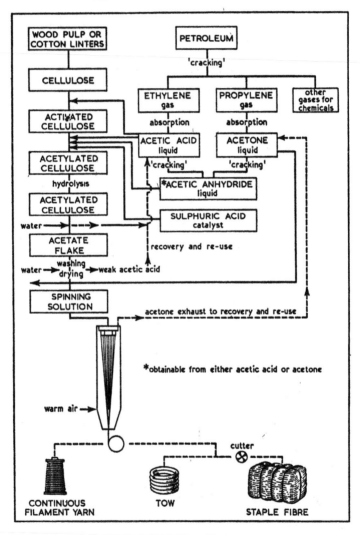

*Acetate Flow Chart*

## Modification of Filament

In common with other man-made fibres, cellulose acetate fibres may be produced in physically-modified forms by manipulation of the spinning process.

Spun-dyed acetate fibres are now being made in great variety by the addition of pigments to the spinning solution prior to extrusion. Carbon black provides black fibres; titanium dioxide is used to modify lustre.

The use of spinneret orifices of appropriate shapes will produce filaments of unusual cross-sections, and many variations were made commercially. Flat filaments reflect the light and yield novelty glitter yarns. X- and Y-shaped cross-sections provide yarns of improved handle and covering power, and yarns which crimp in water. Thick and thin yarns and slub yarns are made by varying the rate of feed of the solution to the spinneret, and by the extrusion of different filaments which are subsequently combined into composite novelty yarns.

## (B) ACETYLATION OF CELLULOSIC FIBRES

In the production of cellulose acetate fibres by the normal technique, the raw material is cellulose in a fibrous form that is not, however, suitable for textile use. Cotton linters are too short to be spun into a satisfactory yarn; wood cellulose is contaminated with natural gums and resins, and the fibres in purified wood pulp are again too short to form a textile yarn. During acetylation, the fibrous structure of the cellulosic raw material is destroyed; the acetylated cellulose forms a solution from which cellulose acetate is precipitated as flakes; the flakes are redissolved and spun into filaments.

It has long been known that cellulose fibres may undergo chemical modification without losing their fibrous form. If a cellulosic textile fibre is acetylated under such conditions, it may be converted into a fibre which is either cellulose triacetate or an intimate mixture of cellulose triacetate and cellulose and is in a state suitable for immediate textile use.

## Cotton

Cotton fibres of textile quality may be acetylated without losing their fibrous form, the modified fibres being cellulose acetate. Practical processes for the production of acetylated cotton have been established by the U.S. Department of Agriculture, using either a batchwise or continuous technique.

*Batchwise Process.* Cotton fibre, yarn or fabric is purified by boiling in dilute caustic soda, washed and dried. It is then allowed to soak in acetic acid for at least 1 hour. Excess acid is squeezed from the cotton, which is then treated with a mixture of acetic acid and acetic anhydride in the presence of a small amount of perchloric acid catalyst. After treatment for 1 hour at 20°C., the cotton is washed and dried.

*Continuous Process.* This process, which is designed for use with fabric, is essentially the same as the above, but presoaking is shortened by carrying it out at 82°C. for 2 minutes. The fabric is cooled and passed continuously through a bath of perchloric acid in acetic acid, followed by treatment with acetic anhydride in an acetylation vessel for 3 minutes at 20°C. The cotton is then washed and dried.

Partially acetylated cotton produced in this way has greatly improved rot- and heat-resistance (see PA cotton, Vol. 1).

### Viscose

During the 1950s, Japanese scientists developed a similar process for the direct acetylation of viscose fibre without destroying its fibrous form. This technique has now become the basis of a commercial process which produces a cellulose acetate fibre direct from a polynosic-type viscose by direct acetylation of the fibre. Fibres carried the trade names 'Alon' or 'Tohalon'.

The acetylation is carried out by soaking polynosic rayon staple in a solution of a catalyst (e.g. sodium acetate, zinc sulphate), partially drying the fibre and then passing it through a chamber containing acetic acid vapour at 110°C. The fibre then passes through an acetylating chamber containing acetic anhydride vapour at 130°C.

Acetylation takes place, to a slightly less degree than in normal secondary acetate. The fibre is washed in water, lubricant is added, and it is dried in warm air. The crimp of the fibre is retained throughout the process.

Acetylation by this technique does not cause degradation of the cellulose to the extent that normal acetylation in solution does. The acetylated viscose fibre is stronger than normal secondary acetate, dry or wet, with lower elongation.

## PROCESSING

### Scouring

Cellulose acetate is sensitive to caustic alkali, which tends to bring about hydrolysis of the acetate groups. Yarns and fabrics should not be scoured under alkaline conditions. Soap and sulphated fatty acid at 60°C., ammonia, or tetrasodium pyrophosphate and surface active agent are effective. Detergents are now widely used.

### Bleaching

Cellulose acetate is a white fibre, and bleaching is seldom necessary. If bleaching is required, alkaline conditions should be avoided. Acid hypochlorite, or a sodium chlorite or hydrogen peroxide bath should be used.

### Dyeing

Cellulose acetate differs in its chemical structure from the cellulosic fibres, such as cotton, viscose and cupro. All but a few of the reactive hydroxyl groups of the cellulose have been acetylated, and acetate will not, as a rule, accept the dyes that are normally used for cellulosic fibres.

When acetate fibre came on the market, its successful commercial development was prejudiced by the fact that available dyestuffs were unsatisfactory for the new fibre. New types of dyestuff were discovered for dyeing acetate, notably the disperse dyes, and acetate can now be dyed satisfactorily in a wide range of shades. Acetate woven fabrics are normally jig dyed between 60 and 98°C at pH 6.0 to 6.5 depending on the choice of disperse dyes. Beam dyeing equipment can also be used to dye woven acetate fabrics, but is normally used to dye acetate tricot.

Boiling water tends to delustre acetate fibre, and dyeing should be carried out if possible at temperatures lower than 85°C. However, acetate of 55 acetyl value is resistant to delustring and does not require dyeing at temperatures lower than 85°C.

*Spun-Dyed Acetate.* A wide range of spun-dyed acetate fibres is now produced.

## Stripping

Treatment with soap solution will bring about partial stripping
of dispersed dyes. Addition of activated carbon will usually
complete the stripping. A stripping bath containing zinc
sulphoxylate formaldehyde and acetic acid may also be used.

## Finishing

The appearance, handle and draping properties of acetate fabrics
are generally excellent, and the dimensional stability of wet
non-textured fabrics is good. Finishing processes designed to bring
about improvements in these respects are not often necessary
with all-acetate fabrics.

Filament yarn fabrics made from acetate tend to suffer from
yarn slippage, and finishes are used to roughen the surface of the
filaments and create a rustle or scroop.

Acetate staple is a constituent of all manner of blended yarns,
especially with viscose staple. The acetate provides drape, soft
handle, dimensional stability and wrinkle resistance. Blended
yarns and fabrics of this type may be subjected to finishing
treatments which are intended primarily to affect the viscose
fibre, and the acetate must be able to withstand the conditions
used. It should be remembered that acetate is sensitive to heat,
water and alkali, and finishing processes should not be used
which subject the fibre to dilute alkali or to water at temperatures
above 80°C. Dry temperatures should not exceed 140°C.

The thermoplastic nature of acetate makes possible the
embossing of acetate and acetate blend fabrics. Patterns may be
embossed on the fabric by passing it, for example, through a
heated calender.

Acetate fibres have a natural sheen which may be destroyed
by incorporating finely-dispersed titanium dioxide in the spinning
solution. Modern acetate fibres are commonly produced as dull
grades in this way. If necessary, a bright acetate may be delustred
by boiling it in water, particularly in soapy water to which a
swelling agent such as phenol has been added.

## STRUCTURE AND PROPERTIES

### Fine Structure and Appearance

The length and fineness of acetate fibres are controlled by the
manufacturer. Continuous filaments can be made to almost any

length. Staple fibre is produced by chopping the continuous filaments into short lengths, which are usually crimped artificially. 38–75 mm (1½–3 in) staple is commonly produced for use on cotton machinery. 75–125 mm (3–5 in) staple is processed on worsted and woollen machinery, and 125–180 mm (5–7 in) staple on spun silk machinery.

Continuous filament acetate is produced commercially in a range of filament counts, usually 1.7–5.6 dtex (1.5–5 den). Staple fibre may be as high as 22 dtex (20 den) or more.

The microscopic appearance of an acetate filament differs from that of the rayons but is very similar to that of triacetate. The filament is marked by longitudinal folds and ridges. The cross-section outline is built up of as many as five or six rounded lobes.

If the acetate filaments have not been dulled artificially by addition of titanium dioxide or other pigments, they are clear and glossy.

### Tensile Strength

Normal acetate fibre, forming the bulk of the output of this type of fibre, has a tenacity of about 9.7–11.5 cN/tex (1.1–1.3 g/den). It does not lose strength so markedly as viscose does when wet; the tenacity falls to 5.7–6.6 cN/tex (0.65–0.75 g/den).

The tensile strength of acetate is about 1260–1540 $kg/cm^2$ (18,000–22,000 $lb/in^2$).

### Elongation

23–30 per cent (Standard); 35–45 per cent (wet).

### Elastic Properties

At 4 per cent elongation, acetate has a recovery of 48–65 per cent. When stretched further, the fibre undergoes plastic flow; it becomes permanently deformed and does not return to its original length when released. At 5 per cent extension, acetate has an immediate recovery of 54 per cent, a delayed recovery of 35 per cent and a permanent set of 11 per cent. The corresponding figures for 10 per cent extension are 27, 32 and 41 per cent.

### Specific Gravity.

1.30.

### Effect of Moisture

Water is held by ordinary cellulose as a result of the attraction between the water-loving hydroxyl groups on the cellulose molecule. In acetate, many of these hydroxyl groups have been replaced by acetate groups; the inherent attraction of acetate for water molecules is therefore less than that of viscose or cuprammonium rayons.

Acetate does not absorb as much water as the rayons. The standard moisture regain is about 6.5 per cent. Immersed in water, acetate will swell by about 6–14 per cent. (Viscose, on the other hand, swells by 35–66 per cent, and cuprammonium by 40–62 per cent.)

### Thermal Properties

Cellulose acetate is a thermoplastic material. It becomes sticky at 190°C. and at 205°C. is soft enough to deform under pressure. It melts at about 232°C.

### *Effect of High Temperature*

The fibre will withstand prolonged high temperatures without serious deterioration. After a week at 120°C., it retains much of its original tensile strength.

### *Flammability*

Cellulose acetate is not readily flammable. Exposed to a naked flame it will melt and burn.

### Effect of Age

There is a slight fall in tensile strength over prolonged periods. The colour of the fibre remains good.

### Effect of Sunlight

Deterioration after prolonged exposure, resulting in some loss of strength. The colour remains good. Retention of tenacity is improved by certain coloured pigments and light-fast titanium dioxide grades.

### Chemical Properties

### *Acids*

Dilute solutions of weak acids do not affect acetate, but the fibres are decomposed by strong acids in concentrated solutions.

Organic acids, including acetic and formic acids, will make acetate fibres swell. At sufficiently high concentrations, aqueous solutions of formic and acetic acids will dissolve cellulose acetate.

*Alkalis*

Alkalis have little effect up to pH 9.5, but strong alkalis cause saponification; the acetate groups are replaced by hydroxyl groups and the cellulose acetate is gradually changed to regenerated cellulose.

*General*

Acetate is attacked by strong oxidizing agents, but is not affected by normal bleaching solutions of hypochlorite or peroxide. (Peroxide degrades acetate on long standing.)

The chemical properties of an acetate fibre depend on the degree to which acetate groups have replaced hydroxyl groups in the cellulose molecule. The more hydroxyl groups there are remaining, the greater is the fibre's 'cellulosic' character. Modern commercial acetate has an acetyl value of 54–55.

**Effect of Organic Solvents**

Acetate swells or dissolves in many solvents, including acetone and other ketones, methyl acetate, ethyl acetate, dioxan, dichloroethylene, cresol, phenol, chloroform, methylene chloride, ethylene chloride. It is insoluble in petroleum chemicals such as white spirit or petrol (gasoline), ethyl ether, benzene, toluene, perchloroethylene, trichloroethylene, carbon tetrachloride, cyclohexanol, xylene.

**Insects**

Moths and other insects do not normally attack acetate; grubs will bite through the fibres in an effort to get at more attractive food.

**Micro-organisms**

Fungi and bacteria may cause surface damage and discoloration, but resistance is generally high.

**Electrical Properties**

Excellent insulator.

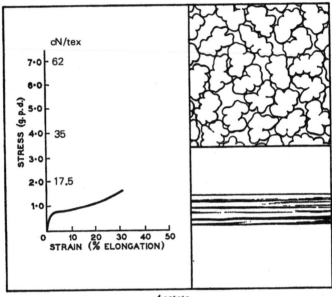

*Acetate*

## Other Properties

Acetate yarns have an attractive natural lustre and a pleasant handle. The surface of the fibre is harder than that of the rayons.

Acetate is a poor conductor of heat.

Acetate yarns and fabrics are non-toxic, and do not irritate the skin.

## ACETATE IN USE

The introduction of acetate into commercial use was an exciting event in the textile world. The chemical structure of the fibre is fundamentally different from that of any natural fibre or of any of the regenerated cellulose rayons. The properties of acetate were, in consequence, quite different from those of any fibre in use at the time of its introduction. It was, in this respect, a step towards the completely synthetic fibre.

The natural attractiveness of cellulose acetate, combined with its useful practical properties, sustained a demand for acetate that increased continuously until 1969-1970.

In general, modern acetate fabrics can be treated in much the same way as natural fibres and the rayons. They can be dyed and finished, washed and dry-cleaned, and will withstand all the conditions that are met in ordinary commercial and domestic use.

It must be remembered, however, that acetate is fundamentally different in its chemical structure from the rayons and natural fibres. It therefore differs in its behaviour in many ways, and these differences must be taken into account in the handling of acetate materials.

Acetate is thermoplastic, for example, and it must not be subjected to high temperatures. It will deform under pressure at about 205°C.

This tendency for acetate fibres to soften on being heated is made use of in the processing of acetate goods. Fabrics can be embossed with patterns that are impressed on the warmed material.

The relatively low moisture absorption of acetate fibres renders acetate less liable to damage by staining with many substances. Fruit juices, ink, food and other water-soluble stains are easily sponged or washed out.

Acetate fabrics dry rapidly, and are particularly suitable for bathing suits, rainwear and umbrellas.

Cellulose acetate does not conduct heat readily; acetate garments are cool in summer and warm in winter.

Acetate has little natural colour; the dyer can produce acetate fabrics in a range of shades varying from a delicate tint to a deep, heavy colour.

The richness and variety of shades, allied with the softness and pliability of the acetate fibre, have helped to make acetate into a 'beauty' fibre. Acetate garments drape well and have an attractive handle; they are soft, and never harsh. They retain their shape if handled with reasonable care and do not easily wrinkle.

Acetate fabrics have an unusually attractive drape. Acetate satins will fall naturally in graceful folds; taffetas retain their crispness under severe conditions of wear.

### Washing

The relatively low moisture absorption of acetate fibres contributes to the good dimensional stability of acetate fabrics when wet compared to viscose fabrics.

Acetate fabrics do not shrink or lose their shape appreciably when washed. They are not affected significantly by boiling water, but some delustring may occur. Lustre is generally restored by ironing. Acetate fabrics will withstand ordinary soap solutions, detergents and bleaches, but alkaline conditions should be avoided.

Washing of acetate garments presents no difficulties, either by hand or machine. Warm (40°C., 104°F.) water and neutral soap or detergent should be used. Agitation should be gentle and garments must not be wrung out or twisted when wet or they may retain creases. They should be kept flat as far as possible.

*Note*

Many fabrics for evening wear, such as poults, satins, brocades and taffetas, are made from acetate. Evening dresses and elaborate styles should be dry cleaned because the construction and decoration of the fabric make them unsuitable for washing at home.

### Drying

Acetate garments dry readily. They should be given a cold rinse prior to spinning, followed by a short spin (15 seconds). Tumbler drying is satisfactory provided that the drier is run cold before stopping. Drying temperatures should not exceed 105°C.

### Ironing

Acetate fabrics should be ironed with a warm iron (HLCC Setting 2). It is preferable to use a damp pressure cloth, and to iron on the reverse side. If temperatures higher than about 120°C. are used, the cellulose acetate may begin to soften and the fibres may be deformed by the pressure of the iron. The surface of an ironed fabric, will become glazed as the plastic filaments are flattened.

In commercial laundering, covered presses at 4.5 kg/cm² (65 lb) steam pressure are satisfactory.

### Dry Cleaning

Acetate fibre is sensitive to many types of solvent (see Structure and Properties), and great care must be taken in bringing any organic solvent into contact with acetate fabrics. Perchloroethylene, trichloroethylene or carbon tetrachloride and petroleum-type solvents (e.g. Stoddard solvent) may be used in dry cleaning.

## END USES

Acetate filament yarns are used in many types of women's dress-wear, from linings, lingerie and gowns to bathing suits and blouses. Staple fibre is used in fuller materials; blended with other fibres acetate staple provides a wide variety of dress fabrics. It provides resilience and resistance to shrinkage in such blends.

In men's wear, acetate filament provides materials for linings and ties, socks and pyjamas. Staple fibre goes into many blends that are spun into suitings, shirt fabrics and materials for sports wear.

Many fine household and furnishing textiles, and materials are made from acetate, and the good electrical insulation properties of the fibre have carried in into the electrical industry. It is used as the insulated covering for electric wires.

# CELLULOSE TRIACETATE FIBRES (TRIACETATE)

## INTRODUCTION

The early attempts to develop textile fibres from cellulose acetate were concerned very largely with the material obtained by complete acetylation of cellulose. This is cellulose triacetate, in which the three available hydroxyl groups of each glucose unit of the cellulose molecule are all acetylated.

Cellulose triacetate is not soluble in acetone, which dissolves cellulose acetate, and few solvents were known for it during the early years of the present century. At that time, when attempts were being made to produce acetate fibres by dry spinning solutions of cellulose acetate, the most satisfactory solvent available for the triacetate was chloroform. And filament yarns were produced experimentally by dry spinning solutions of cellulose triacetate in chloroform.

This experimental work continued up to the outbreak of World War I. In 1914, the Lustron Company, in the U.S.A., began to produce triacetate yarns in quantities of up to 300 lb. per day by dry spinning chloroform solutions. This production continued on a very modest basis until 1927, when it was discontinued. Two factors contributed to failure of this early triacetate fibre project, (1) the use of chloroform as solvent was expensive and dangerous, and (2) the triacetate filaments could not be dyed satisfactorily with the dyestuffs then available.

Despite this setback, interest in cellulose triacetate was kept alive by the success of the closely related secondary cellulose acetate process. The development of special dyestuffs for secondary acetate fibres went a long way towards solving that difficulty for triacetate, and eventually methylene chloride was found to be a satisfactory solvent for triacetate. This solvent was more suitable than chloroform as the basis for a commercial dry spinning process.

With this success against the two biggest barriers to triacetate fibre production, it became apparent that the fibre might have very significant practical advantages over secondary acetate. Cellulose triacetate fibres were found to have a relatively low water imbibition and moisture regain, plus a high degree of chemical inertness. They had a high melting point, below which heating produced irreversible physical changes which improved the chemical and physical stabilty of the fibre. These changes enabled triacetate fibres to be heat set.

After several years' research work, Courtaulds Ltd., U.K., began the commercial development of cellulose triacetate fibres in 1950. The resultant fibre, first referred to as 'JPS', was announced under the trade mark 'Courpleta' in 1954. In that year, British Celanese Ltd. announced that they were to introduce 'Tricel' triacetate fibre. Later, Courtaulds and British Celanese linked their research effort and production experience and now produce a single triacetate fibre, the trade mark 'Tricel' being retained, with British Celanese as the producer.

In 1954, Celanese Corporation of America began production of a triacetate fibre under the trade name 'Arnel', and production of triacetate fibres is now proceeding in several countries.

## NOMENCLATURE

See page 83.

### Note

The information in the section which follows relates particularly to the British fibre 'Tricel', which may be taken as a typical example of a modern triacetate fibre.

## PRODUCTION

Experience gained in the manufacture of secondary cellulose acetate was put to good use in the commercial development of cellulose triacetate. The production of triacetate is, in effect, a stage in the production of the secondary acetate.

### Raw Materials

Cellulose in the form of purified cotton linters or wood pulp.

### Pretreatment

The cellulose is pretreated in acetic acid/water vapour.

### Acetylation

This may be carried out in such a way that the cellulose triacetate either goes into solution as it is formed, or retains the structure of the original cellulose.

### (a) *Solution Process*

Pretreated cellulose is acetylated by treatment with acetic anhydride and sulphuric acid in the presence of acetic acid. As the acetylation proceeds, cellulose triacetate is formed, and it dissolves. The solution is ripened, magnesium acetate and water

101

being added to replace any sulphate groups on the cellulose molecule by acetate groups, and the cellulose triacetate is precipitated by diluting the solution with water. The cellulose triacetate is washed until free of acid, and dried. The triacetate produced in this way has an acetyl value of 61.5 per cent.

A modification of this solution process makes use of methylene chloride as solvent instead of acetic acid. Smaller amounts of sulphuric acid catalyst are needed, and a mild hydrolysis produces a triacetate of 62 per cent acetyl content and higher, up to the maximum.

### (b) *Non-solution Process*

Pretreated cellulose is steeped in benzene or other liquid capable of swelling cellulose triacetate without dissolving it, and is then treated with acetic anhydride and perchloric acid (or other acid) catalyst. Acetylation proceeds, but the cellulose triacetate does not dissolve as it is formed. It retains the shape of the original cellulose.

The triacetate is washed until acid-free and then dried. High molecular weight triacetate may be obtained more easily by this method.

### Dry Spinning

Cellulose triacetate from many batches is blended and dissolved in methylene chloride containing a little alcohol, to form a 20 per cent solution. The solution is then filtered and deaerated, and pumped to the spinneret.

The jets of solution emerge from the spinneret into a vertical spinning tube where they meet a stream of hot air. The methylene chloride evaporates, leaving solid filaments of cellulose triacetate.

The filaments are led over an oiler which applies antistatic lubricant, and are collected in the same way as secondary acetate filaments.

If continuous filament yarn is required, the filaments from the spinneret are collected on to bobbins by a cap or ring spinning mechanism, which applies a twist.

If staple is being produced, the filaments from a number of multi-holed spinnerets are brought together into a tow. This is crimped mechanically and then cut into staple of the required length.

## Wet Spinning

As in the case of secondary cellulose acetate, solutions of the triacetate may be spun into coagulating baths which contain water or other liquids which bring about precipitation of the filaments.

## Melt Spinning

Cellulose triacetate is a thermoplastic material, and molten triacetate may be spun into filaments by melt spinning processes.

## PROCESSING

Triacetate blends readily with other fibres, and the techniques used for acetate are, in general, suitable for triacetate too. Carding, combing, spinning and doubling can be carried out. The increased drag of triacetate makes it advisable to reduce tensions to a minimum in winding continuous filament and spun yarns.

Triacetate yarns are a little more difficult to size, owing to the lower moisture take-up as compared with acetate. Sizes based on polyvinyl alcohol, acrylates, and methacrylates are excellent for continuous filament yarn, and modified starch and polyvinyl alcohol are satisfactory with cotton-spun yarns.

A relatively high humidity facilitates the processing of triacetate yarns, for example in knitting or weaving.

### Desizing

Water-soluble sizes are generally used, and these are removed during scouring. Enzyme treatment will remove starch.

### Scouring

Fabrics need a thorough scouring to remove dirt which is acquired during processing. Surface active agent and trisodium phosphate, with or without soap, may be used effectively at 70°C. (160°F.).

### Bleaching

Triacetate will withstand normal bleaching conditions, and it may be bleached effectively with hypochlorite (acid or alkaline), sodium chlorite, hydrogen peroxide or peracetic acid. Sodium chlorite is recommended. When triacetate is blended with other fibres, fabrics may be scoured and bleached, as a rule, by using techniques suitable for the other components of the blend.

Triacetate fibre withstands alkaline conditions much better than acetate.

### Dyeing

Triacetate fabrics can be dyed with most disperse dyestuffs. In general, higher dyeing temperatures are used, 90—98°C or even up to 110—120°C. Careful selection of dyestuffs is essential when high temperature techniques are used.

The use of swelling agents or pressure-dyeing techniques with Sanderson-type machines yield excellent results.

Triacetate fibre does not stain easily, and vat or sulphur dyestuffs may be used with blends to dye the cellulose component. Where these dyes are used, however, dyeing is done by the two-bath method; the triacetate is dyed by disperse dyes in the second bath. Direct cotton dyes can be used on the cotton or viscose component, cross-dyeing blended or melange fabrics in a single bath.

Blends of triacetate and wool may be dyed similarly by normal cross-dyeing methods. Limitations will be imposed by the fact that wool is very much more sensitive to conditions than triacetate.

Triacetate fabrics can be printed in the same general way as acetate. Disperse colours or vat dyes may be used. The resistance of triacetate to staining means that there is little risk of marking during washing-off.

### Finishing

Triacetate is an attractive fibre with a good handle and excellent draping qualities. It is seldom necessary to apply finishes to bring about improvements in these respects. Softeners are sometimes used, and silicone finishes may be applied to increase water repellency.

Triacetate fibres are often used in blends with other fibres, and finishes may be used to affect the character of other fibres in the blend. It is necessary, therefore, that the triacetate should stand up to conditions used in these finishing treatments. As a rule, the triacetate will cause little difficulty in this respect. Triacetate has a much greater resistance to the effects of hot water and alkaline solutions than has acetate fibre. It is not delustred by boiling soap solutions, and has good dimensional stability when heat set (see below).

Blends of triacetate with cellulosic fibres are commonly treated with resin finishes to reduce moisture sensitivity of the cellulosic fibre and to impart a firmer handle. Careful control of the finishing is necessary in these cases; too much resin can cause poor pleat retention and harsh handle.

## Heat Setting

In the production of cellulose triacetate, the hydroxyl groups of the cellulose molecule have been replaced almost entirely by acetate groups. These bulky, hydrophobic groups have changed the character of the material in several ways. The long molecules are no longer able to pack together as efficiently as the original cellulose molecules, and the powerful forces of attraction between hydroxyl groups is no longer there to hold the aligned molecules tightly together. When cellulose triacetate is heated, especially in the presence of steam, the long molecules are able to move more freely relative to each other; cellulose triacetate is thermoplastic.

Cellulose triacetate has, however, a more symmetrical molecule than secondary cellulose acetate. The triacetate molecule has a succession of large acetate groups attached at regular intervals along the molecule, whereas secondary cellulose acetate has a mixture of acetate groups and hydroxyl groups arranged in random fashion along the molecular backbone. The alignment of molecules into regular order is thus more readily achieved with cellulose triacetate than with secondary acetate.

When cellulose triacetate is heated, the increased freedom of movement of the molecules enables them to adjust their positions with respect to one another. Internal stresses that were caused during spinning are relieved, and the long molecules can align themselves more precisely into crystalline regions.

These changes in the internal structure of the heated triacetate bring about changes in the character of the fibre. The tighter molecular structure is less readily penetrated by moisture, and the amount of water that the fibre will hold is diminished. Moisture regain falls from 4.5% to about 2.5%. As would be expected, this increased resistance to water penetration is accompanied by reduced absorption of dyestuffs, and by greater difficulty in removing dyes which are already in the fibre.

This change in the character of triacetate fibre on heating to temperatures below the melting point is called heat setting. It has become a most important characteristic of triacetate and other thermoplastic fibres, providing a means of stabilizing fabrics and garments, especially of textured yarns, against subsequent deformation during practical use.

If a triacetate fabric is heated under conditions which bring about setting, the molecules of cellulose triacetate will assume those positions that represent a minimum of strain. And they will tend to hold these positions when the fabric is cooled again, even if the fabric is subjected to heat and/or moisture such as it encounters during normal use. The molecules will give up their set positions only if the fibre is heated to temperatures higher than that used in setting. If this should happen, a further adjustment of molecular arrangement might take place.

If a triacetate fabric is heat set whilst being held flat, therefore, it will·acquire a built-in resistance to wrinkling and creasing, and the ability to hold on to its heat set structure under normal conditions of use. It can be made into garments which have easy-care characteristics, and require no ironing.

Heat setting may be taken a stage further by distorting the fabric deliberately into a required shape before the setting is carried out, and then using the setting treatment to hold the fabric permanently in its new shape. If a fabric is folded for example, and then heated to set the triacetate, the molecules in the fibres will take up new positions that relieve the strains set by the distortions caused by folding, and will settle into situations of minimum strain. These positions will then be held unless and until the fibre is heated to a temperature higher than that at which setting took place.

This technique of heat setting is used to set pleats and creases permanently in triacetate garments, and to set three-dimensional shapes in brassieres, collars and the like. Once they have been heat set, these garments retain their new shape during all the conditions of heat and moisture encountered in normal use. Heat set triacetate garments have remarkable dimensional stability.

The temperature used in heat setting is usually some 30 to 40°C. higher than that which might subsequently be encountered, notably in ironing. Fabrics may be subjected, for example, to temperatures of 190–240°C. for periods of 20–30 seconds. The presence of moisture during setting tends to increase the plasticity of the triacetate, and lower temperatures may be used, e.g. 125–130°C. for several minutes.

Heat setting is commonly followed by decatizing, or treatment with steam at atmospheric pressure. This relieves processing strains in the fabric, improving crease-recovery and laundering properties.

*Pleating*

The conditions used in heat setting permanent pleats depend upon the blend and the construction of the fabric. The following examples indicate the conditions that may be used:

100% triacetate: lightweight fabrics: 0.7–1.0 kg/cm$^2$ (10–15 lb/in$^2$) for 15 min; heavyweight fabrics 1.0–1.4 kg/cm$^2$ (15–20 lb/in$^2$) for 30 min;

67% triacetate/33% cotton: 1.4 kg/cm$^2$ (15 lb/in$^2$) for 25–30 min;

60% triacetate/40% wool: 0.7 kg/cm$^2$ (10 lb/in$^2$) for 20 min.

## STRUCTURE AND PROPERTIES

### Fine Structure and Appearance

Triacetate is of bulbous cross-section. The fibres show longitudinal striations.

### Tensile Strength

Tenacity: 10.6–12.4 cN/tex (1.2–1.4 g/den) dry; 6.2–7.1 cN/tex (0.7–0.8 g/den) wet. Ratio wet/dry: 70% approx.
Loop tenacity, dry: 8.8–9.7 cN/tex (1.0–1.1 g/den).
Knot tenacity, dry: 8.8–9.7 cN/tex (1.0–1.1 g/den).

### Elongation

25–30 per cent dry; 30–40 per cent wet.

### Initial Modulus

388.5 cN/tex (44 g/den).

### Specific Gravity

1.32.

**Effect of Moisture**

Regain before heat treatment: 4.5 per cent; after heat treatment: 2.5–3.0 per cent.
The fibre retains 70 per cent of its strength when wet.
Triacetate is not delustred by boiling water, to which it is highly resistant.

**Thermal Properties**

Triacetate is thermoplastic. It differs from acetate in its heat setting characteristics. Heat treatment of triacetate increases the crystallinity and molecular orientation of the fibre. The effect is to set the fibre in a dimensionally stable state; the softening point of the fibre is raised, and its water imbibition and degree of swelling are lowered.

After heat treatment, triacetate has a softening point of 225°C., a fabric glazing point of 240°C. and a melting point of 300°C. Properly set fabrics have a safe ironing temperature of 200°C.

*Effect of High Temperature*

After two weeks exposure at 130°C., triacetate retained 68 per cent of its strength under conditions in which nylon retained 20 per cent and cotton 38 per cent of their tensile strengths.

*Effect of Low Temperature*

Triacetate yarn retains its softness and resiliency at extremely low temperatures.

*Flammability*

Triacetate melts and shrivels to a molten bead when it is ignited. Fabrics will burn as readily as acetate if they are of open weave.

**Effect of Age**

Triacetate is highly resistant to ageing.

**Effect of Sunlight**

Triacetate is highly resistant. On exposure to severe outdoor weathering there is little loss in strength and no yellowing.

**Chemical Properties**

*Acids*

Triacetate is resistant to dilute acids, but is attacked by strong acids in high concentration.

*Alkalis*

Triacetate has an appreciably greater resistance to saponification than acetate. For this reason it cannot be delustred by soap solution or phenol. It resists dilute alkaline solutions such as are commonly encountered in laundering and other wet processing, but is attacked and hydrolyzed by hot strong alkalis.

*General*

Triacetate has a good resistance to the chemicals encountered in normal processing and textile use. It is not affected significantly by common bleaching agents and conditions, including hypochlorites, chlorites, peracetic acid and hydrogen peroxide. Sodium chlorite is recommended as a bleaching agent.

**Effect of Organic Solvents**

Triacetate dissolves in methylene chloride, chloroform, formic acid, acetic acid, dioxan and m. cresol. It is swelled by acetone, ethylene dichloride and trichloroethylene. Triacetate is not affected by methylated spirits, benzene, toluene, xylene, carbon tetrachloride, perchloroethylene and most hydrocarbons.

Trichloroethylene must not be used in dry cleaning, which is preferably carried out with perchloroethylene or petroleum solvents such as white spirit.

**Insects**

Triacetate is not attacked by moths or most tropical insects or larvae which commonly attack textile fibres.

**Micro-organisms**

Triacetate is highly resistant to attack by micro-organisms. Prolonged burial in soil causes no loss of strength, and no microbiological attack can be detected. Does not mildew.

**Electrical Properties**

The electrical resistance of triacetate yarn is very high, and in its unlubricated form it is superior to most textile fibres other than glass, polyesters, polyolefins and fluorocarbons. The antistatic finish which is given to the fibre before processing helps to reduce the effects of static to a minimum in garments and fabrics. Triacetate is very receptive to such finishes.

**Handle**

Triacetate which has been heat set has a crisp, firm handle which is particularly suitable for certain types of fabric including suitings and taffetas. The handle does not match that of acetate fibre for garments to be worn next to the skin, such as lingerie and underwear.

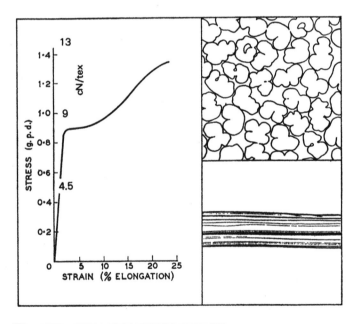

*'Arnel' Triacetate Fibre.*

## TRIACETATE IN USE

The chemical relationship between triacetate and acetate is a close one. Yet it is only in their tensile properties that the two fibres bear any real resemblance to each other.

In many respects, triacetate behaves more like a synthetic fibre than a semi-synthetic fibre. It possesses the thermoplastic properties and the low moisture absorption that we associate with synthetic fibres.

The low moisture absorption of triacetate is reflected in the fact that the fibre retains some 70 per cent of its strength when wet. Fabrics made from triacetate are easily washed and will dry quickly.

The heat setting characteristics of triacetate are of great practical value. When fabrics are heat set they are rendered free from shrinkage, and acquire excellent dimensional stability. They are set permanently in the desired shape. Moreover, heat treatment increases the fastness of the dyestuffs to washing and light if they have been applied prior to heat treatment.

Heat treatment is used to produce permanent effects in fabrics made from triacetate. Permanent pleats are put into woven fabrics of all types, and permanent embossed effects into knitted and woven materials. The permanency of pleats in blended fabrics depends upon the amount of triacetate in the blend. In two-component blends with cellulosic fibres, for example, a minimum of 67 per cent of triacetate is required.

The high melting point of triacetate provides a wide margin of safety in high temperature treatment used in clothing manufacture and laundering.

Triacetate fabrics have little tendency to shrink even before heat treatment. As a result, tighter constructions are required in triacetate and triacetate blended fabrics than in equivalent fabrics of acetate.

Triacetate does not shrink and tighten up the fabric during subsequent dyeing and finishing.

### Washing

Triacetate is not affected by hot water, soap solutions or mild alkalis such as are used in laundering. Garments which have been heat set are dimensionally stable and do not shrink. Laundering of triacetate presents no problems.

Clothes made with triacetate resist soiling and are generally completely washable, quick drying and easy to iron. Fabrics made from triacetate should be washed in warm (40°C., 104°F.) water using detergent or soap flakes. Garments with complicated pleats should preferably be washed by hand, but many garments with simple pleats and most non-pleated garments may be given a minimum machine wash.

Triacetate should never be bleached. Squeezing or wringing should be avoided.

### Drying

Triacetate dries quickly and easily, resembling fully synthetic fibres in this respect. After washing, pleated garments should be given a hand-hot (48°C., 118°F.) rinse and drip dried in their proper shape. Other garments may be drip dried or given a cold rinse and a short spin (15 seconds) followed by line drying. Tumbler drying is recommended for triacetate but it is essential to run the drier cold before switching off.

### Ironing

Triacetate garments should be ironed damp on the reverse side with a warm iron (HLCC Setting 2) or a steam iron.

### Dry Cleaning

Trichloroethylene must not be used. Perchloroethylene or petroleum solvents (e.g. Stoddard solvent) are recommended.

### END USES

Triacetate is established in warpknit garments in underwear and lingerie which retains its shape, and in woven and knitted fabrics which do not shrink or cockle. Triacetate is being used with wool to confer its non-shrink characteristics on the blend, and is blended with cotton and viscose to produce cloths which are completely stable and form permanent pleats.

Triacetate's drip-dry properties, and the fact that many triacetate fabrics need no ironing, have established it as a fibre for use in ease-of-care skirts and dresses. On the other hand, its high melting point permits it to be used in blends or applications where high ironing temperatures are likely to be used, such as mixtures with linen, or in industrial applications.

Triacetate's non-staining properties make it particularly useful for tablecloths and furnishing fabrics.

The permanent pleating effects obtainable in fabrics made from triacetate are of particular interest for applications such as skirts and slacks when blended with viscose rayon staple or with cotton. Blended with wool, triacetate provides fabrics in which the warmth of wool is combined with the heat setting and drip dry properties of triacetate.

## 3. PROTEIN FIBRES

**Introduction**

In producing wool, silk and other animal fibres, Nature makes use of the long-chain molecules of proteins. These are the nitrogen-containing organic substances that play a vital role in the structure and processes of living matter.

Only a minor proportion of the available protein is used for producing natural fibres in this way, and it has long been realized that suitable non-fibrous proteins could perhaps be manipulated to bring them into fibrous form. Proteins have long molecules, the primary requirement of a fibre-forming substance. But the molecules must be brought into some sort of alignment with each other if they are to provide a fibre.

In many proteins, the long molecules are coiled into a compact ball, with coils being linked together in places by chemical bonds. If the molecules of these so-called *globular proteins* are to be

| AMINO–ACID | CHEMICAL STRUCTURE | SYMBOL |
|---|---|---|
| GLYCINE | $NH_2 CH_2 COOH$ | O |
| CYSTINE | $HOOC . CHNH_2 . CH_2 S . S . CH_2 CHNH_2 . COOH$ | ▭ |
| THREONINE | $CH_3 CHOH . CHNH_2 . COOH$ | □ |

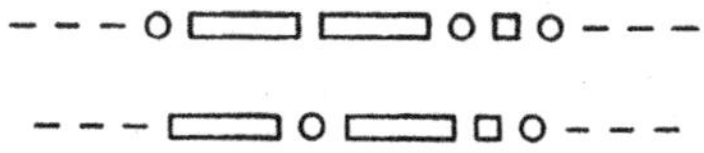

$$- - - O \ \boxed{\phantom{xx}} \ \boxed{\phantom{xx}} \ O \ \square \ O - - -$$

$$- - - \boxed{\phantom{xx}} \ O \ \boxed{\phantom{xx}} \ \square \ O - - -$$

Protein molecules are formed by linking together small amino acid molecules in different proportions and in different sequence. The amino acids shown above, for example, could be linked in innumerable ways, two examples being given.

F                                    115

brought into alignment in such a way as to form a fibre, they must first be subjected to some treatment that destroys the cross-links and permits the molecules to be uncoiled. This process is called *denaturing*.

When a globular protein has been denatured successfully, it may then be possible to dissolve the protein and extrude the solution through the fine holes of a spinneret. As the jets of solution emerge, the protein is coagulated to form solid filaments. In this way, it is possible to make useful fibres from certain types of protein.

If a protein is to be of value as a raw material for making textile fibres, it must be available in adequate quantity, and it must be cheap. A number of proteins satisfy these basic requirements; they are commonly by-products from some industrial process. Among them are casein, zein, arachin and soyabean protein.

Casein is available in the skimmed milk which remains after butterfat has been removed; zein is obtained from maize, and is a by-product in starch manufacture; arachin (groundnut protein) and soyabean protein are left behind after the extraction of oil for margarine and cooking fats.

These proteins have all been used with varying degrees of success for producing protein fibres, but only casein fibres have survived to become a commercially important product.

In general, regenerated protein fibres tend to be weak. The molecules do not align themselves with precision and regularity to form crystalline regions in the fibre, and they cannot hold tightly together to provide the tensile strength that is characteristic of fibres with crystalline structure.

Like wool, regenerated protein fibres will stretch easily; but, unlike wool, they do not have the elasticity that enables them to return to their former length after being stretched.

## Federal Trade Commission Definition

The generic term *azlon* was adopted by the U.S. Federal Trade Commission for fibres of the regenerated protein type, the official definition being as follows:

*Azlon.* A manufactured fibre in which the fibre-forming substance is composed of any regenerated naturally-occurring proteins.

116

## CASEIN FIBRES

### INTRODUCTION

As long ago as 1898, solutions of casein were being spun experimentally to form fibres. Casein solutions were forced through fine jets into hardening baths, forming solid filaments in which the long casein molecules had been given sufficient orientation to hold together in typical fibre form. These early casein fibres were commercially of little value. They were brittle and hard, and lacked the resilience and durability needed for textile use. They swelled to a high degree in water and tended to stick together.

During the early 1930s an Italian chemist, Antonio Ferretti, experimented with casein fibres to try and overcome their drawbacks. He was successful, making casein fibres which were pliable and had many of the properties associated with wool.

Ferretti sold his patents to a large Italian rayon firm – Snia Viscosa – who developed the large-scale manufacture of casein fibres under the trade-name of 'Lanital'. In 1936, the output of 'Lanital' was about 300 tonnes, by the following year it had reached 1,200 tonnes, and in 1939 the production capacity was 10,000 tonnes a year.

Casein fibres have since been produced under various names in a number of countries, e.g. 'Lanital' in Belgium and France, 'Fibrolane' in Britain, 'Merinova' – an improved form of the original 'Lanital' – in Italy, and 'Wipolan' in Poland.

*Note*

Information in the section which follows is based upon 'Fibrolane' produced commercially by Courtaulds Ltd. in the UK. Although production has been suspended, 'Fibrolane' can be regarded as a typical example of a commercial protein fibre.

### PRODUCTION

An outline of the production process is shown on page 119.

#### Raw Material

Casein is obtained by the acid treatment of skimmed milk. The casein coagulates as a curd which is washed and dried, and then ground to a fine powder. 35 litres (7.7 gallons) of skimmed milk produce about 1 kg (2.2 lb) of casein.

### Spinning Solution

Casein is blended to minimize the effect of variations in quality, and is then dissolved in sodium hydroxide solution (caustic soda). The solution is allowed to ripen until it reaches a suitable viscosity, and is then filtered and deaerated.

### Spinning

The spinning solution is wet spun by extrusion through spinnerets into a coagulating bath containing, for example, sulphuric acid (2 parts), formaldehyde (5 parts), glucose (20 parts) and water (100 parts). The jets of solution coagulate into filaments in a manner similar to the coagulation of viscose filaments. They are stretched to some degree during coagulation.

Up to this stage, casein spinning is simpler than that of viscose rayon, as the conditions are not so critical. But subsequent processing may become more involved, as it is necessary to treat the fibre chemically in order to harden it.

The newly-coagulated casein filaments are soft and weak, and will break easily if handled. The spinning process has aligned the casein molecules to some extent, but they are not organized into crystal structures comparable with those of cellulose. Water penetrates readily into the casein filament, pushing apart the long casein molecules and softening and swelling the filament.

The effect of water on untreated casein is such as to render it of little use as a textile fibre. If casein filaments are to be of practical textile use, they must be treated in such a way as to enable the long molecules to hold together in the presence of water, retaining an adequate degree of strength and dimensional stability.

In common with all proteins, casein is a highly reactive material, and it is possible to make use of this activity to create cross-links between adjacent casein molecules. Such cross-links tie the casein molecules together, and prevent them being forced apart by water molecules. Cross-linked casein acquires an increased resistance to the effect of water, retaining a higher degree of tensile strength and resistance to swelling.

Many methods of increasing the water resistance of casein have been developed, and several techniques have been used

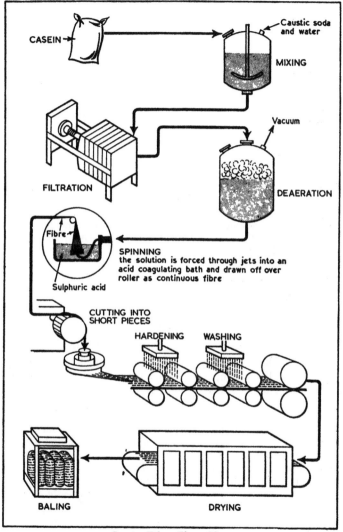

*Casein Fibre Flow Chart*

successfully in practice. The process is commonly described as 'hardening', in that it minimized the softening effects of water. Treatment with formaldehyde forms the basis of many hardening techniques.

In a typical casein fibre production process, bunches of filaments are collected together into a tow as they leave the coagulating bath, and are then steeped in formaldehyde solution. The filaments may be subjected to further stretching at this stage. After treatment, the tow is washed and dried, crimped mechanically, and then cut into staple. The staple may be made into tops for blending with wool, or may be blended during the carding stage.

## PROCESSING

### Spinning

Casein fibre is produced almost entirely as staple, tow or top. A small amount of fibre is used for 100 per cent casein goods, but most casein fibre is blended with wool, cotton, rayon, nylon and other synthetic staple fibres.

Blends containing casein may be spun on all the usual systems.

### Cotton System

For use on the cotton and modified cotton systems, casein fibre is produced, for example, in 3.9 dtex (3.5 den), 50 mm (2 in) or 65 mm (2½ in) staple, and 5.0 dtex (4.5 den), 65 mm (2½ in) staple. Viscose staple is commonly blended with casein for spinning on the cottom system, a typical and highly successful blend being 1/3 casein, 5 dtex (4.5 den), 65 mm (2½ in) and 2/3 viscose staple, 3.3 dtex (3 den), 65 mm (2½ in).

### Woollen System

For use on the woollen system, casein fibre is produced, for example, in 5.0 dtex (4.5 den), 50 and 55 mm (2 and 2½ in) staple, and 10 dtex (9 den), 65 and 100 mm (2½ and 4 in) staple. For carpet blends, heavier deniers are produced, e.g. 20 dtex (18 den) and 33 dtex (30 den), 115 mm (4½ in) staple.

Blended yarns spun on the woollen system commonly contain about ⅓ casein and ⅔ wool or rayon staple. Higher proportions of casein – up to ½ – are used in producing blended carpet yarns containing coarser fibres.

## Worsted System

For use on the worsted system, casein fibre is produced in 3.9, 5.0 and 10 dtex (3½, 4½, 9 den), 100 and 150 mm (4 and 6 in) staple. These fibres are blended with merino or fine cross-bred wools, or rayon staple, for the worsted industry.

## Flax System

Cut tow is commonly used on this system in order to reduce nep formation, e.g. 5.0 and 10.0 dtex (4.5 and 9 den) cut to 150 mm (6 in) staple. This may be blended, for example, with 5.0 dtex (4½ den) cut viscose tow or bright rayon staple and processed into yarn without combing.

Coarse fibres, e.g. 20, 33 dtex (18, 30 den), 200 mm (8 in) staple, may be used on the flax and jute systems of processing in the production of carpet yarns. Fibres of these dimensions are usually blended with 20 and 56 dtex (18 and 50 den) matt rayon staple of a similar staple length, normal staple fibre and not cut tow being used for blends of this type in amounts of up to 50 per cent of casein.

### Sizing

Warps made of staple fibre blend yarns containing casein may be sized satisfactorily from back-beams, either by the 'Cotton Slasher System' or by the 'Rayon Slasher System'. Short warps can be prepared by section warping and sized beams, to beam on a normal multi-cylinder rayon sizing machine.

It is important that the constituents of the size used should be readily removable, i.e. they should be completely removable from the woven fabric by a mild scouring treatment which complies with the conditions suggested.

Among the more common readily removable sizing materials suitable for use with casein blend yarns are:

1. The water-soluble cellulose ethers, and

2. The water-soluble starches (i.e. modified starches, starch ethers and starch esters).

The starch should contain a lubricant, a water-dispersible oil (i.e. one containing a mineral or vegetable oil dissolved or dispersed in a sulphonated oil) being the most satisfactory. Such a lubricant is readily compatible with an aqueous solution of one of the starch or cellulose derivatives referred to above, and is, moreover, easily removed from the woven fabric.

Using a Rayon Slasher, rayon/casein blends yarns may be dried after sizing, at a maximum cylinder temperature of 110°C. Casein blends with acetate or other thermoplastic fibres should be dried at a maximum cylinder surface temperature of 100°C.

Using a Cotton Slasher, the maximum temperatures should be some 5°C. lower in each case.

It is preferable to aim at obtaining a stretch during the sizing operation of not more than 3–4 per cent, and to have some 5–7 per cent of size on the warp yarns for satisfactory weaving.

### Weaving

There are no special difficulties in weaving blend yarns, e.g. of casein and rayon staple containing up to $\frac{1}{3}$ of casein, either as singles – suitably sized, or as unsized folded yarns.

### Desizing

Enzyme products may be used, preferably at pH 4.0 to 6.0. If water soluble sizes have been used, desizing is not necessary.

### Scouring

Synthetic detergents should be used, preferably under acid conditions, e.g. pH 6.0

### Bleaching

In common with all wet processing, bleaching should be carried out if possible under weakly acid conditions, e.g. pH 4.0–6.0, as casein fibres retain maximum strength and minimum swelling under these conditions.

Hypochlorite bleaches should not be used.

Bleaching may be carried out with two volumes hydrogen peroxide buffered to pH 8.0 with 1 kg/24 litres sodium pyrophosphate. Alternatively, an acid stabilizer may be used. Bleaching can usually be effected cold, by steeping overnight.

If alkaline processing is used, it must be followed by careful washing and acidification with acetic acid.

Optical bleaching agents may be applied as for other fibres and blends.

**Dyeing**

Casein absorbs moisture readily and does not have a highly orientated structure. Dyes can penetrate into the fibre without difficulty.

In general, casein can be dyed with dyestuffs used for wool. Acid, basic, direct and disperse dyes are used where good washing-fastness is not a prime essential. Carbolan and Neolan dyes give superior wash fastness.

It is essential to employ modified techniques in the dyeing processes if the desirable properties of casein fibre are to be preserved. In particular, it is necessary to establish careful control over pH and temperature. Buffered systems should be used to keep the pH of the dye liquor between pH 4 and 6.

*Casein Staple Fibre and Continuous Tow*

Casein staple is commonly dyed for eventual use as pressed felts, needleloom carpets, or in woollen blends to be used, for example, in coatings. For these purposes aggregated or metallized acid dyes are mainly used. Selected chrome dyes provide a high standard of fastness.

Continuous tow is dyed for various purposes, and levelling or aggregated dyes will usually provide the required fastness. Heavier deniers require less dye per unit weight than finer deniers, and at the same time are more readily penetrated by dye liquors.

Casein taken from the bale is clean and almost neutral, and will not, as a rule, require scouring or treatment before dyeing. After dyeing, the fibre should be rinsed and given a soft finish, followed by the application of antistatic agent. Continuous tow is usually given the soft finish, but antistatic agent is seldom applied.

*Casein/Wool Blends, Staple Fibre or Tops*

Blends of wool and casein are sometimes dyed in the form of staple fibre in the woollen trade, or as tops for the worsted trade. The procedure is similar to that used with casein alone, but greater care is needed in dye selection to produce a satisfactory solid shade on both fibres in the blend.

It is preferable to convert tops to hanked sliver and dye in machines of the Obermaier type.

In dyeing these blends, it may be necessary to raise the temperature sufficiently to ensure adequate dye fastness on the wool component.

Casein/wool blends are also dyed as yarn or fabric, and as various forms of felt. Dyes commonly used include levelling, aggregated, metallized acid or chrome dyes, and dyeing of yarns is carried out in the Hussong machine or in package form, and piece goods in the winch.

Levelling acid dyes for carpet yarns, certain woollen-type cloths and felts, are selected from those of this class with neutral dyeing properties. Dyeing is carried out at pH 4, for $\frac{1}{2}$ hour at the boil, and thereafter slightly below the boil.

Aggregated or metallized acid dyes, or chrome dyes, provide a higher standard of wet fastness for hosiery yarns, particularly in darker shades, and for some worsted-type cloths. Most of the aggregated or metallized acid dyes are applicable at pH 6, and dyeing is usually carried out at 90–95°C.

Selected chrome dyes are suitable for dark shades of high wet fastness, using either the chromate or the after chrome methods of dyeing.

### Casein/Cellulosic Fibre Blends

Blends of casein with cellulosic fibre may be dyed in the yarn form for the hosiery and carpet trades, but they are more commonly dyed as fabrics.

Yarns are dyed on the Hussong machine, or may be package dyed on cone or cheese; fabrics are usually dyed on a winch.

For most general purposes, these blends are dyed to solid shades with direct dyes and the addition, if necessary, of aggregated acid dyes. As a rule, Class B direct dyes are of greatest interest; they have the least affinity for casein and permit the use of acid dyes on the latter.

Suitable acid dyes are drawn from the aggregated or metallized classes of dye. In general, dyeing at high temperatures will favour the absorption of direct dye by the casein, but the relative rate of dyeing on a cellulosic fibre and casein can be controlled by careful addition of salt and Calsolene Oil HS. It is also possible to produce other attractive effects by dyeing the casein only with aggregated and metallized acid dyes.

Cellulosic blends, being dyed with direct dyes, do not usually require buffered systems, as they have already been processed under slightly acid conditions during scouring. The dye liquor is therefore adjusted to pH 6 with a little acetic acid, the goods receiving a final treatment at pH 4 after dyeing.

### Drying

After dyeing, loose stock and yarns may be centrifugally hydro-extracted before being dried in conventional plant.

Woven fabrics may be hydro-extracted by open width suction machine, or by centrifuging in open width. If basket type extractors are used, excessive running time should be avoided to prevent development of creases and crack marks. High speed water mangles are not recommended.

A recommended drying procedure is either to dry on a slack drier, followed by stentering, or to dry and finish on an over-feed stenter. It is essential to allow an adequate shrinkage from grey to finished dimensions.

### Printing

Casein blend fabrics may be printed very effectively. Good results necessitate thorough preparation. If singeing is necessary, a light treatment with a low burner will suffice. A thorough scour is essential.

Casein fibre is generally white, and bleaching is not usually necessary. If required, however, a mild perborate or peroxide bleach should be used under controlled conditions.

After preparing, the fabric should be dried on the tins, under minimum warp tension, followed by white room stentering to a stable width. Alternatively, the fabric may be dried direct on an enclosed stenter. In all cases, high temperatures and over-drying should be avoided. White room brushing should be unnecessary, but if employed, care should be taken to avoid producing a hairy surface.

Fabrics containing casein may be printed by block, screen, roller, surface roller, and modified paper printer methods. Acid, basic, direct, chrome, mordant, azoic, vat or pigment dyes may be used.

Acid or direct dyes should be applied in a slightly acid or potentially acid paste. In some cases, it may be necessary to modify the viscosity of the printing paste in order to obtain definition similar to that obtained with rayon, wool or cotton fabrics. Casein blends will require a less viscous paste, for example, than 100 per cent rayon fabric, and slightly more viscous paste than that used with 100 per cent cotton.

The minimum amount of alkali should be used on all print mixtures, but vat dyes may be printed by the potassium carbonate formosul method and naphthols may be applied in the usual manner, buffering being advisable. In many cases, it is preferable to use the Rapidogen form of azoic combination. In general, normal printing procedures may be followed, but it may be necessary to modify the printing mixtures slightly in the case of certain dyes, knowledge of which can be obtained from ordinary swatch printing.

Normal ageing and steaming procedures may be followed, but unnecessary and excessive steaming should be avoided, especially in the presence of alkali. After ageing or steaming the fabric should be washed off quickly and not allowed to stand overnight. The fabric should not be finished in an alkaline state.

### Stripping

Partial stripping of blends of casein with wool or rayon staple may be achieved by working the material at approximately 80°C. in a clean liquor containing 5 per cent of Calsolene Oil HS calculated on the weight of material.

If more severe stripping is required, this may be carried out in a liquor containing a neutral solution of sodium hydrosulphite at approximately 50°C.

Initial small-scale experimental strippings should always be carried out.

### Finishing

Crease-resist finishes may be applied to blends containing casein fibre, using temperatures which are preferably not higher than 160°C. (320°F.) for approximately 2¼ minutes. The polymerization may be carried out in a conventional baking chamber, and it should be followed by thorough washing with a neutral detergent.

The final percentage of added resin, calculated on the bone dry weight of fabric, should preferably not exceed 8 per cent.

Handle may be improved by adding proprietary softening agents to the last wet process before drying.

Fabrics containing casein fibre lend themselves readily to raising. Best results are obtained by using a slow rate of raising, and dry raising processes are recommended.

## Carbonizing

Casein will withstand the carbonizing treatment when carried out with the minimum strength of sulphuric acid necessary for the effective removal of vegetable matter. After treatment, the material should be well rinsed and adjusted to pH 4 with sodium bicarbonate.

Carbonizing may be carried out before or after dyeing; if done after dyeing it eliminates the general tendency of the process to cause unlevel dyeing.

## Milling

Casein fibre itself does not display any milling properties, and blends of casein with other non-felting fibres, such as rayon staple or nylon, should not be processed in milling machines.

Blends of casein with wool may be milled, for example, on the Williams-Peace combined scouring and milling machine, or the heavier type of flange roller machine designed only for milling. The stocks can also be used.

It should be remembered that blends of casein and wool will often shrink more quickly than wool itself, and this must be borne in mind in connection with loom settings and the desired finish.

A suitable milling medium is a mixture of 2 parts of soap to 1 part of synthetic detergent. Grease milling may be carried out with only the required amount of free alkali to saponify the oleine, using additional qualities of soap as required.

Acid milling has been used with success for blankets, and is general for most felted structures.

A thorough washing off is essential after scouring or milling, in order to remove soap and alkali. Acetic acid (2 per cent based on weight of goods) is added to the final rinse.

Hat bodies of wool or fur blended with casein, and other felts of various kinds, are generally milled with phosphoric or sulphuric acid at pH 2, preferably at low temperatures. Any type of machine may be used. When felting is complete, washing off should proceed until the pH extract of the material is about pH 4.

## STRUCTURE AND PROPERTIES

### Fine Structure and Appearance

The filaments are smooth-surfaced, with faint striations. Cross-section is bean-shaped to almost round, with a dappled effect due to pitting.

Casein can be spun in the form of fine filaments, with diameters of 20–30μ.

The natural colour is white.

### Tensile Strength

Casein fibre has a tenacity of 9.7–8.0 cN/tex (1.1–0.9 g/den) dry. When wet, the fibres lose much of their strength; tenacity falls to 5.3–2.6 cN/tex (0.6–0.3 g/den).

### Elongation

60–70 per cent, wet or dry.

### Specific Gravity

1.30.

### Effect of Moisture

Casein tends to absorb moisture readily, and the fibres become swollen and soft. They may become plastic and sticky as the temperature is raised. Regain under standard conditions is about 14 per cent (cf. wool).

### Thermal Properties

Casein fibres generally soften on heating, particularly when wet.

### Effect of High Temperature

The fibres become brittle and yellow on prolonged heating at over 100°C. Decomposition is appreciable at 150°C.

### Flammability

Casein fibres burn slowly in air. Flammability is similar to wool.

### Effect of Age

Very resistant.

**Effect of Sunlight**

Very little. Similar to wool.

**Chemical Properties**

*Acids*

Casein is stable to acids of moderate strength under normal conditions. It can be carbonized with cold 2 per cent sulphuric acid solution.

Casein fibre disintegrates in strong mineral acids. It resists dilute mineral acids and weak organic acids, even at elevated temperatures; some loss of strength and embrittlement may occur after boiling for long periods.

*Alkalis*

Like wool, casein is sensitive to alkali. Mild alkalis such as sodium bicarbonate and disodium hydrogen phosphate have little effect at low temperatures. Strong alkalis, such as caustic soda or soda ash cause severe swelling and will ultimately disintegrate the fibre.

*General*

The chemical structure of casein fibre bears some resemblance to that of wool. Both fibres are proteins, but the detailed construction of the protein of casein differs from that of wool. In casein itself there are no sulphur bridges such as there are in wool keratin. Bridges of different chemical types are built into the casein during treatment with formaldehyde or aluminium salts.

Hydrogen peroxide can be safely used as bleach. At high temperatures it will cause some yellowing.

**Effect of Organic Solvents**

Dry cleaning solvents do not cause damage.

**Insects**

Casein fibre is not attacked by moth grubs to the same degree as wool. Damage may be caused, however, when the casein fibre is blended with wool.

## Micro-organisms

Casein fibres are attacked by mildews, particularly when moist.

## Electrical Properties

Dielectric strength of casein fibres is low.

## Other Properties

Casein fibre resembles wool in having a soft warm handle. The fibres are naturally crimped, and yarns have a characteristic warmth and fullness of handle.

Casein fibres provide good thermal insulation. They are resilient, like wool.

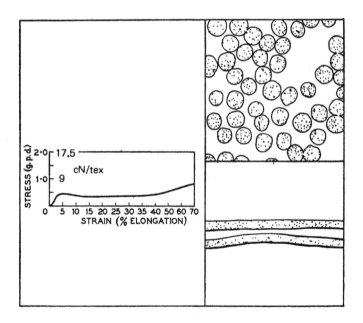

*Casein ('Merinova')*

## CASEIN FIBRES IN USE

The low strength of casein fibre and its sensitivity to water have restricted its use, but it has found a number of applications of importance in certain textile fields.

Casein fibres are produced almost entirely as staple, and are intended primarily as blend fibres for mixture with wool, cotton, rayon, acetate, nylon and other synthetic staple fibres.

In blends with cotton and rayon staple, casein brings warmth, resilience and a full, soft handle. Being of excellent colour, it makes possible the production of good whites, and prolonged wear trials have shown that the whiteness of fabrics made from these yarns is preserved throughout the life of the garment, irrespective of the number of washes it is given.

Blends containing one part casein to two parts rayon staple or cotton have been found particularly satisfactory. Twill and float weaves bring out maximum suppleness in the fabric, and the crease resistance is good. This may be strengthened where required, for example in suitings, by application of a crease-resist finish. Shrink-resist finishes are also used where necessary to improve the shape, stability or resistance to shrinkage on washing.

Most of the casein fibre produced today is used in blends with wool. Casein has a soft handle and warmth that make it particularly suitable for this purpose, and it enables the spinner to produce a yarn of lower cost.

Casein filaments can be spun to very fine diameter, enabling them to blend with the finest qualities of wool. Casein fibre of $20\mu$ diameter is as fine and soft as the 70s wool used for making baby clothes. Casein of $30\mu$ diameter is equivalent to 50s wool.

The proportion of casein used in blends with wool will depend upon the effect required; one third casein is generally satisfactory.

Casein may increase the shrinkage in finishing in some constructions, and allowance should be made for this in the setting of the cloth.

### Washing

Garments containing casein fibre should be washed with care, and treated as gently as wool. Harsh conditions such as cotton will withstand must not be used. High temperatures and strongly acid or alkaline conditions must be avoided. Neutral detergents are preferable for washing.

Fabrics containing casein blended with wool are obviously wool-like in appearance and handle, and they will automatically be treated and washed as wool, causing no difficulties. Blends of casein with cotton will generally be wool-like in handle, and will be treated as wool, but such materials should be suitably labelled to avoid any possibility of harsh treatment during washing.

## Drying

Garments should be dried as wool, care being taken to avoid high temperatures.

## Ironing

The full, soft handle of garments containing casein will be maintained if they are only very slightly damp, or almost dry, before being ironed or pressed. Wool settings should be used, i.e. warm iron (HLCC Setting 2).

## Dry Cleaning

Casein is not affected by dry cleaning solvents, and garments containing casein may be dry cleaned as readily as wool.

## End Uses

### Knitting Yarns

Casein and wool blends are used for fingering and machine knitting yarns, the whiteness of casein lending itself to the production of pastel shades.

### Knitted Fabrics

Worsted spun yarns containing blends of between 30 per cent and 50 per cent casein with wool are very suitable for knitted jersey fabrics where a soft full handle is required, together with a wide range of solid dyed shades, from dark colours to pastels.

Casein/wool blends are used for knitted berets, in which a degree of milling is required to produce the necessary felted structure.

For interlock outerwear, 'T' shirts, cardigans, jumpers etc., interesting effects may be obtained with cotton-spun blends of $\frac{2}{3}$ casein, $\frac{1}{3}$ cotton; $\frac{1}{3}$ casein, $\frac{1}{3}$ rayon staple (spun-dyed), $\frac{1}{3}$ cotton, and similar blends, also introducing nylon. The combination of casein and cotton for knitted fabrics allows great flexibility of garment design and knitted structure.

Casein blended with wool, cotton, rayon staple, nylon and other fibres has many possibilities in the field of circular knitted pile fabrics, Raschel cloths, coatings, blanket fabrics, etc.

*Felts*

One of the earliest uses for casein fibre was in the making of felt for hats. Casein fibres do not have a scaly surface like wool fibres, but they will soften and stick together in warm water, forming a felted mass. Mixed with wool, casein fibres will make the wool felt more readily.

*Pressed Felts*

Blends of casein and wool are made into pressed felts for use as floor coverings. The compact felting of the blend gives high strength and hard-wearing properties necessary for floor covering.

Casein used for this purpose increases the rate of milling, and reduces costs.

*Carpets*

Blends of wool and casein are used in conventional and tufted carpets. The casein contributes a high degree of resilience, good covering power (low fibre density), warmth, excellent soil resistance due to the smooth cross-section, and relatively low price. The casein is obtainable for this purpose in pure white natural colour, or in a large range of spun-dyed colours.

Pile carpets are made with 50 per cent casein blended with wool or rayon staple.

Blends of casein and rayon staple are widely used for needle-loom carpeting. The casein provides softness, bulk and warmth, and 50/50 blends of casein and rayon staple compare favourably with an all-wool carpet in handle, appearance and wearing properties.

*Resilient Fillings and Paddings*

Casein fibre has excellent insulation properties, and good compressibility and resilience. The latter properties depend to a large extent on denier and staple length, and by suitable selection of fibre dimensions the characteristics of finished articles may be obtained to suit requirements.

# GROUNDNUT PROTEIN FIBRE ('ARDIL')

## INTRODUCTION

The natural protein fibres, silk and wool, possess so many attractive properties that they have always served as quality fibres in the textile trade. But animal-derived fibres are, by their very nature, expensive. They are subject to all the uncertainties inherent in anything of animal origin. They are expensive, and vary greatly in quality. Moreover, in the case of wool, their production occupies land that could be devoted to growing food.

The proteins from which these animal fibres are made come, in the first place, from proteins in the plants that are eaten by animals as food. These plant proteins differ from animal proteins in the detailed structure of their molecules. But all proteins are basically similar in chemical design. All protein molecules are in the form of long threads of atoms. Plant proteins, as well as animal proteins, are therefore able to satisfy the first requirement of a fibre-forming material.

The successful production of 'Lanital', 'Fibrolane', and other casein fibres showed that non-fibrous animal protein molecules could be rearranged and aligned to bring them into a fibrous form. There is no reason why the same thing should not be done in the case of protein derived from plants.

## Wet Spinning

In 1935, Professors W. T. Astbury and A. C. Chibnall suggested to Imperial Chemical Industries Ltd., that fibres could be made by dissolving vegetable protein in urea and extruding the solution through spinnerets into coagulating baths. At that time I.C.I. was engaged in research designed to assist in the development of the world's less prosperous areas, with a view to finding new uses for their products. One of the most likely sources of vegetable protein for fibre production was groundnuts, which grow as a staple product in many of the hot, humid regions of the world.

Groundnuts (peanuts, Monkey nuts) are used in large quantities as a source of the arachis oil required for making margarine. The meal remaining after removal of the oil contains a high proportion of protein. This protein was regarded as a potentially suitable source of vegetable protein fibre.

135

Experiments were carried out, and a process was developed for making the peanut protein fibre which became known as 'Ardil'. (The fibre was first made at Ardeer in Scotland.)

By 1938, plans were made for pilot-plant production, but the war held up further progress. Experimental production of 'Ardil' eventually began in 1946. Commercial manufacture followed, a factory being built at Dumfries. By 1951 'Ardil' was in production at the factory, with a planned output of 9 million kg. a year.

Production of 'Ardil' was suspended in 1957.

**Note**

In the section which follows, information on groundnut fibres is based upon the fibre 'Ardil', as it was when production ceased in 1957.

## PRODUCTION

### Raw Material

Groundnuts are the seeds of a sub-tropical annual plant, *Arachis hypogaea L.*, which is cultivated in India, China, West Africa, Borneo and the southern states of U.S.A.

The groundnut plant grows to a height of about 26 cm (10 in). After fertilization the stalk of the ovary elongates, pierces the ground to a depth of 25—75 mm (1—3 in), and the seed-pods ripen underground.

After harvesting, the nuts are shelled or decorticated. The red skins are then removed from the shelled nuts, together with foreign matter such as small stones and nails.

The nuts, which contain about 50 per cent of oil, are crushed and pressed. Some 80 per cent of the available oil is squeezed out, leaving an oily groundnut meal which is reduced in breaker rolls and passed through flaking rolls. The thin flakes pass via a series of buckets on an endless chain into an extraction plant. As they pass through the plant, the buckets of meal are subjected to a thorough washing with solvent (hexane) which removes the remainder of the arachis oil.

The extracted meal is heated under low pressure in steam-jacketed pans to remove residual solvent. It is then cooled, screened, weighed and bagged.

This special technique for removing oil from groundnut meal was devised to provide protein suitable for fibre production.

Extracted meal produced by the normal extraction process used in the production of arachis oil is subjected to too high a temperature, which leads to a deterioration of the properties of the protein.

About 50 per cent of the extracted meal consists of protein, the actual protein content varying according to the maturity of the seeds and the conditions under which they are grown.

The groundnut protein is extracted from the meal by dissolving it in caustic soda solution, the residue after extraction being a valuable cattle food which contains some 3 per cent nitrogen from glutelins insoluble in the caustic soda.

Acidification of the protein solution precipitates the protein, which is the raw material from which fibre is spun.

### Spinning Solution

Groundnut protein dissolves in aqueous urea, ammonia, caustic soda, and solutions of detergents such as alkyl benzene sulphonate. In the manufacture of fibre, dilute solutions of caustic soda were used to dissolve the protein.

A solution of groundnut protein in dilute caustic soda solution is allowed to mature under controlled conditions for 24 hours. During the maturation, the viscosity of the solution increases, probably as a result of the unwinding of folded or coiled molecules of globular protein. At the end of the maturation period, a stable solution of suitable viscosity and spinning characteristics has been produced.

The solids content of the protein solution is between 12 and 30 per cent.

### Spinning

The solution of groundnut protein is filtered and pumped to spinnerets, through which it is extruded at constant rate into an acid coagulating bath. The spinneret holes are typically of 0.07–0.10 mm. diameter.

The coagulating liquor consists of a solution containing sulphuric acid, sodium sulphate and auxiliary substances. It is maintained at a temperature between 12 and 40°C.

As the filament is being spun, it is stretched to increase the alignment of the protein molecules. It coagulates to a filament that is weak and flabby when wet, and brittle when dry. At this stage, the filament dissolves easily in dilute saline solution and in dilute acid and alkali. After leaving the coagulating bath it is treated with formaldehyde to harden and insolubilize it (see Casein Fibre), and it is then dried and cut into staple.

## PROCESSING

### Scouring

Wet processing involving the use of alkali should be carried out at moderately low alkali concentration and temperature. Wool-type scouring conditions are suitable, and processes such as kier boiling should not be used with fabrics containing groundnut protein fibre.

### Bleaching

Sodium hypochlorite and sodium chlorite cause degradation and should not be used. Hydrogen peroxide is the preferred bleaching agent.

### Dyeing

Groundnut protein fibre may be dyed with dyestuffs used for dyeing wool, but the differences in protein structure result in different individual characteristics. In general, the affinity for dyes is higher than that of wool.

## STRUCTURE AND PROPERTIES

### Fine Structure and Appearance

Circular cross-section. Smooth, slightly striated surface.

### Tenacity

6.2–8.0 cN/tex (0.7–0.9 g/den).

### Tensile Strength

8–12 kg/mm$^2$ (11,000–14,000 lb/in$^2$).

### Elongation

40–60 per cent dry; 80 per cent wet.

**Modulus of Torsional Rigidity**

$1.3 \times 10^{10}$ dyne/cm$^2$.

**Specific Gravity**

1.31.

**Effect of Moisture**

Regain: 12–15 per cent (depending upon type).
Expands slightly when wet, drying to original dimensions.
Heat of wetting: 26.6 cal./g.

**Thermal Properties**

Does not soften or melt on heating. Chars at 250°C.

*Flammability*: less flammable than wool.

**Chemical Properties**

*Acids.* High resistance, similar to wool. Withstands carbonizing conditions.

*Alkalis.* Poor resistance, similar to wool.

*General.* Similar to wool. Degraded by sodium hypochlorite and sodium chlorite bleaches.

**Effect of Organic Solvents**

Good resistance, similar to wool. May be dry cleaned without difficulty.

**Insects**

Resistant to attack by moths.

**Micro-organisms**

High resistance to mildews.

**Refractive Index**

1.53.

**Note**

Groundnut protein fibres are generally similar to wool in that they are protein in structure. They do not have the rough scaly surface of wool fibres, and do not undergo felting in the way that wool does. A comparison of properties of groundnut protein fibres ('Ardil') and wool is given in the table below.

Groundnut meal provides a mixture of proteins, the composition of the protein in the filament depending on conditions under which they are produced.

Groundnut protein molecules carry many side chains, and they cannot pack so closely together as the molecules of silk. Groundnut protein yields a relatively weak fibre, which is much more sensitive to moisture than wool.

## COMPARISON OF PROPERTIES –
## GROUNDNUT PROTEIN FIBRE AND WOOL

| Property | Groundnut Protein Fibre ('Ardil') | Wool |
|---|---|---|
| Tensile Strength (kg./mm²) | 8–10 | 12–20 |
| Elongation at Break (per cent) | 40–60 | 30 |
| Young's Modulus (kg./mm²/1 per cent ext'n., at 100 per cent ext'n. per min.) | 1.65 | 2.4 |
| Modulus of Torsional Rigidity (dyne/cm²) | $1.3 \times 10$ | $1.1 \times 10$ |
| Specific Gravity | 1.31 | 1.33 |
| Regain (per cent) | 12–15 | 15 |
| Heat of Wetting (cal./g.) | 26.6 | 26.9 |
| Refractive Index | 1.53 | 1.55 |

## GROUNDNUT PROTEIN FIBRE IN USE

The outstanding characteristic of groundnut protein fibre is its soft, wool-like handle. When it was in commercial production, 'Ardil' groundnut protein fibre was costing about half as much as wool, and it was used largely as a diluent fibre which provided wool-like characteristics at low cost. It was used almost entirely in blends, mostly with wool, but also with cotton and rayon staple.

'Ardil'/wool blends were used for sweaters, blankets, under-wear, carpets and felts. Blends with cotton were used for sports shirts, pyjamas, dress fabrics, and blends with rayon for costume and dress fabrics, tropical clothing, sports shirts and carpets.

## ZEIN FIBRE

### INTRODUCTION

Between 1948 and 1957, a protein fibre 'Vicara' was in production in the U.S. It was made from zein, the protein of maize.

'Vicara' was manufactured by the Virginia-Carolina Chemical Corporation of Richmond, Virginia, in a plant located at Taft-ville, Connecticut. Production was suspended in 1957.

In the natural state, zein molecules are coiled up in the form typical of a globular protein. Before they can be spun into a fibre, the molecules must be uncoiled to permit them to align themselves beside one another. This is done by dissolving the zein in caustic soda, when the neutralization of the acid groups in the protein molecule destroys the attraction between acid and amine groups. The molecules are then able to uncoil and associate together in the positions of alignment necessary for fibre formation.

### PRODUCTION

#### Raw Material

Corn meal is extracted with *iso*propyl alcohol, which dissolves out the zein. After evaporation of the alcohol, the zein is obtained as a pale yellow powder.

#### Spinning Solution

The zein is dissolved in caustic soda solution, which is then filtered and deaerated. The solution is allowed to mature by standing for several hours, when the coiled molecules of zein protein are able to uncoil and straighten themselves out. During the maturation period, the viscosity of the solution increases as these long molecules become associated into larger molecular groups.

### Spinning

When the zein solution has reached the correct viscosity for spinning, it is pumped to spinnerets and extruded into an acid coagulating bath containing formaldehyde. The filaments are stretched, and some cross-linking takes place by reaction between the zein molecules and formaldehyde. After stretching, the zein filaments are subjected to further hardening in formaldehyde, which creates additional cross-linkages between the zein molecules. The filaments are washed, crimped, dried and cut into staple.

## PROCESSING

### Dyeing

'Vicara' could be dyed with most of the normal wool dyes, and with alkaline vat colours. It had excellent resistance to alkali, and the caustic soda used in alkaline vat dyeing had no deleterious effects.

'Vicara' withstood hot water and could be dyed at the boil.

## STRUCTURE AND PROPERTIES

### Fine Structure and Appearance

'Vicara' was made in 2.2, 3.3, 5.6, 7.8, 17 dtex (2, 3, 5, 7, 15 den), and in staple lengths of between 12 and 150 mm (½ and 6 in).

The individual filaments were almost circular in cross-section, and resembled smooth transparent rods. They were crimped mechanically. The colour was golden yellow.

### Tenacity

10.6 cN/tex (1.2 g/den) dry; 5.74 cN/tex (0.65 g/den) wet. Ratio wet/dry: 54 per cent.

### Tensile Strength

1225 to 1365 kg/cm$^2$ (17,500 to 19,500 lb/in$^2$).

### Elongation

25–35 per cent dry; 30–45 per cent wet.

**Elastic Recovery**

96 per cent at 2 per cent elongation; 80 per cent at 5 per cent elongation.

**Specific Gravity**

1.25.

**Effect of Moisture**

Regain 10 per cent.
Moisture imbibition 40 per cent.
Swelling in water 20 per cent.

**Thermal Properties**

Generally similar to wool. Non-thermoplastic, but began to decompose at about 185°C. Melted at 240°C.

*Flammability*: did not burn easily.

**Effect of Age**

None.

**Effect of Sunlight**

Some deterioration on prolonged exposure.

**Chemical Properties**

*Acids.* Highly resistant.

*Alkalis.* Less sensitive to alkali than are wool and other protein fibres. Cold solutions had little effect. Hot solutions of strong alkali caused deterioration.

*General.* Similar to other protein fibres. Good resistance to most chemicals encountered in normal use.

**Effect of Organic Solvents**

Insoluble in most solvents, and could be dry cleaned without difficulty.

**Insects**

Not attacked by moths or other insects that attack wool.

**Micro-organisms**

Resistant to mildews and bacteria.

**Note**

'Vicara' had an attractive handle, and did not cause any irritation to the skin. It had excellent heat-insulating properties, and made up into fabrics that were as warm as wool.

The most significant features of 'Vicara's properties were its comparatively high wet strength and its resistance to alkalis. It could be washed readily and without difficulty.

## ZEIN FIBRE IN USE

'Vicara' offered an unusual combination of attractive properties. It was softer than wool, and made up into fabrics as warm as wool. It was resilient, and gave fabrics a luxurious feel.

The high moisture absorbency made 'Vicara' especially attractive as a clothing fibre, and it could be washed and ironed without difficulty. It did not felt.

'Vicara' was used mostly in blends with cotton, rayon and nylon. It brought excellent handle, resilience, softness and warmth to the mixtures. Mixed with wool, 'Vicara' increased the wear by reducing the tendency to fray.

Blends containing 'Vicara' were used in suitings and clothes, knitted goods, hosiery, blankets and pile fabrics. The 'Vicara' improved crease-resistance and dimensional stability of the garments.

# SOYA-BEAN PROTEIN FIBRE

## INTRODUCTION

Soya-beans have been one of the staple foods of Oriental countries for thousands of years. They are rich in a protein which resembles casein.

In America, soya-beans are now cultivated in great quantity as a source of edible oils and protein. Many attempts have been made to spin this protein into useful fibres. The Ford Motor Company has pioneered in this field; production by this company began in 1939 and reached more than three tons a week by 1942. The fibre was used for making car upholstery. Production was taken over in 1943 by the Drackett Products Co. of Cincinatti, but stopped after a few years.

## PRODUCTION

### Raw Material

Soya-beans have a high protein content (about 35 per cent), and they are grown in abundance in U.S.A. and eastern countries. They provide a cheap and readily available source of protein for fibre production.

The beans are crushed, and the meal is extracted with solvent (hexane) to remove the oil. The protein is dissolved out of the remaining material by dilute sodium sulphite solution, and recovered by acidification of the solution.

### Spinning

Soya-bean protein is dissolved in caustic soda solution, and after being filtered and ripened the solution is pumped to spinnerets. The jets emerge into an acid coagulating bath, and the filaments are stretched, hardened, washed, dried and cut into staple.

## STRUCTURE AND PROPERTIES

### Tenacity
7 cN/tex (0.8 g/den) dry; 2.2 cN/tex (0.25 g/den) wet. Ratio wet/dry: 31 per cent.

### Elongation
50 per cent.

### Effect of Moisture
Regain 11 per cent.

### Note
Soya-bean fibres were of low strength, and were sensitive to moisture to the extent of losing 69 per cent of their tenacity

when wet. They were generally of poor quality compared with other regenerated protein fibres, and had little more than cheapness and availability of raw material to recommend them.

## COLLAGEN FIBRE ('MARENA')

### INTRODUCTION

A protein fibre was made by Carl Freudenberg K.G.a.A., Germany, under the name 'Marena'. Production ceased at the end of 1959.

### PRODUCTION

'Marena' was a collagen fibre, produced from split hides. The split hides were chopped, treated with alkali followed by hydrochloric acid, washed, refined and spun in solution.

The fibres were dried and tanned. They were dope dyed.

### STRUCTURE AND PROPERTIES

**Specific Gravity**

1.32.

**Effect of Age**

None.

**Effect of Sunlight**

None.

**Effect of Acids**

Resistant.

**Effect of Organic Solvents**

Resistant.

**Insects**

Good resistance.

**Micro-organisms**

Good resistance.

146

## COLLAGEN FIBRE IN USE

'Marena' was similar to horse hair. It had good dimensional stability and excellent resistance to dry-cleaning solvents.

It was used largely as brush fibre.


## MISCELLANEOUS PROTEIN FIBRES

Wherever there is a source of cheap, waste protein, there is a potential textile fibre. There are two such sources of industrial waste which have received a great deal of attention, particularly in America. One is the egg albumin remaining behind in the shells at the great egg-drying plants; the other is in chicken feathers.

In the case of the egg albumin, the protein molecules are folded up and globular. The problem is to unfold them and bring them into the extended, fibre-forming state.

Feather protein molecules are already extended, but are highly cross-linked into a network structure. Here, it is a case of breaking down the cross-links without damaging the molecular chains themselves.

The same process has been used experimentally for bringing both these proteins into a spinnable form. Egg or feather protein is treated with a detergent or soap which combines chemically with the protein chains. The detergent has the effect of holding the molecules in the extended position until they have been spun and the fibre coagulated. The detergent is then removed and washed away, leaving the egg or feather protein in the form of extended and aligned molecules. After the usual stretching operation, the molecules are able to pack together and exert their inherent strength.

Other regenerated protein fibres have been made in this way from gelatine and silk waste. In due course, some of these fibres will no doubt find a place in the textile trade.

## 4. MISCELLANEOUS NATURAL POLYMER FIBRES

## ALGINATE FIBRES

### INTRODUCTION

In 1883 an English chemist, E. C. Stanford, discovered that common brown seaweeds contained a substance which served a purpose similar to that of cellulose in land plants. This substance, now called alginic acid, is a polymer of d-mannuronic acid of molecular weight in excess of 15,000. Its structure is as follows:

ALGINIC ACID

Alginic acid accounts for one third or more of the dry weight of many species of seaweed, and is available in virtually unlimited quantities in the millions of tons of weed that litter the world's shoreline.

When alginic acid is treated with caustic soda it is converted into its sodium salt, sodium alginate. Sodium alginate is soluble in water, forming a very viscous solution, and the alginic acid present in seaweed may be extracted by treatment with caustic soda or other alkaline solutions. Alginic acid is precipitated when the sodium alginate solution is acidified.

The molecules of alginic acid and its salts are long and thread-like, and are able to align themselves alongside one another in the manner characteristic of fibre-forming substances. Many attempts have been made to produce commercially useful fibres from alginic acid itself, and from its salts.

A successful development of alginic acid fibres was carried out by Professor J. B. Speakman of Leeds University, England, during World War II. Coarse monofilaments of chromium alginate were first produced. They were green in colour, and were manufactured in some quantity for use in camouflage netting.

In the course of this work, Speakman and his colleagues also made multifilament yarns of calcium alginate. These were of attractive handle and appearance, and were of about the same strength as viscose rayon yarns when dry. They had the added advantage of being flameproof. Unfortunately, calcium alginate yarns proved to be readily soluble in weakly alkaline solutions, including soap and water.

This sensitivity of calcium alginate fibres to weakly alkaline solutions was a serious drawback to the development of these fibres for general commercial use. A fabric made from calcium alginate fibre would dissolve in a scouring bath; a calcium alginate dress might disappear in the wash tub.

Many attempts have been made to spin fibres from other alginate salts, and from alginic acid itself, in the hope of obtaining fibres that would be suitable for normal textile use. Alginic acid fibres are even more readily dissolved in dilute alkali than are calcium alginate fibres. Some metallic alginates, however, are sufficiently resistant to alkali to withstand laundering treatments, and beryllium, chromium and aluminium alginate fibres, in particular, have shown some promise. Aluminium alginate fibres, for example, are washable. Beryllium alginate fibres are particularly stable, but their commercial value is restricted by the toxicity of beryllium and the brittleness of the fibres. Chromium alginate fibres are restricted in their possible applications by their green colour.

As in the case of viscose and other water-sensitive fibres, improved resistance to water penetration might be expected from cross-linking treatments. The alginic acid molecule has reactive acid groupings which could be used to form cross-links, and many techniques for cross-linking alginate fibre molecules have been studied. Cross-linking may be carried out, for example, with formaldehyde and resins, or with hexamethylene diisocyanate.

As yet, it has not been possible to produce an alginate fibre suitable for large scale manufacture and for use in normal textile applications. Calcium alginate yarns are, however, manufactured on a modest scale and are used as speciality yarns in which solubility in weakly alkaline solutions is turned to good effect.

## PRODUCTION

### Raw Material

Seaweed is dried and milled to a fine powder, in which form it may be stored without undergoing bacterial attack. The first step in the production of fibres is to convert the alginic acid in seaweed into sodium alginate. This is done by treatment of the powdered weed with a solution of sodium carbonate and caustic soda.

The solution of sodium alginate is allowed to stand, and the undissolved constituents of the seaweed form a sediment which may be removed. The solution is then bleached, and sodium hypochlorite is added to prevent bacterial attack.

Alginic acid is precipitated from the solution by acidification with hydrochloric acid, and the alginic acid is purified and dried.

### Spinning Solution

The alginic acid is neutralized with sodium carbonate, forming sodium alginate. This is made up into a solution containing about 9 per cent of alginate, which is again sterilized and filtered.

### Spinning

Sodium alginate solution is wet spun into a coagulating bath containing calcium chloride, hydrochloric acid and a small amount of surface active agent. The jets emerging from the spinneret are coagulated into filaments of calcium alginate. These are brought together, washed, oiled, dried and wound.

## PROCESSING

### Dyeing

Basic and direct dyestuffs may be used for dyeing alginate fibres.

### Removal of Yarn from Fabric

Alginate yarns are used primarily as removable linkages, e.g. in the production of hosiery. They may be removed from the fabrics by washing in dilute solutions of sodium carbonate or sequestering agents such as Calgon.

Goods should have sufficient freedom of movement to ensure that the dissolving solution reaches the alginate threads. The following treatments are recommended:

(a) Alginate with Cotton, Viscose Rayon, Linen or Nylon: 20 to 40 minutes at a temperature of at least 50°C. (120°F.) in a bath containing

> Soda Ash 2.5 g./l.
> Common Salt 5 g./l.

(b) Alginate with Wool, Cellulose Acetate or other Alkali-sensitive Fibres: 20 to 40 minutes at a temperature of at least 50°C. (120°F.) in a bath containing

> Lissapol C 2 g./l.
> Calgon 3 g./l.
> Common Salt 5 g./l.

*Note.* Wet steam may cause identification tints to stain goods containing nylon and similar fibres which are being steam set. Excess moisture should be avoided, and when staining occurs it may be removed usually with warm sodium hydrosulphite.

## STRUCTURE AND PROPERTIES

### Fine Structure and Appearance

Alginate fibres are striated length-wise. The surface has a folded appearance; in cross-section the fibres are round to oval with a serrated outline.

### Tenacity

14–18 cN/tex (1.6–2.0 g/den) dry; 4.4 cN/tex (0.5 g/den) wet.

### Elongation

2–6 per cent under normal conditions; 25 per cent wet.

### Specific Gravity

1.779.

### Effect of Moisture

Calcium alginate fibres are insoluble in water but suffer considerable loss of strength when wet.

### Thermal Properties

Alginate fibres are non-flammable. They will not burn even if held in a flame, but will decompose to ash.

### Effect of Alkalis

Calcium alginate fibres will dissolve readily in dilute alkaline solutions, including soap and water.

### Effect of Organic Solvents

No effect.

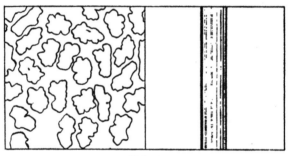

*Alginate*

## ALGINATE FIBRES IN USE

The sensitivity of calcium alginate fibres to dilute solutions of alkali has been a serious drawback to the practical use of these fibres.

Their non-flammability is a most valuable property, and has led to their use, for example, in theatre curtains. A really washable alginate fabric would be of particular appeal in this respect for children's clothes.

The solubility of alginate fibres in alkali has led to a number of specialized applications. The fibres are used, for example, as strength providers in producing loosely-spun wool yarns; the alginate fibres are dissolved away after knitting, leaving a fluffy light-weight fabric which could not have been made by normal methods.

Calcium alginate yarn is of particular interest in the hosiery trade. Socks are linked together by a few courses of alginate yarn, production being continuous. The socks are separated by cutting the alginate yarn, the remains of which are dissolved away. This technique enables perfect welts to be obtained in socks of all types.

For medical use a calcium/sodium alginate yarn provides styptic elastic dressings and dressings which are haemostatic, non-toxic and absorbable in the blood stream. It is used in dental surgery for plugging cavities.

# NATURAL RUBBER FIBRES

## INTRODUCTION

Rubber is a natural polymer obtained by coagulation of the latex produced by certain species of plant, notably *Hevea brasiliensis*, the rubber tree which grows in tropical regions.

In its raw state, rubber is a tough, elastic material which softens on heating, becoming plastic and dough-like. In the processing of rubber, it is kneaded and mixed in powerful mills. This softens the rubber, rendering it more thermoplastic and largely destroying the elasticity of the raw polymer. At the same time, milling provides an opportunity for mixing other materials into the rubber, notably sulphur which takes part in the subsequent process of vulcanization or curing.

When rubber has been softened and mixed on the mill, it is sufficiently thermoplastic to be moulded and shaped by the usual plastics techniques, such as extrusion and compression moulding. If it is then heated in its new shape, the rubber reacts with the sulphur which has been mixed into it, and it sets in its moulded shape. The rubber loses its thermoplasticity and acquires the unusual elasticity we associate with vulcanized rubber.

A cured or vulcanized rubber may stretch to many times its original length, and will return rapidly to that length when the stretching force is removed.

The discovery of vulcanization in 1839 marked the beginning of our modern rubber industry. It provided a method of setting rubber in its moulded form, and of developing the elasticity that was inherent in its molecular structure. Even before the discovery of vulcanization, elastic threads had been made by cutting strips from raw rubber. Many possible applications for these threads had been foreseen, and attempts had been made to produce elastic filaments suitable for textile use. Prior to the discovery of vulcanization, however, it was not possible to produce stable, highly-elastic filaments, and little practical progress had been made.

By 1850, vulcanized rubber threads were in commercial production at the Manchester, England, works of Charles Macintosh. The threads were in the form of fine filaments cut from thin, calendered sheets of vulcanized rubber, a technique which is still used in modified form today.

For the next eighty years or so, cut filaments were the only form of rubber threads available. They came into fairly widespread use for a variety of textile purposes, including fabrics which would 'give' and yet provide support. The square cross-section threads made by cutting sheet rubber were comparatively coarse, however, and they tended to deteriorate in use. They were difficult to incorporate into fabrics by the usual processes of knitting and weaving, and were of poor colour. Elastic threads, prior to the 1930s, played only a modest role in the textile trade.

In the 1930s, a new technique of producing rubber filaments was perfected, in which rubber latex mixed with vulcanizing agents and other materials is extruded through holes in a spinneret. The jets of latex emerge into a coagulating bath to form filaments of rubber which are subsequently vulcanized. This new technique made possible the production of round filaments in very fine counts, and in virtually unlimited lengths.

In the years leading up to World War II, rubber filaments produced by latex extrusion began to make real headway in the textile industry. The old, rigid type of corset and support garment gave way to the lightweight garment in which rubber threads provided a firm but yielding support.

Rubber was virtually the only material available for the

production of elastic filaments at this time, and its shortcomings became more evident as its use became more widespread. It was attacked and degraded by oxygen and by light; it would not withstand prolonged exposure to elevated temperatures; it was attacked by oils and fats, and by perspiration; it tended to discolour, and some of the vulcanizing agents used were liable to stain other fibres.

In the 1950s, rubber filaments really came into their own. Great improvements had been made in the quality of the threads. They could be produced as white filaments which did not discolour unduly in use, and they were more resistant to degradation by oxygen, light and other agents. At this time, also, the introduction of high-speed warp knitting made possible a rapid expansion in the production of elastic fabrics with a two-way stretch, and the demand for lightweight support garments increased by leaps and bounds.

Meanwhile, with the introduction of nylon and other synthetic fibres, finer fabrics of increased strength became available. This created a demand for rubber filaments of finer count and higher modulus, which would be suitable for the production of lightweight fabrics capable of providing powerful support. Rubber filaments met this challenge with the development of high-modulus 'power' threads.

The importance of whiteness in the foundation garment field resulted in continued improvement in this respect. Antioxidants used in earlier rubber filaments had tended to develop a pink colouration during fabric finishing and wear. Also, reaction of constituents of the rubber with traces of dissolved copper salts in water would produce materials that tended to cause a yellowing of nylon fabrics. These problems were solved by modification of the antioxidant and vulcanization systems.

*Spandex Fibres*

In the early 1960s, a new type of elastomeric thread appeared, based upon polyurethanes. These synthetic threads, now known as spandex fibres under the F.T.C. nomenclature (see page xxvi), competed directly with natural rubber threads in the support garment field. And the spandex fibres had obvious advantages over rubber, notably in their higher tensile strengths, higher modulus, and better resistance to oils, fats, perspiration and other organic materials. The elastic recovery of the spandex fibres was better than that of any previous synthetic elastomeric filament, and it seemed at first that natural rubber threads had little future in the support fabric field.

However, production costs of the spandex fibres have been high, and rubber filaments are cheaper. Also, rubber retains some advantages, such as a lower rate of stress decay, which have enabled it to compete more effectively than had been anticipated.

## Counts

The count system used for rubber threads is based upon the diameter of the filament. It is equivalent to the number of filaments that measure 25.4 mm (1 in) when placed side by side. If 62 filaments, for example, make up 25.4 mm when placed side by side, then the filaments are 62s count.

In the case of a round filament, the count is thus the reciprocal of the diameter of the filament, expressed in inches. In the case of square-cut filament, the count is the reciprocal of the width of cross-section expressed in inches.

If the filament is of rectangular cross-section, two adjacent faces are measured, and the width of both is expressed, e.g. $62 \times 40$s count.

## NOMENCLATURE

### Elastomeric Fibre

Natural rubber has an unusual characteristic in that it displays elastic recovery through a very high extensibility, and this property has become known as 'rubber-like' elasticity. Many types of synthetic polymer also display this characteristic to a greater or lesser degree; these polymers, together with natural rubber itself, are described generically as *elastomers,* and fibres made from them are *elastomeric fibres*.

### Natural Rubber Fibre

The term 'natural rubber fibre' indicates a fibre that is made from a specific substance, natural rubber, and it is therefore a descriptive term based upon the chemical constitution of the fibre.

## Federal Trade Commission Definition

*Rubber*

A manufactured fibre in which the fibre-forming substance is composed of natural or synthetic rubber, including the following categories: (1) a manufactured fibre in which the fibre-forming substance is a hydrocarbon such as natural rubber, polyisoprene, polybutadiene, copolymers of dienes and hydrocarbons, or amorphous (non-crystalline) polyolefins, (2) a manufactured fibre in which the fibre-forming substance is a copolymer of acrylonitrile and a diene (such as butadiene) composed of not more than 50% but at least 10% by weight of acrylonitrile units $(-CH_2-CH(CN)-)$.

## PRODUCTION

### Raw Material
Natural rubber is obtained as a latex from the rubber tree, *Hevea brasiliensis*.

### Process

(a) *Cut Rubber Filaments*

The rubber latex is coagulated, and the raw rubber is mixed with vulcanizing agents and other ingredients on a mill. It is then passed through a calender, which produces a thin sheet of very accurately controlled dimensions. The sheet may be cut into filaments before or after vulcanization.

There are a number of techniques for cutting calendered sheets into filaments. In one process, the unvulcanized sheet is cut into a series of flat rings. A number of these are placed one on top of the other, and the pile of rings is vulcanized. The sheets are then mounted on a turntable, and a rotating circular knife in the centre of the rings moves outwards at an appropriate rate so that it cuts away a continuous strip to form a ribbon of filaments.

In another process, the calendered sheet of rubber is vulcanized and then passed over two sets of circular rotating knives, mounted on shafts, one set being below and the other above the sheet. Each pair of knives exerts a scissors-like cutting action on the rubber sheet passing between them, producing filaments which may be collected directly as warps, or passed between pressure rollers. The rollers squeeze the filaments into ribbons in which they adhere lightly together; the individual filaments are readily separated from one another when required.

In the original technique for producing cut rubber thread, which has now almost gone out of use, calendered rubber sheet was vulcanized and then wrapped round a large drum. As the drum rotated, a rotating circular knife moved slowly along the face of the drum, cutting the sheet into a fine spiral of filament. This process produces filaments in comparatively short lengths, up to 180 m (600 ft), and has been superseded by the more modern techniques which produce continuous filaments measuring 1800 m (6000 ft) and more.

Using modern cut filament techniques, it is possible to cut to counts as fine as 85s, and if rectangular filaments are being made the second dimension may be reduced even further by use of calendered sheets of appropriate thickness. A calendered sheet of thickness equivalent to 115s, for example, will provide a filament of average thickness equivalent to 100s count.

The commercial limit for cut filament is usually in the region of 85s.

### (b) *Extruded Rubber Filaments*

Rubber latex is mixed with vulcanizing agents, accelerators, antioxidants, pigments and other materials, and is extruded through glass spinnerets into a coagulating bath, commonly of acetic acid. The jets of latex coagulate, and the filaments are washed, dried and heated to bring about vulcanization of the rubber. The filaments are thus converted into fine, highly-elastic threads, which are dusted with talc to provide a smooth surface which facilitates processing.

The extrusion process produces filaments which are usually of round cross-section, by contrast with the square cross-section of the cut rubber filaments. They are strong and uniform, and can be spun to very fine diameter and in almost any continuous length.

Using this technique, it is possible to produce filaments to counts as fine as 160s.

# PROCESSING

### Scouring

Soaps and detergents may be used for scouring fabrics containing rubber threads. Sodium carbonate may be used if necessary and no significant harm will be done. Caustic soda should not be used.

### Bleaching

Rubber is attacked by strong oxidizing agents, and bleaching must be carried out with care. Hydrogen peroxide and hypochlorite bleaches may be used, the former being preferred. Peracetic acid and sodium chlorite must not be used.

### Dyeing

Rubber filaments are usually coloured where necessary by the addition of appropriate pigments during mixing. If the covering yarn is to be dyed, care should be taken not to use conditions or chemicals which might harm the rubber. High temperatures, oxidizing agents, excessive alkalinity and organic solvents must be avoided. Dyes containing copper in the molecules and dyes which are fixed by the addition of copper salts are not recommended. It is possible to use them without harmful effects, but only by exercising very careful control.

### Finishing

In the finishing of garments containing rubber, precautions should be taken to avoid materials and conditions which might harm the rubber, as in dyeing.

Excessive heat, in particular, may be encountered in many finishing processes, and this will cause degradation of the rubber yarn. Temperatures above 95°C., even for short periods, must be avoided, the maximum permissible total treatment time at this temperature being 1 hour. The permissible time is doubled for each 10°C. drop in temperature.

## STRUCTURE AND PROPERTIES

Natural rubber is a polymer of isoprene in which the isoprene units are arranged in the *cis* configuration; it is *cis*-polyisoprene.

The rubber molecule has a degree of polymerization of about 10,000, i.e. there are some 10,000 isoprene units in the chain.

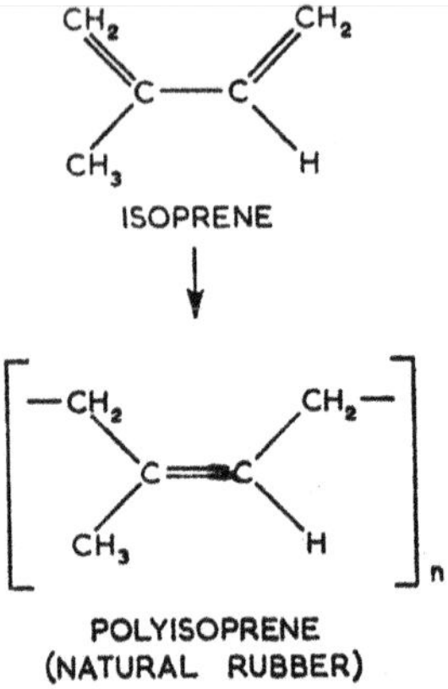

ISOPRENE

POLYISOPRENE
(NATURAL RUBBER)

Natural rubber; *cis*-polyisoprene.

The 'rubber-like' behaviour of rubber is due to the unusual form of its molecules, which are highly-folded. When a piece of rubber is pulled, the molecules tend to straighten out, and if the stretching force is sustained, the molecules may begin to slide over one another and take up new positions. When the stretching force is released, the molecules will return towards their folded state, remaining in their new positions with respect to one another.

Raw rubber is thus a plastic material that displays elasticity to a remarkable degree. If it is heated, the molecules can slide more easily over one another; the rubber is thermoplastic.

When rubber is milled, some breakdown of the long molecules takes place. The entangled molecules cannot hold on to each other as effectively as in the unmilled raw rubber. They are more easily pulled apart, especially when heated. Milled rubber is more thermoplastic than raw rubber.

When milled rubber has been mixed with sulphur and other vulcanizing agents, and then moulded into a new shape, the sulphur undergoes a chemical reaction with the rubber, creating cross-links which hold adjacent molecules together. The folds in the molecules allow them still to be pulled into new positions in response to a force (assuming there are not too many cross-links), but a limit is reached at which the links will no longer permit further movement. If the stretching force is now released, the rubber molecules return to the positions that correspond to the shape in which the rubber was vulcanized.

POLYSULPHIDIC LINK
($x \geq 2$)

MONOSULPHIDIC LINK

CARBON – CARBON LINK

*Vulcanization of Rubber.* When natural rubber is vulcanized, the long chain molecules are linked chemically to adjacent molecules at intervals along their lengths. The nature of the cross-links varies with the conditions under which vulcanization takes place. They may, for example, be polysulphidic, monosulphidic or carbon-carbon cross-links.

**Range of Properties**

The properties of vulcanized rubber may be varied over a very wide range by suitable choice of vulcanization conditions. A lightly-vulcanized rubber may be soft and possess a very low initial modulus, and it could stretch to 8 or 9 times its original length. A highly-vulcanized rubber, on the other hand, may have so many cross-links tying its molecules together that it is a hard, unyielding solid which will not stretch to any noticeable degree.

Rubber filaments can thus be made to a wide range of property specifications, and it is difficult to generalize in considering rubber filaments as a whole. Natural rubber has, however, characteristic properties which are inherent in the nature of the material, and the information which follows may be of use in this respect.

Rubber filaments may be used bare, or covered by other textile yarns of virtually any type. The nature of these composite rubber threads is influenced greatly by the nature of the covering yarn, and by the manner in which the composite yarn is constructed. The information on properties relates only to the rubber filament itself, the term 'rubber' being used always to refer to vulcanized natural rubber.

### Fine Structure and Appearance

Rubber filaments are produced usually as cut filaments of square (or sometimes rectangular) cross-section, or as extruded filaments of round cross-section.

The count of rubber filaments ranges usually from 30s to 125s. Composite yarns are generally covered with two oppositely-wound spirals of textile covering yarns. The gauge of covered yarns ranges from 0.009 in. to 0.040 in. The covering yarns may be of virtually any textile fibre, including cotton yarns in counts of 100s to 24s used singly or as wrappings of 2, 3 and 4 ends, acetate and viscose rayon yarns, 110, 133, 167 dtex (100, 120, 150 den) and nylon and polyester yarns 33–110 dtex (30–100 den).

### Tensile Strength

The breaking stress of a vulcanized rubber in tension is in the region of 1–2 tons/in$^2$. (140–280 kg./cm$^2$.) calculated on the original cross-sectional area. Calculated on the cross-section at break, it may be as high as 15 tons/in$^2$.

A typical rubber filament will have a tensile strength of 385 kg/cm$^2$ (5500 lb/in$^2$). A comparable spandex filament has a tensile strength of 490–700 kg/cm$^2$ (7,000–10,000 lb/in$^2$).

### Tenacity

4.0 cN/tex (0.45 g/den) (cf. spandex: 6.2 cN/tex (0.7 g/den).

### Elongation

700–900 per cent (cf. spandex: 700–800 per cent).

### Elastic Recovery

Rubber has an immediate recovery from stress, under normal circumstances, of 100 per cent.

*Stress Decay*

When rubber is held under stretch, the stresses set up show a gradual decrease with increased time, as the network of cross-linked molecules adjusts to an equilibrium condition.

Stress decay is an extremely difficult property to measure, the fibres being very sensitive to test conditions and to the method of preparation of the samples. The following table shows values obtained by testing production batches of rubber and spandex filaments, the loads being measured at an extension of 200 per cent, using a Scott I.P.2 inclined plane tester, the specimens being first conditioned and then pre-stretched to 500 per cent, the values being read on the second cycle of loading.

### Load/Extension Ratio, Rubber and Spandex Fibres

*Load at 200 per cent extension; psi*

| | |
|---|---|
| **Rubber Thread** | |
| 60 gauge | 235 |
| 100 gauge | 267 |
| **Spandex Fibre** | |
| 280 denier | 714 |
| 420 denier | 635 |
| 840 denier | 600 |

*Note*
These values for load were obtained on a Scott I.P.2. inclined plane tester, the specimens being first conditioned and then pre-stretched to 500 per cent, the values being read on the second cycle of loading. The samples were taken from production batches that were all within specification for this particular property *Courtesy Natural Rubber Producers' Research Association.*

The figures show a much greater reduction in the load/extension ratio during the first and second loading cycle in the case of the spandex fibre than in the case of the rubber. The values for the spandex fibre, however, remain higher. This would appear to give the spandex fibre a considerable advantage over the natural rubber in this important property. It is equally important, however, that the load/extension ratio should be maintained throughout the life of a garment, with the fibre held in a highly-stretched state.

The following table shows the results of further tests comparing the stress decay for natural rubber and spandex fibres. The tests were made using an 'Instron' tester in a conditioned atmosphere, the specimens being extended to 300 per cent extension and maintained there for the stated times. The rubber thread is of heavier gauge than the spandex so that the stress loss was measured at comparable values of stress. Samples of finer rubber thread, however, were tested in the same way and it was found that the percentage stress loss was the same for all counts tested. The results show that the stress decay for the spandex fibre is about twice that for the rubber fibre.

Comparison of Stress Decay, Rubber and Spandex Fibres

|  | Stress Loss (per cent) | |
|  | Spandex Fibre | Rubber Thread |
|  | 622 dtex (560 den). | 60 gauge |
|---|---|---|
| After  5 mins. | 46.6 | 23.4 |
| After 10 mins. | 50.2 | 25.0 |
| After 20 mins. | 53.2 | 27.1 |
| After 40 mins. | 55.2 | 28.0 |
| After 80 mins. | 86.0 | 28.9 |

*Note*
This comparison of stress decay was made on an 'Instron' tester in a conditioned atmosphere, the specimens being extended to 300 per cent extension and maintained there for the stated times. The rubber thread is of heavier gauge than the spandex so that the stress loss was measured at comparable values of stress. Samples of finer gauges of rubber thread, however, were tested in the same way and it was found that the percentage stress loss was the same for all counts tested. *Courtesy Natural Rubber Producers' Research Association.*

*Permanent Set*

When the load is removed from a stretched rubber fibre, recovery is almost 100 per cent if the time of stretching is small.

If the fibre is kept in a stretched condition, as in the stress decay tests above, the relaxation of stress is accompanied by an incomplete return to the original length when the fibre is relaxed. The amount of permanent set in the case of rubber filaments is small, and most of it takes place during the initial period of stretch.

As would be expected from the results of the tests on stress decay, the permanent set of rubber filament is much less than with spandex fibre. The following table compares the two fibres in this respect. In the experiments from which these values were obtained, 25.4 cm (10 in) samples of thread were used, measurements of extended length being taken 5 minutes after elongations of 100, 200 and 300 per cent had been retained for the times stated and the samples removed from the stretching frame. Thus, 5 minutes was allowed for recovery in each case.

Comparison of Permanent Set, Rubber and Spandex Fibres

| | *Extended Length (in.)* | | | | | |
|---|---|---|---|---|---|---|
| | *100% Elong'n.* | | *200% Elong'n.* | | *300% Elong'n.* | |
| | *Spandex* | *Rubber* | *Spandex* | *Rubber* | *Spandex* | *Rubber* |
| After | | | | | | |
| 5 min. | 10.40 | 10.30 | 10.45 | 10.30 | 10.90 | 10.35 |
| 10 min. | 10.70 | 10.30 | 10.60 | 10.40 | 11.15 | 10.50 |
| 20 min. | 11.00 | 10.35 | 11.00 | 10.45 | 11.50 | 10.65 |
| 40 min. | 11.10 | 10.40 | 11.10 | 10.50 | 11.70 | 10.80 |
| 80 min. | 11.10 | 10.50 | 11.30 | 10.65 | 11.85 | 10.85 |

*Note*
10 inch samples of thread were used in these experiments, spandex 622 dtex (560 den) and rubber 60 gauge. The measurements of extended length were taken five minutes after elongations of 100, 200 and 300 per cent had been retained for the times stated and the samples removed from the stretching frame. Thus, five minutes was allowed for recovery in each case. *Courtesy Natural Rubber Producers' Research Association.*

The data show that the permanent set of the natural fibre is much less than that of the spandex. Thus, for an initial extension of 300 per cent for 80 minutes, the rubber has stretched from

10 in (25.4 cm) to 10.85 in (27.6 cm) whereas the spandex fibre has stretched from 10 in (25.4 cm) to 11.85 in (30 cm), i.e. more than twice as much. Moreover, the rubber shows little increase in stretch after the first 5 minutes, whereas the spandex fibre continues to stretch as time goes on.

This difference in permanent set between rubber and spandex fibres is aggravated at higher temperatures. The figures on pages 167 and 168 show the difference in heat ageing behaviour of rubber and spandex fibres.

### Modulus

52.5–70 kg/cm² (750–1000 lb/in²) at 500 per cent extension. (cf. spandex: 175 kg/cm² (2500 lb/in²) at 500 per cent extension).

### Specific Gravity

0.960–1.066.

### Effect of Moisture

Negligible.

### Thermal Properties

*Effect of High Temperature*

At temperatures approaching 140°C., further vulcanization of the rubber may take place, causing increased hardness and decreased strength.

At temperatures of 350°C. and above, rubber softens as molecular breakdown occurs, and then becomes hard and brittle.

Oxygen plays a major role in the degradation of rubber at elevated temperatures.

In general, temperatures above 95°C. even for short periods should be avoided. The maximum permissible time for treatment during fabric finishing is 1 hour at 95°C., the permissible time being doubled for each 10°C. drop in temperature.

*Effect of Low Temperature*

Rubber stiffens with decrease in temperature from about 70°C. down to −20°C. At about −60°C., it becomes glass-like and brittle. If maintained at −25°C., rubber crystallizes and loses its elasticity, but this returns on heating.

*Flammability.* Rubber will burn, but it does not readily propagate flame.

*Specific Heat.* 0.41–0.45.

*Thermal Conductivity.* 0.25–0.31 (relative to water).

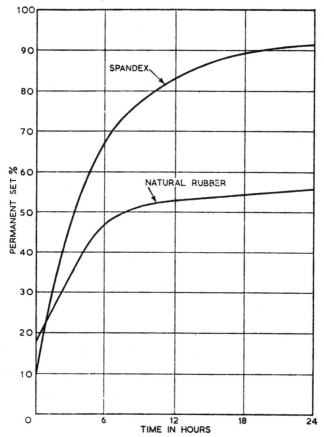

*Heat Ageing, Rubber and Spandex Fibres*
A wet heat ageing test in which thread at 200 per cent extension was immersed in water heated to 70°C.—*Natural Rubber Producers' Research Association.*

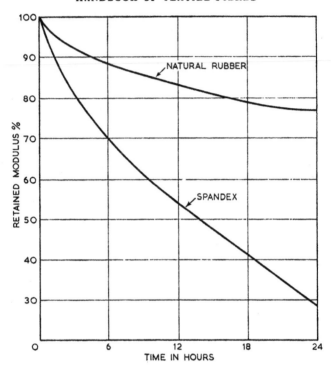

*Heat Ageing, Rubber and Spandex Fibres*
A dry heat oven ageing test carried out at 100°C. at 200 per cent extension—*Natural Rubber Producers' Research Association.*

### Effect of Age

In the absence of excessive heat, sunlight, oils and greases, and catalysts such as copper and manganese, the ageing effects are very slow.

The general effects of age are loss of modulus and extensibility.

### Effect of Sunlight

Light has both a discolouring and a deteriorating action on rubber which is proportional to the intensity of the light and the time of exposure.

In a covered thread, the rubber core is largely protected from the effects of sunlight.

## Chemical Properties

*Acids.* Natural rubber is resistant to most inorganic acids, but is attacked by concentrated sulphuric acid and by oxidizing acids such as nitric and chromic acids.

Mineral acids and the stronger organic acids, unless cold and very dilute, react with the pigment used in white rubber core threads, causing some change of shade.

*Alkalis.* Resistance to alkalis is generally good.

*General.* Oxygen and, more especially, ozone attack rubber, causing degradation which is made apparent by surface cracking. Antioxidants are incorporated in rubber during mixing, and

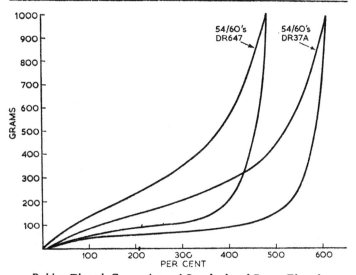

*Rubber Thread. Comparison of Standard and Power Threads.*

These stress-strain diagrams of DR647 Power 'Lactron' and DR37A standard 'Lactron' rubber threads illustrate the increased modulus of the Power thread.—*Lastex Yarn and Lactron Thread Ltd.*

these will give adequate protection for most normal textile applications.

Rubber is attacked by chlorine and other halogens. Hydrogen peroxide and hypochlorite bleaches may be used, but peracetic acid and sodium chlorite bleaches must be avoided.

Copper salts cause degradation, and must be kept out of contact with rubber.

### Effect of Organic Solvents

Natural rubber is swelled to a small extent by acetone, alcohols and vegetable oils such as castor oil. It is attacked by hydrocarbons, oils and fats. Dry cleaning fluids should be avoided.

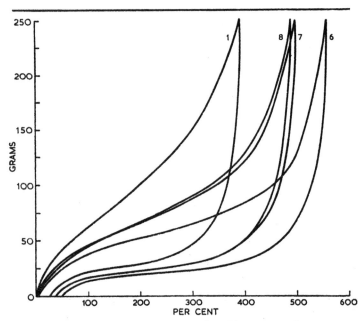

*Stress-Strain Diagram, Spandex Fibre ('Vyrene')*

'Vyrene' 110's count. First and sixth cycles of extension. Seventh cycle, after 6 hours at rest, eighth cycle after 24 hours at rest.—*Lastex Yarn and Lactron Thread Ltd.*

**Insects**

Resistant.

**Micro-organisms**

Rubber is attacked only under very warm and damp conditions.

**Electrical Properties**

Electrical resistivity: $1.7 \times 10^{16}$ ohms/cm. cube.
Dielectric Constant: 3–15.
Power Factor: 0.002–0.1.

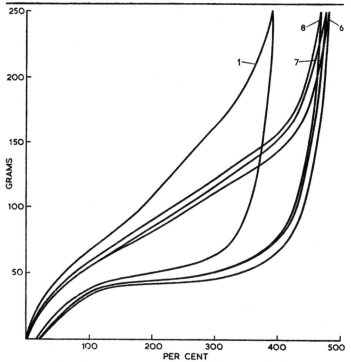

*Stress-Strain Diagram, Rubber Thread ('Lactron')*

'Lactron' 75's count. First and sixth cycles of extension. Seventh cycle after 6 hours at rest, eighth cycle after 24 hours at rest.—*Lastex Yarn and Lactron Thread Ltd.*

## NATURAL RUBBER FIBRES IN USE

**Covering**

Rubber threads are used in knitting and weaving either in the bare form, or covered with spirally-wound textile yarns.

Covering is carried out by drawing a rubber thread through the centre of a rotating hollow spindle. The rubber is stretched to a desired extent as it is emerging from the spindle.

The covering yarn is held on a cheese or bobbin which is mounted on the spindle, and as the bobbin rotates the yarn passes over-end from the bobbin and wraps spirally round the rubber filament as it emerges from the spindle.

In recent years, spindle speeds have increased greatly, largely owing to the use of air-drag to tension the covering yarn. Flyer or ring and traveller mechanisms were previously employed. Using the latter, spindle speeds were in the range of 5,000 to 10,000 rev./min.; modern spindles rotate at about 20,000 rev./min.

There are many reasons for covering rubber threads in this way. The covering yarn protects the rubber from light and other degradative influences, and from materials such as fats and greases which would attack the rubber. Covering also permits the rubber to be held in a partially stretched state in the composite yarn, enabling the covered yarn to exert greater restraining power when used in a support garment. The covering yarn will also restrict the total stretch of the composite yarn; the rubber alone could stretch to 9 or 10 times its original length, but this is not generally desirable in a fabric.

**Knitted, Woven and Braided Fabrics**

Rubber threads are used on all types of conventional equipment in the production of knitted, woven and braided fabrics.

In relatively dense fabrics such as woven webbings or braids, bare rubber filaments may be incorporated in the fabric at high extension, and held in the extended form by the fabric structure, thus providing a fabric with high power.

In open fabrics, it is more difficult to use' the structure of the fabric itself to maintain the rubber thread in its extended state, and in fabrics of this type covered rubber threads are commonly used. The covering yarns of the composite threads hold the rubber core at a controlled degree of stretch when the fabric is relaxed. Composite rubber threads have the added advantage that the rubber is protected from light and from contamination by oils and other degradative materials. They are also easier to control in knitting and weaving processes.

The performance of an elastomeric fibre such as rubber in practical use is influenced greatly by this technique of using the elastomeric threads in a partly extended form. The ability of the stretched material to sustain its power under these conditions during the life of the garment is a vitally important factor.

Elastomeric fibres used for support garments are usually materials capable of total extension of at least 600 per cent. During the initial stages of manufacture of the fabric, the elastomeric thread is prestretched to about 300 per cent. It is then held at a stretch of perhaps 100 to 200 per cent either by the fabric structure or by the covering yarn, so that this degree of extension obtains when the fabric is relaxed. The fabric thus has a high initial resistance to stretching. When the garment is put on, it has to be stretched, and it must then contract on to the body so as to exert sufficient force to achieve the degree of control that is required.

When the manufacturer is designing a support garment, he aims to produce a garment which will provide precisely the right degree of support, and will maintain this support during the life of the garment. If the restraining force is too great, the garment will be uncomfortable; if the restraining force is too small, the necessary control will not be achieved.

During its active life, the support garment will be stretched by body movement, and will return when relaxed to its normal shape, with the elastomeric threads in their positions of controlled extension.

If the elastomeric fibre tends to 'creep' and undergo stress decay when held under tension, it may lose a significant amount of its initial tension before and during its practical use. The designer will have to allow for this loss of tension by tightening up the construction of the garment, so adding to the general discomfort of the wearer, both at rest and during movement.

The following requirements are important, therefore, in an elastomeric fibre:

(1) An extension at break of about 600 per cent, to allow the thread to operate in the middle of its extension range.

173

(2) A high ratio of load to extension (modulus) for extensions in the region of 200–300 per cent.

(3) The ability to retain a high load/extension ratio after continued stretching, and after being held for an indefinite period at an extension of about 200 per cent.

(4) Adequate tensile strength. Very high tensile strength is not needed, as the fibre is never stretched beyond about half its total range, either during manufacture or in subsequent use. The rigid fibres of the fabric act as a buffer which prevents extension beyond a certain point.

Rubber threads satisfy these requirements, and in some respects are superior to other elastomeric materials, including spandex fibres. The stress decay of rubber threads is generally lower than that of spandex fibres. Garments made from natural rubber threads retain their elasticity to a much greater extent, both during the manufacturing and finishing processes of the fabric, and throughout the life of the garment. Thus, natural rubber thread makes it possible to achieve the degree of control required with a much more gentle action which will be retained over a longer period.

Rubber threads are generally less heat sensitive than spandex fibres. They will withstand the conditions encountered in processing without being affected to the extent that spandex fibres are.

### Washing

Garments containing rubber threads should be washed by hand in warm soapy water, and thoroughly rinsed. The surplus water should then be squeezed out, not wrung, and the garments dried with a minimum of heat.

### Ironing

It is generally unnecessary to iron garments containing rubber threads. When ironing is employed, only low temperatures should be used.

### Dry Cleaning

Special care should be taken not to rub or stretch fabrics containing rubber yarns during dry cleaning. It is generally better to avoid dry cleaning, where possible.

**End Uses**

Fabrics designed with rubber threads to improve the fit of garments or to provide support are made in great diversity. By using the appropriate type of rubber yarn and associated textiles, a wide range of fabric handle, from relatively stiff batistes and nets to soft and supple swimwear and underwear fabrics can be obtained. Over the whole range, these fabrics provide exceptional comfort.

The type of garment for which rubber threads are used are as follows:

| *End-Use* | *Fabrics* |
|---|---|
| Corsetry | Woven batistes, satins, brocades and lenos with rubber yarns as weft.<br>Knitted garments (roll-ons, etc.).<br>Narrow woven and braided fabrics used as trimming.<br>Warp-loom knitted elastic nets and laces with rubber yarn as warp. |
| Swimwear | Woven batistes, jacquards and satins with rubber yarn as weft.<br>Knitted garments and fabrics.<br>Ruched elastic fabrics made by sewing rubber yarn on to the back of a rigid woven fabric. |
| Footwear | Woven elastic fabrics for bonding with leather to give the elasticized leather which is widely used in women's footwear. |
| Surgical hosiery | Knitted surgical stockings, anklets, etc. |
| Men's and Children's Hosiery | Rubber yarn is knitted or laid-in the tops of men's and children's socks to make them self-supporting. |
| Ladies', Gents' and Children's Underwear and Outerwear | Rubber yarns are used in many different ways either to perform the function of garment support or to maintain garment shape. Rubber yarns are also used extensively for the ruching of all types of rigid materials used in lingerie, dresses, gloves, etc. |

## SILICATE FIBRES

Fibres spun from mineral silicates or mixtures of minerals containing silicates.

### INTRODUCTION

Most of the fibre-forming substances made available to us by Nature are organic materials, in which the element carbon forms the backbone of a long polymer molecule, as in cellulose and the proteins. There are, however, many naturally-occurring inorganic substances which are capable of being formed into fibres, and some of these have become of commercial importance. They include many minerals and mixtures of the silicate type.

Fibres are spun from a variety of silicate-type minerals and mixtures of minerals, including basalt, wollastonite, dolomite/clay mixtures, etc. The fibres produced from them are generally mixtures of silicates, e.g. calcium, aluminium and magnesium.

### NOMENCLATURE

Mineral silicate fibres are included in various commonly-used fibre groupings. They are *inorganic fibres* and *mineral fibres.* Their resistance to high temperature puts them into the *high temperature fibre* classification.

These fibres are usually described as *mineral silicate fibres,* a term which indicates that they are silicates of natural origin, as distinct from silicates which are made synthetically, e.g. glass.

### PRODUCTION

#### Raw Material

Basalt, wollastonite, dolomite/clay mixtures and the like.

#### Production of Fibre

The processes used in making silicate fibres vary in detail depending upon the nature of the minerals used as raw materials, but the general principles are alike in most cases. The raw material is mixed with coke and fed into a form of blast furnace through which an air jet is blown. The coke burns, releasing heat that melts the charge of siliceous raw material. The molten mineral flows to the bottom of the furnace, from which it emerges to meet a blast of high pressure steam. The molten mineral is disintegrated into small droplets which are attenuated into viscous filaments as they are blown along by the steam jet. The filaments solidify as they cool, forming fibres which are collected on a moving belt as a blanket or batt of matted fibre. A binder of resin may be sprayed on to the batt, and the resin cured by passing the batt through a curing oven. Alternatively, the batt may be sprayed with oil and then broken up into loose fibre.

## STRUCTURE AND PROPERTIES

### Fine Structure and Appearance

Mineral silicate fibres are generally smooth-surfaced and glass-like in appearance, and are of almost round cross-section.

### Tensile Strength

High.

### Specific Gravity

2.8–2.9.

### Effect of Moisture

Negligible.

### Thermal Properties

Mineral silicate fibres are generally resistant to temperatures of, for example, 850–900°C. for prolonged periods.

*Thermal conductivity*. Low.

*Flammability*. Non-flammable.

### Chemical Properties

Mineral silicate fibres are chemically inert, and will resist most of the chemicals encountered in normal use. They resemble glass in their reaction to most chemicals.

### Effect of Organic Solvents

Nil.

H                                     177

**Insects**

Completely resistant.

**Micro-organisms**

Completely resistant.

## MINERAL SILICATE FIBRES IN USE

Mineral silicate fibres are used largely as thermal and acoustical insulation materials, especially where resistance to elevated temperature is an important factor, e.g. in boiler and steam-pipe insulation, and the sound-proofing of buildings.

## SILICA FIBRES

Fibres spun from silicon dioxide (silica), which may or may not contain minor amounts of other materials.

$$SiO_2$$

## INTRODUCTION

The element silicon is the most abundant in the earth's crust, occurring in combination with many other elements as silicates, and in the form of its oxide, silica.

Silica occurs as agate, amethyst, chalcedony, cristobalite, flint, jasper, onyx, opal, quartz, rock crystal, sand and tridymite. It is typically a hard, transparent, glassy mineral, with a very high melting point. Quartz, for example, softens at about 1,500°C., and melts at 1,710–1,755°C. It is chemically extremely stable, resisting the attack of almost all common chemicals.

In 1838, a French scientist, M. Gaudin, discovered that fused quartz could be drawn out into fine filaments, and these were subsequently used as springs in torsion balances. They are still used for this purpose today, providing springs that have almost perfect elasticity, and do not undergo deterioration from fatigue or corrosion.

In more recent times, quartz filaments have assumed a new importance as high-temperature-resisting fibres, and they are now being produced in considerable quantity for applications which require high-temperature-resistance and corrosion resistance.

The outstanding properties of silica in these respects has stimulated the development of other routes to silica fibres. They are now being produced indirectly from glass filaments which are treated to remove constituents other than silica.

A modification of the latter technique consists in spinning a viscose filament containing a high proportion of finely-dispersed silica; the organic materials are then burned away to leave a porous filament of silica.

## TYPES OF SILICA FIBRE

Silica fibres made by the routes indicated above are similar in that they are all basically silica, but they differ in certain characteristics which derive from differences in their methods of production. The fibres are best considered separately, therefore, as different types of silica fibre:

(1) Quartz Fibres.
(2) Silica-from-Glass Fibres (Silica (G) Fibres).
(3) Silica (Viscose Process) Fibres (Silica (V) Fibres).

## NOMENCLATURE

Silica fibres are *inorganic fibres*, and they come within the category of *high-temperature fibres*. They are commonly described also as *ceramic fibres*.

## (1) QUARTZ FIBRES

Fibres spun from naturally-occurring silica in the form of quartz.

### PRODUCTION

#### Raw Material

Quartz is an almost pure form of silica, which occurs naturally in massive form, e.g. as rock crystal, and as silica sand.

#### Spinning

The melting point of quartz is in the region of 1,710 to 1,756°C., and the fabrication of quartz into fibres has been made possible by the development of techniques capable of operating at such high temperatures.

Quartz fibres of 5–10μ diameter are made by softening quartz rods in an oxy-hydrogen flame, and drawing the rods out into filaments of about 0.1 mm. diameter. These are then passed through a series of oxy-hydrogen jets, the quartz being blown forward by the jets to form fine fibres which collect on a rotating drum.

Continuous filaments may be made by softening quartz rods in an oxy-hydrogen flame, and then drawing out the fused quartz into filaments of diameter 0.7μ and less.

## STRUCTURE AND PROPERTIES

### Fine Structure and Appearance

Smooth-surfaced, glass-like filaments of near-round cross-section.

### Tensile Strength

Quartz fibres are strong. Tensile strength is typically in the region of 650 kg./mm$^2$.

### Elastic Recovery

Quartz is an almost perfectly-elastic material, recovery from deformation being almost 100 per cent.

### Specific Gravity

2.6.

### Effect of Moisture

Nil.

### Thermal Properties

*Softening Point.* 1,500°C.

*Melting Point.* 1,700–1,756°C.

### Effect of High Temperature

Quartz fibres may be heated for prolonged periods at temperatures up to 1,400–1,600°C. without undergoing deterioration.

*Flammability.* Non-flammable.

*Thermal Conductivity.* Low.

### Chemical Properties

*Acids*

Highly resistant to all acids except hydrofluoric acid, even after prolonged treatment at normal temperatures. Attacked by phosphoric acid at elevated temperatures.

*Alkalis*

Highly resistant to most alkalis, even after prolonged treatment at normal temperatures. Caustic potash at 90°C. may cause some loss of strength.

*General*

Highly resistant to almost all common chemicals.

QUARTZ FIBRES IN USE

### General Characteristics

Quartz is a strong, almost perfectly elastic and corrosion-resistant fibre, which is capable of withstanding temperatures up to some 1,600°C. This combination of properties gives it a number of important applications in the industrial field.

### End Uses

Quartz fibres are widely used as filtration materials, especially where resistance to corrosive substances and/or high temperature is essential These fibres are also used as insulation materials, serving at temperatures above those where mineral silicate fibres are normally used. Examples of applications include rockets and missiles, jet aircraft, nuclear power plants and industrial furnaces.

## (2) SILICA-FROM-GLASS FIBRES (SILICA (G) FIBRES)

Fibres obtained by the treatment of glass fibres to remove constituents other than silica.

INTRODUCTION

The demand for high-temperature-resisting fibrous materials was stimulated by the development of gas turbines and jet engines during World War II. The exhaust ducts of these engines, operating at temperatures in the region of 700 to 800°C., required insulation from the airframe and other structural parts of the aircraft, and the insulant had to be reasonably fuel-tight, light in weight and able to withstand the prolonged high frequency vibration associated with jet engines.

Silica was obviously a suitable material for this purpose, and a new route to silica fibre production was discovered to provide the fibre in quantity and at reasonable price. This consisted in spinning glass fibre and then leaching out all the constituents other than the silica. Fibres produced in this way were over 96 per cent pure silica, and they proved admirable for the purpose for which they were intended. Since the end of the war, they have developed into one of the most important types of high-temperature-resisting fibres, with a wide range of industrial applications.

## FORMS AVAILABLE

Silica (G) fibres are commonly available in the form of bulk fibre, batt (felt), cloth, tape, sleeving, braided rope, seamless tubular cloth, cord and yarn.

## NOMENCLATURE

Silica fibres made from glass fibres may be described conveniently as *silica (G) fibres* to distinguish them from quartz and from silica fibres produced by the viscose process (silica (V) fibres).

### Note

The information which follows relates primarily to 'Refrasil', produced by The Chemical and Insulating Co. Ltd., U.K. This may be taken as a typical example of a modern silica fibre made by the leaching of glass fibre.

## PRODUCTION

### Glass Fibre

This is produced in the usual way, as described on page 649.

### Silica Fibre

The conversion of glass fibre or textile to silica fibre or textile is in theory a straightforward extraction process, but in practice there are difficulties which must be overcome. These are (a) loss of strength, (b) increased brittleness and (c) shrinkage in length and diameter of fibres after processing the glass fibre or textile.

A typical manufacturing process follows three stages:

(a) Chopped glass fibre is leached with hydrochloric acid until most of the non-silica material has been removed.
(b) The leached fibre is washed and then felted into blankets or batts.
(c) The batts are dried and heat-treated, the fibres bonding together to increase the strength of the material.

## STRUCTURE AND PROPERTIES

### Chemical Structure

Silica fibres produced by the leaching of glass fibre are almost pure silicon dioxide, commonly over 98 per cent. The composition of a typical modern silica (G) fibre of this type ('Refrasil') is as follows:

$$SiO_2 \qquad 98.7 \text{ per cent}$$
$$Fe_2O_3 \qquad 0.1 \text{ per cent}$$

$$\left.\begin{array}{l} Al_2O_3 \\ TiO_2 \end{array}\right\} \qquad 0.4 \text{ per cent}$$

$$CaO \qquad 0.3 \text{ per cent}$$
$$MgO \qquad 0.1 \text{ per cent}$$

The balance consists of combined water and traces of other elements.

### Fine Structure and Appearance

Smooth-surfaced fibre of near-circular cross-section. Fibre diameters range from 0.01 mm ($5\mu$) to 0.02 mm ($10\mu$). Fibre length is about 19 mm (¾ in).

Silica (G) fibres are white. Batt and bulk fibre are like raw cotton in appearance; textiles resemble glass textiles.

Fibre length of batt (felt) and bulk fibre may be varied between 6 and 50 mm (¼–2 in).

Very fine filament batt is also manufactured with a fibre diameter of 0.75μ.

**Tensile Strength**

Silica (G) fibre textiles have adequate strength for the majority of applications. Minimum breaking strength of a typical yarn, 5950 m/kg (3000 yd/lb) is 0.454 kg (1 lb); 3570 m/kg (1800 yd/lb) is 0.908 kg (2 lb).

**Elongation**

Very low.

**Elastic Recovery**

Silica (G) fibres have almost perfect elasticity.

**Specific Gravity**

2.1.

**Effect of Moisture**

Silica is not affected by moisture, the properties of the fibre being unchanged when wet.

**Thermal Properties**

*Softening point.* 1,500°C.

*Melting point.* 1,700–1,756°C.

*Effect of High Temperature*
Silica (G) fibres remain unchanged after continuous and prolonged exposure to temperatures of 1,000°C., and will withstand much higher temperatures for short periods. Resistance to thermal shock is high. Fibres may be heated to 1,000°C. and then quenched in cold water without appreciable change.

*Flammability.* Non-flammable.

*Thermal Conductivity*

The thermal insulation afforded by silica (G) fibre textiles is extremely high.

*Specific Heat*

| Temp (°C.) | B.T.U.s/lb./°F. | Cals./g./°C. |
|------------|-----------------|--------------|
| 260 | 0.19 | 0.19 |
| 538 | 0.21 | 0.21 |
| 815 | 0.23 | 0.23 |
| 1095 | 0.26 | 0.26 |
| 1370 | 0.27 | 0.27 |
| 1650 | 0.28 | 0.28 |

## Chemical Properties

### Acids

Silica (G) fibres are very stable to acids, other than hydrofluoric and phosphoric acids.

### Alkalis

Silica (G) fibres are stable to mild alkalis at low or high temperatures, but strong alkalis will cause some loss of tensile strength and related properties.

### General

Resistance to most common chemicals is excellent.

### Insects

Not attacked.

### Micro-organisms

Not attacked.

### Electrical Properties

Silica (G) fibres are extremely good high temperature electrical insulators, and are virtually the only really flexible materials available in textile form that will serve as electrical insulants at temperatures up to 1,000°C.

The table below shows values for the insulation resistance of silica (G) fibre cloth at various temperatures as determined by The National Physical Laboratory, U.K.

| Temp. (°C.) | Insulation Resistance (ohms/cm²) |
|---|---|
| 20 | $5 \times 10^7$ to $5 \times 10^8$ |
| 200 | $5 \times 10^{12}$ to $2 \times 10^{13}$ |
| 500 | $2 \times 10^9$ to $5 \times 10^9$ |
| 900 | $2 \times 10^7$ to $5 \times 10^7$ |

*Note.* Lower figures relate to plain weave, 0.45 mm (0.018 in) thick. Higher figures relate to 8-end satin weave, 0.90 mm (0.036 in) thick.

### Effect of Radiation

Silica (G) fibres of low boron content may be used in regions of high neutron and gamma flux without appreciable radiation heating effects. The following figures were obtained by exposing silica (G) fibre materials to neutron bombardment in the GLEEP reactor at U.K.A.E.R.E., Harwell, England.

| Material | Macroscopic Nuclear Absorption Cross Section (cm² per gram) |
|---|---|
| Refrasil tape T–RF–3 | 0.016 |
| Refrasil yarn Y–RW–445 | 0.016 |
| Refrasil fibre F–RF–75 | 0.019 |

These low values mean that the materials are completely satisfactory for reactor use, especially since the thermal properties remain substantially unaffected by neutron bombardment.

### SILICA (G) FIBRES IN USE

### General Characteristics

Silica (G) fibres combine light weight with low thermal conductivity, flexibility, high temperature resistance and high resistance to thermal shock. In addition, they have the chemical resistance that is characteristic of silica, and excellent high temperature electrical insulation properties.

Silica (G) fibre batt has a bulk density in the region of 4 lb./ft³. (64 kilo./m³.). It can be compressed to a bulk density of about 9 lb./ft³. Batt consists of a close self-bonded felt of fibres interlocked in random orientation. As no bonding agent is used there is no distortion or loss of strength at high temperatures.

Silica (G) fibre textiles have adequate strength for most applications, but where heavy abrasion is likely to be encountered, e.g. in the machine braiding of yarns and cord, it is preferable to treat the goods with a finishing agent that will increase their resistance to abrasion. This permits the braiding or other operation to be carried out successfully, and the finish is removed quickly when the silica goods are first heated in use. The application and removal of the coating has no effect at all on the final thermal and electrical properties of the silica (G) fibres.

### End Uses

The industrial applications open to silica (G) fibres are many and varied, but they fall mainly into three groups:

     (1) High temperature electrical insulation.
     (2) High temperature thermal insulation.
     (3) Chemical engineering.

#### High Temperature Electrical Insulation
Silica (G) fibre cloths, sleevings and tapes are used for a variety of applications in the electrical industry, e.g. as supporting carrier for the heating element in many domestic appliances, especially electrical convector heaters, electrical furnace insulation, sleeving for thermocouples, high frequency heating coils.

#### High Temperature Thermal Insulation
As a light-weight high temperature thermal insulator, silica (G) fibre is used in furnace, nuclear and chemical plant applications. It is used in the following forms: (a) loose fill (batt or bulk fibre), (b) silica fibre cloth covered blankets (cloth and batt), (c) metal clad blankets (metal foil and batt), and (d) Refrasil Cloths which are fabricated into furnace curtains for the control of heat, protection of expensive equipment during welding operations and also to protect personnel in high temperature processes.

#### Chemical Engineering

In high temperature chemical engineering operations Refrasil is increasingly being used. In steelmaking, silica sleeving and tape are used to wrap hoses subject to radiant heat and splashes of molten steel. Refrasil cloth is being used increasingly for furnace

187

curtains in conveyor-fed annealing and normalising furnaces and also draped over large castings to control the rate of cooling. In continuous steel casting processes gaskets precision cut from thick Refrasil cloth are used in special valves controlling the flow of molten metal. Refrasil Ropes are used in gaskets for furnace door seals. Many other applications involving newer forms of Refrasil materials are continually being found in high temperature chemical engineering and engineering processes.

In atomic energy applications, the low boron content of silica (G) fibres is advantageous in that there is no heating effect when the silica is irradiated with neutrons.

## (3) SILICA (VISCOSE PROCESS) FIBRES; SILICA (V) FIBRES

Fibres produced by dispersing silica or derivatives in viscose solution, spinning by the usual viscose techniques, and then removing the combustible materials to leave fibres consisting substantially of silica.

### INTRODUCTION

The production of fibres from a high-melting material such as silica presents considerable technical difficulties. Filaments may be melt-spun by a direct process, as in the case of quartz, but this is not readily adapted to the production of textile grade filaments on a large scale.

Filaments may, however, be produced by indirect methods, as in the production of silica (G) fibres, in which the fibre is produced by spinning a lower-melting material which is subsequently converted to the high-melting material.

This indirect method of production may be taken a stage further, as in the production of silica (V) fibres. The silica, in this case, is dispersed in a solution of viscose, which is spun into rayon in the usual way. The cellulosic material of the viscose may then be burnt away at any stage of processing, to leave a fibre formed from the silica residue.

188

The technique of producing fibres from inorganic materials in this way has been under investigation for a considerable time. Early forms of inorganic fibre were weak, however, and the high loadings of inorganic material precluded the manufacture of anything other than coarse filaments, e.g. of $50\mu$ diameter.

Modern research, notably by Avtex Fibers Inc., has made possible the production of very fine filaments of silica and other high-melting materials by the viscose technique. The basis of the process is to disperse the inorganic materials in a very fine form in the viscose solution.

Many inorganic substances are soluble in alkali, and will dissolve in the viscose solution. They are subsequently regenerated as the viscose is coagulated in the coagulating bath, providing a fine dispersion of the inorganic material in the viscose filament.

Silica has been made into filaments successfully by this viscose technique, and rayon fibres and fabrics containing silica alone or mixed with carbon and other substances are on the market ('Avceram').

## NOMENCLATURE

Silica fibres made by the viscose technique may be described conveniently as *silica (V) fibres* to distinguish them from quartz and from silica fibres produced by the glass fibre leaching technique (silica (G) fibres).

### Note
The information which follows relates primarily to 'Avceram' produced by Avtex Fibers Inc., U.S.A. This may be taken as a typical example of a modern silica fibre made by the viscose technique.

## PRODUCTION

### Spinning
Production of silica (V) fibres by Avtex Continuous Ceramic Fiber-Process makes use of the conventional viscose process (see page 11). Sodium silicate is dissolved in the viscose solution prior to spinning, and this is converted to hydrated silicic acid during the coagulation of the fibre in the acid coagulating bath.

The technique used is, in this respect, different from the long-established technique of incorporating particles of pigment or delustrant in spinning dopes. In the latter case, the resulting fibre contains essentially the same particle size inorganic material as was supplied in the spinning dope. In the Avtex process, inorganic material is dissolved in the spinning dope, and is thus present in a molecularly dispersed form. As the cellulose has been regenerated in the acid coagulating bath at the same time as the inorganic material is precipitated, the resulting hybrid fibre has few, if any, domains of either cellulose or silica, but is essentially a continuum of both these materials.

### Treatment of Fibre

Fibres produced as above may be further processed at any subsequent stage to produce a variety of products, including a silica fibre.

Pyrolysis of the rayon-silica material under reducing conditions will produce a carbon-silica fibre. If this carbon-silica fibre is oxidized, the carbon may be burned off and a silica filament formed. Alternatively, pyrolysis of the rayon-silica fibre may be carried out under oxidizing conditions, resulting in the direct formation of a silica fibre, without going through the carbon-silica fibre stage.

The carbon-silica fibres produced in this way are unique in that both the carbon and the silica are present as fibrous structures. This may be shown by the fact that oxidation of the carbon-silica fibre results in a fibrous silica, while chemical extraction of the silica from the carbon-silica composition results in a fibrous carbon product.

Since the carbon and silica are indistinguishable as to structure, even using high magnification electron photomicrography, the tendency for these materials to react to form silicon carbide is enhanced. Treatment of the carbon-silica fibres in an inert atmosphere produces continuous silicon carbide filaments.

Two products, based on rayon-silica and pyrolysed rayon-silica are produced commercially by this process, under the trade names 'Avceram'–RS (rayon-silica) and 'Avceram'–CS (carbon-silica).

## STRUCTURE AND PROPERTIES

('Avceram'–RS and 'Avceram'–CS).

### Fine Structure and Appearance

*'Avceram'–RS* yarn is a composition of regenerated cellulose and

silica. The composition may be varied over a wide range up to a silica content of about 60 per cent by weight; the commercial product contains approximately 35 per cent silica and 65 per cent regenerated cellulose.

The fibres are superficially identical with unmodified rayon; filament is 1.3 dtex (1.15 den).

'Avceram'–CS contains 35 per cent carbon and 65 per cent silica.

The fibres are similar in appearance to typical carbon yarns; filament is 0.69 dtex (0.62 den).

**Tensile Strength**

|  | 'Avceram'–RS | 'Avceram'–CS |
|---|---|---|
| Tenacity cN/tex (g/den) | 15.9 (1.8) | 53 (6.0) |
| Tensile Strength kg/cm$^2$ (lb/in$^2$) | 3500 (50,000) | 11,200 (160,000) |

## SILICA (V) FIBRES IN USE

Rayon-silica yarns may be woven, knitted, felted, chopped into staple, or made into felts or papers. Controlled pyrolysis to produce the desired type of silica, silica-carbon or silicon carbide materials may be carried out at any stage.

The fabrics produced in this way used primarily as reinforcement in the manufacture of high temperature resin systems for use in rocket and missile technology, e.g. re-entry system heat shields and rocket nozzles.

# B: SYNTHETIC FIBRES

**Introduction**

The manufacture of synthetic fibres is a wonderful example of the way in which industrial chemistry is contributing to modern life. We now accept synthetic fibres such as nylon and 'Terylene', 'Acrilan' and 'Courtelle' as everyday materials, using them as casually as we use the wool and cotton, flax and silk that have provided man's fabrics for thousands of years.

Yet until the 1930s, synthetic fibres existed only as a few experimental filaments that showed little sign of serving any useful purpose in the textile trade. Who would have dreamed that in twenty years or so the production of synthetic fibres would have become one of the world's great industries?

During the early years of the present century chemists became interested in the unusual materials we now know as plastics. In time, it became realized that the strange properties of plastics were a consequence of their molecular structure – their molecules were long and thread-like, made up of atoms strung together like miniature strings of beads.

These materials with thread-like molecules became known as *polymers,* and during the 1920s and 1930s great progress was made in the study and understanding of polymers of different types. It was found that rubbers and fibres were polymers, their unique properties deriving from the particular shape of long molecule they possessed. Fibres, for example, were polymers in which the long molecules could pack together alongside one another like sticks in a bundle of faggots.

In the years after World War I, chemists carried out a great deal of research on methods of linking up atoms and groups of atoms in such a way as to create long molecules. Many synthetic polymers were made, some of them being plastics, others rubbers. Many of these polymers would dissolve in solvents, and it was natural for people to try extruding the polymer solutions through spinneret holes to find out whether filaments could be made in the same way as rayon.

As long ago as 1913, German chemists were producing an experimental fibre from polyvinyl chloride in this way. But the fibre was of little interest as a potential textile fibre. In 1928, fibres were spun in Germany from a copolymer of vinyl chloride and vinyl acetate, and in 1936 a fibre was being made commercially from chlorinated polyvinyl chloride. This was the fibre Pe Ce, which many claim as the first synthetic textile fibre.

Pe Ce was only of limited value as a textile fibre, and it was never of great commercial importance. The real beginning of the synthetic fibre industry was to stem from the work of Wallace H. Carothers on polyesters and polyamides, which led to the fibre we know as *nylon*.

Today, polymer chemistry is one of the most fruitful branches of scientific research, and many classes of polymer are now being spun into textile fibres of immense commercial value.

## 1. POLYAMIDE FIBRES

### INTRODUCTION

Polyamides are polymers which contain recurring amide groups as integral parts of the main polymer chains. Naturally-occurring polyamides include the protein fibres, e.g. silk and wool. Synthetic polyamide fibres form one of the most important of all classes of textile fibre, which we know today as nylon (see Nomenclature, page 207).

Synthetic polyamides are made by a condensation reaction taking place between small molecules, in which the linkage of the molecules occurs through the formation of amide groups. The reactant molecules are selected to yield linear molecules of polyamide after reaction has taken place, and the linear polyamides from which modern nylon fibres are spun are typically of two structural types, which may be represented as follows:

(1)   $H_2NRNH(COR'CONHRNH)_nCOR'COOH$

(2)   $H_2NRCO(NHRCO)_nNHRCOOH$

In each case R and R' represent chains of atoms separating the functional groups, and 'n' represents the number of recurring units in the polymer molecule, i.e. the degree of polymerization.

Synthetic linear polyamides of these two types are produced by one or other of the two following routes:

(a) The type of polymer shown in (1) above is produced by the interaction of a diamine and a dibasic acid, e.g. hexamethylene diamine and adipic acid:

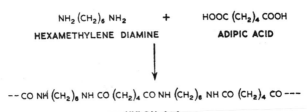

$$NH_2(CH_2)_6NH_2 \quad + \quad HOOC(CH_2)_4COOH$$

HEXAMETHYLENE DIAMINE          ADIPIC ACID

$$--CO\,NH(CH_2)_6\,NH\,CO\,(CH_2)_4\,CO\,NH\,(CH_2)_6\,NH\,CO\,(CH_2)_4\,CO---$$

NYLON 6:6

POLYHEXAMETHYLENE ADIPAMIDE

(b) The type of polymer shown in (2) above is produced by the self-condensation of an amino acid, or a derivative such as a lactam:

$$CH_2(CH_2)_4 CO NH \longrightarrow --NH(CH_2)_5 CO NH (CH_2)_5 CO NH (CH_2)_5 ---$$

**CAPROLACTAM**                      **NYLON 6**

Fibres have been spun experimentally from thousands of polyamides produced by one or other of these condensation reactions. But only a very few have attained real commercial importance in the modern textile industry. One of them, which we know now as nylon (or, more precisely, nylon 6.6 – see Nomenclature, page 207) was the first commercially successful synthetic textile fibre.

### Nylon 6.6

The story of nylon's discovery is a romance of modern scientific research. It is the research of fiction come to life; the chance discovery that led to a world-wide industry.

In 1927, the management of one of America's largest chemical firms, E. I. du Pont de Nemours & Co., decided that the outlook of their industry was becoming too restricted. More and more, they were depending on research to keep them up to date in their processes and in their development of new products. But to a large extent it was 'directed' research – research that was related to the products already being manufactured, and to established scientific fields.

Backed by this research, du Pont was flourishing; but in spite of this, insufficient was being done to broaden the scope of the industry during the years ahead. It was decided, therefore, that research should begin along lines which did not necessarily bear any direct relation to the bread-and-butter activities of the firm itself. The research chemists would be given a free hand to explore at will in any new scientific territory to which their fancy took them. They would not be expected to contribute anything of direct commercial value to the firm; nor would they have to convince the management of any useful purpose in the research they chose. It was to be research for research's sake, carried out in the conviction that if anything was discovered, it would be something entirely new. In 1928, the work began. Responsibility for leading the team of chemists fell to Wallace H. Carothers, a comparatively young and little-known organic chemist who had served as instructor for four years at Illinois and Harvard Universities.

Carothers, during the next few years, was to lead his team with a brilliance that had rarely been surpassed in scientific research. Had he lived, Carothers would have become one of the great scientists of our time. But he died at the age of forty-one, before the results of his work were to come to fruition.

## Polymer Research

Carothers and his team, in 1928, chose for their research the study of the long chain molecules which give us our rubbers, plastics and fibres. At that time the chemistry of polymers was little understood, and much work had to be done before an adequate background could be built up in the new scientific field. Much of our present understanding of the behaviour of long chain molecules is a result of the work carried out by Carothers and his colleagues.

The first stage of the research included a study of the techniques by which short molecules could be linked together into long polymer molecules. Carothers concentrated his attention upon the production of polymers by condensation reactions, rather than by addition reactions involving unsaturated monomers (see page xxiv). And in the initial experiments, he made use of the esterification reaction between alcohol and acid groups to provide the link that joined small molecules together.

In an esterification reaction taking place between two mono-functional substances, e.g. methyl alcohol and acetic acid, the two materials link to form an ester, and water is eliminated:

$$CH_3OH + CH_3COOH \rightarrow CH_3COOCH_3 + H_2O$$

Carothers and his colleagues carried out this reaction between difunctional substances, i.e. glycols, containing two hydroxyl groups, and dibasic acids, containing two carboxyl groups. They selected reactants which would not readily form ring molecules, in the anticipation that the esterification process would link the glycol and dibasic acid molecules alternately end-to-end to produce a long polymer molecule.

$$HO-R-OH + HOOC-R'-COOH \rightarrow$$
$$-O-R-OOC-R'-COO-R-OOC-R'-COO-$$

The esterification took place as expected, and polyesters were formed. Initially, these had molecular weights reaching into the range 2,000 to 5,000. As the experimental techniques became more refined, polyesters of molecular weight 10,000 and higher were obtained. These high molecular weight polymers were given the name superpolyesters. They were commonly clear, viscous liquids which solidified on cooling to horny, opaque solids.

During this early period, Carothers and his colleagues studied the relationships between the properties of polyesters and their molecular structures. This essential but unspectacular phase of the research increased their understanding of polymer science, but yielded little of outstanding practical interest.

In April 1930, an unusual characteristic of superpolyesters was observed, which was to prove of immense importance. It was discovered that if the molten polyester was touched with a glass rod, and the rod was then drawn away, the viscous material stuck to the rod, forming a fine strand linking the rod and the molten mass. As soon as the strand left the hot vessel, it met the cold air and solidified to form a long continuous solid filament. This was quite flexible, and strong enough to be wound on to a bobbin.

*Drawing*

Here, then was something new – a molten synthetic polymer that could be formed into a filament resembling the long continuous filaments of viscose rayon or natural silk.

A further observation was then made. It was found that the polymer filaments could be stretched or drawn out when cold, until they were several times their original length. When stretched the filaments showed no tendency to return to their original length in the way that stretched rubber does. They simply extended until a point was reached at which further extension was resisted, and they remained in their new length.

Moreover, the physical characteristics of the filaments underwent a dramatic change after being cold-drawn. In the undrawn state, they were dull and opaque, with little resistance to tensile stress; in the drawn state, the filaments became transparent and lustrous, and displayed greatly increased tenacity, toughness and elasticity.

These two phenomena, i.e. the formation of filaments from molten polymer, and the cold-drawing of the filament to change its properties, had resulted from the production of linear polyester molecules of great length. Condensation between the glycol and the dibasic acid had resulted in polymers that were essentially linear; there were no side branches or large pendant groups to prevent the molecules packing close alongside each other, and the long molecules were able to form regions of crystallinity in which the forces of attraction between them could develop effectively. Also, the length of the polyester molecules was such as to link the crystalline regions together into the structure that is typical of a fibre (see page xi).

X-ray diffraction studies of undrawn superpolyester filaments showed that the long molecules were indeed packing together into crystalline regions. But in the undrawn fibre, these crystalline regions were not in alignment with each other, or with the long axis of the filament. They were in a state of random orientation.

When the filaments were cold-drawn, however, the X-ray diffraction patterns showed that the crystalline regions were pulled into alignment until they were oriented parallel to the fibre axis. At the same time, drawing brought about an increased degree of alignment of the molecules in the amorphous regions of the filament. This permitted more of the molecules to pack together into crystalline regions, so that the degree of crystallinity was increased.

*Superpolyamides*

The physical properties of some of these early cold-drawn polyester filaments showed certain advantages over established man-made filaments, such as viscose rayon, notably in tenacity and elastic properties. And it became evident that superpolyesters produced by condensing glycols and dibasic acids might provide a route to the production of true synthetic fibres of real commercial value. Carothers and his team began a systematic investigation of superpolyesters with the aim of producing such a fibre.

At an early stage, it became apparent that the superpolyesters then available had deficiencies which precluded their being of value as commercial textile fibres. In particular, the melting points were too low, and the fibres would not withstand the conditions which would be encountered in normal textile use. It was decided, therefore, that work should be concentrated on the study of superpolyamides, which would be expected to provide filaments with properties more satisfactory to textile use.

Initially, polyamides were produced by self-condensation of 9-aminononanoic acid. Polymers were obtained which melted at about 195°C., and fibres spun from this polymer were comparable in strength and flexibility with those of natural silk. This was a promising start.

A wide range of superpolyamides was then made, using both routes, i.e. the self-condensation of amino-acids and lactams, and the condensation of diamines with dibasic acids. At the same time, the method of spinning filaments from the molten polymers came under intensive development. Instead of spinning fibres by touching the molten polymer and pulling away a strand, the chemists began extruding the molten polymers through fine orifices in spinnerets. In principle, this technique resembled that used in spinning viscose rayon, but it involved the extrusion of a molten mass held at high temperature instead of a solution of polymer at normal temperature as in the case of rayon.

This technique of melt spinning presented many practical difficulties. The molten polymer had to be held at a high temperature without decomposing during the spinning operation. Pumps, filters spinnerets and other equipment had to be designed to operate at these high temperatures; the first spinneret used in melt spinning molten polyamides was an electrically heated hypodermic needle.

Following an intensive research campaign during the early 1930s, a superpolyamide was chosen for further development as a potential textile fibre. This was polyhexamethylene adipamide, made by condensation of hexamethylene diamine and adipic acid. The polymer was of suitable melting point (about 250°C.), and the fibres spun from it had properties that made it attractive as a textile fibre. Equally important, this polymer offered the most attractive proposition from the point of view of raw materials and manufacturing costs.

Once this decision had been made, a full-scale development campaign was put in hand, with the object of producing a polyamide fibre on a commercial basis. The task was to prove one of great complexity. There was no background of industrial experience on which the chemists and technologists could draw. The process of producing polyamides was new. The technique of spinning a molten polymer at high temperature had never been attempted before.

Processes had to be devised for the manufacture of the two main raw materials, hexamethylene diamine and adipic acid. The basic source, originally, was phenol, and adequate supplies of this had to be ensured.

Altogether, some eight years of high pressure work by chemists, physicists, engineers and textile technologists were required before production of the new fibre could begin. In April 1937, the first pair of stockings was made from polyamide fibre produced in the experimental plant. The process for making the fibre was then established on a pilot plant scale, and on 27 October 1938 the news of this first synthetic textile fibre was announced to the world. It was to be called 'nylon', a name that had been coined as a generic term for synthetic polyamide fibres.

## Commercial Production

By the end of 1939, the first factory for commercial production of nylon was in operation at Seaford, Delaware. And by May 1940, nylon stockings were being sold to the American public. The end of 1941 saw nylon in production at Seaford to the extent of 3.6 million kilograms per year. But this was insufficient to meet the demand, and a second factory was opened at Martinsville, Virginia. From these two factories, nearly 11.3 million kilograms of nylon per annum were turned out by 1942.

In Britain, the first nylon spinning factory was in operation by January 1941, and the second by December 1942. Canada, also, was in production. During the war, all the yarn produced by these factories was used for essential war purposes. Most important of all was the manufacture of parachutes, which needed strong, light and elastic fibre – a job for which nylon was admirably suited.

After the war, nylon production underwent an enormous expansion. The new fibre was soon being manufactured in many countries, and in many forms; multifilament yarns, monofilament, high tenacity yarns and staple. It was accepted enthusiastically by the general public, and has now come into use in almost every branch of the textile industry.

**Nylon 6**

During his early researches into the production of polyamides, Carothers made a polymer by the self-condensation of caprolactam. He described, in 1932, the fibres that could be spun from this polymer, and polycaproamide was one of the many polyamides considered for possible development as a commercial textile fibre. Following the decision to concentrate on the production of a polyhexamethylene adipamide fibre, however, the polymer made from caprolactam was no longer a candidate for immediate development by du Pont.

In Europe, considerable research had been carried out on polyamides during the 1930s, notably by the German firm of I.G. Farbenindustrie. In 1937, fibres were being spun experimentally from polycaproamide, but it was not until 1939 that these fibres were being produced on a commercially practicable basis.

During World War II, polycaproamide fibres were manufactured in Germany, and used under the name 'Perlon L'. Chemically, these nylon 6 fibres (see Nomenclature, page 207) were similar to the nylon 6.6 fibres produced by du Pont, the slight differences in molecular structure (see page 203) being reflected in some differences in the physical properties of the fibres.

In Germany, as in the U.S.A. and Britain, nylon fibres were used almost entirely for essential war purposes, notably in the production of parachute fabric. After the war, many of the 'Perlon L' factories were dismantled, but production of the fibre began again in Germany in 1948.

**Polyamides Today**

*Nylon 6.6 and Nylon 6*

Since the end of World War II, nylon 6 and nylon 6.6 have established themselves on the world market as two of the most important of all synthetic textile fibres. Production of both types increased at an impressive rate in the immediate postwar years. Almost every industrial country now has its nylon factories, and nylon fibres are making their way into an ever-widening field of textile and industrial applications.

Nylon 6 and nylon 6.6 together account for almost the entire production of polyamide fibres. The range of properties offered by the two fibres makes for great versatility, and there are few textile fields in which nylon does not compete effectively with other fibres. This versatility of nylon 6 and nylon 6.6 has increased steadily over the years as research has introduced all manner of modifications and improvements to the basic types of fibre. The production of trilobal filaments and textured yarns in comparatively recent times, for example, has opened up vast new fields of application for nylon fibres.

Since the end of World War II, continuous expansion of the synthetic fibre market has built up the production of nylon 6 and nylon 6.6 into one of the world's great new industries. An enormous capital investment has been made in the nylon producing industry; vast sums have been expended on the research that is necessary to ensure maximum efficiency of production and continuous improvement of the product. The result has been to create an industry that is operating with great efficiency, and producing two closely-similar types of polyamide fibre for which there is a sustained and indeed increasing demand.

If either of these two polyamide fibres, nylon 6 and nylon 6.6, suffered from serious deficiencies with respect to their suitability for widespread use in the general textile field, the way would be open for the development of a more satisfactory polyamide fibre. Like any other textile fibre, nylon 6 and nylon 6.6 have shortcomings with respect to their use in specific applications. But, by and large, the two nylon fibres provide a range of properties that enables them to compete effectively throughout almost the entire field of general textile applications.

This situation provides little incentive for the introduction of a new polyamide fibre to compete in the general textile field. The prospects are made even less attractive by the cost situation. Nylon 6 and nylon 6.6 are now being produced by highly-efficient processes that are established and thoroughly understood. The economic conditions that influenced the choice of these two polyamides for development in the early days are still valid today. The raw materials in each case are substances with 6 carbon atoms linked in linear fashion in the molecule, and these substances are inherently attractive as raw materials; they may be made by opening out the 6-membered ring of aromatic substances available from coal, oil and other sources.

Under present circumstances, therefore, it is difficult to see how any other polyamide fibre could establish itself in direct competition with nylon 6 and nylon 6.6 as a general purpose textile fibre. These two fibres seem destined to continue as the two general-purpose polyamide fibres for as long as we can see ahead. Competition will develop not from a new contender, but between the two established fibres which already share the polyamide fibre field between them.

## Similarities and Differences

The chemical structures of nylon 6 and nylon 6.6 are virtually identical, differing only in the arrangement of the atoms in the amide groups, as follows:

$$- NH(CH_2)_5CONH(CH_2)_5CO - \qquad - NH(CH_2)_6NHCO(CH_2)_4CO -$$
$$\text{Nylon 6} \qquad\qquad\qquad\qquad \text{Nylon 6.6}$$

There are differences also in the molecular weight distribution and the average molecular weights of the two polymers, resulting from the differences in polymerization techniques used in manufacture.

These structural relationships between nylon 6 and nylon 6.6 are reflected in the relationships between the two fibres spun from the polymers. In general terms, the characteristics of the two fibres are similar, and they fulfil similar roles in the textile industry. But there are differences in the properties of the fibres, which vary in practical significance according to the requirements of particular end-uses.

Some of the more important differences between nylon 6 and nylon 6.6 are as follows:

(1) Nylon 6 has a lower melting point (about 215°C.) than nylon 6.6 (about 250°C.). For most practical textile purposes, this is not a significant factor. The melting point of nylon 6 is generally high enough to meet all normal requirements, but there are circumstances in which the higher melting point of nylon 6.6 is advantageous.

The lower melting point of nylon 6 has been claimed to result in lower fuel costs during manufacture. Less heat is needed to keep the polymer molten during spinning.

(2) Nylon 6 has a greater affinity for certain dyestuffs than nylon 6.6. Dyed together with acid dyes in the same dyebath, nylon 6 will dye to a shade several times deeper than that attained

by nylon6.6. However, colourfastness is generally better on nylon 6.6 because the dye is more closely combined with the fibre. Two-tone (or multi-tone) dyeings may be obtained by dyeing fabrics constructed from both fibres; similar effects can be more readily achieved by combination of different varieties of the same fibre, e.g., bright and dull yarns.

Blends of other fibres with nylon, notably wool blends, may often be dyed more advantageously when nylon 6 is used rather than nylon 6.6.

(3) Both nylon 6 and nylon 6.6 are sensitive to ultraviolet light, and tend to undergo degradation and yellowing to varying degrees after prolonged exposure to sunlight. Resistance is affected greatly by titanium dioxide and additives used to stabilise yarns against UV radiations.

(4) Nylon 6.6 has a better resistance to the effect of high temperatures inasmuch as it has a higher melting point, and can withstand higher temperatures without rapid loss of tensile and other properties. At temperatures well below the melting point of nylon 6, however, both fibres will undergo degradation on prolonged heating, and resistance of nylon 6 to this degradation is greater than that of nylon 6.6.

This increased resistance to heat degradation, coupled with the lower melting point, makes it possible for nylon 6 to be held in a molten condition without undergoing noticeable degradation. Nylon 6.6, on the other hand, tends to degrade fairly rapidly when held in a molten condition. This difference between the two polymers is important in its influence on the manufacture of fibre. It has proved simpler to spin filaments direct from molten nylon 6 after production of the polymer, without intermediate cooling and solidification, than in the case of nylon 6.6.

When setting fabric on a stenter, nylon 6 is at a disadvantage in that it must be set within a closer temperature range ($6^\circ$C) than nylon 6.6 ($20^\circ$C).

(5) Nylon 6 is claimed to have better elastic recovery and fatigue resistance than nylon 6.6. Nylon 6.6 tyre cords, however, have superior fatigue resistance to nylon 6 tyre cords when each is made to optimum conditions.

(6) Nylon 6 filaments blend more readily than those of nylon 6.6. This is an advantage when a soft full hand is required, but is disadvantageous when a crisp hand is required.

(7) Nylon 6 is more suitable for R.F. welding than nylon 6.6.

## Direct Competition

Nylon 6 and nylon 6.6 developed into full-scale industries under war-time conditions, one becoming the polyamide fibre of the Axis Powers, and the other of the Allied Powers. When the war ended each fibre was established in its own region of the world, to the exclusion of the other.

If the differences between the two nylons had been so significant as to give one an impressive advantage over the other, or if the production costs of one had been much less than those of the other, one type of nylon would no doubt have established clear ascendency after the war ended. But with little to choose between nylon 6 and nylon 6.6 in either respect, there was no incentive for either side to switch production from the fibre it had established.

During post-war years, both nylon 6 and nylon 6.6 have gone ahead, each in its own part of the world. The differences between them were insufficient to warrant a preferential development of either; such differences as there were, in fact, contributed to the maintainance of the status quo by making dyers and other processors reluctant to change from one fibre to the other when little technical advantage was to be gained.

In the early development years, therefore, each type of nylon was busy establishing its position largely on the basis of historical development. The situation changed, however, during the 1960s and 1970s, when both types of nylon made headway in territory that was previously regarded as the almost exclusive preserve of the other. In the U.S.A., for example, there was no production of nylon 6 in 1954, but by 1962 this fibre was accounting for some 10 per cent of the total U.S. output. And by 1964, nylon 6 made up about 25 per cent of the total output. Meanwhile, production of nylon 6.6 had been making similar progress in Germany; by 1964, this fibre was accounting for about 33 per cent of the total nylon production.

This pattern has been followed in other countries too. Nylon 6 production is now established in the U.K., which was previously a nylon 6.6 preserve.

With nylon 6 and nylon 6.6 now competing directly with each other in major world markets, the technical differences between them are becoming of greater significance than before. The costs of production, too, are being scrutinized with the greatest care.

Despite the handicap of its lower softening point, nylon 6 appears to be making headway against nylon 6.6 in many countries, including the U.K. and U.S.A. Its protagonists point to the deeper dyeing characteristics, increased light and heat stability as described above. These factors have given it the edge on nylon 6.6 in specific applications, such as tyre cord. The softer hand of nylon 6, too, has been of advantage in some applications, such as tricot fabrics, and the production of fabrics from false-twisted yarns.

With only one raw material to be made, compared with two in the case of nylon 6.6, it would seem that nylon 6 production should be simpler and cheaper than that of nylon 6.6. A new route to caprolactam from toluene has increased the advantage enjoyed by nylon 6 in this respect. On the other hand, it is claimed that diamines and diacids are cheaper to produce than amino acids, giving nylon 6.6 an advantage in costs.

The extra heat stability of nylon 6, allied with its lower melting point, have made the development of continuous processes simpler. Nylon 6 is now being spun direct from molten polymer as it is produced. The high melting point of nylon 6.6, and its tendency to decompose if held in a molten condition, have made the development of continuous spinning processes more difficult.

With so many technical and economic factors involved, it is difficult to predict what the long-term result of competition between nylon 6 and nylon 6.6 will be. In the meantime, it can be forecast with some certainty that nylon 6 and nylon 6.6, between them, will be the mainstay of polyamide fibre production for some decades to come.

### Other Polyamide Fibres

With nylon 6 and nylon 6.6 so well established, and with the economics weighted so heavily against the introduction of new general purpose polyamide fibres, the prospects of developing new polyamide fibres are obviously restricted. The opportunities must lie very largely in the direction of producing specialized types of polyamide fibre which can serve in particular applications for which the regular nylons (and other fibres) are inadequate.

All fibres establish a place for themselves in the textile industry by offering a range of properties that differs to a greater or lesser degree from the property ranges of other fibres. The user assesses the characteristics of every fibre and selects that which offers him the range of properties best suited to his particular needs, at the price he can afford to pay.

Inevitably, the characteristics of any individual fibre will suit some applications better than others, and for certain applications the range of properties it offers will be entirely unsuitable. The demand for any fibre depends upon its price, and upon the breadth of the spectrum of textile applications for which its characteristics are satisfactory.

In the case of the polyamide fibres, nylon 6 and nylon 6.6, the inherent characteristics of the fibres meet the needs of a wide range of important textile applications, and the fibres are produced on a large scale to meet the resulting demand. There are, however, specific applications for which the properties of nylon 6 and nylon 6.6 are inadequate, and important potential markets may be closed on this account. In such cases, it may well be that the fibre fails to meet requirements by only a single factor; the end-use may require, for example, that the fibre should retain its tensile characteristics at a temperature higher than the melting point of the fibre.

If the end-use, in an example such as this, represents an important potential market for the fibre, there is obviously a case for producing a specialized form of polyamide in which the deficiency in the properties is made good.

It is against this type of background that new polyamide fibres are being developed, and the approach is being made from two directions. On the one hand, the established nylons may be modified chemically or physically to meet specific needs. On the other hand, new types of nylon are being produced, which differ chemically from nylon 6 and nylon 6.6.

## NOMENCLATURE

The term 'nylon', as coined by du Pont, was defined as 'a generic term for any long-chain synthetic polyamide which has recurring amide groups as an integral part of the main polymer chain, and which is capable of being formed into a filament in which the structural elements are oriented in the direction of the axis'.

207

Prior to the introduction of polycaproamide fibres ('Perlon L') in Germany, the polyhexamethylene adipamide fibre developed by du Pont in the U.S.A. was the only fibre of commercial importance that came within this definition of 'nylon'. Throughout the wartime years, the du Pont fibre was known simply by this name.

The development of 'Perlon L', however, brought another polyamide fibre on to the market, and this fibre too could be described as 'nylon' under the terminology established by du Pont. Clearly, some method of differentiating between these two polyamide fibres, and others which have since appeared, became necessary.

This can be done most effectively and precisely by referring to the polyamide fibre in terms of its chemical structure; the original nylon, for example, is polyhexamethylene adipamide, and 'Perlon L' is polycaproamide. This terminology is too cumbersome, however, for everyday use, and a simple method of nomenclature was devised which retained the accepted term of 'nylon', but distinguished between the different forms of polyamide.

Using this nomenclature, the number of carbon atoms in the constituents of the nylon are indicated by appropriate figures, the diamine being considered first in the case of a polyamide made by condensing diamine and dibasic acid. Thus, the original nylon, made from hexamethylene diamine and adipic acid, is nylon 6.6 – the diamine and dibasic acid both contain 6 carbon atoms. The nylon made from hexamethylene diamine and sebacic acid, likewise, is nylon 6.10.

When the polyamide is made by self-condensation of a single constituent, e.g. an amino acid, the nature of the polyamide is indicated by a single figure representing the number of carbon atoms in the molecule of the original constituent. Nylon 6, for example, is the fibre made by self-condensation of caprolactam, i.e. 'Perlon L' (now known simply as 'Perlon').

## Copolymers

In describing copolymers, the major component is named first, followed by the minor components in order of decreasing proportions. The percentages of each component are written in parentheses.

In the copolymerization of hexamethylene diamine, adipic acid and sebacic acid, for example, a resultant polyamide described as nylon 6.6/6.10 (90:10) would be a copolymer containing the 6.6 and 6.10 components in the proportions (by weight) of 90:10.

## Federal Trade Commission Definition

*Nylon.* A manufactured fibre in which the fibre-forming substance is a long chain synthetic polyamide in which less than 85% of the amide (–CO–NH–) linkages are attached to two aromatic rings.

*Aramid.* A manufactured fibre in which the fibre-forming substance is a long-chain synthetic polyamide in which at least 85% of the amide linkages (–CO–NH–) are attached directly to two aromatic rings.

## TYPES OF POLYAMIDE FIBRE

Two types of fibre – nylon 6.6 and nylon 6 – dominate the polyamide fibre field. As already described, these are the general-purpose fibres that represent the bulk of polyamide fibre production. In recent years, however, a number of new polyamide fibres have assumed commercial importance. In some cases, these have already come into commercial production; in other cases, they are still under development, but show prospects of achieving real importance in due course.

In this section of the Handbook, the commercially important types of polyamide fibre are discussed as follows:

(1) Nylon 6.6

(2) Nylon 6

(3) Nylon 11

(4) Nylon 6.10

(5) New types of Polyamide Fibre.

## (1) NYLON 6.6

Nylon 6.6 fibre is spun from polyhexamethylene adipamide, a polyamide made by condensation of hexamethylene diamine and adipic acid:

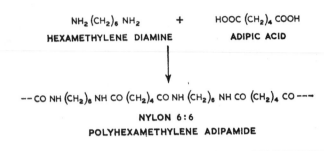

$$NH_2 (CH_2)_6 NH_2 \quad + \quad HOOC (CH_2)_4 COOH$$

HEXAMETHYLENE DIAMINE          ADIPIC ACID

$$-- CO\ NH\ (CH_2)_6\ NH\ CO\ (CH_2)_4\ CO\ NH\ (CH_2)_6\ NH\ CO\ (CH_2)_4\ CO ---$$

NYLON 6:6

POLYHEXAMETHYLENE ADIPAMIDE

## TYPES OF NYLON 6.6 FIBRE

Nylon 6.6 is produced as multifilament yarns, monofilaments, staple and tow, in a wide range of counts and staple lengths to suit virtually all textile requirements.

The fibres are available in bright, semi-dull and dull lustres, and with additives such as optical bleaches for specialized end-uses.

The properties of nylon 6.6 fibres vary over a range which is limited by the inherent characteristics of the polymer, each manufacturer controlling his process to produce fibres that will meet specific requirements. In general, commercial nylon fibres fall into two main classes, (a) regular tenacity and (b) high tenacity.

Nylon 6.6 is a thermoplastic fibre, and lends itself well to physical modifications associated with this property. Crimped and textured yarns of all the familiar types are available.

### Modified Fibres

Nylon 6.6 filaments are commonly produced in round cross-section, but fibres of special (e.g. multilobal) cross-section are now available from several manufacturers. Bicomponent fibres of various types have been introduced, and many others are under development.

These radical modifications of the basic nylon 6.6 fibre have introduced new fibres which have special characteristics differing from those of the normal nylon fibres.

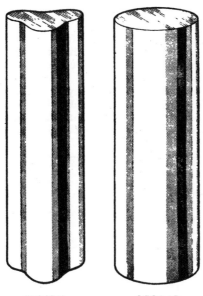

TRILOBAL        CIRCULAR

A nylon 6.6 fibre of trilobal cross section. The physical and chemical characteristics of the fibre are identical with those of normal cross-section, but the trilobal cross section results in fabrics with a drier hand, increased covering power and greater bulk. Fabrics show a unique, rich, three dimensional highlight effect, and prints on both woven and knitted fabrics have unusual clarity and definition.

## PRODUCTION

**Reactant Synthesis**

Nylon 6.6 polymer is made by condensation of two substances: (a) adipic acid, and (b) hexamethylene diamine.

These starting materials are synthesized usually via one or other of three routes.

### (1) *Cyclohexanol Route*

This is the original route used in making the starting materials for nylon 6.6, and it is still the route by which much of the nylon 6.6 is made today. The stages in the synthesis are shown below.

Originally, cyclohexanol was made from phenol, which was, in turn, obtained from the benzene distilled from coal tar or petroleum. The phenol is reduced to cyclohexanol by hydrogenation in the presence of a catalyst (1).

Much of the cyclohexanol used today is produced by a more direct route from benzene, which is reduced to cyclohexane (2); the latter is then oxidized by air in the presence of catalyst, forming a mixture of cyclohexanol and cyclohexanone (3).

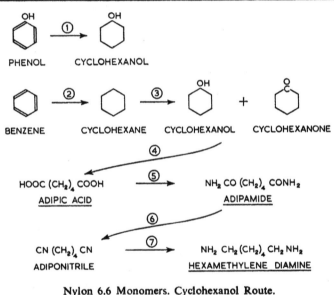

Nylon 6.6 Monomers. Cyclohexanol Route.

Cyclohexanol, or the mixture of cyclohexanol and cyclohexanone produced by the second route, is oxidized to adipic acid (4).

Hexamethylene diamine, the second starting material, is made from adipic acid by the following route:

(a) Adipic acid is reacted with ammonia to form adipamide (5).

(b) Adipamide is dehydrated to adiponitrile (6).

(c) Adiponitrile is reduced to hexamethylene diamine with hydrogen in the presence of a catalyst (7).

## (2) *Butadiene Route*

Butadiene is a basic raw material of synthetic rubber manufacture in the U.S.A. and it is produced in great quantity from petroleum. It is made into adiponitrile by the following route (see below):

(a) Butadiene is chlorinated with chlorine gas, to form dichlorobutene (1).

(b) Dichlorobutene is treated with hydrocyanic acid, forming 1, 4-dicyanobutene (2).

(c) Dicyanobutene is hydrogenated in the presence of catalyst, to form adiponitrile (3).

Adiponitrile is then converted into adipic acid or hexamethylene diamine by hydrolysis or reduction respectively.

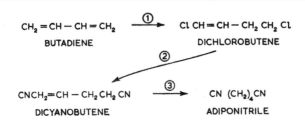

Nylon 6.6 Monomers. Butadiene Route.

## (3) *Furfural Route*

Furfural is produced in large quantities from corn cobs and oat hulls. It is converted to adiponitrile by the following route (see below):

Furfural is converted to furan (1) and this is reduced to tetrahydrofuran (2). Treatment with hydrochloric acid converts

213

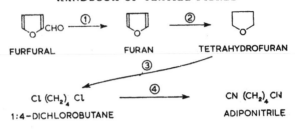

FURFURAL        FURAN       TETRAHYDROFURAN

Cl $(CH_2)_4$ Cl                   CN $(CH_2)_4$ CN

1:4–DICHLOROBUTANE            ADIPONITRILE

Nylon 6.6 Monomers. Furfural Route.

tetrahydrofuran to 1:4-dichlorobutane (3) which is converted to adiponitrile by treatment with sodium cyanide (4).

Hexamethylene diamine or adipic acid may then be made from the adiponitrile as described above.

### Polymerization

The condensation reaction that results in the formation of nylon polymer takes place between the amine groups on each end of the hexamethylene diamine, and the carboxyl groups on each end of the adipic acid. If the two reactants are mixed in exact stoichiometric quantities, the reaction could theoretically continue until all the small molecules had linked together into one huge molecule. This could not, of course, take place in practice, as the opportunities for amine end groups and carboxyl end groups to meet and react diminishes as polymerization proceeds, and the mobility of the polymer molecules is reduced.

If the two reactants are mixed together in quantities which are not stoichiometrically balanced, the condensation reaction will proceed in the normal way. But a point will be reached at which all the end groups of one type have been reacted, and the end groups of the polymer chains are now all of the type present in the component that was used in excess. The polymerization will then stop.

The manner in which this imbalance of components affects polymerization is best illustrated by considering an extreme case. If condensation is carried out, for example, using 1 molar proportion of diacid to 2 molar proportions of diamine, the polymerization will result in the production of a 'polymer' containing only three component residues, with amine groups on each end of the molecule:

$$H_2N(CH_2)_6NHCO\ (CH_2)_4CONH(CH_2)_6NH_2$$

If the proportions of the two components are, say, 1 molar proportion of diacid to 1.25 molar proportions of diamine, i.e. 4 diacid molecules to 5 diamine molecules, then the polymer formed would contain 9 component residues. The polymer molecule would again have amine groups at each end, and further condensation would be impossible.

Two important points are evident from this effect of varying the balance of the components used in the polycondensation reaction;

(a) a high degree of polymerization will be attained only by ensuring that the balance of components is adequately controlled,

(b) the degree of polymerization attained may be controlled by using components in carefully calculated non-stoichiometric proportions, representing the required degree of imbalance.

*Stabilization*

In the production of nylon 6.6 polymer, it is necessary to allow the polymerization to proceed until an adequate degree of polymerization has been attained. Polymers below a molecular weight of about 5,000 will form fibres only with the greatest difficulty; polymers of molecular weight between approximately 5,000 and 10,000 will form fibres which are generally too weak for practical use. It is not until the molecular weight is greater than about 12,000 that fibres of adequate strength are produced. It is necessary, therefore, that the polymerization conditions should be such as to allow this degree of polymerization to be reached.

As the degree of polymerization increases still further, however, new difficulties arise. The polymer becomes intransigent and difficult to melt and spin. In practice, it is necessary to control the polymerization to provide a polymer of average molecular weight in the region of 12,000 to 22,000 the actual figure being determined by the fibre characteristics that are required.

It is apparent that the polymerization reaction can be controlled in this way by using extremely highly purified components in very carefully calculated proportions. By suitable choice of the balance of components, the polymerization may be stopped at any desired degree of polymerization.

This technique is, in fact, used in practice. A common modification is to create the necessary imbalance by using stoichiometric proportions of the two components, and adding a small proportion of a monofunctional ingredient which serves as a chaingrowth stopper in the same way as the extra proportion of a component. Acetic acid, for example, is added to the mixture of hexamethylene diamine and adipic acid used in producing nylon 6.6 polymer. The amount of acetic acid is calculated to block the ends of the polymer chains after the desired average molecular weight has been reached. This technique is called 'stabilization'.

*Polycondensation*

If the hexamethylene diamine and adipic acid are pure, they may be mixed in stoichiometric quantities directly in aqueous solutions, the equivalence being determined by electrometric titration. This solution is then used directly for the polymerization to nylon 6.6 polymer.

The correct stoichiometric balance between the two components may also be obtained by reacting the two materials together to form a salt, hexamethylene diammonium adipate, in which one molecule of each component is present. This is commonly called 'nylon salt'.

$$(\overset{+}{N}H_3(CH_2)_6\overset{+}{N}H_3)(\overset{-}{COO}(CH_2)_4\overset{-}{COO})$$

Nylon salt is prepared by neutralizing solutions of the two components in methanol. The salt is relatively insoluble in methanol, and it crystallizes out as the solution cools. The crystals are separated by centrifuging, washed and dried.

Nylon salt produced in this way is extremely pure. In it, the two nylon 6.6 components are present in exact stoichiometric proportions. The salt is dissolved in water to form a 60 per cent solution, and acetic or adipic acid is added in amount calculated to stop polymerization at the desired stage.

Condensation of the aqueous solution of nylon salt as carried out in a stainless steel pressure vessel, using an inert atmosphere of nitrogen or hydrogen to ensure that oxygen is excluded. Nylon polymer is extremely susceptible to decomposition in the presence of oxygen at the temperatures used in condensation, and it is essential that all traces of oxygen should be kept out of the vessel.

Condensation is commonly carried out in two stages:

(1) The solution is heated at 220–230°C. for up to 2 hours, at a pressure of about 17.5 kg/cm² (250 lb/in²).

(2) The temperature is raised gradually, and steam is allowed to escape from the vessel, the pressure being at about 17.5 kg/cm² (250 lb/in²). When the temperature has reached 275–280°C., the molten material is held at this temperature at atmospheric pressure, or under vacuum, until the desired degree of polymerization has been reached.

The molten polymer is then extruded through a slit in the base of the vessel, the ribbon of viscous material falling on to a slow-moving wheel which is cooled by water. The polymer is immediately chilled and solidifies to a tough ribbon of horn-like nylon 6.6 polymer (polyhexamethylene adipamide), which is typically about 30 cm (12 in) wide and 6 mm (¼ in) thick.

The ribbon is passed into a machine which chops it into small pieces or chips.

### Spinning

The spinning of nylon differs fundamentally from techniques that are used in rayon and acetate manufacture. Viscose rayon, for example, is spun by extruding cellulose xanthate solution into a coagulating bath which regenerates insoluble cellulose (wet spinning); acetate is spun by extruding a solution of cellulose acetate in volatile solvent into a stream of hot air, the solvent evaporating to leave a solid filament (dry spinning). Nylon, on the other hand, is melt spun. The polymer is heated until it melts, and the molten material is then forced through holes in spinnerets. As the jets of molten nylon emerge, they are cooled and solidified by contact with a stream of cold air, forming solid filaments.

In the melt spinning of nylon 6.6, great care must be taken to avoid the risk of decomposition which is always present when the polymer is molten and at a high temperature. As in the polymerization process, the polymer is always protected during spinning by maintaining an inert atmosphere of nitrogen or other protective gas above it. The spinning technique is such as to maintain a small amount of polymer in the molten state at any time, so that the opportunity for decomposition is at a minimum. Details of the spinning operation are shown in the figure on page 219.

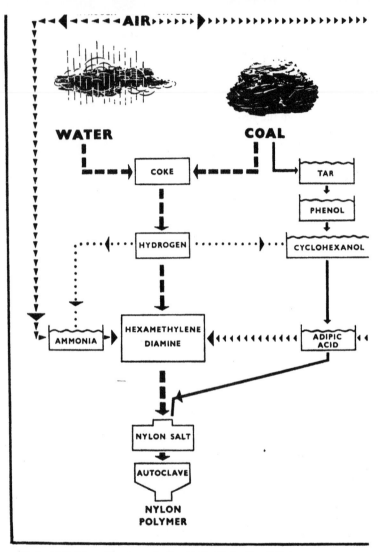

AIR

WATER

COAL

COKE

TAR

PHENOL

HYDROGEN

CYCLOHEXANOL

AMMONIA

HEXAMETHYLENE
DIAMINE

ADIPIC
ACID

NYLON SALT

AUTOCLAVE

NYLON
POLYMER

*Nylon 6.6 Flow Chart.* Left: production of polymer.
Right: Spinning nylon yarn from polymer
– I.C.I. Fibres Ltd.

218

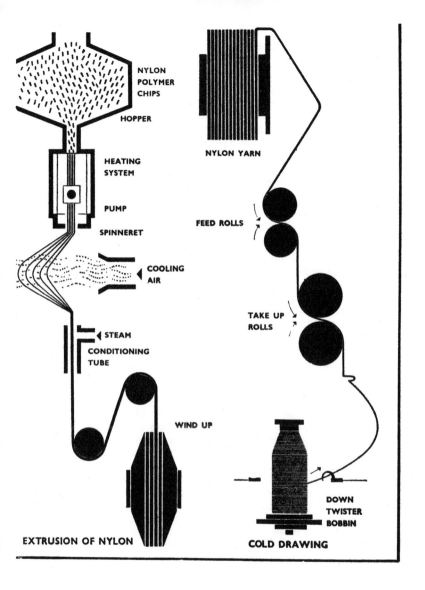

NYLON POLYMER CHIPS

HOPPER

HEATING SYSTEM

PUMP

SPINNERET

COOLING AIR

STEAM

CONDITIONING TUBE

WIND UP

EXTRUSION OF NYLON

NYLON YARN

FEED ROLLS

TAKE UP ROLLS

DOWN TWISTER BOBBIN

COLD DRAWING

219

The yarn emerges from the cooling chamber at the rate of some 1,200 metres per minute. It has been cooled to about 70°C. If the dry, warm yarn were wound directly at this stage, it would subsequently absorb moisture from the air and increase slightly in length. This would create instability in the package. After leaving the cooling chamber, therefore, the nylon filaments enter a conditioning tube through which steam is passed. This moistens the yarn and allows it to reach a state of equilibrium with respect to moisture it would otherwise absorb from the air.

The filaments emerging from the conditioning tube are brought together and given a slight twist before being wound on to the package.

### Drawing

At this stage, the long molecules of nylon polymer are folded and in a state of random orientation. The filaments are weak and opaque. In order to develop the inherent lustre and strength of the nylon, it is necessary to bring the molecules into alignment with respect to the long axis of the fibre. The filaments are therefore stretched or drawn.

Drawing is carried out by a technique similar to that used in the production of rayon. The undrawn yarn is passed round a pair of feed rollers which control the speed at which it leaves the package. It then passes several times round a second roller which is rotating such that its surface speed is four or five times faster than that of the first feed rollers. The filaments of nylon in the yarn are thus stretched to four or five times their original length as they pass from the first to the second rollers, the molecules in the filaments being drawn into alignment.

The drawn nylon is lustrous and strong. It may now be heat-set in boiling water before being wound, or it may be wound up immediately after drawing.

During drawing, the diameter of the filaments has been reduced, and they have acquired great tensile strength. The appearance of the filaments has changed; they are now translucent and lustrous, whereas the undrawn yarn was dull and opaque.

The physical properties of the nylon yarn produced in this way will depend to some degree on the degree of orientation of the molecules. This, in turn, depends upon the drawing or stretching to which the filaments have been subjected. The characteristics of the fibre can thus be controlled during manufacture.

# PROCESSING

### Scouring

Most processing agents and many types of dirt and soil are removed easily from nylon by scouring. Some types of soil, however, including graphite and certain oils and greases may be difficult to remove, especially if the soiled nylon is subjected to a heat-setting treatment before scouring.

Everything should be done, therefore, to keep nylon clean during mill processing, and nylon fabrics should be scoured before heat-setting to remove substances that might contribute to yellowing during heat-setting.

## *Method*

Nylon should be scoured with mild agitation at a moderate temperature, e.g. 50–60°C. Suggested scouring formulas are given in the table on page 222.

High temperatures will usually increase the effectiveness of the scour, and cause partial setting of the fabric. Wrinkles or creases set in the fabric during a high temperature scour may be difficult to remove in subsequent operations.

Yarns or fabrics carrying a polyacrylic-type size must be scoured before heat-setting, as it will be impossible to remove the polyacrylic acid after heat-setting.

The removal of graphite becomes more difficult with time, and nylon lace should be scoured before storage.

## *Equipment*

### *Jig Scouring*

The jig may be used for scouring woven fabrics made from filament yarns. It is advisable to remove creases by steam framing before the fabrics are loaded into the jig.

### *Beck Scouring*

The beck is used for scouring most spun woven fabrics, some warp knit fabrics, and some light and loosely woven filament fabrics.

Beck scouring is not recommended for most heavy or tightly woven filament fabrics because of wrinkling and subsequent streakiness.

SCOURING FORMULAS (NYLON 6.6)

| Materials and Conditions | Rawstock and Top | Yarn and Fabric Light Soil | Fabric Medium Soil | Fabric Heavy Soil | Coning Oil on Yarn | After Dyeing | Graphite and Grease |
|---|---|---|---|---|---|---|---|
| 'Duponol' RA Surface Active Agent | 1% | 1% | 2-3% | 3% | — | 1% | — |
| 'Alkanol' HCS Surface Active Agent | — | — | — | — | 1% | — | — |
| Ammonium hydroxide | 1% | — | — | — | — | — | — |
| Tetrasodium pyrophosphate | — | 1% | 2-3% | — | 1% | — | — |
| Caustic soda | — | — | — | 3% | — | — | — |
| 'Varsol' hydrocarbon solvent or xylol | — | — | — | 0.5-1.5 oz./gal. | — | — | — |
| Sodium carboxymethyl cellulose | — | — | — | — | — | — | 0.133 oz./gal. (0.1% sol'n.) |
| 'Triton' X-100 concentrate | — | — | — | — | — | — | 1.33 oz./gal. (1.0% sol'n.) |
| 'Eccoterge' L-46 concentrate | — | — | — | — | — | — | 1.33 oz./gal. |
| Stoddard Solvent | — | — | — | — | — | — | 26.6 oz./gal. (20% sol'n.) |
| Time (mins.) | 15 | 30 | 30 | 45-60 | 30 | 15 | 45 |
| Temp (°F.) | 140 | 120-140 | 180 | 180-212 | 120-140 | 140 | Room |

Note: Stains and yarn tints that resist normal scouring treatments can usually be removed by reducing or stripping agents such as sodium hydrosulphite or zinc sulphoxylate formaldehyde, or with bleaching agents.

222

The beck should not be overcrowded with fabric, in order to prevent wrinkling.

### Rotary Drum or Paddle Machine Scouring

This method is used mainly for scouring knit goods such as socks and sweaters, which are usually placed in bags before scouring.

### Rope Soaper Scouring

This machine is used for scouring severely soiled fabrics of selected constructions which have a high resistence to creasing or wrinkling and to slippage of yarns.

### Beam-Dyeing Equipment Scouring

Warp knit fabrics and sheer open weave fabrics are commonly scoured on beam-dyeing equipment. Higher temperatures may be used than in beck scouring, where creases and rope marks tend to be set in the fabric at high temperatures.

Anti-foaming agents are generally necessary to eliminate excessive foaming when scouring with synthetic detergents on beam type equipment.

## Bleaching

Nylon is a white fibre as produced by the manufacturer, and seldom requires bleaching. If a bluish-white cast is required, this may be achieved by tinting with a small amount of a suitable blue or violet dye, such as 'Latyl' Blue RB.

Bleaching may become necessary when fabrics are stained or discoloured during processing. Scouring should always precede bleaching in such cases, as bleaching alone is not always effective in removing certain types of stains, especially oils and greases. In some cases, bleaching may set the stains in a nylon fabric and make complete removal impossible. It may even be necessary, therefore, to spot or dry clean the fabric prior to scouring and bleaching. Nylon should be bleached only when the discoloration resists removal by scouring.

### 100 per cent Nylon Yarns and Fabrics

#### Acid Sodium Chlorite Bleaching

100 per cent nylon is bleached most effectively by means of the acid sodium chlorite method. This method is useful as a

bleach for (1) maximum whiteness, (2) removing yarn tints, and (3) whitening fabrics that have been yellowed by exposure to air at high temperatures. Fabrics should be treated in open width form to prevent wrinkles from being set by the hot bleaching bath.

### Peracetic Acid Bleaching

This technique produces a good white with nylon when only a light or moderate coloration must be reduced. It has the added advantage of being less corrosive to equipment than the acid sodium chlorite technique, and there are no toxic fumes.

### Hydrogen Peroxide; Sodium Hypochlorite

These are not effective bleaches for nylon, and they should not be used in high concentration or at elevated temperatures for long periods of time. Some degradation of nylon may take place under such conditions.

Blends of other fibres with nylon may require the use of these bleaching materials; in such cases, the conditions of time, temperature and concentration should be kept as moderate as possible.

### Blends of Nylon and Rayon

Bleach as for 100 per cent nylon.

### Blends of Nylon and Polyester Fibres

Bleach as for 100 per cent nylon.

### Blends of Nylon and Acrylic Fibres

Bleach as for 100 per cent nylon.

### Blends of Nylon and Acetate Fibres

Bleach as for 100 per cent nylon, except that temperature should not exceed 77°C.

### Blends of Nylon and Wool

The acid sodium chlorite bleach recommended for nylon will damage wool, and cannot therefore be used for blends of nylon and wool. A bleach is commonly used, such as a hydrogen peroxide/tetrasodium pyrophosphate bleach at 50°C. This will bleach the wool, but will have little effect on the nylon component.

## Blends of Nylon and Cotton

These blends can be bleached with (1) acid sodium chlorite, (2) hydrogen peroxide in a continuous unit or kier, or (3) sodium hypochlorite.

Acid sodium chlorite is used to remove yellowing from nylon which results from heat-setting, as neither of the other two bleaches will be effective. Heat-setting is generally required when a blended fabric contains 50 per cent or more of nylon.

### Dyeing

### (1) 100 per cent Nylon

Nylon has affinity for many classes of dyes, and may be dyed successfully with a very wide range of dyestuffs. For most purposes, nylon is dyed with disperse and acid dye classes, the latter including neutral-dyeing premetallized and chrome dyes. Selected direct dyes are used, and vat dyes may sometimes be applied, commonly on blends of nylon and cotton.

Some manufacturers have introduced nylons with special dyeing characteristics. These have made possible the production of multicolour effects in single-bath dyeing of 100 per cent nylon fabrics.

## Disperse Dyes

These are especially suited to the dyeing of nylon, and they are widely used in dyeing yarns and fabrics. They provide a method of dyeing that is both simple and practical.

As in the case of acetate and polyester fibres, the dyeing mechanism is one of solid solution. The dyes have excellent levelling or transfer properties, and produce level, well-penetrated dyeings. Fastness to light and washing is only moderate.

Shade build-up allows light, medium and some dark shades.

## Acid Dyes

Acid, neutral-dyeing premetallized, acid-dyeing premetallized, chrome and selected direct dyes are used where maximum fastness is required, or where depth of shade precludes the use of disperse dyes. Most of these dyes have good fastness to washing, being superior to disperse dyes in this respect. Some have particularly good fastness to light. Chrome dyes are, in general, fast to soaping at the boil.

Acid dyes do not transfer as well as disperse dyes, and the selection of dyes and dyeing procedure should be made with great care if maximum levelness is to be obtained. This is particularly true when filament materials are being dyed.

Retardation of the dye rate will generally give good results in achieving maximum levelness. This may be done by temperature control, pH control, and the use of additives in the dyebath.

Acid dyes are absorbed more rapidly as the dyebath temperature is increased, or as the pH is lowered. The dyeing rate may be controlled, therefore, by initiating dyeing at or near the neutral point, and then gradually raising the temperature and lowering the pH.

Dyebath additives may be of either the anionic or cationic types. The anionic additives are colourless surface-active agents which provide anions which compete with the dye anions for the amine groups in the fibre molecule. The cationic additives provide colourless cations which form complexes with the anions of the dyestuff. As the dyeing temperature is approached, the complexes dissociate to release the dye anions in low concentration, a condition which favours level dyeing of the fibre.

Levelling of most acid dyes is enhanced by increasing the dyeing temperature to 121°C. in pressurized equipment. This technique also results in increased penetration and improved wetfastness.

### Direct Dyes

Some dyestuffs of this class are used successfully for dyeing nylon; others are of little use. The useful colours are, in general, those which are similar in chemical structure to the acid dyes, and they may be applied to nylon by methods similar to those used for acid dyes.

Direct dyestuffs provide shades of good fastness to washing and to perspiration. The washfastness of these dyestuffs on nylon is better than that of the same dyes on cotton or wool.

### Neutral-Dyeing Acid Dyes

All acid dyes may be applied to nylon at low pH, but some of these exhaust well at neutral or slightly alkaline pH; these are designated as neutral-dyeing.

The neutral-dyeing (non-metallized) acid dyes generally have fair to good transfer properties and good wetfastness.

### Neutral-Dyeing Premetallized Dyes

These dyes are recommended for dying raw stock and top, and in continuous pad-dyeing systems. They can, however, be applied to nylon yarns in either skein or package form, and to spun nylon piece goods. The use of these dyes on filament yarns is limited by their poor transfer properties, even at high temperatures.

### Acid-Dyeing Premetallized Dyes

The acid-dyeing premetallized dye 'Chromacyl' Black W is used to produce full black shades having very good lightfastness and good wetfastness on either filament or spun nylon. It is applied usually from a dyebath containing formic acid.

### Chrome Dyes

'Pontachrome' Black TA and 'Solochrome' Black WDFA are examples of chrome dyes used with nylon. They produce full black shades having excellent fastness to light, to washing and to all wet processing.

### Vat Dyes

These dyes are used successfully to provide high wet and light fastness in a good range of shades. Dyeing is carried out in alkaline solutions, using temperatures higher than those used for cotton.

Vat dyes are frequently used on blends of nylon and cotton when outstanding fastness is desired. Vat dyes alone give a reasonable union in light and medium shades in well blended fabrics.

### Differential Dye Effects

Multicoloured effects may be obtained from a single dyebath by making use of nylon yarns with different dyeing characteristics. I.C.I. Fibres Ltd., for example, produce the following classes of yarn: standard dyeing; deep dyeing (acid dyeing); basic dyeable (acid—dye—resist).

By using combinations of these classes of nylon, which differ only in their dyeing characteristics, the dyer can obtain colour and white, two-colour, three-colour and tone-on-tone effects. The most versatile of combinations is obtained by using deep dye and basic-dyeable yarns on which complementary colours, as well as colour and white effects, may readily be obtained. The shade on the basic-dyeable nylon should be restricted to medium depth, in order to prevent cross-dyeing of basic dyes on to the deep dye yarn. Also, the easiest fabric to dye is one containing 50 per cent deep dye and 50 per cent basic-dyeable yarn; the greater the deviation from these proportions, the greater will be the restrictions on shade combinations that can be achieved.

## (2) *Blends*

Nylon is often used in blends with other fibres, to provide fabrics with properties not attainable by using a single fibre.

The following points should be considered when blends or combinations containing nylon are to be dyed:

- (A) Desired Effect
    - (1) Union dye
    - (2) Cross dye
    - (3) Leaving one fibre white.
- (B) Type of Operation
    - (1) Batch
    - (2) Continuous.
- (C) End-Use Fastness Requirements.

### *Blends of Nylon and Wool*

Nylon and wool have affinity for the same types of dyes, but their absorption rates are different. This factor makes every change of shade or fibre proportion a different dyeing problem. The blocking of dye sites by one or more dyes in a mixture may also be a difficulty which must be solved by careful dye selection. For these reasons, it is essential to test every colour-combination before carrying out the dyeing.

Cross-dyeing or dyeing one fibre and leaving the other undyed is not generally a commercial proposition with nylon/wool blends, as both fibres take up the same dyes. In some cases, it is possible to treat the nylon with a dye-resist before blending with the wool.

### (a) *Neutral-Dyeing Acid Dyes*

These are commonly used for dyeing nylon/wool blends, producing satisfactory unions in pale to heavy shades, and building up well. They have good wetfastness and fair to good lightfastness.

### (b) *Level-Dyeing Acid Dyes*

Many of these dyes produce a good union on nylon and wool blends in light to medium shades.

### (c) *Acid-Dyeing Premetallized Dyes*

These dyes are commonly used on nylon/wool fabrics that have been carbonized. They produce unions that have lightfastness in pale shades of 15 to 40 sun hours ('Chromacyl' dyes).

When dyed neutral or with formic acid, they do not level as well as with sulphuric acid, but this must be used with great care as it can damage nylon. The dyebath pH should not be lower than 3.0, or degradation of the nylon may result.

### (d) *Chrome Dyes*

As a class, chrome dyes offer the best lightfastness on wool. Except in a few cases, they have a good lightfastness on nylon, and they are therefore used to produce selected shades on nylon/wool blends.

The blends may be difficult to chrome, as the wool absorbs a large proportion of the chrome from the bath, leaving insufficient for the nylon.

### (e) *Neutral-Dyeing Premetallized Dyes*

These dyes are not generally recommended for the union-dyeing of blends of nylon and wool, as they show preferential affinity for the nylon. Selected dyes will, however, produce solid shades on 50/50 blends if the dye affinity of the nylon has been reduced with a resistant pretreatment.

## *Blends of Nylon and Cellulosic Fibres* ·

Several methods of dyeing nylon/cellulosic fibre blends may be used. Union-dyeing, cross-dyeing and single-fibre dyeing are all possible, the choice depending on the fastness requirements of the fabric, the available equipment and the proportion of fibres in the blend.

Usually, the nylon portion is dyed with acid dyes, and the cellulosic fibre with selected aftertreated direct dyes. When moderate washfastness is satisfactory, disperse dyes are put on the nylon and direct dyes on the cellulosic fibre.

Vat dyes may be applied to nylon/cotton blended fabrics in light and medium shades, using a one-step procedure if the percentage of nylon in the blend is about 25 per cent or less. For heavy shades, or blends containing more than 25 per cent nylon, a two-step procedure is usually necessary. Acid or pre-metallized dyes are applied to the nylon on the first pass, and vat dyes applied to the cotton on the second pass.

### Blends of Nylon and Acetate Fibre

These may be dyed readily with disperse dyes. Acid dyes may be used to dye the nylon and leave the acetate white, but the reverse is not possible.

Most of the disperse dyes produce strength and shade on nylon which differ from those produced on acetate. Only a limited number of disperse dyes produce good unions.

Subtle cross-dye effects may be obtained by selecting dyes which have more affinity for one fibre than the other.

### Blends of Nylon and Silk

When dyeing these blends, a careful colour selection is required for union shades. Cross-dyeing is not possible.

### Blends of Nylon and Polyester (PET) Fibre

White reserve effects may be produced on nylon/polyester piece-dyed warp-knitted fabrics. The nylon component is dyed with selected Procion or acid dyestuffs, leaving the polyester filament component white. The brightness of the polyester may be enhanced by the subsequent application of a fluorescent brightening agent to produce a white comparable with that obtainable on 100 per cent polyester fabric. Level shades on 100 per cent filament nylon fabrics are difficult to obtain using Procion dye-stuffs, but where the effect is broken as in a striped fabric, the evenness of dyeing is perfectly satisfactory.

Both components of woven or warp-knitted nylon/polyester

mixture fabrics may be dyed using a single-bath process which is shorter and cheaper than the two-bath technique commonly used to obtain cross-dyed effects.

The preferred method is based on the use of the high-temperature beam-dyeing machine, since it has been found that under these conditions selected disperse dyes give a stain on the nylon which is weaker than, or at least no heavier than, the shade on the polyester fibre component. When these selected dyes are used, the fastness of the stain on the nylon is adequate for a wide range of apparel outlets. While the presence of this disperse-dye stain on the nylon exerts some influence on the final shade which can be obtained, a wide range of attractive colour combinations can be produced on the finished fabric by cross-dyeing the nylon with Nylomine and acid milling dyestuffs.

Three-component mixtures such as nylon/polyester/cotton may also be handled by the one-bath process, but the problems of shade control in bulk scale working make it advisable to leave the cotton undyed. Where the cotton is in an intimate blend with the nylon or polyester fibre, care should be taken in fabric design and choice of colour to avoid excessive contrasts which would show as patches of changing colour as the cotton component is preferentially abraded away in wear.

Certain textured fabrics, such as seersuckers, which are unsuitable for beam-dyeing, may be handled on the winch machine.

### Printing and Surface Effects

Many types of nylon fabrics are printed commercially. Nylon flat-woven and nylon tricot fabrics are printed by conventional roller- and screen-printing techniques to meet the needs of a great variety of end-uses.

Nylon carpets, hosiery, speciality fabrics and yarns are also printed, but special techniques and equipment are often required.

A wide selection of dyes is available for printing nylon, including many acid, direct, premetallized, vat and disperse types. Resin-bonded pigments are also used. The selection of dyes from any of these classes should be based on the end-use of the fabric to achieve adequate colourfastness and brightness of shade.

Fabrics of multilobal cross-section nylon and Schreiner-calendered nylon tricot fabrics are especially suited to printing.

The increased cover and smoothness of these fabrics give clear definition of printed patterns, good colour register, and more desirable opaque backgrounds.

Specific formulae for printing pastes depend upon the type of dye to be used, the method of application, and the nature of the fabric. Pastes to be used in screen printing are generally thicker than those for roller work. Fabrics of spun yarn require a thinner paste than sheer filament constructions.

### Acid and Direct Dyes

These are widely used in printing piece goods for dresswear. They provide the brightest shades that will meet minimum fastness requirements for dress fabrics. They build up readily to full shades, and are used for attractive combination shades.

### Neutral-dyeing and Acid-dyeing Premetallized Dyes

These types are duller than many acid and direct dyes, but exhibit very good fastness on nylon fabrics. They are used where maximum lightfastness and wetfastness are required, as in carpets, car upholstery and swimwear.

### Vat Dyes

Selected vat dyes print easily and produce shades with outstanding fastness to washing. Lightfastness is commonly inferior to that of the same dyes on cotton.

### Disperse Dyes

These dyes are economical to use with nylon, and possess good affinity, build up and levelling properties. The shades are duller than those produced with acid dyes, and are inferior to the acid dyes in wetfastness. They must be processed with care to avoid staining of whites during rinsing for removal of thickener.

### Resin-Bonded Pigments

Pigments are applied to nylon fabrics usually from an emulsion containing sufficient resin to bind the colour mechanically to the fabric. The emulsion may be of the water-in-oil or oil-in-water types. The pigments are fixed to the fabric by curing the resin.

Pigments are used in printing sheer nylon fabrics and in the production of opaque white-on-white effects which cannot be produced by any other method. The print may be stiffened to some extent, but pigment printing is usually satisfactory where the coverage is small enough to leave the hand of the fabric unimpaired.

## Discharge Printing of Nylon

The discharge printing of nylon has been limited by the following factors:

(a) Few dyes are satisfactorily dischargeable on nylon.

(b) Dyes which are dischargeable do not usually have sufficient levelness to meet market requirements.

(c) Discharged areas have a tendency to discolour with age.

Best results have been obtained with pastel shades, discharging to white or to a colour with vat dyes in the paste.

### Stripping

Disperse or acid dyes may be removed from nylon by a stripping procedure based on a reducing action. The following is an example of the type of stripping technique which may be used:

1. Add the following to the stripping bath, based on weight of goods:

| | |
|---|---|
| Zinc sulphoxylate formaldehyde | 5 per cent |
| Acetic acid (56 per cent) or formic acid for heavy shades | 10.0 per cent |
| 'Duponol' D Paste or 'Lissapol' N surface active agent | 0.5 per cent |

2. Strip for 45 minutes at 88 to 99°C.
3. Drop the bath and rinse well.
4. Scour with 'Duponol' D Paste, 'Duponol' RA, or 'Lissapol' N.

If this procedure does not remove acid dyes completely, the following additional treatment may be used:

1. Run at the boil for about 30 min. – raise 3°C./min. in a solution containing:
   Acetic acid (56 per cent) 0.13 oz./gal. (0.8 g/l; 0.1% soln).
   Sodium chlorite 0.07 oz./gal. (0.4 g/l; 0.05% soln) pH to 6.0 to 7.0.
2. Drop the bath and rinse well.

3. Scour with:
   'Duponol' RA 0.06 oz./gal. (0.36 g/l; 0.05% soln).
   Sodium bisulphite pyrophosphate 0.133 oz./gal. (0.8 g/l; 10% soln).

## Chemical Finishing Treatments

Fabrics and yarns of nylon may be made softer, stiffer or repellent to oil and water by treatment with the appropriate type of finishing agent. Antistatic properties may be imparted to nylon by use of an antistatic finishing agent.

The dermatological behaviour of finishing agents should always be checked with great care, as some may cause skin irritation or sensitization.

### Modification of Fabric Hand

Nylon fabrics often undergo finishing treatment designed to provide increased softness and improved draping properties, together with increased yarn pliability and surface lubricity.

'Avitex' NA and 'Cirrasol' AD softeners are examples of the type of finishing agent used for this purpose. If more lubricity is required, an equal part of 'Methacrol' Lubricant K or 'Lubrol' W may be added in addition to the 'Avitex' NA or 'Cirrasol' AD. These finishes are applied by padding on a quetsch or by exhaustion from a dilute bath. A finish level of 0.5 to 2.0 per cent is usually desirable.

These two finishing agents are also used to improve the sewing qualities of nylon fabrics on high speed sewing machines. The amount of finish applied is in the region of 2.0 per cent.

Some nylon fabrics are treated with finishing agents to increase stiffness of handle. Thermoplastic resin dispersions, such as 'Elvacet' 81–900 polyvinyl acetate and 'Methacrol' FNH resin dispersion are examples of finishes used for this purpose. They are usually padded on the goods, to produce concentrations of 0.5 to 1.0 per cent for satins and taffetas, and 1.0 to 3.0 per cent for marquisettes (based on the weight of fabric).

If a more durable finish is required, a thermosetting resin such as 'Aerotex' M3 may be used.

### Repellents

Good water repellency is obtained by the application of 'Zelan' S to nylon fabrics. The water repellency is durable and resistant to laundering if properly applied. 'Zelan' S is applied by padding from an aqueous bath, followed by drying and curing at elevated temperature.

Application of 'Zepel' B fabric fluoridizer to fabrics of nylon and its blends with other fibres provide water- and oil-repellent and stain-resistant finishes which are durable to both laundering and dry-cleaning. Typical applications involve the use of 2.5 to 3.0 per cent 'Zepel' B with water-repellent adjuvants and/or resins selected to meet individual finishing requirements. The finish is applied by padding, followed by drying and curing at 150°C. for 2 to 4 minutes. It is not necessary to after-wash unless the formulas used contain resins or other products which require an after-wash.

### Antistatic Finishes

An antistatic finish which is durable to laundering with soap is attained by applying 'Zelec' DP in the range of 1 to 2 per cent to nylon. The effect is neutralized by exposure of the treated fabric to synthetic detergents, but it may be regenerated by rinsing with warm soap solution and then with water.

Temporary antistatic finish is provided by the application of agents such as 'Avitex' NA or 'Cirrasol' AD. Yarns with 'built-in' anti-static properties are also available.

### Flame Retardants

Clean, undyed, finish-free nylon fabrics have low flammability, and in this respect are satisfactory for many practical purposes. The flammability may be increased, however, by certain resin-finish applications or by certain dyes. Such fabrics may be treated with flame-retardant finishes to reduce their flammability. Examples of these finishes are 'Pyróset' Fire Retardant N2 and N10, and 'Fi-Retard' NBX.

These flame-retardants have an adverse effect on the light and gas-fading fastness of dyed nylon. The stiff finish which is often produced by a retardant may be modified by a softener.

### Mechanical Finishing Treatments

Modification of the hand and surface appearance of nylon fabrics may be achieved by various mechanical treatments. The effect desired in the finished fabric determines the type of equipment used.

## 1. *Cold Calendering*

Calendering is used to polish and smooth a fabric surface, or to minimize the mark-off of fabrics treated with resins or water repellents. The rolls of the calender are normally unheated, and pressures of 30 to 50 tons are used.

## 2. *Flat Calendering – Tricot Fabrics*

Flat or Schreiner calendering of tricot nylon produces a durable finish which increases the covering power of the fabric and changes its hand and appearance. The change in appearance results from a compaction of the fabric stitch and from diffusion of light reflected from the embossed fabric surface. A change in the fabric hand results from a decrease in thickness of the fabric, and a change in the frictional properties of its surface. Fully dried tricot, calendered at zero tension, produces the optimum finished effect.

Schreiner calendering can be used to:

(a) produce thin, lightweight fabrics with a high degree of opacity, which retain good porosity and good bursting strength;

(b) obtain excellent whiteness with a fully delustred appearance;

(c) obtain a more effective print base by increasing cover and smoothness;

(d) produce a two-fold increase in covering power of the fabric;

(e) increase the styling potential of tricot by changing its normal jersey appearance.

In the Schreiner calendering of nylon tricot, the fabric is passed through an engraved (diagonal grooves 118–142/cm; 300–360/in) Schreiner calender at temperatures of 193 to 205°C. at 12–27 m (13–30 yd) per minute. On a 122 cm (48 in) width machine, good results are obtained at roll pressures of 80 to 100 tons.

Calendering may be carried out at any stage during finishing, but the effects obtained will depend upon the stage at which it is introduced into the finishing sequence.

Schreiner calendering results in excellent opacity, thinness, uniformity of appearance, durability, defect coverage and hand. With scoured or dyed tricot, calendering flattens the yarn and closes the stitch, but with less shrinkage than occurs with gray goods. Durability and hand are excellent in this procedure, with opacity and thinness rated as good.

Calendering has the least effect on fully finished nylon tricot, as the goods have already been heat-set and resist shaping. This is the most economical way of Schreiner calendering nylon tricot, as intermediate drying and framing are unnecessary. The effects on the fabric, however, are only fair to good, and the calendering may bring out undesirable lustre which must then be broken up by subsequent wet working.

### 3. *Embossing*

Nylon 6.6 may be durably embossed without the use of resins. This is achieved by embossing at elevated temperatures with close control of the moisture content of the fabric, dwell time and roll pressure.

In most cases, it is not necessary to heat-set shallow embossed patterns to obtain a lasting effect. Deep patterns, however, require heat-setting. The fabric should be completely relaxed during heat-setting to prevent loss of the three dimensional effect.

### 4. *Napping and Sueding*

Napping and sueding equipment is used to produce raised surfaces on nylon fabrics, the conditions necessary to obtain a desired effect on a particular fabric being best determined by experiment. Knit fabrics should be introduced loop first to achieve maximum effect in any raising operation.

Nylon has a higher tenacity than natural fibres, and the clothing normally used on nappers for wool or cotton fabrics may not be of adequate strength. Stiffer napper clothing is usually required to obtain a good napped surface on a nylon fabric. A softener acting as a lubricant, wet or dry, assists in raising fibres to the surface by reducing inter-fibre friction.

Best results are usually obtained in sueding by using the finer grades of sandpaper.

### 5. *Shearing*

Fabrics of spun nylon and blend fabrics containing nylon may be sheared to remove excess surface fibre. In the case of pile fabrics, shearing is used to cut the pile to a uniform height. Shear setting varies, depending upon fabric construction, pile density and the depth of cut required.

Fuzz which is deeply embedded in the fabric construction cannot be removed by shearing, and it may be necessary to remove this by singeing. This is preferably done after dyeing, to avoid deep-dyeing of melt balls.

### 6. *Semidecaters and Palmers*

Smooth, pressed fabrics are obtained by semidecating or by treating in a palmer. Conditions will vary depending upon the finished effects required.

### 7. *Pleat-Setting*

Pleats may be set in nylon fabric by exposure of the pleated item to saturated steam at a pressure of 0.7 kg/cm² (10 lb/in²) for 20 minutes. Several commercially-available machines will set pleats in nylon by exposing the fabric to a hot roll. It is essential that the fabric should not be heat-set before the pleating operation.

#### Heat-Setting

Nylon 6.6 may be heat-set before or after dyeing. If this is done effectively, it will have the following results:

(1) Residual shrinkage will be removed, ensuring dimensional stability of fabrics during processing and wear.

(2) Resistance of fabrics to wrinkling during processing and wear will be increased.

(3) Yarn twist will be stabilized.

(4) Edge curl of fabrics will be prevented.

(5) Hand of fabrics will be softened.

Nylon 6.6 fabrics may be heat-set using either (1) dry heat, (2) saturated steam under pressure, or (3) hot water (121°C.). The degree of set is determined by both the duration and the temperature of the treatment, and the nature of the setting medium.

Nylon 6.6 tends to shrink during setting, developing a force of about 3.5 cN/tex (0.4 g/den). A force equal to this will prevent shrinkage; a greater force will stretch the yarn.

Wrinkles or creases formed during heat-setting or during dyeing of an unset fabric will be almost impossible to remove.

The selection of a method for heat-setting nylon depends on the form of the fibre and on the properties desired in the heat-set product. The following methods are commonly used:

(1) Dry Heat
    (a) Hot Air
    (b) Radiant Heat (Infra Red)
    (c) Hot Roll

(2) Saturated Steam Under Pressure

(3) Hot Water.

## STRUCTURE AND PROPERTIES

### Fine Structure and Appearance

Nylon 6.6 fibres are smooth-surfaced, with no striations. In microscopic appearance they are as featureless as glass rods. The fibres are commonly of round cross-section, but special types of nylon 6.6 of multilobal cross-section are now produced.

### Tenacity

The tenacity of nylon may be varied within limits by adjusting the manufacturing conditions. Filament produced for stockings has a tenacity of 40.6–51.2 cN/tex (4.6–5.8 g/den), dry; 35.3–45.0 cN/tex (4.0–5.1 g/den), wet. High tenacity nylon may have a tenacity of 79.5 cN/tex (9.0 g/den), dry; 68.0 cN/tex (7.7 g/den), wet. Nylon staple has a tenacity of 36.2–39.7 cN/tex (4.1–4.5 g/den), dry; 31.8–35.3 cN/tex (3.6–4.0 g/den), wet.
Loop tenacity: about 90–95 per cent of normal.
Knot tenacity: about 85 per cent of normal.

### Tensile Strength

Regular filament: 4,550–5,950 kg/cm² (65,000–85,000 lb/in²).
High tenacity filament: 6,300–9,100 kg/cm² (90,000–130,000 lb/in²).
Staple: 4,200–4,620 kg/cm² (60,000–66,000 lb/in²).

### Elongation

Regular filament: 26–32 per cent (30–37 per cent, wet).
High-tenacity filament: 19–24 per cent (21–28 per cent, wet).
Staple: 37–40 per cent (42–46 per cent, wet).

**Elastic Recovery**

Nylon is a highly elastic fibre, in that it will recover its original dimensions after being deformed by the application of a stress. Standard filament has an elastic recovery of 100 per cent at up to 8 per cent extension; high tenacity filament has a recovery of 100 per cent at up to 4 per cent extension.

In this respect, nylon has a resemblance to rubber, but it does not recover or snap back as quickly as rubber after the release of tension. Like rubber, however, it tries to return to its original length when held in a stretched condition, and until allowed to contract it exerts a force that resists the restraining influence.

If nylon is stretched for several days, and then allowed to relax, it will recover some 50 per cent of its stretch almost immediately. The rest of the recovery takes place more slowly. During the first 24 hours, nylon will recover about 85 per cent of the total stretch, but it may take 2 weeks to recover completely. The rate of recovery is increased by increase in temperature or relative humidity.

**Initial Modulus**

Regular filament: 353.2–530.0 cN/tex (40–60 g/den).

**Average Stiffness**

Regular filament: 159 cN/tex (18 g/den).
High tenacity filament: 282.6 cN/tex (32 g/den).
Staple: 97.1 cN/tex (11 g/den).

**Average Toughness**

Regular filament: 1.08.
High tenacity filament: 0.77.

**Flex Resistance**

Excellent.

**Abrasion Resistance**

Excellent.

**Specific Gravity**

1.14.

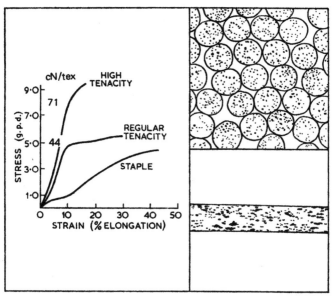

*Nylon 6.6*

### Effect of Moisture

Nylon absorbs only a small amount of moisture compared with most natural fibres. It has a regain of 4–4.5 per cent.

The tenacity of thoroughly wet nylon is 80 to 90 per cent of its dry (conditioned) tenacity. The elongation of wet nylon is 5 to 30 per cent greater than that of dry (conditioned) nylon.

The filaments do not swell appreciably in water; the diameter increases only by about one-fiftieth.

### Thermal Properties

*Melting point.*

Approximately 250°C. The range of temperature over which nylon softens and melts is very narrow.

## Effect of Low Temperature

Nylon retains its strength well at low temperatures. After several hours at $-40°$C., a nylon rope does not lose strength, and the strength is retained after the rope is reconditioned at normal temperatures.

The tenacity of regular and high tenacity yarns increases slightly with a slight decrease in elongation when they are chilled to about $-80°$C.

## Effect of High Temperature

Nylon can withstand temperatures up to about 150°C. for many hours without undue loss of strength. It turns slightly yellow after 6 hours at 150°C.

Prolonged exposure in air at elevated temperatures causes deterioration of nylon, as evidenced by permanent losses in breaking strength, breaking elongation and toughness. The fibre discolours (yellows) to some extent, and under many conditions of exposure there is also an increase in the resistance to initial stretch.

Apart from this effect of prolonged exposure, there is an instantaneous and reversible change in the fibre properties with increase in temperature, resulting in a decrease in tenacity and an increase in elongation.

Anti-oxidants are added to nylon yarns for some (generally industrial) end-uses to confer a high degree of heat resistance.

## Flammability

Nylon is less flammable than cotton, rayon, wool or silk. If a flame is applied to a nylon fabric, the material melts and tends to drop away. The fabric does not normally support combustion on its own, but its flammability may be increased by the presence of certain chemical finishes and dyes.

Ignition temperature: 532°C.

## Heat Capacity and Heat of Fusion

The specific heat of nylon at 20°C. is 0.4 calories/gram/°C. At 230°C., it is 0.6 calories/gram/°C.

The heat of fusion of nylon is 22 calories/gram.

## Thermal Conductivity

The thermal conductivity of nylon polymer is 1.7 B.T.U./hr./ ft$^2$./°F. for 1 inch of thickness.

## Thermal Expansion and Contraction

Preshrunk nylon yarn decreases slightly in length with an increase in temperature and increases slightly in length with a decrease in temperature. In the temperature range from 25 to 130°C. the change in length is about 0.007% per °C (0.004% per °F), when the moisture content of the yarn remains constant.

## Shrinkage Properties

When nylon yarn is removed from the bobbins and allowed to relax with no tension, it may tend to contract or shrink by about 2.0 to 2.5 per cent, the actual amount depending upon the previous treatment of the yarn. This is relaxation shrinkage.

The unrelaxed residual shrinkage of nylon yarn is the amount of shrinkage that takes place in boiling water immediately after the yarn is removed from the bobbins. A typical value for this residual shrinkage is in the range 8 to 12 per cent, commonly 9 to 10 per cent. This figure includes the relaxation shrinkage, and if this has been allowed to take place before placing the yarn in boiling water, the residual shrinkage is reduced accordingly.

When completely preshrunk yarn is immersed in water it will gain in length. This gain may be as much as 3 per cent. It is a reversible effect and should not be confused with true shrinkage.

### Effect of Age

Negligible.

### Effect of Sunlight

In common with most other textile fibres, nylon is affected by prolonged exposure to light. There is a gradual loss of strength, but little or no discolouration.

The degree to which nylon resists deterioration by sunlight or ultraviolet light in a particular range of wavelengths depends upon a number of factors, as follows:

1. Lustre. Bright nylon is considerably more resistant to light than Semidull or Dull.

2. Amount of surface exposed.

3. Size or diameter of exposed filaments. High denier per filament nylon is more resistant to light.

4. Length of time of exposure and intensity of light.

5. Time of year and geographical location. Nylon deteriorates more rapidly in summer.

6. Temperatures during exposure.

7. Location of exposed nylon, indoors or outdoors.

Stabilizers may be incorporated in nylon to give improved resistance to light and heat for specialized applications.


### Chemical Properties

*Materials Having No Permanent Effect on Nylon Yarn*

Most compounds of the following general types have little or no effect on the tenacity and elongation of nylon yarn under ordinary conditions of exposure:

Alcohols

Aldehydes

Alkalis

Dry cleaning solvents

Ethers

Halogenated hydrocarbons

Hydrocarbons

Ketones

Soaps and synthetic detergents

Water, including seawater

The results of tests carried out on nylon yarn, bristle and fishing line are shown in the following table. The nylon was subjected to treatment as indicated in the table, and strength measurements were made before exposure and on the washed and dried sample after exposure. None of these treatments caused a significant loss (less than 5 per cent) in the strength of the samples.

## Materials Having No Permanent Effect on Nylon 6.6

| Chemical | Concentration (per cent) | Temp. (°C.) | Time (hr.) |
|---|---|---|---|
| Acetone | | 25 | 60 |
| Acetic acid | 3 | 100 | 3 |
| Benzene | | 25 | 60 |
| Carbon tetrachloride | | 25 | 60 |
| Cotton seed oil | | 82 | 120 |
| Dichlorodifluoromethane | | 25 | 172 |
| Ethyl alcohol | | 25 | 60 |
| Formic acid | 3 | 100 | 3 |
| Hydroxyacetic acid | 3 | 100 | 3 |
| Lard | | 82 | 120 |
| Methyl alcohol | | 25 | 60 |
| Potassium carbonate | 19 | 25 | 2 months |
| Potassium hydroxide | 10 | 65 | 3 |
| Sea water (Florida) | | | 4 weeks |
| Sodium cyanide | 10 | 25 | 168 |
| Sodium hydroxide | 10 | 85 | 16 |
| Stoddard solvent | | 25 | 60 |
| Tetrachloroethylene | | 25 | 60 |
| Trichloroethylene | | 25 | 60 |

## Materials Having a Permanent Effect on Nylon Yarn

Hydrochloric acid and sulphuric acid in 5.0 per cent concentrations at room temperature cause about 25 per cent loss in strength of nylon yarn in 11½ days. Strength losses also occur with lower concentrations of these acids. The rate of deterioration increases with the concentration and temperature.

Oxalic acid in 3.0 per cent concentration at room temperature causes some deterioration of nylon yarn. The rate of deterioration increases rapidly with temperature. At 100°C., about 35 per cent loss of strength occurs in 3 hours.

Some deterioration of nylon yarn takes place during treatment with bleaches of the ordinary types. The degree of deterioration depends upon a number of factors including the type and concentration of the bleaching agents, the pH of the bleach bath, and the time and temperature of the treatment.

Air (oxygen) at high temperatures causes deterioration of nylon (see Thermal Properties).

*Solvents for Nylon*

1. Concentrated formic acid at room temperature (27°C.). (The solubility falls off rapidly as the concentration of formic acid is decreased.)

2. Phenolic compounds such as phenol, cresols, xylenols and chlorinated phenols (27°C.).

3. Calcium chloride in methanol (saturated solution at 27°C.).

4. Concentrated nitric acid, concentrated sulphuric acid and concentrated hydrochloric acid. These agents appear to dissolve the fibre but in fact break down the molecular structure of the polymer.

5. Hot solutions of calcium chloride in (a) glacial acetic acid, (b) ethylene chlorohydrin, and (c) ethylene glycol.

6. Hot solutions of zinc chloride in methanol.

7. Benzyl alcohol at the boil.

*Swelling Agents for Nylon*

The following agents cause an appreciable increase in the diameter of nylon filaments. The effect of these agents on the strength of the yarn was not determined.

| Chemical | Concentration (per cent) | Solvent or Diluent |
|---|---|---|
| Acetic acid | 100 | — |
|  | 2 | Methanol |
| Adipic acid | 2 | Methanol |
| Aniline | Saturated | Water |
| Benzene sulphonic acid | 20 | Water |
| Benzoic acid | Saturated | Water |
| Boric acid | Saturated | Methanol |
| Chloral hydrate | 20 | Water |
| Chloroacetic acid | 20 | Water |
| Formic acid | 20 | Water |
| Glycol | 100 | — |
| Lactic acid | 100 | — |

| Chemical | Concentration (per cent) | Solvent or Diluent |
|---|---|---|
| Lithium bromide | Concentrated | Methanol |
| Metacresol | 2 | Methanol |
| | 2 | Water |
| p. Hydroxybenzoic acid | 20 | Methanol |
| Phenol | 2 | Water |
| Phosphoric acid | 20 | Methanol |
| | 20 | Water |
| Zinc chloride | Saturated | Water |

### Perspiration

Exposure to synthetic perspiration, both acid and alkaline, has little effect on the tenacity of nylon. After 2 hours at 100°C., exposure to an acid perspiration solution resulted in only an 8 per cent loss in tensile strength. Under less severe conditions of overnight exposure at room temperature there is no loss of strength.

### Acids

Dilute acids have little effect on nylon under the conditions encountered in practical use. Hot mineral acids will, however, decompose nylon. The fibres disintegrate in boiling hydrochloric acid of 5 per cent strength, and in cold concentrated hydrochloric, sulphuric and nitric acids.

### Alkalis

Nylon 6.6 has excellent resistance to alkalis. It can be boiled in strong caustic soda solutions without damage.

### Effect of Organic Solvents

Concentrated formic acid, phenol and cresol are solvents for nylon. The fibre is not attacked by solvents used in dry-cleaning.

### Insects

Nylon cannot serve as food for moths or beetles.

### Micro-organisms

Nylon is not weakened by moulds or bacteria.

**Electrical Properties**

The low moisture absorption of nylon encourages the accumulation of static electricity, but the effects may be overcome by use of antistatic finishes and static eliminators.

*Volume Resistivity*:

$4 \times 10^{14}$ ohms-cm. at 18 per cent r.h.
$5 \times 10^9$ ohms-cm., wet.

*Breakdown strength*:

Film, unrolled, 9 mils.; 1,300 v./mil.
Film, rolled, 2 mils.; 3,000 v./mil.

*Dielectric Properties*

|  | Power Factor (per cent) | Dielectric Constant |
|---|---|---|
| 1,000 cycles; 18 per cent r.h.; 22°C. | 5.0 | 4.0 |
| 1,000 cycles; wet; 22°C. | 11.0 | 20.0 |
| 60 cycles; dry; 33°C. | 1.8 | 3.8 |
| 60 cycles; dry; 90°C. | 13.0 | 7.0 |

**Allergenic Properties**

Nylon is chemically inert and will not cause irritation to the skin.

**Refractive Index**

In axial direction: 1.547
In transverse direction: 1.521

NYLON 6.6 IN USE

**General Characteristics**

Nylon fibres offer a range of properties which have enabled it to become one of the most successful of all synthetic textile fibres.

*Mechanical Properties*

The outstanding mechanical properties of nylon include:

(a) High strength-to-weight ratio.
(b) High breaking elongation.

248

(c) Excellent recovery from deformation.

(d) High abrasion resistance.

(e) High flex resistance.

The tenacity of nylon may be adjusted to provide for most of the requirements of textile applications, including high tenacity fibres for industrial use.

The high elongation of regular nylon, associated with its excellent elasticity, makes nylon the ideal fibre for form-fitting garments, such as ladies' hose, which will regain their shape after being stretched.

Some items made from nylon have shown phenomenal abrasion resistance in use. This property is related to the inherent toughness of the fibre in terms of resistance to flexing, its natural pliability and its recovery from deformation. Because of nylon's toughness and recovery after deformation its flex resistance is excellent.

The mechanical properties of nylon are affected only to a low degree by moisture, and fabrics remain strong and stable when wet.

The high strength, associated with low specific gravity, give nylon a high strength to weight ratio. The low specific gravity suggests that yarns should be bulky and give higher coverage than many other fibres, but this is offset to some degree by the manner in which the smooth, regular, round filaments are able to pack together into compact yarns. The effect of this is well illustrated by comparison of nylon with silk. Although boiled-off silk has a specific gravity of 1.25, nylon (specific gravity 1.14) hose of the same weight and gauge are much more sheer.

Although nylon has a high breaking tenacity, its initial modulus is relatively low. This means that it is particularly sensitive to stretching under low loads. The lower end of the stress-strain curve rises less steeply than the middle portion, and in this lower region a small change of load produces considerable elongation. This lower end of the curve represents the general working area for tensions in textile processing, so that care must be exercised to establish tensions as low as possible and maintain them as evenly as possible.

The history of the yarn being used has a considerable influence on its sensitivity to stretching. In general, any treatment (such as heat relaxing) which increases elongation, also increases the tendency to stretch under low loads. By the same token, normal tenacity yarn stretches more easily than high tenacity yarn.

Humidity is also a factor which influences nylon's resistance to stretching. In one series of tests, for example, wet nylon stretched 1 per cent with only 24.2 per cent of the load required to stretch bone dry yarn 1 per cent. Nylon conditioned at 50 per cent r.h. required 64.6 per cent of the load needed at 0 per cent r.h.

## Heat Setting

Nylon fibre, yarn or fabric may be heat set readily in either dry heat or steam. The threshold temperature for dry heat is in the region of 205°C., but much lower temperatures will bring about heat setting in the presence of water, e.g. in the region of 120°C.

Goods which have been heat set by either the dry or wet process will subsequently hold their geometrical configuration through processing in boiling water, and will exhibit wrinkle recovery characteristics commensurate with the severity of the setting operation.

Fibre which is heat set under dry conditions suffers a compacting of the internal structure, which results in decreased dye receptivity or dye rate. Fibre heat set under wet conditions results in a more open structure, which dyes deeper.

These properties imparted to nylon by heat setting have been a major factor in its acceptance in many consumer uses. It is fortunate that nylon 6.6 heat sets readily at temperatures in the range of 205°C., which is well below its melting point of about 250°C. This allows a good operating margin in heat setting, hot drawing, calendering, pleating and yarn texturing operations.

## Texturing

The thermoplastic nature of nylon has made possible the successful development of a wide variety of permanent bulking and textured effects on continuous filament yarn. Many processes are used, which are based generally on twisting, crimping, looping or curling (deforming) with some sort of heat setting. In addition to providing additional bulk and texture to the filament yarn, these processes may or may not impart a certain degree of stretch and recoverability.

The textile industry has made good use of this aspect of nylon processing, by developing an extensive variety of end uses, particularly in knit articles and garments. Examples include men's women's and children's hose, bathing suits, underwear, gloves, sweaters and dresswear.

In addition to the properties of loftiness and stretch, these yarns and the fabrics made from them are characterized by good cover, light weight, soft hand, dullness, increased absorption, durability and freedom from pilling. Each of the processes is capable of producing modifications of the bulked yarn by changing the basic characteristics, providing flexibility and adaptability to different end use requirements.

### Heat Resistance

Nylon 6.6 has adequate resistance to heat for practically all apparel and home furnishing uses. Anti-oxidants are added to some yarns to confer a high degree of heat-resistance for some industrial uses.

### Flame Resistance

The flammability of the fibre itself is only one aspect of flammability of the fabric constructed from it. Other factors are involved, including construction, type of surface (brushed or napped, for example, as compared with clean), the types of dyes and chemicals used in finishing, and the moisture content of the material.

Clean, undyed fabrics of nylon have a good resistance to burning and are considered highly satisfactory for practically all textile purposes. When nylon melts, it forms a melt ball which drips away. Usually, when the igniting flame is removed, the flame extinguishes itself. Dyed and finished nylon, however, will in some cases support the spread of flame. In principle, anything that will tend to hold the structure together and retard or prevent dripping of the molten polymer will help to support the spread of flame. For this reason, premetallized and chrome dyes, thermosetting resins and certain flameproofing compositions developed for other fibres may tend to increase the capacity of nylon fabrics to support the spread of flame.

### Moisture

The rate with which nylon comes to equilibrium with the surrounding air depends upon the available surface as well as density and type of packing. For this reason, the only safe conditioning treatment should be controlled by daily weighings until constant weight is reached. A sheer fabric will come to equilibrium in less than an hour if loosely draped over a rack. The same fabric when tightly rolled may take 10 days. Shipping bobbins and warps sometimes require 20 days or more.

## Static

The relatively low moisture absorption of nylon is conducive to the accumulation of static electricity. This may cause a problem in nylon processing, encouraging ballooning in yarns and the inability to lay nylon fabric straight on tables and trucks. Static may also cause a yarn or fabric to pick up soiling materials. It is necessary in the processing of yarns and fabrics, therefore, to ensure that static is kept to a minimum.

Static may be controlled by modification of the polymer and the use of antistatic agents. A high humidity in processing will also prevent the accumulation of static electricity. If humidity is too high (above 75 per cent, it may make the finish tacky and create tension difficulties. High humidity may also produce uneven stretch and recovery due to the tendency of the nylon to stretch more easily as the humidity increases.

Static eliminators are very efficient when positioned effectively. In warping for tricot, for example, the eliminators should be placed at each drop wire bank on the creel just after the eye board and as the yarn goes on the beam or spool.

Nylon should always be allowed to condition for at least 24 hours open to the atmosphere in which it is to be processed.

## Light

Most fibres undergo degradation when exposed to light, resulting in loss of strength and other properties. The relative merits of fibres in respect of light resistance can be assessed only against identical concurrent exposures, and some data comparing nylon with other fibres is given in the following table.

The results of this test indicate that nylon rates below the fluorocarbon yarn and acrylic fibre both in the open and under glass. Behind glass, the polyester is second only to the fluorocarbon yarn. Some improvement occurs in nylon durability by placing it behind glass.

The wavelengths that degrade nylon range from 3,200

## Fibre Degradation by Light

| Fibre | Average Percentage Strength Retained | |
|-------|----------|----------|
| | Outdoors | Behind Glass |
| Fluorocarbon yarn ('Teflon' TFE) | 100 | 100 |
| Acrylic fibre ('Orlon') | 57 | 61 |
| Nylon yarn (bright; 6.7 dtex/f) | 38 | 50 |
| Polyester (bright, semidull, dull) ('Dacron') | 16 | 72 |
| High tenacity rayon | 8 | 23 |

Note: Continuous exposure, Miami, Florida; facing south at 45° angle.

Ångströms into the visible blue slightly above 4,000 Ångströms. For this reason, the effect is less pronounced on nylon than on the polyester when the two fibres are placed under glass.

The light durability of nylon is extended by certain premetallized dyes.

### Washing

Coloured nylon goods are best washed in warm water at 50°C., and white nylon goods at 60°C. Blends of nylon and cotton may be washed at rather higher temperatures, up to about 70°C. White and light-coloured fabrics may pick up tints from coloured fabrics, and the two should be washed separately.

Bleaching is generally unsatisfactory, and should be avoided.

### Drying

Nylon fabrics absorb very little water, and they dry quickly. Drip drying or cold tumble drying are preferred.

### Ironing

Nylon has excellent qualities of crease resistance and shape retention, especially when it has been effectively heat set. A minimum amount of ironing is required; this may be carried out with a dry iron at 'synthetic' setting, or by using a steam iron or steam press. Ironing temperatures above about 150°C. should be avoided. Glazing may occur at higher temperatures, and sticking may occur between 205 and 240°C. Excessive exposure to ironing temperatures may cause yellowing.

Previous treatment of the fabric has a considerable effect on its reaction to pressing conditions. Certain dyes may change colour at elevated temperatures. If a change of form is desired, e.g. in creases, pleats or wrinkles, the conditions of temperature and moisture under which the form was established must be exceeded to produce a permanent change.

**Dry Cleaning**

Nylon is not affected by any of the solvents commonly used in dry cleaning, and there are no difficulties involved.

**End-Uses**

*Home Furnishings*

The major uses of nylon in this field are in carpets and upholstery.

*Carpets*

It is apparent that several of the characteristics of nylon are of value in carpets. Among these are appearance retention, which is a function of abrasion resistance, texture retention, recovery from crushing, durability and dye fastness.

A major proportion of all carpets and rugs is now made on tufting machines, and three general types of nylon yarns are used:

(1) Frieze and Nubby.

(2) Velvet.

(3) BCF (Bulked Continuous Filament).

All these three types depend upon heat setting to provide their maximum function in carpets.

*Frieze Yarn.* In its relaxed, kinked form, a frieze yarn looks as though it would not pass through a tufting needle. But the low forces needed to elongate nylon (low initial modulus) are advantageous in this respect, as the yarn appears unkinked when wound on cones. When unwound, the nylon does not recover instantaneously, and adequate time is thus provided in tufting before the yarn asserts its elastic recovery.

*Velvet Type Yarns.* These tend to elongate on the package, but 'bloom' when converted into carpet pile.

*BCF Yarns.* These are bulky in carpets, but compact while on the shipping package. They pass readily through the creel and tufting machine, developing subsequently the maximum bulk and cover in the carpet, where tensions are no longer present.

An important property of all three types of heat-set nylon yarns is their ability to pass through piece-dyeing operations and retain their appearance and physical characteristics.

BCF yarns have gained wide acceptance because of the economy provided by this type of yarn in comparison with spun yarns; many handling and processing operations are avoided while still providing a yarn for carpets with excellent performance characteristics in terms of tufting efficiency, piece dyeability, bulk, cover, wear, styling and appearance retention.

By altering the end-groups balance and the physical properties of nylon yarn, its dye rate and dye capacity can be controlled. This technique is used in producing tri-dye effects which bring improved styling of both BCF and spun yarns. This principle is used in carpets, upholstery and apparel end uses.

Further styling effects in carpets and upholstery are achieved by combining yarns of mixed count, e.g. 20, 17 and 16.7 dtex/f (18, 15, 6 den/f) with the cross-dyeable yarns and including three plies and two twist levels along with two different weight levels. By combining these factors, as many as 380,000 style combinations are obtainable, excluding colour and pattern.

Adding to this the possibilities obtained by using the different staple and filament products, the number of combinations extends into the millions for tufted carpets.

The use of space dyeing, space resist treatments prior to piece dyeing, yarn dyeing and printing introduces other possibilities for stylng in the carpet field.

### Upholstery

Nylon's combination of strength, dyeability, abrasion resistance, high light fastness and lustre ('Antron') have opened up an important market in the upholstery field. Nylon provides fabrics which combine beauty, performance and practicability.

## Apparel

The following characteristics, among others, are important in fabrics to be used in wearing apparel:

(1) Protection – Cover, warmth
(2) Aesthetic appeal – Texture, colour, hand, drape
(3) Ease of care
(4) Durability – Abrasion resistance and strength
(5) Appearance retention – Wrinkle recovery and soil resistance
(6) Comfort – Weight, openness, tactile property.

Many of these characteristics are influenced greatly by the yarn size, denier per filament, lustre, fibre cross-section, fabric construction and weight. But within the framework of yarn and fabric construction, some fibres perform better than others, and in this respect nylon competes most effectively.

Nylon lends itself to most types of fabric construction, including the following types which are important in the apparel field:

Surah, tricot (plain and brushed), simplex, linings, hosiery (circular and fully-fashioned), taffeta, crepe, satin, reinforced twill and sateen cotton, ski-wear fabric, velvet fleece, brocade, matlasse, circular knit single, circular knit double, full fashion knit, tape and ribbon, lace, sheer, organza and seersucker.

These fabrics are used for all manner of apparel end-uses, including the following:

Lingerie, swimwear, ladies' outerwear, men's outerwear, stretch sportswear, hunting apparel, uniforms, hosiery, sweaters, dresses, (knit), skirts (knit), socks, gloves.

The introduction of textured yarns of nylon in recent years has opened up a vast new field of application in the production of garments that perform because of either bulk with low weight or bulk with elastic recovery for better fit. In addition, special constructions such as bouclé effects are obtainable and different tactile properties are provided.

### Hosiery

Nylon made an immediate appeal as a hosiery fibre, and from the day of its appearance on the market it began to oust silk from its dominant position in the quality hosiery field. Once pre-boarding was adopted, there were no real problems associated with the use of nylon in ladies' hosiery. Setting of the fabric configuration was essential to permit passage of the hose through the dyebath without the formation of severe wrinkles. In addition, the boarding operation shapes the hose to a desirable dimension and imparts elastic memory and recovery to the fabric. This is particularly important to the appearance of hose in use.

Post boarding was introduced at a later stage.

The development of nylon stretch yarns was followed by a rapid increase in the use of nylon in men's hose. The acceptance of these yarns resulted from comfort, appearance retention, elasticity and excellent durability.

Cotton hose have better wash fastness than nylon, but the fibrillation characteristics of cotton during wear cause it to become whitened after a few repeated home laundry cycles. Nylon, by contrast, will withstand repeated laundering for very long periods without undergoing deterioration in its appearance.

The excellent abrasion resistance of hose made of nylon make them much less susceptible to hole formation and wear.

### Lingerie

Tricot is one of the most important of the warp knit fabrics, and nylon tricot has made impressive headway in recent years.

Nylon's acceptance in this use is undoubtedly related to its easy care properties, such as quick drying and adaptability to automatic home laundering. Equally important in the development of nylon tricot, however, was the ability of lingerie producers to style day wear fabrics which are light in weight with adequate 'see through cover', and sheerer sleepwear fabrics requiring less cover. These factors were of importance in the development of nylon tricot before its easy care properties were recognized.

For optimum shrinkage control and wrinkle recovery, it is essential to heat-set nylon tricot. This may be done at one or other of two stages during the production of the fabric, and two different processes are now recognized. They are referred to broadly as (a) the single-pass process, and (b) the double-pass process.

In the single-pass process, the fabric is scoured, dyed, dried and heat-set, whereas in the double-pass process it is scoured, dried and heat-set, followed by dyeing and drying. In the single-pass process, the severity of heat-setting is restricted by fabric discolouration, dimensional stability and smooth appearance after laundering. In the two-stage process, a more complete degree of heat-setting is achieved, which actually yellows the fabric and necessitates subsequent bleaching, but provides excellent dimensional stability and smooth appearance after laundering.

Using the AATCC Wash and Wear Performance scale, a fabric finished by the single-pass process rated 2.5, whereas fabric produced by the two-pass process rated 3.5. There was less relaxation shrinkage in the second fabric.

The wrinkle recovery properties of properly prepared nylon tricot are excellent. This is due in part to the mobility of the knit structure, but also to a great degree to the excellent work recovery of the fibre. Nylon has a high work recovery under conditions of high deformation, whereas polyester yarns have superior work recovery under conditions of low deformation.

In tricot fabric, the mobile structure works to the advantage of the fibre having the best recovery properties under conditions of high deformation, and nylon is excellent in this respect.

### Gloves

Nylon has been used successfully in ladies' gloves made from sueded simplex fabrics. A leather grain appearance may be obtained by embossing a 'crush'-like pattern on to simplex. The pattern is durable to hand laundering.

### Outerwear

Nylon taffeta is used in shell cloth in jackets and wind-breakers. These garments are generally lined with fleeced nylon fabrics or fiberfill, or are laminated to other materials. The colour fastness of these garments is acceptable, and they have excellent appearance retention and easy care properties.

Nylon taffeta and tricot have found an important outlet in blouses and dresses, where the same combination of properties is advantageous.

### Staple

Nylon staple is widely used as a component of blends with other fibres, notably of cotton. A high strength, high modulus type fibre will convey significantly higher yarn and fabric strength when blended with cotton, than can be obtained with yarn and fabric made from 100 per cent cotton. Such blends are commonly made with 15 to 50 per cent of nylon in the blend. As the percentage of nylon increases, fabric flex life, tearing and breaking strength and abrasion resistance increases in such a way that a blend containing 50 per cent nylon staple with carded cotton will have a 30 per cent greater fabric breaking strength. There will also be a threefold improvement in abrasion resistance and a sixfold increase in flex life as compared with a fabric of 100 per cent cotton.

Nylon staples of this type are used to reinforce cotton fabrics such as twills, sateens, denims, whipcords, drills and ducks. The following table illustrates typical improvements in strength and abrasion resistance when 25 per cent nylon staple is added in the warp only.

**Effect of Nylon Staple on Fabric Properties**

| Fabric Property | Twills | | Denims | |
| --- | --- | --- | --- | --- |
| | 100% cotton | 75% cotton/ 25% nylon warp | 100% cotton | 75% cotton/ 25% nylon warp |
| Fabric weight (g/m²) | 257.6 | 264.4 | 381.4 | 381.4 |
| Breaking strength (kg) | 48.1 | 54.5 | 78 | 84.4 |
| Stoll abrasion (cycles to destruction) | 260 | 1100 | 1600 | 4000 |
| Accelerated abrasion (min. to destruction) | 27 | 50 | 20 | 36 |

In the latent cure type finished fabric, nylon is used to impart strength and wear life. The nylon component commonly amounts to some 15 per cent.

Blends with cotton containing high percentages (e.g. 50 per cent) of nylon have great resistance to wear, abrasion and flex life, and good resistance to thermal radiation.

### Industrial

The high strength, elasticity, abrasion resistance and other characteristics of nylon have opened up a wide field of application in industrial fabrics. The following are a few of the more important industrial end-uses:

Air springs, belting, filter fabrics, fish net, twine, hose, wash nets, press covers, ironer covers, paddings, parachutes, webbing, sewing thread, cordage coated fabrics, body armour or ballistic cloth, tents, aerial targets, screening, felts, reinforced plastics, ropes, blend paper, papermakers' felts, tyres, non-woven fabrics, back gray fabric.

## Tyre Cord

The high strength and abrasion resistance of nylon 6.6, coupled with other properties such as resistance to moisture and heat, good fatigue resistance and high work of rupture, enabled nylon to make rapid headway in the tyre cord market.

## Conveyor Belting

The use of nylon as reinforcement in heavy-duty conveyor belting is rapidly gaining ground. Nylon offers the following advantages in this application:

(1) High tensile strength.

(2) High impact strength and resistance to shock maintained over long periods

(3) Good adhesion to rubber.

(4) Resistance to rotting.

(5) Excellent flex and fatigue resistance.

(6) Low specific gravity compared with natural fibres used for rubber reinforcement.

## Hose

Nylon is widely used in many types of industrial hoses. It offers a high strength-weight factor, good wet strength retention, excellent flex life and loop-strength ratio. Nylon-reinforced hose is made with knitters, vertical and horizontal braiders, Wardwell braiders, Chernak looms; nylon is also used in mandrel-built wrapped hose.

## (2) NYLON 6

Nylon 6 fibre is spun from polycaproamide, a polyamide made by the self-condensation of 6-amino-caproic acid or its lactam, caprolactam:

$$CH_2(CH_2)_4 CONH \longrightarrow --NH(CH_2)_5 CONH (CH_2)_5 CONH (CH_2)_5 ---$$

**CAPROLACTAM**                    **NYLON 6**

### TYPES OF NYLON 6 FIBRE

Nylon 6 is produced as multifilament yarns, monofilaments, staple and tow, in a wide range of counts and staple lengths to suit virtually all textile requirements.

The fibres are available in bright, semi-dull and dull lustres, and with special additives such as optical bleaches for specialized end-uses.

Like nylon 6.6 fibres, the fibres of nylon 6 may be varied in properties over a range which is limited by the inherent characteristics of the polymer, each manufacturer controlling his process to produce fibres that will meet specific requirements. In general, commercial nylon 6 fibres fall into two main classes, (a) regular tenacity and (b) high tenacity.

261

Nylon 6 is a thermoplastic fibre, and lends itself well to physical modifications which are associated with this property. Crimped and textured yarns of the familiar types are available.

**Modified Fibres**

As in the case of nylon 6.6, nylon 6 fibres are commonly produced in round cross-section, but fibres of special (e.g. multilobal) cross-section are now available from several manufacturers.

## PRODUCTION

### Reactant Synthesis

The caprolactam used in producing nylon 6 polymer is made by one of several routes, of which the following are important:

### (1) *Cyclohexanone Route*

This is the route by which caprolactam is commonly produced for nylon 6 manufacture (see page 263).

Cyclohexanone may be made from benzene via one of several routes, including the following:

(a) Benzene is chlorinated to chlorobenzene (1), which is then converted to phenol (2). Phenol is reduced to cyclohexanol (3), which is oxidized to cyclohexanone (4).

(b) Benzene is nitrated to nitrobenzene (5), which is then reduced to aniline (6). The aniline is then converted to cyclo-hexanol (7), which is oxidized to cyclohexanone.

(c) Benzene is hydrogenated to cyclohexane (8), which is then oxidized to cyclohexanone (9).

The cyclohexanone produced by any of these routes is reacted with hydroxylamine (in the form of its sulphate $NH_2OH.H_2SO_4$), forming cyclohexanone oxime (10).

Cyclohexanone oxime is treated with sulphuric acid, and undergoes the Beckmann transformation to form caprolactam (11).

Caprolactam Production. Cyclohexanone Route.

## (2) *Cyclohexane Route*

Benzene may be hydrogenated to cyclohexane (1), which is then nitrated (2). The nitro-compound is reduced, forming cyclohexanone oxime (3), which is converted to caprolactam as above.

Caprolactam. Production. Cyclohexane Route.

## (3) *Cyclohexylamine Route* (upper diagram, page 264).

Aniline is hydrogenated to convert it to cyclohexylamine (1). Hydrogen peroxide is reacted with this to form an addition compound, which is then converted to cyclohexanone oxime by treatment with ammonium tungstate solution (2). The cyclo hexanone oxime is converted to caprolactam as above.

ANILINE     CYCLO-HEXYLAMINE     CYCLOHEXANONE OXIME

Caprolactam Production. Cyclohexylamine Route.

## (4) *Hexahydrobenzoic Acid Route*

This is a process patented by Snia Viscosa for the production of caprolactam from toluene which is available from petroleum refining. Toluene is oxidized to benzoic acid (1) which is hydrogenated to hexahydrobenzoic acid (2). Treatment of this with nitrosyl sulphuric acid in the presence of oleum produces caprolactam (3).

TOLUENE     HEXAHYDRO-BENZOIC ACID     CAPROLACTAM

Caprolactam Production. Hexahydrobenzoic Acid Route.

## Polymerization

The polymerization of caprolactam is carried out usually by one or other of two processes, either (a) a non-aqueous process, or (b) an aqueous or hydrolytic process.

### (a) *Non-aqueous Process*

In this process, the carprolactam is heated in the presence of catalysts (e.g. alkali metals and their salts) at temperatures of up to 280°C. Polymerization proceeds by the opening of the caprolactam rings and the linking of the opened rings into polymer molecules. The reaction is rapid, and high molecular weight polycaproamide may be produced, e.g. with a degree of polymerization in the region of 200. The polymers are highly crystalline. They are generally superior in physical properties to polymers made by the alternative method, but the process is not used, as a rule, in the production of fibres. It is of particular interest in the production of cast polyamide plastics.

### (b) *Hydrolytic Process*

This is the technique commonly adopted in the production of polycaproamide for fibre manufacture. The process is usually operated on a continuous basis. Caprolactam, mixed with about 10 per cent of its weight of water, together with dulling agents (where required), acid catalyst and acid chain-stopper, are fed continuously into the top of a stainless steel column, which may be 6 m (20 ft) high and 45 cm (18 in) diameter.

The column is heated to 250–270°C., and as the caprolactam flows downwards through the column it undergoes polymerization to polycaproamide. An equilibrium condition is reached, the material at the base of the column containing about 89–92 per cent of polymer and 11–8 per cent of caprolactam. The amount of caprolactam in the equilibrium mixture depends upon the temperature; at 260°C., there is about 11 per cent.

The polymerization of caprolactam takes place via two routes, (a) a polycondensation reaction, and (b) a polyaddition reaction.

#### *Polycondensation Reaction*

The water added to the caprolactam as it enters the reaction vessel acts as a hydrolytic agent, reacting with some of the caprolactam to form 6-amino caproic acid.

$$\left[\begin{array}{c}(CH_2)_5\\ CO-NH\end{array}\right] \longrightarrow NH_2\,(CH_2)_5\,COOH$$

Polycondensation of this acid then takes place, setting free water which forms more amino caproic acid from caprolactam. This undergoes condensation, contributing to the polymerization, and so the process goes on.

#### *Polyaddition Reaction*

The polyaddition reaction takes place through the opening of caprolactam rings, and the linking together of the opened molecules. There is no intermediate formation of amino acid, and the reaction does not involve the liberation of water or other small molecules. It is an addition reaction, and not a condensation reaction.

Polyaddition takes place alongside the polycondensation reaction, both contributing to the creation of polymer molecules. Polyaddition predominates over the polycondensation.

Polymerization continues until the amino end groups are blocked by the organic acid which was added to serve as a polymerization stabilizer, as in the case of nylon 6.6 production. Polycaproamide made in this way has a degree of polymerization which is commonly in the region of 120–140.

The molten polycaproamide may be spun directly at this stage, without any intermediate isolation of solid polymer (see Direct Spinning, below). More commonly, it is extruded from the pressure vessel as a thick macaroni-like strand which is cooled by a water spray or by falling on to a cooled metal band. The solidified polymer is chopped into small chips of maximum diameter about 6 mm (¼ in).

The chips are washed in demineralized water, which dissolves out the bulk of the caprolactam; they are then centrifuged and dried in vacuo at a temperature below 85°C. The washed and dried chips contain about 1 per cent of caprolactam.

### Spinning

Polycaproamide (nylon 6 polymer) melts at 215–217°C., i.e. about 35°C. lower than polyhexamethylene adipamide (nylon 6.6 polymer). Molten polycaproamide is comparatively stable, and may be held (under an inert atmosphere) at 250°C. for 16–24 hours without deterioration. In this respect, polycaproamide is much less sensitive to spinning conditions than polyhexamethylene adipamide, and the spinning of polycaproamide direct from the polymerization vessel may be carried out much more readily than that of polyhexamethylene adipamide.

Direct spinning of polycaproamide is now established on a commercial scale, and there are thus two spinning techniques in operation for the production of nylon 6 fibres, (a) Spinning from Polymer Chips, and (b) Direct Spinning from Polymerization Stage.

The use of one or other of these two processes is dictated largely by the degree of uniformity required in the filaments that are produced.

In the production of multifilament yarn, a high degree of uniformity is essential, as variations in filament denier will tend to show up in the finished fabric. In the production of staple fibre, on the other hand, uniformity is not such a critical factor, and greater latitude is permissible. Variations in fibre denier, within reasonable limits, will tend to lose themselves during subsequent mixing and processing of the staple fibre.

The molten polymer in the polymerization vessel will contain up to 11 per cent of unchanged caprolactam, the actual amount varying with polymerization conditions. This comparatively high proportion of a volatile constituent is a variable factor that is difficult to control during spinning of the molten material, and filaments spun directly from the melt are not readily held within limits that are acceptable for multifilament yarns. They are, however, satisfactory for the production of staple fibre.

When polycaproamide is extruded and isolated as chips, on the other hand, the excess caprolactam is removed by water washing, and the spinning of fibre from polymer chips may be controlled to provide a much more uniform product. The chip spinning process, therefore, is preferred in the production of multifilament yarns.

### (a) *Spinning Multifilament Yarn from Polymer Chips*

In the production of multifilament yarns, dry chips of poly-caproamide are allowed to fall into the melting zone of a stainless steel spinning apparatus. This may consist, for example, of an electrically-heated grid maintained at about 250–260°C.

As the chips melt, the molten polymer flows through the grid into a conical-shaped sump leading to a metering pump. The molten polymer is at all times maintained under a blanket of nitrogen to prevent oxidation and decomposition.

From the pump, the molten polymer is forced through filters consisting of layers of sand or graded gauze, and from the filter it flows to the spinneret. This is a stainless steel disc perforated with holes in number and shape depending on requirements.

As the jets of molten polymer emerge into the cold air, they solidify. During the spinning process, the heating of the polymer has tended to bring about regeneration of some free caprolactam, and the filaments contain a higher percentage (some 3 per cent) than was present in the dry polymer chips. The presence of this free caprolactam precludes the use of a steam conditioning chamber such as is used in the production of nylon 6.6 filaments; the filaments of nylon 6 would become sticky in the hot steam, and would stick together and to any surfaces they met.

The nylon 6 filaments are conditioned, therefore, by passing over rollers which moisten them with water, solutions of antistatic agents and lubricant before being wound on to the package. Throughout the take-up region, the atmosphere is controlled to sustain a relative humidity of 45–55 per cent and a temperature of 18–20°C. This ensures subsequent stability of the package.

Take-up speeds of more than 1,000 meters per minute are used, and the speed of winding is co-ordinated with that of the metering pump to ensure that the filament diameter is accurately controlled. Packages may contain about 1 kilogram of yarn.

### Stretching or Drawing

Nylon 6 filaments are drawn by passing over two sets of rollers, as in the case of nylon 6.6, the second set moving with some 4 times the surface speed of the first set. The yarn is thus stretched to a ratio of about 1 to 4. The stretching operation is carried out under rigidly controlled conditions, e.g. at 15°C. and 55–65 per cent r.h. Drawing may be carried out at higher temperatures to produce filaments of higher tenacity.

### Twisting

The multifilament yarn at this stage has little or no twist. If added twist is required, the yarn may be up twisted, using conventional machinery, on to perforated bobbins. The degree of twist is determined by the end use for which the yarn is destined. A hosiery yarn, for example, may be given a twist of 1,182 turns/m (30 turns/in) whereas a yarn for weaving or knitting may be given only 276 turns/m (7 turns/in).

### Washing and Twist-setting

The caprolactam formed during spinning is still present in the filaments of nylon 6, and for some purposes this is better removed. The yarn is therefore washed in hot water which dissolves out the caprolactam. At the same time, the yarn undergoes some heat-setting which sets the twist.

*Coning*

The yarn is then dried, and a size is applied, prior to a final winding on to cones. The yarn may also be lubricated during this final winding stage.

## (b) *Direct Spinning of Staple Fibre from Polymerization Stage*

Molten polymer is pumped from the polymerization vessel to a series of spinnerets. These contain a large number of holes, and the filaments from many of them are brought together and wound on to large bobbins. A cable of 5,500 dtex (4,950 den) may be built up in this way, providing a package weighing several kilograms.

A number of these bobbins are then run together to produce a tow of 500,000 dtex (450,000 den) or more.

*Stretching of Tow*

The tow is passed over a series of heavy-duty feed rollers which stretch the filaments in two stages. The first stage stretches the tow some 2.5 to 3.0 times, and the stretched tow then passes forward to the second stretching stage where it is heated and stretched to bring the total stretch to 3.5 to 4.0 times.

*Washing, Crimping and Cutting*

The stretched tow is passed through hot water, which dissolves out the bulk of the caprolactam. It is dried in warm air and may be treated with various finishes. Finally, the tow is crimped mechanically and moves forward to the cutter which chops it into the required staple length.

## PROCESSING

### Scouring and De-Sizing

Nylon 6 yarns, in common with other synthetic fibre yarns, are commonly sized with polyacrylic acid, glue, polyvinyl alcohol, oils or waxes. All these sizes, together with dirt, spinning oil and other contaminants are readily removed by scouring.

The agents used in scouring vary depending on the nature of the material to be removed. When spinning finish only is to be removed, a scour with soap or detergent at 50°C. may be adequate. If an acrylic size is present, however, it is necessary

269

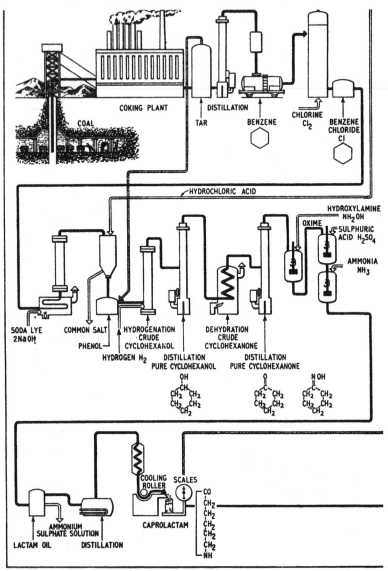

*Nylon 6 Flow Chart*

270

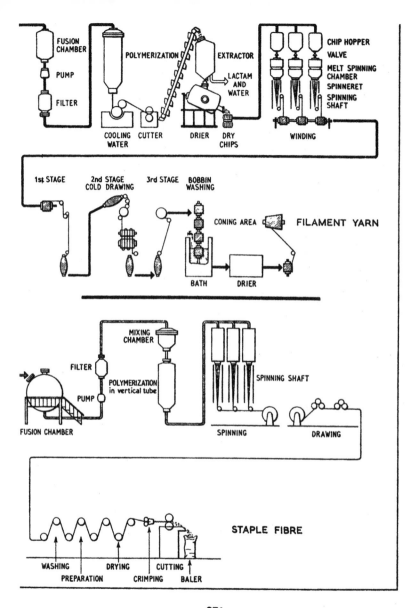

FUSION CHAMBER

POLYMERIZATION

EXTRACTOR

LACTAM AND WATER

CHIP HOPPER

VALVE

MELT SPINNING CHAMBER

SPINNERET

SPINNING SHAFT

PUMP

FILTER

COOLING WATER

CUTTER

DRIER

DRY CHIPS

WINDING

1st STAGE

2nd STAGE COLD DRAWING

3rd STAGE

BOBBIN WASHING

CONING AREA

FILAMENT YARN

BATH

DRIER

MIXING CHAMBER

FILTER

POLYMERIZATION in vertical tube

PUMP

SPINNING SHAFT

FUSION CHAMBER

SPINNING

DRAWING

STAPLE FIBRE

WASHING

PREPARATION

DRYING

CRIMPING

CUTTING

BALER

271

to use a scour liquor containing soda ash, at a temperature of 60–70°C.

For heavier soiling, an emulsified solvent detergent, a stronger alkali, or both, are recommended.

In some cases, solvent scouring in pure or charged perchloroethylene is preferred, particularly in the case of nylon 6 weft-knitted fabrics and garments.

Many commercial deep dye and basic dyeable nylon 6 yarns are tinted with a special sighting colour and, in this case, it is necessary to destroy the dyestuff chemically by addition of hydrosulphite to the scour bath.

### Bleaching

Polyamide fibres in general are naturally white, and chemical bleaching is unnecessary except in special circumstances. Bleaching may be necessary, for example, if heat-setting has caused discolouration, or if the nylon is blended with other fibres that need bleaching.

Although recommended and used in the past, strong oxidative bleaches such as peracetic acid, sodium chlorite and sodium hypochlorite must not be used for the bleaching of nylon 6 as they can cause excessive oxidation and a resultant reduction in dyeability. Moreover, the resistance of nylon 6 to photodegradation is often drastically reduced after such oxidative treatments resulting in unacceptable strength losses after exposure to light. When the use of an oxidative bleach is unavoidable, e.g. for nylon 6/cellulosic blends, hydrogen peroxide is the preferred agent; very little oxidation occurs in this case provided the correct conditions are employed.

### Optical Bleaching

Usually bleaching of nylon 6 is required only for optically-brightened shades and in this case a reduction bleach using stabilized hydrosulphite is employed. It is often possible to apply the optical brightening agent from the bleach bath making a separate bleaching treatment unnecessary. When post setting is to be carried out, the heat sensitivity of the whitening agent should be considered.

## Dyeing

### (1) *100 per cent Nylon 6*

The dyeing characteristics of nylon 6 are essentially similar to those of nylon 6.6, but dyestuffs tend to diffuse more readily into nylon 6 fibre than into nylon 6.6 fibre. This means that with many dyes, nylon 6 requires less dyestuff to attain a given shade, and its rate of exhaustion exceed that of nylon 6.6.

Nylon 6 has a marked affinity for many classes of dyestuff. The classes with the widest applications are the disperse, acid, premetallized, direct and selected reactive dyestuffs combinations.

As in the case of nylon 6.6, basic dyes and vat colours are rarely used. The basic dyes tend to have poor light fastness; the vat dyes are costly and difficult to apply, have poor light fastness and low migration power.

As nylon 6 is heat-set at a maximum temperature of 193°C., the heat stability of the dyestuffs used is not usually critical. Care should be taken, however, in choosing the dyes. When setting precedes dyeing, it is essential that an even treatment is given, or there may be some difficulty in obtaining level shades.

### *Methods and Preparation*

The techniques used in dyeing nylon 6 are similar to those used in dyeing nylon 6.6. It is often desirable to dye from a slightly higher dyebath pH and to raise the dyebath temperature more slowly to the final level, in order to prevent too rapid exhaustion of the dyestuff, which will result in uneven dyeing.

It is most important that the goods should be thoroughly scoured before they are dyed.

### *Disperse Dyestuffs*

These dyes present few difficulties in application, and have excellent migration and levelling properties, even when applied at low temperatures. They offer a wide range of colours and are recommended particularly for light and medium shades. Their covering ability is very good and increases with dyeing time and temperature.

Although generally insoluble in water, the disperse dyes are easily dispersed with the aid of a suitable detergent; some new dyes of this class are self-dispersing.

L                                           273

Disperse colours are of two types, (a) normal disperse dyes, and (b) dyes which are diazotized and developed on the fibre. The latter group are sometimes used for heavy shades, exhibiting good wet fastness with somewhat poorer light fastness. The converse is true with the normal disperse dyes, which have moderate light fastness and less satisfactory wet fastness.

Some of these dyes have a tendency to sublime on to white goods particularly when exposed to heat. Many in commercial use, however, will resist saturated steam at 120°C. and higher.

Disperse dyes are universally used for hosiery dyeing, but are only suitable for the lightest shades on warp-knit, weft-knit and woven fabrics. The wash fastness of the majority of these dyes is insufficient for dyeing medium or heavy shades. Light fastness properties are, on the whole, adequate for most end uses.

## Acid Dyestuffs

These dyestuffs are used generally for medium to deep shades and usually show good light fastness, wash fastness and resistance to sublimation. They are classified conveniently into two classes, (a) Acid Levelling Dyes and (b) Acid Milling Dyes.

### (a) Acid Levelling Dyes

These have good levelling power and show good coverage of affinity variations between nylon 6 fibres. They are dyed generally from a fairly acidic dyebath, e.g. pH 3–5. The wash fastness of acid-levelling dyes is inferior to that of other acid dye classes and they are therefore used only for pale and medium shades or where fastness to wet treatments is not of prime importance.

### (b) Acid Milling Dyes

This class of dyestuff shows a higher affinity for nylon 6 and greater care in application is needed. This is achieved by a slow rise in dyebath temperature, a higher dyebath pH and by use of selected anionic and cationic levelling agents. Because of the higher affinity of these, levelling and coverage of affinity variations is much poorer than with acid levelling dyes but fastness to wet treatments is much higher. Because of their good all-round fastness, acid milling dyestuffs are suitable for most end uses.

Careful selection of dyestuffs is important as blocking may occur if monosulphonated and polysulphonated dyes are present in the same bath. The monosulphonated dye, with its higher affinity for nylon 6, will dye preferentially and sometimes displace the polysolphonate. It is important, therefore, to select the dyes carefully and not to use incompatible mixtures of dyestuffs when dyeing compound shades.

## Premetallised Dyestuffs
These are special dyestuffs complexed with chromium, nickel or cobalt. They have a very high affinity for nylon 6 and therefore control of dyeing parameters is very important; for pale shades dyeing is often carried out from a slightly alkaline dyebath to slow down the rate of dyeing. These dyes have good wet fastness properties (comparable with acid milling dyes) but are characterised by their very high light fastness. Because of this they are nearly always used where very good fastness to light is required, e.g. curtains, upholstery, etc. Unfortunately, most of the dyestuffs give relatively dull shades.

## Chrome Dyestuffs
This class of dye is only of interest for specific end uses where economic heavy shades of high light and wash fastness are required. Chrome dyes are applied in a similar manner to acid dyes except that, after dyeing, the dyestuff is complexed in a fresh bath containing potassium dichromate and reducing agent. This results in a strong chromium/dye complex of high fastness being formed "in situ" in the fibre. As the true shade is developed only after this chroming process, colour matching is difficult as sample swatches have to be afterchromed before matching to the standard pattern can be carried out. Another problem is that free chromium salts in finished apparel can cause localised skin irritation.

## Direct Dyestuffs
Direct dyes are chemically similar to acid dyes and are used primarily for the dyeing of cellulosic fibres. However, selected dyes may be used for dyeing nylon 6; application conditions and fastness properties are similar to acid milling dyes.

## Reactive Dyestuffs
These dyestuffs are designed for application to cellulosics and wool. Certain selected dyes may be used for dyeing nylon 6 and,

because they form a covalent link with the nylon 6 molecule, very high fastness to washing can be achieved. These dyestuffs find only limited usage because of their poor levelling power and limited shade range.

## Improvement in Fastness

The fastness of dyed nylon 6 to wet treatments can often be improved significantly by the use of a syntan (synthetic tanning agent) or full back tan. They should, however, be used only when necessary as they can alter the shade, especially on subsequent washing, and can also break down on post heat setting.

## Deep Dye and Basic Dyeable Yarns

Dye variant nylon 6 yarns are becoming increasingly common and can be used to give novel colour and white, tone in tone and two colour effects. The effects produced are very much dependent on the dyeing parameters used and particular attention should therefore be made to dyebath pH, dyeing auxiliaries and dyestuff selection so that the desired effect may be achieved.

## (2) Blends of Nylon 6 and Wool

The warmth of wool and the strength and hard-wearing properties of nylon 6 combine to produce highly-desirable blends. Dyeing of these blends can present some difficulties, but they are overcome by the use of carefully selected acid, premetallized or chrome dyestuffs together with suitable retarding agents.

Dyes generally have a greater affinity for nylon 6 than for wool, and they will dye the nylon more rapidly and to a greater depth in light to medium shades. As dyeing continues and the shade deepens, the difference in tone between the two fibres diminishes but is still distinguishable. The use of retarding agents, however, reverses this state of affairs, as the nylon is retarded more than the wool.

In heavy shades, the nylon 6 will be completely saturated with dye and the excess colour will be available for the wool, dyeing it to a deeper shade in darker colours.

In addition to the proper selection of dyestuffs, the type of wool and the percentage of the two fibres in the mixture must be taken into consideration in preparing the dye bath.

## (3) *Blends of Nylon 6 and Cellulosic Fibres*

For union shades of nylon 6 and cellulosic fibres, direct and disperse colours are generally used in the same bath. The direct colours tend to stain the nylon 6, but this can be minimized by the use of a resisting agent. The direct and disperse colours should be prepared and introduced into the bath separately.

### Heat-Setting

Nylon 6 may be heat-set effectively at temperatures lower than those used in heat-setting nylon 6.6. This gives an added flexibility to the setting process, permitting the use of dyestuffs, optical whiteners and other agents prior to heat-setting which would be sensitive to the temperatures employed with nylon 6.6.

Heat-setting of nylon 6 follows the pattern common to thermoplastic fibres generally. Strains which have been established in the fibres at any stage of processing are relaxed, and the fibres are set in the positions they occupy during setting. This gives stability and shape permanence to the goods.

Setting may be carried out at any stage during processing, from fibre to finished garment. The later the stage at which setting is carried out, the greater will be the effect on the stability of the final product; strains introduced during processing that follows setting will not, of course, be affected.

Knitted and woven fabrics of nylon 6 are commonly heat-set. During setting, the cloth must be smooth and crease-free, but relaxation must be allowed if residual shrinkage is to be reduced to an acceptable level. This is usually achieved by over-feeding the fabric on a pin stenter.

An adequately set nylon 6 fabric will not crease to any significant extent in normal washing operations.

### Method

Two methods of heat-setting are in common use, i.e. (a) dry heat, or (b) wet heat (steam).

#### (a) *Dry Heat Setting*

Hot air or gas at 190–193°C. may be used, the exposure time being 20–30 seconds, depending on the fabric being processed.

Where steam injection is available, less degradation and yellowing of the nylon 6 occurs and heat setting temperatures can be raised to 195°C. Infra-red setting may also be used on similar equipment, the fabric being passed under specially arranged infra-red units. Very short setting times are possible with infra-red, owing to the high penetration achieved.

Pin stenters are the most suitable equipment for heat-setting nylon 6, either by hot air or infra-red. Clip stenters are not suitable, owing to their inferior control of the fabric, as they do not overfeed and may cause dyeing variations at the selvedge. Hot roll machines using heated cylinders are suitable for stable fabrics, but there is insufficient width control to permit less stable constructions to be processed. This latter method is thus unsuitable in most cases.

### (b) *Wet Heat Setting*

Steam setting is carried out at 105–115°C., depending on the fabric and yarn. This technique is most suitable for weft-knit fabrics, as penetration is good with this type of construction. It is difficult to set tubular weft fabrics in any other way.

When batching on a mandrel for steaming, the greatest care must be used to get even tension from inside to outside, and to avoid steam escapes through the edges of the batch. Quality of steam is also important, as excess condensed moisture must be avoided if even dyeing properties are to be achieved. The steam must be saturated but not superheated. A vacuum cycle before steaming is of great value in obtaining even setting, and should be used whenever possible.

Slight variations in treatment may result in noticeable dyeing differences and it is essential that setting treatments should be uniform between steaming batches. Similarly, the size of the fabric roll should be limited to ensure complete and uniform penetration of steam throughout the piece. In some cases, particularly with warp-knitted fabrics processed on the beam, it is possible to obtain a good degree of set by treating in hot water at the boil or 110°C. This is known as hydrosetting. Unfortunately, fabric shrinkage is often excessive and special fabric constructions to allow for shrinkage are necessary when setting by this method.

The stage in processing at which setting is carried out depends upon a number of factors. From the economic point of view, it is desirable to set as the final operation, as this reduces the total number of operations. This may be impossible, however, because of instability of the fabric, or the possibility of damage to dye or finish, and setting might be carried out in the loomstate or after a scour.

Setting in the loomstate requires great care, or stains may be fixed resulting in discolouration. Setting after scouring has the drawback of increasing the number of operations necessary, and difficulty may be experienced in obtaining the required yield from a grey fabric unless allowance has been made in the construction.

### Pleating

Pleats may be heat-set in nylon 6 fabrics, and will withstand the effects of wear and washing. Nylon 6 monofilament (e.g. 22 dtex; 20 den) locknit fabric, for example, is often pleated for lingerie applications.

Before pleating, the cloth must be scoured well at 60°C., optically bleached or dyed, and then dried on a stenter at a maximum temperature of 130°C. Pleating should be carried out at 170–180°C., depending on the pleating speed and size of pleat required.

When optical brightening agents or dyes are applied before pleating, care must be taken to select those that are not heat-sensitive at the temperatures used.

### Texturing

In common with other thermoplastic yarns, nylon 6 may be deformed by heat treatment to produce bulked, stretch or torque yarns. The false twist technique, for example, is commonly used.

Nylon 6 may be used to produce false twist yarns for all crimp yarn outlets, and garments made from correctly processed nylon 6 yarns have a different character from those made from nylon 6.6.

Nylon 6 has a lower bending modulus than nylon 6.6, and this results in an unusually soft yarn, the softness being apparent in the garment. The effect is particularly noticeable in the case of garments produced on medium gauge knitting units.

In the production of false twist textured yarns, lower temperatures are used with nylon 6 than with nylon 6.6.

## Filament and Yarn Denier

In the production of false twist textured yarns, the filament count is selected with the end-use of the yarn in mind. Filaments of high count will have a high retractive power in yarn and fabric, and they are suitable, therefore, for form-fitting garments such as swimwear, slacks and trews, stockings, etc. Low filament counts will have less retractive power, producing a softer and fuller hand in a fabric; they are more suitable, for example, for outerwear garments.

## STRUCTURE AND PROPERTIES

### Fine Structure and Appearance

Nylon 6 fibres are smooth-surfaced, with no striations. In microscopic appearance they are similar to nylon 6.6 and are as featureless as glass rods. The fibres are commonly of round cross-section, but special types of nylon of multilobal cross-section are now produced.

### Tenacity

The tenacity of nylon 6 may be varied within limits by adjustment of the manufacturing conditions. In general, the greater the degree of stretch during drawing, the higher the tenacity and the lower the elongation.

Tenacities of typical nylon 6 yarns are as follows:

Standard: 39.7–51.2 cN/tex (4.5–5.8 g/den), dry; 36.2–45.0 cN/tex (4.1–5.1 g/den), wet.

High tenacity: 66.2–73.3 cN/tex (7.5–8.3 g/den), dry; 47.7–62.7 cN/tex (5.4–7.1 g/den), wet.

Staple: 33.6–48.6 cN/tex (3.8–5.5 g/den), dry; 30.9–41.5 cN/tex (3.5–4.7 g/den), set.

### Tensile Strength

Standard: 5,110–5,880 kg/cm² (73,000–84,000 lb/in²).
High tenacity: 7,700–8,400 kg/cm² (110,000–120,000 lb/in²).

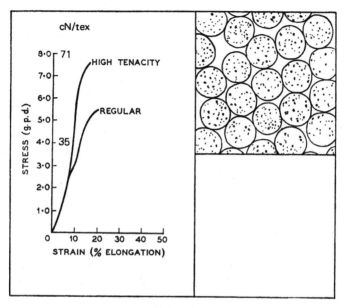

*Nylon 6*

### Elongation

Standard: 23–42.5 per cent, dry; 27–34 per cent, wet.
High tenacity: 16–19 per cent, dry; 19–22 per cent, wet.
Staple: 23–50 per cent, dry; 31–55 per cent, wet.

### Elastic Recovery

Like nylon 6.6, nylon 6 is a highly elastic fibre, in that it will recover its original dimensions after being deformed by the application of a stress. Standard filament has an elastic recovery of 100 per cent up to an extension of 6–8 per cent. Recovery from 10 per cent extension is about 85 per cent.

Recovery from extension also follows the pattern set by nylon 6.6, only part of it being instantaneous, and the remainder taking place over several hours.

## Initial Modulus

Nylon 6 has a low initial modulus, i.e. it stretches readily when subjected to a stress.

Regular filament: 309–441.5 cN/tex (35–50 g/den).

## Average Stiffness

Regular filament: 203.1 cN/tex (23 g/den)
High tenacity filament: 388.5 cN/tex (44 g/den).

## Average Toughness

Regular filament: 0.67.
High tenacity filament: 0.68.

## Flex Resistance

Excellent.

## Abrasion Resistance

Like nylon 6.6, nylon 6 has an outstanding resistance to abrasion. Several factors contribute to this, including inherent toughness, natural pliability, and high flex resistance.

## Specific Gravity

1.14.

## Effect of Moisture

Regain: 4–4.5 per cent.
The moisture absorption increases to a maximum of about 9 per cent at 100 per cent r.h.
The tenacity of wet nylon 6 is 80 to 90 per cent of its dry (conditioned) tenacity. The elongation of wet nylon 6 is some 5 per cent greater than its dry (conditioned) elongation. There is a slight amount of lateral swelling of nylon 6 fibre in water, but the length remains almost unaffected.

## Thermal Properties

*Melting point*: 215°C.

282

## Effect of Low Temperature

Nylon 6 retains its strength well at low temperatures. There is, in fact, a slight gain in tensile strength after exposure of nylon 6 at low temperatures ($-17°C$.). The effect is reversible, the nylon 6 regaining its original strength when returned to room temperature.

## Effect of High Temperature

Nylon 6 loses strength with increase in temperature. The effect is a condition of the thermoplastic nature of the fibre, and is not permanent. The tensile strength is regained when the fibre is returned to room temperature.

Prolonged exposure of nylon 6 to air at elevated temperatures causes deterioration, with permanent loss in breaking strength, elongation and toughness. The fibre discolours (yellows) to some extent.

## Flammability

Nylon 6 resembles nylon 6.6 in flammability. It melts when heated above $215°C$., and the molten droplets tend to fall away. A nylon 6 fabric does not normally support combustion on its own, but its flammability may be increased by the presence of certain chemical finishes and dyes.

## Effect of Age

Negligible.

## Effect of Sunlight

Nylon 6 suffers some loss of strength on prolonged exposure to light, with superficial yellowing and a general deterioration of other fibre properties.

The degree of deterioration due to light is affected by many factors, of which the following are important:

(1) The transparency or opacity of the fibre. Bright nylon 6 is more resistant than dull.

(2) The count of the yarn. Yarns of higher count are more resistant than those of lower count.

(3) **Dyes** and finishes used on the yarn. These may have a considerable effect on light sensitivity.

### Chemical Properties

In general, nylon 6 is highly resistant to chemical degradation, and is similar in this respect to nylon 6.6.

*Materials Having No Permanent Effect on Nylon 6 Yarn*

Under ordinary conditions, nylon 6 yarn is not adversely affected by compounds of the following types:

| | |
|---|---|
| Alcohols | Esters |
| Aromatic Hydrocarbons | Thiols |
| Aliphatic Hydrocarbons | Alkalis |
| Ketones | Soaps and |
| Ethers | Synthetic Detergents |

Conditions and chemical agents used for some tests on nylon 6 yarns are listed below. Strength and elongation determinations were made on untreated yarn and on treated yarn after washing, drying and conditioning.

Under the stated conditions, test results indicated no significant loss in strength or elongation for yarn samples as follows:

1. Immersed in 10 per cent aqueous solution of sodium hydroxide at 85°C. for 16 hours.

2. Immersed in 10 per cent aqueous solution of potassium hydroxide at 65°C. for 3 hours.

3. Immersed in 3 per cent aqueous solution of acetic acid at 99°C. for 3 hours.

4. Immersed in 3 per cent aqueous solution of formic acid at 99°C. for 3 hours (50 per cent formic acid solution at 80°C. dissolves nylon 6 after 15 seconds).

5. Immersed in methyl alcohol, ethyl alcohol, acetone, carbon tetrachloride and benzene for 72 hours at room temperature.

6. Exposed in a nitrogen atmosphere to temperatures of 150, 175 and 200°C. for 3 hours.

7. Exposed to steam atmosphere at 99°C. for 6 days.

*Materials Having a Permanent Effect on Nylon 6 Yarn*

A 3 per cent solution of oxalic acid in water at 99°C. for 3 hours causes a loss of almost 30 per cent in strength and elongation in nylon 6 yarn. Higher concentrations, higher temperature and longer exposure times cause a rapid increase in chemical degradation.

Nylon 6 yarns heated in dry air for 5 hours at 150°C. undergo deterioration, losing brightness and becoming yellow.

Most of the common bleaches, other than optical agents, cause some degradation in nylon 6.

*Solvents for Nylon 6*

The following will dissolve nylon 6:

1. Concentrated formic acid (a 50 per cent formic acid solution at 80°C. will dissolve nylon 6 rapidly).

2. Concentrated hydrochloric, nitric and sulphuric acids.

3. A 25 per cent solution of zinc chloride in methanol at 50°C.

4. Phenol and phenolic compounds.

*Acids*

Dilute acids have little effect on nylon 6 under the conditions encountered in practical use. Hot mineral acids will, however, decompose nylon 6. The fibres disintegrate in boiling hydrochloric acid of 5 per cent strength, and in cold concentrated hydrochloric, sulphuric and nitric acids.

*Alkalis*

Nylon 6 has excellent resistance to alkalis. It can be boiled in strong caustic soda solutions without damage.

**Effect of Organic Solvents**

Concentrated formic acid, phenol and cresol are solvents for nylon. The fibre is not attacked by solvents used in dry cleaning.

### Insects
Nylon 6 cannot serve as food for moths or beetles.

### Micro-organisms
Nylon 6 is not attacked by moulds or bacteria.

### Electrical Properties

*Surface resistance:* 2.0 x $10^{12}$ megohms.

*Specific resistance*: $2.6 \times 10^8$ megohms.

*Power factor*: 0.04 at 1 megacycle; 0.07 at 1 kilocycle; 0.20 at 50 c.p.s. (All values determined at 60–70 per cent r.h.)

*Dielectric strength*: 90 kV/cm.

### Allergenic Properties
Nylon 6 is absolutely free of all toxic properties, and is chemically inert. It will not cause irritation to the skin.

### Identification of Yarn as Nylon 6 or Nylon 6.6
The following test is used to distinguish between nylon 6 and nylon 6.6

#### Preparation of Solution
A 50 per cent formic acid solution is prepared by dilution of the 90 per cent formic acid solution commonly available, e.g. 1 litre of acid is diluted with enough cold water to bring the total volume to 1,800 c.c.

#### Procedure
The 50 per cent formic acid solution is heated carefully to 80°C. Several pieces of yarn or individual filaments are dropped into the solution. Nylon 6 will shrivel or ball up and dissolve almost immediately, very little agitation being necessary. Nylon 6.6 will float in the solution and appear not to be affected.

286

The temperature control is very important. At temperatures several degrees lower than 80°C., neither nylon 6 nor nylon 6.6 will appear to be destroyed. If the temperature is at about 90°C., both nylon 6 and nylon 6.6 will disintegrate.

When making this test, it is advisable to carry out preliminary tests first on known samples of nylon 6 and nylon 6.6.

## NYLON 6 IN USE

### General Characteristics

Nylon 6 offers a range of properties that are generally similar to those of nylon 6.6. The characteristics of nylon 6.6, as they influence the applications of the fibre, apply equally well to nylon 6, apart from the effects of the differences between the two polyamide fibres as outlined on page 203. The following points, in particular, are important in this respect:

### (1) *Melting Point*

The lower melting point of nylon 6 (about 215°C.) as compared with nylon 6.6 (about 250°C.) means that greater care must be taken with nylon 6 in all processes involving the use of elevated temperatures. Careless ironing, for example, will damage nylon 6 more readily than nylon 6.6.

The lower melting point of nylon 6 influences the optimum temperature which may be used in processes such as heat setting. Nylon 6 is heat set at temperatures which are lower than those used for nylon 6.6.

### (2) *Affinity for Dyestuffs*

The increased affinity of nylon 6 for some types of dyestuff makes for greater versatility in dyeing, with the possibility of producing brighter, deeper prints (see page 272).

The extra dye affinity of nylon 6 is advantageous also in the production of wool/nylon blends.

### (3) *Light Sensitivity*

Despite the increased resistance to the effect of light which is shown by nylon 6, it is still not generally recommended for those uses in which exposure to light is an important factor, e.g. curtains.

287

### (4) *Heat Resistance*

The lower melting point of nylon 6 means that it may not be used at elevated temperatures where nylon 6.6 is still effective. At temperatures below its melting point, on the other hand, nylon 6 has a somewhat better resistance to the effect of prolonged heating. This can be an important factor in some applications, e.g. tyres, where the fibres must withstand elevated temperatures.

### (5) *Fatigue Resistance*

Nylon 6 has a better fatigue resistance than nylon 6.6, and this is important too in applications such as tyres, where the fibre is subjected to repeated stresses.

### (6) *Hand*

In general, nylon 6 goods have a softer hand than those made from nylon 6.6. This may be advantageous in applications where a soft, full hand is desirable, e.g. in tricot and in fabrics made from textured yarns. Nylon 6 has been particularly successful in the tricot market, where its softness has proved an attractive feature.

In applications where nylon 6.6 has benefited from its greater filament rigidity, e.g. in textured yarns which have a crisp hand associated with maximum stretch and recovery, it is advisable to use nylon 6 yarns containing filaments of increased diameter and in fewer numbers. Small increases in diameter produce large increases in rigidity, and it is possible in this way to bring nylon 6 yarns more into line with nylon 6.6 yarns of similar total denier. The increase in filament denier also increases stretch and recovery.

### (7) *Affinity for Softening Agents*

Nylon 6 has a greater affinity for softening agents than nylon 6.6. Cationic softeners are absorbed more rapidly and completely from solution, and there is a greater softening effect from a given amount of softener. This makes necessary the use of finishing procedures that ensure slow enough absorption of finish to provide a uniform absorption of softener.

## (8) *Toughness*

Nylon 6 has a breaking tenacity in the same range as that of nylon 6.6, but the elongation at break of nylon 6 is generally greater than that of a nylon 6.6 of similar tenacity. This means that nylon 6 has a greater toughness than nylon 6.6

## (9) *Modulus*

The initial modulus of nylon 6 is lower than that of nylon 6.6, which means that nylon 6 deforms more readily under a load. This is a factor in giving nylon 6 its softer hand, as high modulus tends to result in increased stiffness.

The low initial modulus of nylon 6 can make for difficulties in processing, as the yarn will stretch easily at relatively modest tensions. In general, however, nylon 6 may be handled at tensions similar to those used for nylon 6.6

## (10) *Shrinkage*

Other things being equal, nylon 6 fibres tend to shrink more in boiling water than similar yarns of nylon 6.6. This is an important factor in the processing of nylon yarns. In the production of hosiery, for example, it has the following consequences:

(a) It is necessary to knit hose rather looser than when nylon 6.6 is used, in order to obtain proper widthwise stretch. Depending on the mill handling conditions, this can result in increased snagging in the greige hose.

(b) To offset the above disadvantage, the stitch in finished hose of nylon 6 is somewhat tighter, and so fewer picks will be encountered after boarding.

(c) A high shrinkage monofilament may be advantageous in hosiery manufacture. In the production of tubular stockings, for example, the finished shape is obtained entirely by shrinking the knit tube to the boarding form, and the high shrinkage coupled with high elongation is of definite advantage in producing a stocking with proper fit.

In certain non-run styles, machine limitations make it necessary that the ankle be knit rather loose. A high shrinkage nylon is again advantageous in producing a stocking with satisfactory fit.

## Washing

Nylon 6 goods are washed preferably in hand-hot water (up to 48°C.); blends of nylon 6 and cotton may be washed at rather higher temperatures, up to about 70°C. White and coloured fabrics may pick up tints from coloured fabrics, and the two should be washed separately. Many garments are machine-washable.

Bleaching is generally unsatisfactory, and should be avoided.

## Drying

Nylon 6 fabrics absorb very little water, and they dry quickly. Drip drying or cold tumble drying are preferred. Garments should be given a cold rinse and a short spin (15 seconds) followed by line drying. It is advisable to avoid overloading the spin drier with too many garments.

## Ironing

Nylon 6 has excellent qualities of crease resistance and shape retention, especially when it has been effectively heat set. A minimum amount of ironing is required; if necessary, this should be carried out with a warm iron (HLCC Setting 2), and the garment must be damp. Ironing temperatures above 150°C. should be avoided. Glazing may occur at higher temperatures, especially if the garment is dry, and sticking may occur at temperatures between 180 and 200°C. Excessive exposure to ironing temperatures may cause yellowing.

Previous treatment of the fabric has a considerable effect on its reaction to pressing and ironing conditions. Certain dyes may change colour at elevated temperatures. If a change of form is desired, e.g. in creases, pleats or wrinkles, the conditions of temperature and moisture under which the form was established must be exceeded to produce a permanent change.

*Note.* Certain garments such as rainwear and anoraks will often have been treated with special water-repellent finishes which will require less severe washing treatment. The same may apply to quilted and lined garments, and these should be washed in warm water and ironed where necessary with a cold iron (HLCC Setting 1).

## Dry Cleaning

Nylon 6 is not affected by any of the solvents commonly used in dry cleaning, and no difficulties are involved.

## End-Uses

The end-uses of nylon 6 are generally similar to those of nylon 6.6.

*Note.*

*Tyre Cord.* Like nylon 6.6, nylon 6 has made great headway in the tyre cord field. It is claimed that nylon 6 has certain advantages over nylon 6.6 in this application, notably the following:

(a) Nylon 6 has greater thermal stability at the temperatures encountered in tyres.

(b) The adhesion of nylon 6 to rubber is stronger than that of nylon 6.6.

(c) The flex resistance of nylon 6 is greater than that of nylon 6.6.

## (3) NYLON 11

Nylon 11 fibre is spun from polyundecanamide, made by the self-condensation of 11-amino-undecanoic acid:

$NH_2(CH_2)_{10}$ COOH $\longrightarrow$

OMEGA AMINO UNDECANOIC ACID

-- NH $(CH_2)_{10}$ CONH $(CH_2)_{10}$ CONH $(CH_2)_{10}$ CO ---

NYLON 11

### INTRODUCTION

For some years, a nylon-type polymer has been made in France under the name polyamide 11, polyundecanamide or 'Rilsanite', primarily for use as a plastic. The polymer is made by self-condensation of 11-amino-undecanoic acid, and under the nomenclature scheme used for polyamide fibres (see page 207) it is nylon 11.

Nylon 11 may be melt spun into fibres, and the production of nylon 11 fibres was developed in France by Organico S.A., with the cooperation of the Italian firm Snia Viscosa.

### TYPES AND SIZES OF NYLON 11

Nylon 11 has been made as multifilament yarns, monofilaments, staple and tow, in a range of deniers and staple lengths. It was produced under the trade name 'Rilsan', for example, in the following sizes:

Filament: 12/1, 12/2, 18/3, 29/10, 45/16, 57/20, 90/32, 145/50, 290/100.

Staple: 1.4/32, 37, 62 mm.; 2.9/37, 62, 95, 112 mm.; 6/40, 70, 100, 120 mm.

Tow: 110,000 dtex (100,000 den).

## PRODUCTION

### Reactant Synthesis

The 11-amino-undecanoic acid used in production of nylon 11 may be made by three commercially-important routes; (a) from castor oil, (b) from ethylene and carbon tetrachloride, and (c) from dodecane.

### (a) *Castor Oil Route*

Nylon 11 has been produced in France and Brazil from 11-amino-undecanoic acid made from castor oil (see below). The oil is obtained from castor beans; it is extracted by the combined action of pressure and organic solvents. The oil contains 85 per cent of triglyceryl ricinoleate.

Triglyceryl ricinoleate is converted to methyl ricinoleate by treatment with methyl alcohol.

Methyl ricinoleate is pyrolyzed at high temperature, yielding heptaldehyde, methyl undecylenate and a small amount of fatty acids (1). Pure heptaldehyde and methyl undecylenate are isolated by fractional distillation.

Methyl undecylenate is hydrolyzed to undecylenic acid (2).

Undecylenic acid is aminated by reaction with ammonia, to form 11-amino-undecanoic acid (3).

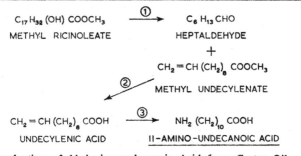

Production of 11-Amino-undecanoic Acid from Castor Oil.

Castor oil is available in very large quantities, and the process for producing 11-amino-undecanoic acid is relatively simple and straightforward.

Important by-products are produced in this process, including glycerol, heptaldehyde and residual oils from the cracking stage. These are an important factor in the economics of nylon 11 production by the castor oil route.

Glycerol is used in innumerable industries, and is always in demand. Heptaldehyde is converted into heptyl alcohol and heptanoic acid, which are raw materials in the plastics industry, and are a useful source of organic chemicals containing a chain of seven carbon atoms. The residual oils are used for the manufacture of detergents.

The economic development of the castor oil route to nylon 11 is influenced, therefore, by a number of factors, and it remains to be seen how successful the process can become. Not least among the many considerations that must be taken into account is the fact that castor oil is an agricultural chemical. It is subject to all the variations and fluctuations that are inherent in the production of a natural product.

### (b) *Ethylene and Carbon Tetrachloride Telomerization*

Telomerization is, in effect, a polymerization reaction which is stopped after only a few monomer units have linked together. This is achieved by carrying out the polymerization in the presence of a relatively large amount of a chain-stopping material, which provides radicals which block the active ends of the growing polymer chain and so prevent further polymerization.

When ethylene and carbon tetrachloride are heated together at high temperature in the presence of a catalyst (e.g. benzoyl peroxide), the ethylene undergoes polymerization until such time as the ends of the polymer chain are blocked by Cl and $CCl_3$ radicals formed from the carbon tetrachloride. Conditions may be adjusted so that this occurs after only 1, 2, 3, 4 or 5 ethylene molecules, for example, have linked together (1).

The product obtained by telomerizing ethylene and carbon tetrachloride is a mixture of compounds of general structure $Cl(C_2H_4)_nCCl_3$, where n is a small number, e.g. 1 to 5. One of the products, which may be separated by distillation, is 1-chloro-11-trichloroundecane i.e. $Cl(CH_2)_{10}CCl_3$.

Hydrolysis of this by aqueous sulphuric acid yields 11-chloroundecanoic acid (2), which is reacted with ammonia to produce 11-amino-undecanoic acid (3).

$$n\,C_2H_4 + CCl_4 \xrightarrow{\text{①}} Cl\,(C_2H_4)_n\,CCl_3$$

$$n = 1,\ 2,\ 3,\ 4,\ 5,\ etc.$$

$$Cl\,(C_2H_4)_5\,CCl_3 + H_2O \xrightarrow{\text{②}} Cl\,(CH_2)_{10}\,COOH$$

11–CHLORO – UNDECANOIC ACID

$$\text{③}$$

$$NH_2\,(CH_2)_{10}\,COOH$$

11–AMINO – UNDECANOIC ACID

Production of 11-Amino-undecanoic acid from Ethylene and Carbon Tetrachloride (Telomerization).

---

### (c) *Dodecane*

A process for the production of undecanolactam from dodecane has been developed.

**Polymerization**

Polymerization of 11-undecanoic acid is carried out in three stages, the monomer being fed as an aqueous suspension into the reaction vessel.

*Stage 1.* Water is removed, and the 11-undecanoic acid is melted. The temperature is raised to 215°C., and polycondensation begins.

*Stage 2.* Polycondensation is allowed to proceed until the desired degree of polymerization has been reached.

*Stage 3.* The molten polymer is held for a time at 215°C. to allow the molecular weight distribution to attain a satisfactory state.

The molten polymer is passed to a storage tank, from which it may be fed directly to the spinnerets.

*Note.* The basic reaction which occurs during nylon 11 production is as follows:

$$nH_2N(CH_2)_{10}COOH \rightarrow H(HN(CH_2)_{10}CO)_nOH + n-1\ H_2O$$

**Spinning**

Molten nylon 11 polymer is very stable at the temperature used

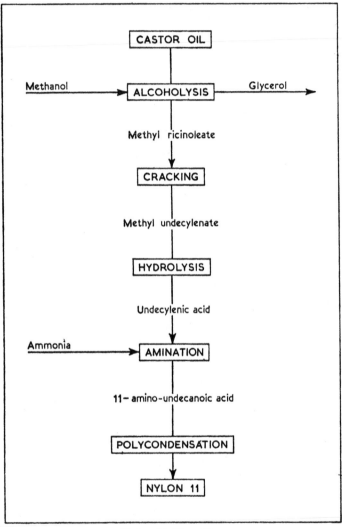

*Nylon 11 Flow Chart*

during melt spinning (about 215°C.), and it may be stored for long periods without deterioration. The usual precautions are taken to prevent oxidation by maintaining molten polymer under an inert atmosphere.

There is little tendency for 12-membered ring formation to occur during polymerization, and the amount of low molecular weight material in the polymer is very small. The polymer is commonly spun direct after production, without any intermediate chip-production stage.

Spinning is carried out in a manner similar to that used in nylon 6 production, and the filaments are drawn to a degree depending on the type of fibre required.

## PROCESSING

### Dyeing

Nylon 11 has a lower moisture absorption than nylon 6 or 6.6, and this caused some dyeing difficulties during the early development stages. Satisfactory dyeing techniques have now been established.

Metallized and acetate dyes give good results, with excellent levelling properties. Many acid and naphthol dyes yield shades of excellent fastness. o-Phenyl phenol and other dyeing assistants are used with these dyes.

Nylon 11 has an excellent resistance to acid solutions, and acid dyeing baths of pH as low as 2 can be used effectively. Nylon 11 can be spun-dyed effectively, with the help of a wide range of colouring matters. The low spinning temperature permits of the use of many organic dyestuffs for this purpose.

## STRUCTURE AND PROPERTIES

The properties of nylon 11 are essentially similar to those of nylon 6.6 and nylon 6, inasmuch as they are all polyamides. The differences between nylon 11 and nylons 6.6 and 6 are those that would be anticipated from the lengthening of the chain of carbon atoms separating the amide groups (see page 309). The moisture absorption has decreased, and the specific gravity is lower, making nylon 11 the lightest textile fibre other than the polyolefins.

The melting point of nylon 11 is lower than that of nylon 6, in accordance with the tendency for melting point to fall with increase in the number of methylene groups between the amide groups. But as the methylene groups are an even number, nylon 11 falls into the higher of the two series of polyamides. For this reason, the melting point is not very much lower than that of nylon 6, which has 5 methylene groups between amide groups, and therefore comes in the 'odd-number' series.

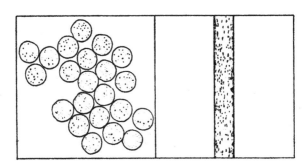

*Nylon 11*

### Fine Structure and Appearance

Nylon 11 filaments are similar in appearance to those of other nylons. They are smooth surfaced, and commonly of circular cross-section.

### Tenacity

Fibres may be produced in a range of tenacities by varying the drawing and other conditions during production. Tenacities are commonly in the range 44.15–66.23 cN/tex (5.0–7.5 g/den). The tenacity is virtually unaffected by moisture.

### Tensile Strength

Fibre of tenacity 66.23 cN/tex (7.5 g/den) has tensile strength in the region of 6,860 kg/cm² (98,000 lb/in²).

### Elongation

Regular filament: 25 per cent wet or dry.

**Elastic Recovery**

100 per cent at 6 per cent elongation.

**Initial Modulus**

Nylon 11 has a higher initial modulus than either nylon 6 or nylon 6.6. It is typically about 441 cN/tex (50 g/den).

The higher rigidity of nylon 11 gives it a better dimensional stability to repeated strains and a better resistance to creep than other polyamides.

**Flex Resistance**

Excellent.

**Abrasion Resistance**

Excellent.

**Specific Gravity**

1.04.

**Effect of Moisture**

Nylon 11 absorbs less water than nylon 6 or 6.6. It has a regain of 1.18 per cent. The moisture absorption does not increase as rapidly with increasing humidity as in the case of nylon 6 and 6.6.

Water absorption at 20°C. and 65 per cent r.h., 1.3 per cent. Nylon 6 and 6.6 under similar conditions absorb 4 and 3.6 per cent respectively.

**Thermal Properties**

*Melting Point*: 189°C.

*Softening Point*: 170°C.

*Effect of High Temperature.* The behaviour of nylon 11 on prolonged heating is similar to that of nylon 6 or 6.6. The fibre yellows in dry air at 150°C.

Decomposition takes place at about 300°C. in an inert atmosphere. Molten nylon 11 can be maintained for weeks at spinning temperature (215°C.) without undergoing decomposition.

**Effect of Sunlight**

Similar to nylon 6.6 (special sunlight-resisting grades are available).

**Chemical Properties**

*Acids.* Nylon 11 is fairly resistant to dilute mineral acids. There is slow deterioration at high temperatures and high concentrations.

*Alkalis.* Nylon 11 has a high resistance to alkalis.

*General.* Nylon 11 has a higher resistance than other polyamides to oxidizing agents. It is inert to common reagents.

**Effect of Organic Solvents**

Nylon 11 has a high resistance to common organic solvents, and is generally similar in this respect to nylon 6 and 6.6. It is soluble in phenols and 100 per cent formic and acetic acids.

**Insects**

Nylon 11 is not attacked by moth grubs or beetles.

**Micro-organisms**

Nylon 11 is not attacked by mildews or bacteria.

**Electrical Properties**

Nylon 11 has excellent electrical insulation properties, which are retained under conditions of high humidity.

**Handle**

Nylon 11 has a pleasant, soft handle.

NYLON 11 IN USE

**General Characteristics**

Nylon 11 resembles nylon 6 and 6.6 in many of its important properties, but it possesses features which affect its potential applications in several ways.

The melting point of nylon 11 (189°C.) is on the low side for general textile use, and great care must be taken in ironing and other elevated-temperature treatments.

The initial modulus of nylon 11 is higher than those of the other nylons, resulting in increased stiffness and rigidity. This is advantageous in applications such as brush bristles, and it also makes for easier processing. Nylon 11 yarns do not stretch so easily as nylon 6 or 6.6 yarns when subjected to physical processing such as winding.

The high initial modulus of nylon 11 suggests that this is a useful nylon for the huge tyre cord market. Tyres reinforced with nylon 11 would not be subject to flat-spotting to the extent that nylon 6 and nylon 6.6 reinforced tyres are.

The low moisture absorption of nylon 11 enables it to retain its excellent insulation properties at high humidities; this is a useful characteristic in electrical applications.

Nylon 11, with a specific gravity of only 1.04, is a very light fibre, with much greater covering power than the other polyamides.

### Washing

Like other polyamide fibres, nylon 11 is easy to wash, using conditions similar to those used for nylon 6 and 6.6. Fabrics may be washed repeatedly without yellowing.

### Drying

Nylon 11 has a low moisture absorption, and dries very rapidly. It should be dried in the same way as nylon 6 and 6.6 using temperatures as low as possible.

### Ironing

Nylon 11 fabric may be ironed safely at temperatures in the region of 80–100°C. Great care must be taken to avoid using temperatures which might soften the fibre.

### Dry Cleaning

Nylon 11 may be dry cleaned without difficulty. It is not affected by the usual dry cleaning solvents.

### End-Uses

Nylon 11 is used in a great variety of textile applications. It is made into tricot knitted lingerie and underwear, hose and woven fabrics. It serves in the same sort of fields as nylon 6.6 and nylon 6. Fabrics of nylon 11 are hard wearing and comfortable, with all the ease-of-care characteristics associated with nylon 6.6 and 6.

Heat-setting processes are used in the same way as with other thermoplastic fibres. Permanent pleats and creases may be obtained effectively.

## (4) NYLON 6.10

Nylon 6.10 fibre is spun from polyhexamethylene sebacate, made by the condensation of hexamethylene diamine and sebacic acid:

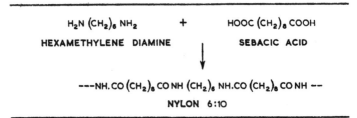

$$H_2N (CH_2)_6 NH_2 \quad + \quad HOOC (CH_2)_8 COOH$$

HEXAMETHYLENE DIAMINE      SEBACIC ACID

$$---NH.CO (CH_2)_8 CO NH (CH_2)_6 NH.CO (CH_2)_8 CO NH --$$

NYLON 6:10

### INTRODUCTION

Polyhexamethylene sebacate is produced in relatively small amounts, largely for use in plastics and in the manufacture of synthetic bristle. Fibres have been spun from it, and they have a number of interesting characteristics. Sebacic acid, however, is expensive, and there seems little likelihood of nylon 6.10 fibre being manufactured on a major scale for textile applications.

### PRODUCTION

Sebacic acid is produced from castor oil. It is condensed with hexamethylene diamine, the process being similar to that used in the production of nylon 6.6. The product, polyhexamethylene sebacate, may be melt spun without difficulty.

### STRUCTURE AND PROPERTIES

Nylon 6.10 fibres are similar to those of nylon 6 and 6.6 in many respects. The moisture absorption of nylon 6.10 is lower than that of either nylon 6 or 6.6, and nylon 6.10 fibres are unusually resilient.

### Regain

2.6.

### Melting Point

214°C.

## (5) NEW TYPES OF POLYAMIDE FIBRE

### Introduction

Polyamide fibres in widespread use today, i.e. nylon 6 and 6.6, have a range of characteristics which serve them well in the general textile field. These fibres have high tensile strength, which is associated with a high degree of elasticity; they have excellent bending strength and outstanding resistance to abrasion. Nylon fabrics have good dimensional stability which is enhanced by heat setting.

The polyamide fibres have a relatively low specific gravity, making for lightweight fabrics. Their moisture absorption is low enough to permit of easy washing and drying.

This combination of properties, together with those detailed in the sections dealing with the individual nylons, has created for polyamide fibres a vast market which spans almost every field of textile application. But it is inevitable that there are applications for which the range of properties offered by nylon is inadequate. Characteristics advantageous in one application may be less desirable in other applications, for which a different balance of properties is required.

The relatively low moisture absorption of nylon, for example, contributes to its ease-of-care characteristics. But it also encourages the accumulation of static electricity, which may be undesirable or even dangerous in certain circumstances. There are applications in which it would be advantageous to have a nylon with higher water absorption.

Other properties too have proved inadequate in specific applications. Polyamide fibres generally have a sunlight resistance which is adequate for normal textile uses, but they are attacked too readily by ultra-violet light to make them the preferred choice in applications that must withstand continuous sunlight. Degradation by light is commonly less serious in the lustrous types of nylon fibre than in fibre which has been dulled by addition of titanium dioxide. Much can be done to improve light resistance by the use of chemical additives, but nylon's tendency to undergo degradation in sunlight has told against it in curtains and similar applications.

When nylon is washed in water containing iron salts, it tends to absorb the dissolved materials, and acquires a yellow colour which is difficult to remove.

The initial modulus of nylon is low, and the fibre thus extends readily at low loadings. This makes for difficulty in processing, and in certain applications. It is a factor in the flat-spotting which occurs when nylon is used for reinforcing tyres.

Nylon is sensitive to bleaching agents and to acids, and it tends to lose strength when heated for prolonged periods, e.g. at temperatures above 150°C.

These characteristics of nylon fibres are of little significance over a vast range of nylon's textile applications. But in certain applications, they are sufficient to make nylon less competitive than it might be. And in specific applications, they may render nylon altogether inadequate.

For most purposes, for example, nylon has an adequate resistance to heat, and the melting point – especially of nylon 6.6 – is high enough for normal textile applications. In recent years, however, developments in space travel, supersonic flight and other fields have created a demand for fibres which can withstand temperatures higher than those encountered in everyday applications. The normal nylons offer a range of properties which is attractive for these specialized applications, but in many cases their melting points are too low to permit of their use.

In many electrical fields, similarly, the characteristics of polyamide fibres are generally attractive, but there are specific applications in which they would be more satisfactory if the moisture absorption was lower.

It is apparent, therefore, that nylon would be able to play an important role in many specialized fields of application if certain characteristics were changed to suit specific needs. Some of these fields, though specialized in the sense that they may require particular properties in the fibre, are of great importance and

304

absorb great quantities of fibre. Tyre cord, for example, is a 'specialized' application which is a major outlet for nylon and other fibres.

During the 1960s and 1970s there was an increasing awareness of the need for developing polyamide fibres with properties that would enable them to compete more effectively in specialized fields and a number of new types of polyamide fibre came on the market. Some of these are designed to serve in comparatively narrow specialized fields. Others have extended the range of general textile applications and brought improvements in the characteristics of nylon fabric generally.

The development of polyamide fibres with properties different from the 'standard' nylon fibres has followed three lines:

(1) Physical modification of existing nylon types.

(2) Chemical modification of existing nylon types.

(3) Production of new types of polyamide.

## (1) PHYSICAL MODIFICATION OF EXISTING NYLON TYPES

**Modification of Polymer**

The polyamides from which nylon 6 and nylon 6.6 fibres are spun have been subjected to intensive study over a very long period. A great deal is known about the extent to which the properties of the fibre may be modified by suitable control of the molecular structure of the polymer.

The physical properties of nylon 6 and nylon 6.6 may be modified within limits by influencing the average molecular weight, molecular weight distribution, degree of crystallinity and degree of orientation of the polymers. Full advantage is now taken of techniques such as drawing which allow the manufacturer to control tenacity, flexibility and other mechanical properties.

The extent to which the properties of the fibre may be varied in this way is limited by the fundamental chemical structure of the polyamide. The melting point of polyhexamethylene adipamide (nylon 6.6 polymer) may be increased, for example, by increasing the molecular weight. But a point is soon reached at which further increase in molecular weight has no effect; the optimum melting point for this particular polyamide has been reached.

The possibility of effecting dramatic changes in properties, beyond those already known, by adjustment of the physical state of the polymer is thus remote.

**Modification of Fibre**

*Multilobal Cross-Section*

It has long been recognized that the cross-sectional shape of a fibre has an important influence on many important characteristics, and nylon – in common with other synthetic fibres – is being produced by some manufacturers in a variety of non-circular cross-sections. Multilobal cross-section nylons, for example, are now in large scale use.

The advantages claimed for multilobal nylon fibres include: (1) increased cover, (2) crisp, silk-like, firm handle, (3) reduced pilling in spun yarn fabrics, (4) increased bulk, (5) a sparkle or highlight effect, (6) resistance to soiling, especially in carpet yarns. The increase in surface area of a multilobal filament, on the other hand, means that more dyestuff is required to achieve a particular shade, and the wash-fastness is reduced.

*Textured and Bulked Yarns*

In common with other thermoplastic fibres, nylon may be textured and bulked by the various processes in common use. A wide variety of yarns modified in this way is now available.

*Bicomponent Fibres*

A number of bicomponent polyamide fibres are now on the market, in which two filaments of different constitution have been brought together during spinning to form a single bicomponent filament. The two components of the twin filament display different shrinkage properties on application of wet or dry heat, and the application of heat to the bicomponent fibre causes differential shrinkage which produces a crimp.

*Heterofil; Melding*

Heterofil fibres and filaments are bicomponent fibres with a core/sheath structure. Nylon heterofil fibres may be used for making non-woven fabrics (e.g. I.C.I. Fibres 'Cambrelle') by a technique called melding (a combination of melting and welding). The heterofil fibres may have a core of relatively high melting point polymer and a sheath of relatively low melting point polymer which is designed to flow on heating. Fibre webs are melded by the application of controlled heat and pressure, the fibres being bonded at cross-over points.

## (2) CHEMICAL MODIFICATION OF EXISTING NYLON TYPES

The polyamides from which nylon 6 and 6.6 are spun are chemically active materials, and the structure of these polymers may be modified by carrying out chemical reactions on them. This provides an opportunity for modification of the characteristics of existing nylon fibre types.

During the production of nylon fibres, the polymer molecules come together in places to form regions of crystallinity. Elsewhere, the long molecules remain in a more or less random arrangement, forming regions of amorphous polymer. The crystalline regions are much less readily penetrated by chemical reagents than the amorphous regions, and it is much easier, therefore, to bring about chemical modification of the polyamide in the amorphous regions. Chemical modification thus tends to have a more significant effect on those properties of the fibre which depend chiefly on the amorphous regions, e.g. dyeability and moisture absorption. The mechanical properties, such as tenacity and flexibility, which derive primarily from the crystalline regions, are less readily influenced.

### Cross-Linking

The long chain molecules of nylon may be linked together by reacting them with chemicals carrying an active group at each end of the molecule. The isocyanate group, for example, will react readily with amine or carboxylic groups, such as may be present at the ends of polyamide molecules. Reaction of polyamides with a di-isocyanate, therefore, would be expected to link up adjacent polyamide molecules.

This technique has been used successfully in attempts to modify nylon fibres with a view to improving their flat-spotting characteristics when used in tyres. Exposure of nylon 6 brought about increased initial modulus and lower extensibility, with a significant improvement in flat-spotting characteristics.

### Graft Polymerization

The grafting of polymers and other substances on to the sides of polyamide molecules in another chemical technique which can produce substantial results. Acrylic acid grafts on nylon 6.6 provide sodium salts which have an attraction for moisture. Fibre treated in this way has high wet-crease recovery properties.

Calcium salts of acrylic acid grafts tend to raise the melting point of the nylon, e.g. to 360°C. or higher depending upon the degree of grafting.

Moisture absorption of nylon fibres may be increased by the graft polymerization of ethylene oxide on to the fibre. This also improves flexibility.

### (3) NEW TYPES OF POLYAMIDE

#### Introduction

During the 40–50 years that have passed since the intensive investigation of polyamide fibres began, many thousands of polyamides have been prepared and examined as potential fibre-forming materials. Every conceivable dibasic acid, diamine, amino acid and lactam has been polymerized, and the polymer spun into fibre.

Many of these polyamides have provided fibres which excel in particular properties by comparison with nylon 6 and nylon 6.6. But none has yet been developed commercially to the extent of presenting a serious challenge to the position of the two established fibres.

The reason is primarily an economic one. None of these polyamide fibres could be produced in the necessary quantity and at a price which would be competitive with that of the established nylons. Nor does it seem likely that this position could change in the foreseeable future.

Nylon 6 and nylon 6.6 are both produced from raw materials containing 6 carbon atoms, and under present economic conditions these are materials which can be produced cheaply and in quantity from available raw materials. In addition, the production of nylon 6 and 6.6 has now been developed to a stage of high efficiency, and there is an enormous capital investment in these two fibres throughout the world. If another polyamide fibre was to compete effectively against them, it would have to offer convincing improvements in general textile characteristics, and be capable of production in comparable quantity at a cheaper price.

Despite the bleak outlook which faces challengers to nylon 6 and 6.6 in the general field, there remain great opportunities in the production of new types of polyamide for specialized applications. A number of such polyamide fibres are already in commercial production, and others are under development.

The most obvious line of approach to the development of these new polyamide fibres has been to examine the effect of using diamines, diacids, amino acids and lactams containing more or less than the 6 carbon atoms present in the monomers of nylon 6 and 6.6. As would be anticipated, the polyamides prepared in this way are basically similar in properties to nylon 6 and 6.6, but the changes in the spacing between the amide groups in the polymer chain affects certain properties in important ways.

The melting point of polyamides in a homologous series, for example, tends to be lowered as the number of methylene groups between the amide groups increases. Plotted as graphs, the melting points follow two curves, one representing an odd number of methylene groups, and the other an even number of methylene groups. The even numbers are higher than the odd numbers.

Nylon 11, for example, which has an even number of methylene groups between the amide groups, has a melting point which is only slightly lower than that of nylon 6, with 5 methylene groups separating the amide groups.

The melting point of a polyamide is increased to a significant extent by introducing phenylene rings into the molecule, especially when substitution is in the para position.

Moisture absorption is another important property which is affected by the distance between the amide groups in the polymer chain. As the number of methylene groups increases, the moisture absorption is lowered.

By adjustment of the segment of polymer chain that separates the amide groups in a polyamide it is possible to modify the properties of the polyamide in significant ways. This technique has been used in the production of polyamides which are now serving in specialized fields for which the normal nylon fibres are unsuitable. High-temperature resisting polyamides (e.g. 'Nomex') and polyamides of high dielectric strength (e.g. nylon 11) are cases in point.

## NYLON 3

Many polyamides have been made by self-condensation of the lactams of $\beta$-amino acids. Polymers of high molecular weight may be obtained, using polymerization techniques similar to those used with nylon 6. The ease of polymerization decreases with increase in the degree of substitution of the monomer.

Fibres spun from these nylon 3 polymers are highly crystalline, with melting points in the region of 300°C. Decomposition tends to occur at the high temperatures needed in melt spinning.

Nylon 3 polymers are difficult to dissolve, but solutions have been prepared using mixtures of methanol and calcium thiocyanate, and fibres have been wet-spun from these solutions.

Nylon 3 fibres are characterized by their high melting points and excellent resistance to oxidation. The affinity for dyes is lower than that of nylon 6.

## NYLON 4

Nylon 4 fibres are spun from polypyrrolidone:

$$[-(CH_2)_3CONH(CH_2)_3CONH-]_n$$

### INTRODUCTION

Polymerization of pyrrolidone to polypyrrolidone or nylon 4 has been studied by a number of firms. Fibres spun from nylon 4 polymer have characteristics which make them interesting as speciality polyamide fibres.

## PRODUCTION

2-pyrrolidone is polymerized in the presence of alkali catalyst. The polymer has a higher melting point than that of nylon 6, and it tends to decompose readily at temperatures above 265°C. Melt spinning may be carried out with difficulty.

## PROPERTIES

Nylon 4 fibres have a high melting point (273°C). Tenacity is 40.0 cN/tex (4.5 g/den). Water absorption is high; regain 8%. This is advantageous in that it reduces the tendency for build-up of static electricity and improves fabric comfort. Nylon 4 is more susceptible than nylon 6 to oxidation and is sensitive to hypochlorite bleaches. Fabrics are easy to launder and dry easily.

## NYLON 5

Nylon 5 fibres are spun from polyvalerolactam:

$$[-NH(CH_2)_4CO-]_n$$

## INTRODUCTION

Polyvalerolactam (nylon 5) fibres have been examined by a number of firms, including I.C.I., du Pont, and Tennessee Eastman. Fibres of high quality have been obtained, with properties generally similar to those of nylon 6.6. Further development of nylon 5 fibres on a commercial scale will depend primarily on the economics of monomer production.

## PRODUCTION

### Reactant Synthesis

Valerolactam may be made from cyclopentadiene, (see page 312). Cyclopentadiene is hydrogenated to cyclopentane (1), which is then oxidized to cyclopentanone (2). Conversion to the oxime (3) is followed by the Beckmann transformation which produces valerolactam (4).

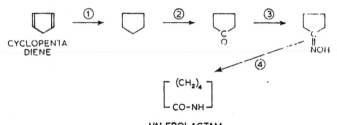

CYCLOPENTA
DIENE

$(CH_2)_4$

CO-NH

VALEROLACTAM

Production of Valerolactam

## Polymerization

Valerolactam may be polymerized in a manner similar to caprolactam (see page 264) at a temperature of about 280–290°C. The polymer produced in this way (molecular weight about 15,000–16,000) is in equilibrium with some 15 per cent of low molecular weight cyclic oligomers.

## Spinning

Polyvalerolactam may be melt spun in the same way as nylon 6.

## STRUCTURE AND PROPERTIES

The properties of nylon 5 fibres are generally similar to those of nylon 6.6.

## Tenacity

42.4–44.0 cN/tex (4.8–5.0 g/den). Wet is about 90 per cent of dry. Fibres of up to 83.9 cN/tex (9.5 g/den) may be produced.

## Elongation

20–28 per cent.

## Initial Modulus

Higher than nylon 6.6.

## Creep Characteristics

Nylon 5 fibres have low creep at elevated temperatures, and are an improvement on nylon 6.6 in this respect.

**Specific Gravity**

1.13.

**Effect of Moisture**

Regain 4.1.

**Thermal Properties**

*Melting Point*: 250–260°C.

**Chemical Properties**

Excellent resistance to most common chemicals and solvents.

# NYLON 7

Nylon 7 fibres are spun from polyheptanoamide (polyoenantha-mide or polyenanthamide):

$$[-(CH_2)_6CONH-]_n$$

## INTRODUCTION

Polyheptanoamide (nylon 7) fibres have been developed in the U.S.S.R. under the name 'Enant'. The physical properties of these fibres are generally similar to those of nylon 6 and 6.6, but there are differences in certain characteristics which could be of commercial significance.

## PRODUCTION

Polyheptanoamide is made by self-condensation of either 7-amino-heptanoic acid or its lactam.

### Reactant Synthesis

The monomer may be made by the following routes:

(a) *Telomerization*

The steps in this synthesis are shown below. Telomerization of ethylene in the presence of carbon tetrachloride is carried out as already described under nylon 11 (see page 294). One of the products is 1-chloro-7-trichloro-heptane (1).

Hydrolysis of this material by aqueous sulphuric acid (2) yields 7-chloro-heptanoic acid. Treatment of this with aqueous ammonia (3) forms 7-amino-heptanoic acid.

7-chloro-heptanoic acid may also be reacted with anhydrous ammonia (4) to form heptano-lactam.

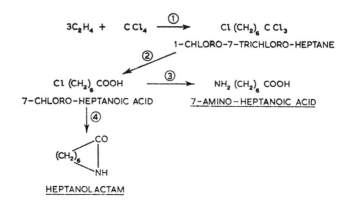

Production of Heptanolactam. Telomerization Route.

(b) *Cyclohexanone Route*

The steps in this synthesis are shown below. Cyclohexanone is oxidized to caprolactone (1), which is converted to 6-chloro-caproic acid by treatment with hydrochloric acid and zinc chloride (2).

6-chloro-caproic acid is converted to 6-cyano-caproic acid by treatment with cyanide (3), and the ester of this acid is then hydrogenated to the ester of 7-amino-heptanoic acid (4). Hydrolysis of this yields 7-amino-heptanoic acid (5).

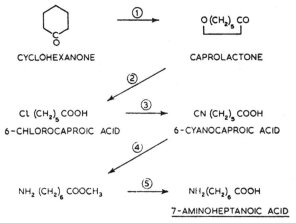

CYCLOHEXANONE      ① →      CAPROLACTONE

$O(CH_2)_5 CO$

② 

③ →

$Cl(CH_2)_5 COOH$    $CN(CH_2)_5 COOH$

6-CHLOROCAPROIC ACID      6-CYANOCAPROIC ACID

④

⑤ →

$NH_2(CH_2)_6 COOCH_3$      $NH_2(CH_2)_6 COOH$

7-AMINOHEPTANOIC ACID

Production of 7-Aminoheptanoic Acid from Cyclohexanone.

---

### Note

The telomerization process is used in the U.S.S.R., and is potentially an economic route to nylon 7. The raw materials are cheap and abundant, and the process itself presents no great difficulties. Telomerization produces a mixture of tetrachloro-alkanes, and the successful commercial development depends upon the economic use of tetrachloro-alkanes other than 1-chloro-7-trichloro-heptane.

### Polymerization

Self-condensation of the monomer is carried out in a manner similar to that used in making nylon 6, to yield a nylon 7 polymer of molecular weight as high as 30,000.

### Spinning

In the absence of oxygen, nylon 7 is thermally stable up to about 300°C. As it melts at about 225°C., melt spinning may be carried out without difficulty.

Under equilibrium conditions, nylon 7 contains only a very small proportion (about 1.5 per cent) of monomer and other low molecular weight materials. There is no necessity for extraction of either the polymer or the fibre.

## STRUCTURE AND PROPERTIES

The physical properties of nylon 7 are, in general, intermediate between those of nylon 6 and nylon 6.6.

### Tenacity
37.1 cN/tex (4.2 g/den) dry; 35.3 cN/tex (4.0 g/den) wet.

### Elongation
35 per cent.

### Initial Modulus
Higher than that of either nylon 6 or nylon 6.6 at 40–60°C.

### Specific Gravity
1.10.

### Effect of Moisture
Nylon 7 absorbs less moisture than either nylon 6 or nylon 6.6. Regain: 2.9.

### Thermal Properties
*Melting point*: 220–230°C.

*Effect of High Temperature.* Nylon 7 has a higher resistance to the effects of elevated temperatures than either nylon 6 or nylon 6.6.

### Effect of Sunlight
Better resistance than either nylon 6 or nylon 6.6.

## NYLON 7 IN USE

The higher melting point (by comparison with nylon 6) and the low moisture absorption could be important advantages for nylon 7 in certain applications. Coupled with the increased initial modulus, the low moisture absorption of nylon 7 makes for superior wash-and-wear characteristics.

316

In general, it would seem that the uses of nylon 7 will fall into line with those of nylon 6 and 6.6.

## NYLON 8

Nylon 8 fibres are spun from polycaprylamide:

$$[-(CH_2)_7CONH-]_n$$

## PRODUCTION

### Reactant Synthesis

#### (A) *Butadiene Route*

Dimerization of butadiene provides cyclooctadiene (1), which is hydrogenated to cyclooctane (2). This may be converted to capryl lactam by routes similar to those used in converting cyclohexane to caprolactam.(see page 263).

Capryl Lactam Production. Butadiene Route.

#### (B) *Acetylene Route*

Polymerization of acetylene produces cyclooctatetrene (1), which is partly hydrogenated to cyclooctene (2). This is oxidized to cyclooctene epoxide (3) and transformed to cyclooctanone (4). This is converted to the oxime (5), which undergoes the Beckmann transformation to capryl lactam (6).

317

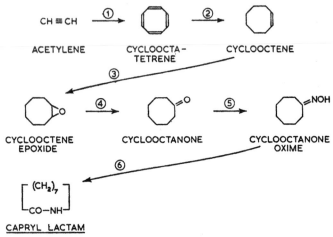

Capryl Lactam Production. Acetylene Route.

### Polymerization

Capryl lactam polymerizes readily to nylon 8 polymer, in a manner similar to the polymerization of caprolactam. The polymer contains only a very small proportion of low molecular weight material (e.g. 0.5–2.5 extractable with water). Molecular weights of 16,000 to 23,000 are readily obtained.

### Spinning

Nylon 8 polymer melts at about 200–205°C., and may be melt-spun without difficulty.

## STRUCTURE AND PROPERTIES

A typical nylon 8 polymer, of molecular weight in the region 16,000 to 23,000, has properties which are generally similar to those of nylon 6.

### Tenacity

37.1 cN/tex (4.2 g/den) dry; 36.2 cN/tex (4.1 g/den) wet.

**Elongation**

38 per cent.

**Specific Gravity**

1.09.

**Effect of Moisture**

Regain 2.9.

**Thermal Properties**

Melting Point: 200–205°C.


# NYLON 9

Nylon 9 fibres are spun from polynonanoamide:

$$[-(CH_2)_8CONH-]_n$$

These fibres have been produced commercially in the U.S.S.R. as 'Pelargon'.

## PRODUCTION

### Reactant Synthesis

9-amino-nonanoic acid is one of the products of the telomerization reaction (see page 294). In any large scale production of other amino acids by this route, 9-amino-nonanoic acid would be a readily available by-product.

### Polymerization

Self-condensation of 9-amino-nonanoic acid takes place readily, and polymers of molecular weight in the region 20,000–25,000 are made without difficulty. The polymer contains only a very small proportion of low molecular weight material at equilibrium (about 0.5–1.5 per cent extractable with water).

### Spinning

Nylon 9 polymer melts at 210–215°C., and may be melt spun as readily as nylon 6. The molten polymer is thermally stable so long as it is protected from atmospheric oxygen.

## STRUCTURE AND PROPERTIES

The characteristics of nylon 9 fibres are generally similar to those of related polyamide fibres. The water absorption is lower than that of nylon 6, nylon 6.6 or nylon 7. The following properties are typical of a polymer of molecular weight in the region 20,000–25,000.

### Tenacity

37.1 cN/tex (4.2 g/den) dry; 36.2 cN/tex (4.1 g/den) wet.

### Elongation

40 per cent.

### Specific Gravity

1.09.

### Effect of Moisture

Regain: 2.5.

### Thermal Properties

Melting point: 210–215°C.

# NYLON 12

Nylon 12 fibres are spun from polylaurylamide:

$$[-NH(CH_2)_{11}CO-]_n$$

## INTRODUCTION

These fibres have created considerable interest. Development of nylon 12 fibres has taken place in France, Germany and the U.S.A., and there are prospects of production on a commercially important scale as a speciality polyamide fibre. Nylon 12 is inherently expensive, however, and it is unlikely that it could become of importance as a general purpose polyamide fibre.

## PRODUCTION

### Reactant Synthesis

Stages in the synthesis of lauryl lactam from butadiene are shown below.

Butadiene is trimerized to 1, 5, 9-cyclododecatriene (1), which is converted to cyclododecanone (2). This is converted to the oxime (3), which undergoes the Beckmann transformation to 12-amino-dodecanoic acid (4). The lactam of this is lauryl lactam (5).

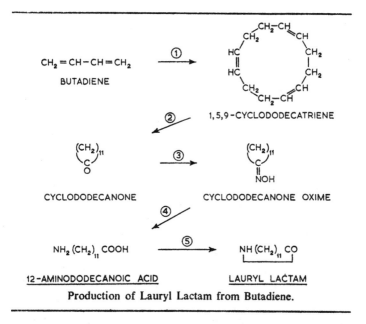

Production of Lauryl Lactam from Butadiene.

### Polymerization

Polycondensation of the monomer is carried out in the usual way, and polymers of molecular weight in the region 22,000–25,000 may be obtained without difficulty. Nylon 12 polymer contains only a very small proportion of low molecular weight material (0.75 per cent extractable with water).

**Spinning**

Nylon 12 polymer melts at 180–190°C., and is readily melt-spun into fibres.

## STRUCTURE AND PROPERTIES

Nylon 12 fibres are similar in many characteristics to nylon 6 and 6.6. The moisture absorption is low, and dielectric properties are excellent. The melting point (180–190°C.) is rather low for general textile applications.

The following properties are typical of a nylon 12 fibre of molecular weight in the region of 22,000.

**Tenacity**

33.6 cN/tex (3.8 g/den) dry or wet.

**Elongation**

40 per cent.

**Specific Gravity**

1.08.

**Effect of Moisture**

Regain: 2.2.

**Thermal Properties**

Melting point: 180–190°C.

## QIANA

This is the trade name for a fibre spun by E.I. duPont de Nemours from polymer produced by condensation of a *trans, trans*–di(4–aminocyclohexyl) methane with a dibasic acid containing 8–14 carbon atoms, e.g. dodecanedioic acid.

## STRUCTURE AND PROPERTIES

Trilobal cross-section. Tenacity 26.5–31.0 cN/tex (3.0–3.5 g/den). Elongation 20–30%. Good recovery from stretching. Sp. gr: 1.03. Moisture regain: 2.0–2.5. Melting point 275°C. Chemical stability excellent. Dyes well to fast shades.

## QIANA IN USE

A lustrous silky fibre with many characteristics similar to those of nylon 6 and 6.6. Fabrics made from Qiana have excellent drape and form retention, and the fibre has found a ready market notably in the field of high fashion.

## BICONSTITUENT NYLON—POLYESTER

Fibres consisting of microfilaments of polyester dispersed in a nylon 6 matrix possess unusual optical and dyeing characteristics. They have a higher tenacity than nylon 6.6 and lower regain.

## BICOMPONENT NYLON-POLYESTER

Fibres consisting of a polyester core sheathed in nylon 6.

## AROMATIC POLYAMIDES

It has long been known that the melting point of a polyamide may be raised by introducing aromatic rings into the polymer molecule. These may be incorporated into the diamine, diacid or both.

Many aromatic polyamides have been made experimentally, and some are now of commercial importance as special purpose (commonly high-temperature resistant) polyamide fibres.

The effect of introducing aromatic rings is particularly marked when phenylene groups are incorporated in the molecule through the para positions. Meta substitution provides polymers of lower melting point, intermediate between that of the para substituted polymers and those of normal straight-chain constitution.

### Polymetaxylylene Adipamide (Nylon MXD–6)

Condensation of meta xylylene diamine and adipic acid produces polymetaxylylene adipamide, which is commonly known by the name nylon MXD–6. Its structure (1) is shown on page 324.

Nylon MXD–6 has been examined by a number of firms, notably by Celanese Corporation, and it has a number of attractive features. In particular, its flat-spotting characteristics are a great improvement on those of the normal nylons, and it has given excellent results as a tyre cord.

Nylon MXD–6 is very susceptible, however, to the effects of heat and moisture, and its value as a textile fibre is restricted. This weakness is presumably due to the presence of methylene groups between the aromatic rings and the amide groups. These permit free rotation of the aromatic ring with respect to the amide group. The polarity of the amide group is low.

### Polyhexamethylene Terephthalamide (Nylon 6T)

Condensation of hexamethylene diamine with terephthalic acid produces polyhexamethylene terephthalamide, or nylon 6T. Its structure (2) is shown on page 325).

This structure does not possess the weakness inherent in nylon MXD–6, as the amide group is linked directly to the aromatic ring. The molecule is stiff, and the amide groups are highly polar. Nylon 6T has been studied by a number of firms, including Celanese Corporation, and it has a range of interesting properties. The melting point (370°C.) is too high to permit of effective melt spinning, and fibres are spun from solutions of the polyamide in concentrated sulphuric acid. Also, special polymerization techniques are necessary.

The structure of nylon 6T resembles a combination of the structures of nylon 6.6 and a polyethylene terephthalate fibre such as 'Terylene' or 'Dacron'. As would be anticipated, the fibres spun from nylon 6T combine many of the properties of nylon 6.6 and the polyester fibre. They have the low density, moderate moisture regain, high abrasion resistance, easy dyeing, high alkali resistance and excellent elastic properties associated with nylon, and the high initial modulus, especially at elevated temperatures, of the polyester fibres. Nylon 6T has an outstanding resistance to stretch, and a very high recovery at high temperatures. The flat-spotting characteristics of the fibre are similar to those of polyester fibres. In addition, nylon 6T has good heat, light and chemical resistance, good dye fastness and resistance to the effects of moisture that are characteristic of polyester fibres. Nylon 6T retains its full strength after 5 hours at 185°C. It discolours after 1 hour at 220°C.

## Fully-Aromatic Polyamides; Aramids

The maximum effects of introducing aromatic rings into the polyamide molecule are obtained by condensing monomers in which, in each case, the functional groups are separated by phenylene groups. Aromatic diamines, for example, condensed with terephthalic acid provide polyamides with exceptional resistance to high temperatures. The intermolecular bonding and chain stiffness are such as to confer high thermal stability on the polymer molecules.

When all the phenylene units in the polyamide are para-substituted, the optimum effects are obtained, and the polymers have melting points or decomposition points in the region of 555°C. With all phenylene units in the meta-substituted position, the polymers melt or decompose at about 410°C.

These fully aromatic polyamides may be prepared with the para- and meta-substituted units in any desired proportions, and the properties of the polyamides may be selected accordingly.

The polyamide shown as formula 3 below, for example, is the basis of a fibre developed by Chemstrand Corporation.

Fibres formed from polyamides in which at least 85% of the amide linkages are attached to aromatic rings are known as *aramids* (F.T.C. Definition). Examples are 'Nomex' and 'Kevlar' produced by E.I. duPont de Nemours & Co Inc.

① NYLON MXD-6

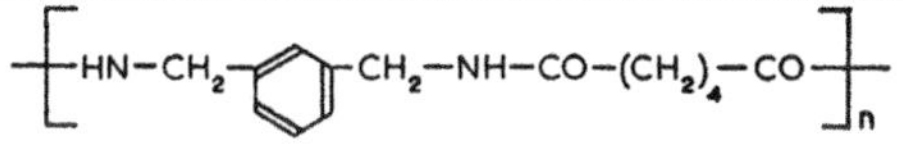

② NYLON 6T

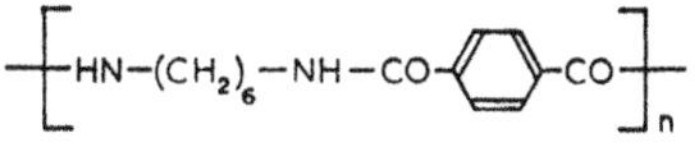

③ Aromatic Polyamides.

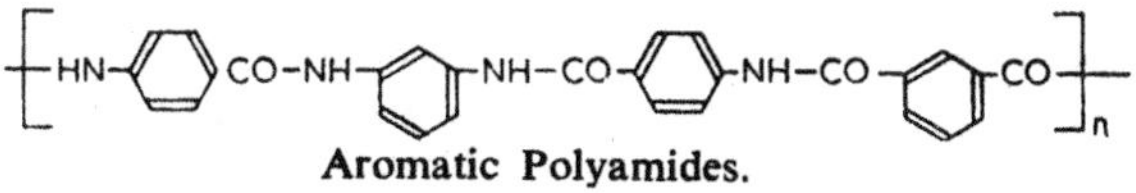

## PRODUCTION

Aramids are produced by reaction of aromatic diacid chlorides with aromatic diamines in a solvent such as N,N-dimethylformamide (DMF). Polymers are dry spun from solvent solution into a hot air stream or wet spun into a coagulating bath, followed by stretching.

## STRUCTURE AND PROPERTIES

Aramid fibres are commonly round or dog-bone in cross-section. They are pale yellow to cream before bleaching. Tenacities are high, e.g. 38–190 cN/tex (4.2–21.5 g/den) dry and 28–159 cN/tex (3.2–18 g/den) wet. Elongation decreases with increasing fibre tenacity , from 3–30% dry, 4–27% wet. Recovery from low stretch is almost 100%. Aramid fibres are stiff; resiliency and recovery from bending are excellent, and abrasion resistance is high. Sp. gr: 1.38–1.45. Regain 3.5–7.0. Resistance to chemicals is high, and aramids are unaffected by most household chemicals. They are attacked by concentrated acids and oxidising agents at high temperatures. Heat resistance is high, degradation occurring on prolonged heating in air at temperatures above 370°C. The fibres have very low flammability and are self-extinguishing. Sunlight causes some discoloration and a slight loss of tenacity . Aramid fabrics provide effective screens against high-energy nuclear radiation. A good range of fast colours may be obtained by using disperse dyes.

## ARAMID FIBRES IN USE

Fabrics made from aramid fibres such as 'Nomex' and 'Kevlar' are lustrous and attractive, with good draping characteristics and handle. Shape retention and wear resistance are excellent, but the high strength of the fibres can cause pilling. Like other nylon fabrics, aramid fabrics wash easily and drip-dry quickly. Dry cleaning does not present problems. Ironing may be carried out safely  at temperatures up to 300°C.

Applications for aramid fibres are commonly those which take advantage of the high strength, low flammability and excellent heat resistance, e.g., specialised astronautical and military uses, industrial and protective clothing, heat and electrical insulation materials, filtration cloths and tyre cords.

# B: SYNTHETIC FIBRES

## 2. POLYESTER FIBRES

### INTRODUCTION

Polyesters are polymers made by a condensation reaction taking place between small molecules, in which the linkage of the molecules occurs through the formation of ester groups.

Polyesters are commonly made by interaction of a dibasic acid with a dihydric alcohol:

$$HOOC—X—COOH + HO—Y—OH$$
$$\rightarrow - - - OC—X—COO—Y—OCO—X—COO—Y—OCO - -$$

The formation of polyesters was studied by Wallace H. Carothers of du Pont during the investigation of polymers which led eventually to the discovery of nylon. Development of the polyesters was overshadowed, however, by the polyamide research, and it was not until 1941 that a valuable polyester fibre was discovered. In that year, J. T. Dickson and J. R. Whinfield of the Calico Printers' Association in England made a synthetic fibre from polyethylene terephthalate by condensing ethylene glycol with terephthalic acid (see page 332).

After the war, development of the fibre was carried out under licence by I.C.I. Ltd. in the U.K. and by du Pont in the U.S.A., resulting in the fibres known respectively as 'Terylene' and 'Dacron'.

Today, polyethylene terephthalate fibres are being made in many countries, and modified forms of this fibre are also produced. Other polyesters have been produced and spun into fibres, some of which have become of commercial importance (cf. 'Kodel').

### TYPES OF POLYESTER FIBRE

In the years since World War II, polyethylene terephthalate fibres of the 'Terylene' and 'Dacron' type have established a dominating position in the polyester fibre field. Other types of polyester have, however, been spun into fibres with varying degrees of practical success, and a few of these have become of commercial importance.

The position has now been reached where it is preferable to consider polyester fibres as specific types, based upon their chemical structures. The differences between them are too significant to permit of their being considered simply as 'polyester' fibres.

For the purposes of the Handbook, polyester fibres are subdivided into the following types, based upon their chemical structures, and the abbreviations shown are used in referring to the fibres:

(1) Polyethylene Terephthalate Fibres (PET Polyester Fibres).

(2) Poly-1, 4-Cyclohexylene-Dimethylene Terephthalate Fibres (PCDT Polyester Fibres).

(3) Other Types of Polyester Fibre.

## NOMENCLATURE

The first polyester fibres to be introduced on a commercial scale (i.e. 'Dacron' and 'Terylene') were spun from polyethylene terephthalate. This chemical term was, of course, too complex for everyday use, and the fibres became known simply as 'polyester' fibres. They are still known generally by this term today.

The term 'polyester' is, however, a specific chemical name which refers to any polymer in which the linkage of small molecules takes place through the formation of ester groups. It refers equally well, for example, to polyethylene adipamide made by condensing ethylene glycol with adipic acid, or to polypropylene terephthalate made by condensing propylene glycol with terephthalic acid.

In the selection of 'official' names for synthetic fibres of different types, the U.S. Federal Trade Commission included the term 'polyester' and specified that this must be used for fibres of the polyethylene terephthalate type (see below). This was an unfortunate choice, as this definition of 'polyester' is at variance with the chemical meaning of the term. Confusion could have been avoided if a completely new, non-chemical term (cf. nylon) had been coined instead of using a term with a well-defined chemical meaning.

Under the present F.T.C. regulations, it is possible for a

polyester fibre to be marketed which does not meet the F.T.C. definition, and could not, therefore, be called a polyester fibre!

## Federal Trade Commission Definition

The generic term *polyester* was adopted by the U.S. Federal Trade Commission for fibres which satisfy the following official definition:

*Polyester.* A manufactured fibre in which the fibre-forming substance is any long-chain synthetic polymer composed of at least 85 per cent by weight of an ester of a substituted aromatic carboxylic acid, including but not restricted to substituted terephthalate units $p(-R-O-CO-C_6H_4-CO-O-)$ and para-substituted hydroxybenzoate units $p(-R-O-C_6H_4-CO-O-)$.

## (1) POLYETHYLENE TEREPHTHALATE FIBRES (PET POLYESTER FIBRES)

Fibres spun from polyethylene terephthalate:

$$---O(CH_2)_2OCO\langle\bigcirc\rangle COO(CH_2)_2OCO\langle\bigcirc\rangle COO(CH_2)_2O---$$

**POLYETHYLENE TEREPHTHALATE**

## INTRODUCTION

The discovery that potentially-valuable textile fibres could be spun from polyethylene terephthalate (PET) (see page 334) was made by Dr. J. T. Dickson and Mr. J. R. Whinfield in 1941, in the laboratories of the Calico Printers' Association Ltd. in Lancashire, England. Development of the PET fibre was held up during the war, but in 1947 the world rights to manufacture the new fibre, with the exception of those for the U.S.A., were purchased by Imperial Chemical Industries Ltd. (E.I. du Pont de Nemours and Co. Inc. bought the rights for the U.S.A.).

In Britain, the manufacture of PET fibre began on a pilot plant scale in 1948, the fibre being marketed under the name 'Terylene'. Since then, production of 'Terylene' has expanded rapidly. At Wilton in Yorkshire, I.C.I. Ltd. have spent millions of pounds in the construction of 'Terylene' plants. In January 1955, a large plant came into production, with an annual capacity of 5 million kg divided almost equally between filament yarn and staple.

In 1956, a second unit of similar capacity began production, and this was followed by a 'Terylene' plant in Northern Ireland.


In the U.S.A., the du Pont company began producing PET fibre on an experimental basis in 1950. The fibre was known in its early days as 'Fiber V', but was subsequently given the trade name 'Dacron'.

PET polyester fibres are now being produced in many countries.


## TYPES AND SIZES

PET polyester fibres are produced as multifilament yarns, monofilaments, staple fibre and tow, in a wide range of counts and staple lengths to suit virtually all textile requirements.

The fibres are available in bright, semi-dull and dull lustres. The properties of the fibres may be modified over a range which is limited by the inherent characteristics of the polymer, each manufacturer controlling his process to produce fibres that will meet specific requirements. In general, commercial PET polyester fibres fall into two main classes, (a) regular tenacity and (b) high tenacity.

PET polyester fibres are produced commonly in round cross-section, but fibres of special (e.g. triangular) cross-section are now available from a number of manufacturers.

PET polyester fibres are thermoplastic, and lend themselves well to physical modifications associated with this property. Crimped and textured yarns of all familiar types are available.


## PRODUCTION

Polyethylene terephthalate is made by the condensation of

terephthalic acid, or a derivative such as dimethyl terephthalate, with ethylene glycol.

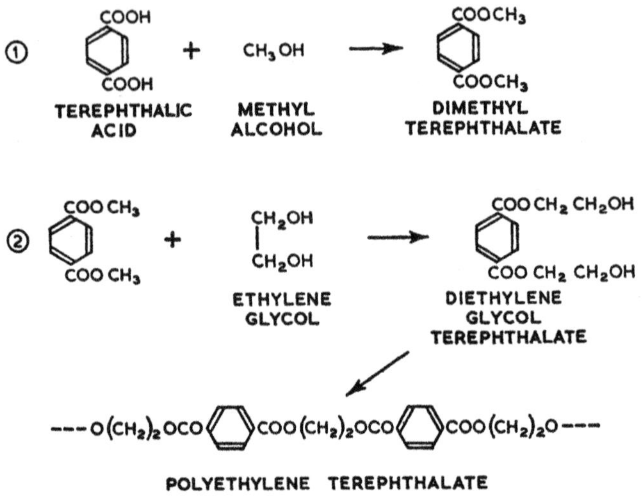

---

POLYETHYLENE TEREPHTHALATE

Production of polyethylene terephthalate

---

**Reactant Synthesis** (see diagram page 333).

### (a) *Ethylene Glycol*

This is made by the catalytic oxidation of ethylene, which is obtained from petroleum cracking. Ethylene oxide is produced (1). Hydration of this yields ethylene glycol (2).

### (b) *Terephthalic Acid; Dimethyl Terephthalate*

*Para*-xylene obtained from petroleum is oxidized (3), for example with nitric acid or with air in the presence of a catalyst.

Terephthalic acid is esterified with methyl alcohol (4) to form dimethyl terephthalate.

(a)

CH₂=CH₂  →①  ethylene oxide  →②  CH₂OH / CH₂OH

$$CH_2{=}CH_2 \xrightarrow{①} \underset{CH_2}{\overset{CH_2}{\diagdown}} O \xrightarrow{②} \begin{array}{l} CH_2OH \\ | \\ CH_2OH \end{array}$$

ETHYLENE    ETHYLENE OXIDE    ETHYLENE 'GLYCOL

(b)

p. XYLENE    TEREPHTHALIC ACID    DIMETHYL TEREPHTHALATE

PET Polyester Fibres. Production of monomers

---

### Polymerization (see diagram page 334).

Polyethylene terephthalate is made by condensing ethylene glycol with either terephthalic acid itself or with dimethyl terephthalate.

Condensation of ethylene glycol with terephthalic acid (1) is an esterification reaction, water being eliminated as the reaction takes place. Condensation of ethylene glycol with dimethyl terephthalate (2) is an ester interchange reaction, methyl alcohol being eliminated as the reaction takes place. The polymer obtained in this way would be expected to have an ester end group instead of the carboxylic acid end group in the case of the polymer obtained by the terephthalic acid route.

In either case, the condensation is carried out by heating the ethylene glycol and terephthalic acid or dimethyl terephthalate and removing the water or methyl alcohol *in vacuo*. When the desired degree of polymerization has been reached, the clear, colourless polyester is extruded through a slot on to a casting wheel. The polymer solidifies into an endless ribbon, which is fed to a cutter and cut into chips in the form of cubes with 3–6 mm (1/8 – 1/4 in) sides.

The chips are despatched to the spinning room via a suction pipe.

Production of polyethylene terephthalate

---

### Spinning

Polyethylene terephthalate melts at about 260°C., and the molten polymer is stable so long as oxygen is rigorously excluded. Every care is taken during melt spinning, as in the polymerization process, to prevent air coming into contact with the molten polymer.

In the spinning building, the chips of polymer are dried to remove traces of moisture, and then passed to storage hoppers. From the hoppers the chips are fed as required to the spinning machines.

Spinning is carried out in a manner similar to that used for polyamide fibres, the molten polymer being pumped through holes in a spinneret. As the filaments emerge, they solidify and are wound into packages of undrawn yarn.

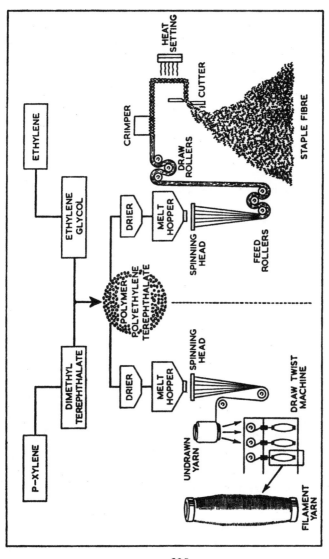

*PET Polyester Fibre Flow Chart*

The undrawn yarn is stretched to about five times its original length on draw-twist machines, the stretching being carried out usually at elevated temperature. If high tenacity yarn is being made, the filaments are drawn to a higher degree than in the manufacture of regular tenacity yarn.

It is normal practice for PET polyester yarns to be drawn hot, as this gives a more uniform product than cold drawing. The stretching of heavy denier yarns and monofilaments may, however, be carried out at room temperature, as the poor heat conductivity of the fibre makes for irregularities in thick filaments which are drawn hot.

### Staple Fibre

Staple fibre is produced by spinning a great number of filaments and bringing them together to form a heavy tow. This is drawn and then crimped mechanically, and the crimp is set in the fibre by heat treatment. The tow is then cut into staple of the desired length.

### PROCESSING

The basic finishing processes for 100 per cent PET polyester filament yarn fabrics may be arranged in the following three sequences:

(1) Scour – heat-set – dye
(2) Heat-set – scour – dye
(3) Scour – dye – heat-set

Loom stains and other forms of contamination are difficult to remove from cloth which has been heat-set, and it is preferable to scour before heat-setting. Also, goods which have not been heat-set before dyeing will tend to shrink during dyeing, and the dyed goods will be subjected to high-temperatures after dyeing. For these reasons, sequence (1) is the most generally useful.

Sequence (2) eliminates a drying process and is suitable for fabrics which are perfectly clean in the loomstate. It is, however, rarely used for PET polyester fabrics except in the case of curtain nets.

If sequence (3) is used, some stiffening of the fabric is likely to occur following heat-setting. The degree of stiffening depends on the construction of the fabric and on the tensions which are developed during setting. In other respects, the sequence is attractive, and it is limited only by the fastness of the available dyestuffs to sublimation under heat-setting conditions.

### Scouring

PET polyester fibres are supplied in a high state of cleanliness, and it is generally unnecessary to scour the material prior to dyeing. If, however, a batch of fibre should become so dirty as to require scouring, the bath may be set with

| | |
|---|---|
| Water | 1,000 parts |
| Soda ash | 2 parts |
| Detergent, such as textile soap, or Lissapol C, D, NC or ND | 1 part |

The temperature is raised to 70°C. for 15–30 minutes and, after scouring, the goods should be rinsed thoroughly to eliminate all traces of alkali. A small quantity of acetic acid may be added to the final rinse.

PET polyester goods will often acquire dirt and stains during manufacture, including after-waxing agents applied to sized warp yarns, loom stains and other forms of soiling. If goods are heat-set before removal of these stains, subsequent cleaning of the goods will become difficult and perhaps impossible. It is preferable, therefore to scour goods before heat-setting. A bath of the following composition is commonly used:

| | |
|---|---|
| Water | 1,000 parts |
| Soda ash | 2–3 parts |
| or caustic soda (flake) | 0.5 parts |
| Detergent, such as textile soap, Lissapol C, D, NC or ND | 1–2 parts |

Fabrics of relatively open structure, such as voiles, marquisettes and lenos may be scoured in a shallow winch at temperatures below 60°C.

### Bleaching

The natural white colour of PET polyester fibres is usually satisfactory for most purposes, and bleaching is unnecessary. When fabrics are to be finished white a slightly improved colour may be obtained by bleaching in a bath set with 2–3 parts of sodium chlorite per 1,000 parts of water at pH 3–4. Other common bleaching agents have little or no effect on the colour of PET polyester fibres.

The best results are obtained by making use of fluorescent brightening agents. Those used with PET polyester fibres commonly have the following characteristics:

(a) A relatively high dyeing temperature is required. One result of this is that they are virtually non-effective as additives in domestic washing powders.

(b) High wet fastness. This ensures little loss under normal washing conditions, obviating the need for restoration of the white during washing.

(c) Lightfastness is high enough for all apparel uses, and in most cases is suitable also for furnishings and curtains.

Fluorescent brightening agents are applied by the methods used in dyeing with disperse dyes, i.e.

(a) Application at the boil, without carrier.

(b) Application at or near the boil, with carrier.

(c) High temperature application (130°C.), without carrier.

(d) Application by pad-thermofix techniques.

The method selected will depend upon the type of machinery available, and the structure of the fabric. Most fabrics may be handled on a beam-dyeing machine at or above 100°C., while a jig may be used for the more stable woven fabrics.

Rope processing in the winch is often used, especially for warp-knitted fabrics. There is some slight risk of rope marking, but this can usually be removed during subsequent stentering.

The selection of a particular fluorescent brightener depends upon the fluorescent tone and fastness required, and on the method of application. The amounts of brightener and carrier are determined by the strength and effectiveness of the products, and the manufacturers' instruction should be followed.

Most fluorescent brightening agents suitable for PET polyester fibres may be applied from a sodium chlorite bleach liquor, so that a one-stage chemical and optical bleach is possible. Carriers of the ortho-phenyl-phenol class must not be used. however. either during or after chlorite bleaching.

Techniques recommended for obtaining a good white finish on PET polyester and blends are outlined below.

## 100 per cent PET Polyester and PET Polyester/Nylon Blends

### Warp-knitted Fabrics

An excellent white is obtained by treating the heat-set fabric with acidified sodium chlorite, followed by the application of a fluorescent brightening agent at 130°C., or 100°C. with a compatible carrier, on a beam-dyeing machine.

With most brightening agents, a good white can be obtained by a single-bath method in which the fluorescent brightening agent and compatible carrier are included in the sodium chlorite liquor.

### Weft-knitted Fabrics

Weft-knitted fabrics may be handled on the winch, using similar techniques.

### Woven Fabrics

The beam-dyeing machine is recommended for bleaching woven fabrics. Alternatively, the jig may be used, the concentration of chlorite bleaching chemicals being increased by some 20 per cent to compensate for the shorter liquor.

### Pad-Thermofix Application of Fluorescent Brightening Agent

Padding and baking the fluorescent brightening agent, without chemical bleaching, is often employed, especially for warp-knitted or woven curtain nets. The process is cheaper than those described above, but the results are usually inferior.

The baking stage may be combined with heat-setting, and also with chemical finishing where this is carried out, providing a very economical finishing procedure. The padding liquor contains sufficient brightening agent, based on the expression of the mangle, to give the required amount on the fabric. This varies with the product and the effect required, and the manufacturers' instructions should be followed. No carrier is needed, but a wetting agent may be incorporated if desired.

After padding and drying, the fabric is heated on a stenter at 220°C. for 10–30 seconds depending on the characteristics of the stenter. Excessive exposure to high temperature should be avoided in order to avoid impairing the brightening effect.

### PET Polyester/Cellulosic Fibre Blends

Chemical bleaching should be employed, especially when the cellulosic component is cotton. For best results, fluorescent brightening agents should also be applied for both components of the blend.

Several methods of chemical bleaching may be used, the choice depending on the result required and the equipment available. Acidified sodium chlorite provides a most effective single-stage bleach, and a high white is obtained from a chlorite bleach followed by a peroxide bleach.

A good result is also obtained from a hypochlorite bleach followed by alkaline hydrogen peroxide. Hypochlorite or peroxide used alone as a single stage process will give a white suitable for many purposes, e.g. as ground for subsequent dyeing. Processes excluding chlorite are generally more suitable for colour-woven goods.

Depending on the weight and construction of the fabric, it may be possible to heat-set and then process in rope form at temperatures up to 100°C. without undue risk of rope marking. Lighter weight fabrics, and those of tight construction generally, must be handled in open width on the beam machine or jig.

PET polyester/flax blends may be bleached according to these general principles, but more intense treatment is required.

### PET Polyester/Wool Blends

The PET polyester component cannot be bleached with chlorine compounds in the presence of the wool component. Chemical bleaching of PET polyester/wool blends is virtually restricted, therefore, to the bleaching of the wool. Hydrogen peroxide is the preferred agent, and bleaching is carried out after a fluorescent brightening agent has been applied to the PET polyester component. This corrects any yellowing of the wool which may have occurred during the application of the agent to the PET polyester fibre.

#### Pre-Setting

Unset PET polyester filament yarns will shrink when allowed

to relax in boiling water, commonly by some 7 per cent. At 130°C., shrinkage is of the order of 10 per cent.

If unset yarns in package form are subjected to any process, such as dyeing, involving elevated temperature, the shrinkage will bring about consolidation of the package. In dyeing, this restricts the even flow of dye liquor and causes unlevel dyeing.

Before dyeing, PET polyester filament yarns are wound on to collapsible paper tubes and relaxed for 20 minutes in saturated steam, preferably at a temperature 5°C. higher than the maximum temperature which will be reached in the dyeing.

Precautions should always be taken to ensure that conditions in the steaming chamber are as uniform as possible. Variation in the steam setting conditions can cause variation in dyeing properties, and it is preferable to steam the whole of each dyeing batch at the same time.

### Heat-Setting

The purpose of heat-setting is to stabilize the fabric to the effects of any heat treatment which it may receive in subsequent finishing processes, in making-up, or in use. The selection of heat-setting conditions is controlled, therefore, by the intended end-use of the fabric and by the thermal history of the yarns from which it is constructed.

For PET polyester fabrics, the heat setting temperature should be higher than the temperatures which are to be used in pleating, embossing or calendering, so as to eliminate the possibility of uncontrolled shrinkage in these processes. It should be emphasized that the ability of PET polyester fibre to accept a permanent set is not influenced by the temperature of prior heat-setting. In this respect, PET polyester fibres differ from some other synthetic fibres.

### *Steam Setting*

Pressure steaming is not recommended for the setting of PET polyester fabrics, because of variations between the inside and the outside of the batch in the degrees of shrinkage and restraint which are produced.

### *Hot-Air Setting*

A hot-air stenter is commonly used in heat-setting PET polyester filament fabrics. The pin stenter is usually preferred for the setting of woven fabrics, but for fabrics in which a pin-marked selvedge is unacceptable, the clip stenter may be used in association with cylinder setting (see below).

There are difficulties associated with the control of warp shrinkage in clip stenters. The clips tend to mark the selvedges of the cloth under the tensions which are developed in heat-setting, and tight selvedges may be torn from the body of the fabric. Local cooling at the selvedges is also more severe in the clip than in the pin stenter, and may produce marked unlevelness in subsequent dyeing.

The high dry-heat shrinkage of PET polyester filament fabrics necessitates accurate control in stentering in order to avoid fabric distortion. The selvedges should be pinned accurately, without excessive overfeed.

## Cylinder Setting

Cylinder or cylinder-and-blanket setting machines are used principally in the stabilization of heavy industrial PET polyester fabrics, where stenters could not withstand the high tensions developed in setting. This method is also used with fabrics such as sail-cloths which require consolidation during finishing. In conjunction with stentering, it is also used to produce the exceptionally uniform finish which is required for PET polyester base fabrics which are to receive a plastic coating.

## Heat-Setting Conditions

PET polyester filament fabrics are commonly set by exposure for 10–30 seconds in the hot zone of the setting stenter. The extremes of this range correspond with lightweight fabrics of less than 68 $g/m^2$ (2 oz./sq.yd.$^2$) and with heavy fabrics of up to 203 $g/m^2$ (6 oz./yd.$^2$). The following table suggests setting temperatures and shrinkage allowances suitable for the flat finishing of a wide range of PET polyester filament fabrics:

| Fabric Class | Method | Setting Temp. (°C.) | Shrinkage Allowance (%) | |
|---|---|---|---|---|
| | | | Warp | Weft |
| 1 | Pin stenter | 210–220 | 3–5 | 3–5 |
| 2 | Pin stenter | 200–210 | 3–5 | 3–5 |

| Fabric Class | Method | Setting Temp. (°C.) | Shrinkage Allowance (%) Warp | Weft |
|---|---|---|---|---|
| 3a | Pin stenter | 150–160 | 0 | 0 |
| 3b | Pin stenter | 150–200 | 0–5 | 0–5 |
| 4 | Clip stenter plus | 150–160 | 0 | 0 |
| | cylinder | 200 | Free | Free |
| 5 | Cylinder | 210 | 10 approx. | Free (15 approx.) |

*Class 1* includes leno curtain fabrics and similar very open structures.

*Class 2* includes the majority of PET polyester filament woven fabrics (taffetas, twills, satins, etc.).

*Class 3* includes colour-woven fabrics. The majority of fabrics made from high-temperature or carrier dyed yarns are stable to making-up and to mild laundering conditions. Fabric properties such as handle and crease recovery are, however, generally improved by the mild setting treatment suggested for Class 3a.
Fabrics made from yarns dyed by other methods or which contain unset white yarns fall into Class 3b, and should be heat-set at the highest temperature permitted by the sublimation of the dyestuffs used.

*Class 4* includes fabrics for which completely free shrinkage is not permissible, but which must be finished free from pin or clip marks.

*Class 5* includes fabrics in which shrinkage is desirable, such as sailcloths, or which are too heavy for stenter processing.

## Additional Finishing of Heat-Set Fabrics

Harshness or 'paperiness' produced by heat-setting may be removed by subsequent wet processes, such as jig-scouring, which give some mechanical action. The rate of softening is accelerated by the presence of caustic soda in the scouring bath. Fabrics which have become excessively harsh should be processed at the boil on an enclosed jig in a bath set with 5 parts of caustic soda (flake) per 100 parts of water. The process produces a loss in fabric weight of the order of 0.5 per cent. The effect produced is quite different from that of the conventional 'caustic soda softening' process in which 6–10 per cent of the fabric weight is lost.

### Dyeing

PET polyester fibres are hydrophobic, and dyeings of useful depth are obtained by using those classes of dyestuffs which are substantially insoluble in water. These include the disperse dyestuffs, azoic dyestuffs (applied by a modified technique), and a limited number of vat dyes.

Disperse dyestuffs provide build-up and colour fastness that is adequate for most purposes, and a wide range of colours is available. Azoic dyestuffs produce a range of bright reds and maroons, a black and a navy blue, all of which have good fastness to severe washing. The usefulness of the azoic reds and maroons is limited, however, by their tendency to be dulled and to lose rubbing fastness after steam-setting or heat-setting processes.

Vat dyestuffs yield a limited range of shades, one or two of which are very bright, but they build up to only medium depths.

PET polyester fibres have a high affinity for disperse dyestuffs, but the rate of diffusion of these dyestuffs in the fibre is relatively low. The rate of dyeing may be raised to a commercially acceptable level either by working at the boil in the presence of an accelerating agent or carrier, or by dyeing under superatmospheric pressure at temperatures in the region of 130°C. The high temperature dyeing method is preferable and should be used where possible, in order to take advantage of the excellent levelling action obtainable under these conditions. This technique also avoids the adverse effect of certain carriers on the light fastness of some dyestuffs.

The full range of available disperse dyes can be applied at high temperatures, whereas some of the newer disperse dyes do not build up well when dyed at the boil in the presence of a carrier.

Disperse and selected vat dyestuffs may also be applied to piece goods by a pad-bake technique, using dry heat at temperatures in the range 190–215°C.

## (1) *100 per cent PET Polyester Goods*

### Loose Stock and Slubbing

PET polyester loose stock and slubbing are commonly dyed by the high-temperature process, using disperse or azoic colours. The carrier process may also be used, but it may then be necessary to repack the materials during dyeing in order to obtain a level result.

In handling slubbing, great care should be taken to avoid the insertion of 'real' twist, as this may become 'set-in' during dyeing and lead to difficulties in subsequent drafting processes.

### Staple Fibre Yarns

Most 100 per cent PET polyester staple fibre yarns may be wound directly on to cheeses or cones for dyeing. Some yarns, however, especially those spun from stretch-broken tows (e.g. yarns spun on the Schappe or Stains systems), have a high potential shrinkage, and these should be stabilized by pressure steaming prior to winding the dyeing package.

### Filament Yarns

Twist must be inserted in PET polyester filament yarns to obtain a dyeing package of sufficient permeability. The minimum twist levels vary with yarn deniers, but as a guide it is recommended that 167 dtex (150 den) be thrown to at least 235 turns/m (6 turns/in). The thrown yarns should be wound on collapsible tubes and relaxed in steam. This eliminates the potential shrinkage, which would otherwise make the package consolidate in dyeing. The relaxed yarns are rewound into suitable packages for dyeing. When undyed yarns are to be used in the same fabric as dyed yarns, it is essential to stabilize the undyed yarns to prevent puckered effects during cloth finishing.

### Bulked Yarns

Special handling techniques must be used in dyeing bulked PET polyester filament yarns. In general, carrier dyeing produces a rather inferior handle to that obtained in high-temperature dyeing at 120°C. Dyeing at temperatures higher that 120°C. tends to cause loss in bulk.

*Piece Goods*

Woven PET polyester fabrics are handled usually on jigs or in beam-dyeing machines Some lightweight fabrics of open structure, however, such as lenos and marquisettes, may be handled on winches. Knitted fabrics, laces and the heavy type of staple fibre goods may also be dyed on the winch machine as well as on the beam. Hose, half-hose and knitted garment lengths are dyed in paddle machines.

High-temperature equipment is available for both jig and beam dyeing, but the beam machine is preferred for all but the most closely woven fabrics, because of its superior levelling action. Minor changes in sett can make even very closely woven fabrics suitable for beam dyeing without affecting their suitability for raincoats, overalls and similar end-uses. Designers should therefore bear in mind the possibility of beam dyeing when designing new fabrics.

## (2) *Blended Goods*

### *PET Polyester/Wool Blends*

The PET polyester component of these blends is dyed with disperse dyes, and the wool component with selected milling-acid, chrome or neutral-dyeing premetallized colours, The presence of the wool restricts the dyeing temperature to 105°C., and the blend must be dyed, therefore, by the carrier process. Piece goods are usually winch dyed at 95–98°C.

Most disperse dyestuffs will stain the wool component of the blend, but by careful selection of the dyestuffs, carrier and dispersing agents it is possible to dye up to medium shades by a one-bath process. A two-bath process is recommended for deep shades, so that the wool can be 'cleared' after the PET polyester has been dyed. The best combination of fastness properties on the two components is obtained by dyeing the fibres separately before blending.

A large proportion of PET polyester/wool goods are used in 'wash-and-wear' end-uses, and dyes for the wool component should be selected carefully to give the best combination of colour fastness to light and washing. The wet fastness of the wool component is particularly important in dyed yarns, which must withstand the conditions encountered in worsted or woollen finishing routines. Finishers should ensure that yarn dyed goods spend a minimum of time piled in the wet state; such fabrics should not be allowed to stand wet overnight.

PET polyester fibre has a much greater abrasion resistance than wool, and with cross-dyed styles there is a danger that changes in shade may occur during wear. Small differences in shade or in tone between the two components are, however, tolerable. Where the style demands the production of a cross-dyed effect, it is preferable that the PET polyester component should carry the darker shade; a deepening in shade is usually less noticeable than a change in the opposite direction.

### PET Polyester/Silk Blends

The problems encountered in dyeing these blends are similar to those of PET polyester/wool blends. Disperse dyes will usually stain silk appreciably, but the stain may be cleared and the silk dyed in a second bath. The silk may be dyed to fast, bright colours with reactive dyes, as well as with the dyes which are traditionally associated with this fibre.

### PET Polyester/Cellulosic Fibre Blends

Blends of PET polyester fibres with cellulosic fibres such as cotton, flax and viscose rayon may be dyed by using disperse dyes on the PET polyester component, and vat, Procion or direct cotton dyes on the cellulose.

The use of vat dyes necessitates a two-bath process, but it gives the best colour-fastness properties. Procion dyestuffs also require a two-bath process, and give good colour fastness together with bright, clear shades. Selected direct cotton dyestuffs may be applied from the same bath as disperse dyes, but an aftertreatment is needed to confer reasonable wet fastness. Even when aftertreated, direct cotton dyestuffs are satisfactory only for light shades, except in those end-uses where garments are washed infrequently.

As in the case of PET polyester/wool blends, the production of cross-dyed effects on PET polyester/cellulosic fibre blends is limited preferably to outlets where the different abrasion effects during wear will be small. In outlets where severe wear is anticipated, only very slight differences in shade between the components can be tolerated.

PET polyester/cellulosic fibre yarns for effect threads should be dyed with disperse dyes and with vat dyes selected to withstand the proposed finishing routine. Disperse dyes may be selected to withstand heat-setting at 180°C. in dark-medium shades, but few are able to withstand chlorite-bleaching conditions. Other methods of bleaching must be chosen, therefore, for colour-woven materials.

In addition to the batch methods which are available, it is possible to dye PET polyester/cellulosic fibre piece goods continuously by a pad-bake process, using mixtures of disperse and Procion dyestuffs. A semi-continuous pad-bake-jig process is also available, using mixtures of disperse and vat dyestuffs, or using only vat dyestuffs selected to dye both fibres. Pastel shades may be produced by a padding technique using the solubilized vat dyes; these dye the cotton or viscose rayon, but produce only a persistent surface stain on PET polyester fibres.

## PET Polyester/Triacetate Blends

Disperse dyestuffs are used for dyeing both components of this blend, and dyeing is restricted, therefore, to the production of tone-in-tone and solid effects. The production of solid shades requires careful selection of dyestuffs. Piece goods may be dyed usually on the winch with a carrier, but the best results are obtained on the high-temperature beam-dyeing machine.

## PET Polyester/Nylon Union Fabrics

Knitted and woven filament fabrics based on this union may be handled on the types of equipment used for the corresponding 100 per cent PET polyester fabrics. The production of contrasting shades is limited by the high affinity of most disperse dyes for both fibres. The depth of shade produced on the nylon component may be reduced, however, by careful selection of dyestuffs and the use of high-temperature dyeing machinery. Contrasting primary colours are therefore difficult to obtain, and it is advisable to produce primary colours on the PET polyester with harmonizing secondary colours on the nylon. As a general guide the weaker, or where they are of similar strengths, the brighter shade should be on the PET polyester component.

When white reserve effects are required, these should always be produced on the PET polyester component. This is the easier fibre to reserve, and its white can also be enhanced by the application of a fluorescent brightening agent. A wide range of shades can be produced on the nylon component without staining the PET polyester.

If yarn-dyed PET polyester is also incorporated in the fabric, multi-coloured striped or check effects can be produced by cross-dyeing. Shades and dyestuffs for these dyed yarns should be selected carefully; very deep shades should be avoided where possible.

### Printing

#### *100 per cent PET Polyester Woven Fabrics*

Fabrics woven from 100 per cent PET polyester fibre yarns are usually printed with disperse dyestuffs, although a limited number of vat dyestuffs may be used. The disperse dyes provide a wide range of bright and deep shades, but the choice of individual dyestuffs is governed not only by the projected end-use, but also by the method to be used in fixing the colours on to the fibre.

#### *Methods of Colour Fixation*

Three methods are commonly used for the fixation of printed colours on to PET polyester fabrics:

(a) Prolonged steaming (1–2 hours) at atmospheric pressure, using a carrier to promote dyestuff migration.

(b) Steaming for 20–30 minutes in a high pressure star-frame steamer, using high-temperature steam ($1.4–2.0$ kg/cm$^2$; $20–28$ lb./in.$^2$).

(c) Fixation by treatment for 30–90 seconds in dry heat (190–200°C.), using a stenter or baking oven.

High-temperature steaming and the dry-heat process both give excellent colour yields. Low pressure steaming is the least effective of the three methods, but it makes use of conventional equipment.

#### *PET Polyester/Cellulosic Fibre Blended Fabrics*

The most widely applicable method for the printing of PET polyester/cellulosic fibre blended fabrics makes use of mixtures of disperse and reactive dyes. This gives a wide range of solid shades with good build-up and fastness properties. For the highest colour-fastness requirements, mixtures of disperse and vat dyes may be applied, using a two-stage fixation process.

A rather more restricted range of shades having high light and good washing fastness, but only moderate build-up and rubbing fastness, can be obtained very simply by making a minor modification to the 'all-in' vat process. Alcian 'X' dyes may be printed alongside the vat dyes to extend the available range or shades.

Unfixed dyestuff residues are normally removed from 100 per cent PET polyester materials by a reduction-clearing process. The use of such a process on PET polyester/cellulosic fibre blenas would, however, strip the dyestuff from the cellulosic component. It is essential, therefore, that the 'soaping' and rinsing treatments should be made as intensive as possible. The use of a good modern soaping and rinsing range, fitted with spray pipes and devices for inducing turbulence in the soaping liquors, is recommended.

When 'soaping' and rinsing are carried out in rope form on a winch, 'soaping' should continue for at least 10 minutes. Disperse/reactive prints should preferably receive a washing-off treatment with a mixture of a non-ionic detergent and polyvinyl pyrrolidone prior to the 'soaping' process.

#### Hydroextraction

Almost all the moisture in a wet PET polyester fabric is held mechanically in the inter-fibre spaces, and most of it can be removed in hydroextraction. The quantity of water absorbed by the fibre itself is very small.

#### Drying

PET polyester fabrics are readily dried at 120°C., and are normally handled on pin or clip stenters. Care must be taken to ensure that the fabric is not stentered under heavy tension. If it is necessary to remove scouring creases, the damp fabric should be stentered out, without overfeed, to a width which exceeds the scoured width by 7–14 mm per m (¼–½ in. per yd.), and then dried at 140–150°C. Relaxed drying methods are preferred for caustic soda softened fabrics and some yarn-dyed fabrics.

**Stripping**

Imperfect dyeings may result from errors in handling or from faulty machine operation. These imperfections will show as poor colourfastness, poor dye penetration, streaky or uneven dyeing, shading from selvedge-to-selvedge or end-to-end, dye spots, carrier spots, or resist spots. The causes of such problems, discounting machine failure, may usually be traced to one or other of the following:

1. Too rapid strike of dye caused by incorrect carrier selection, excessive rate of temperature rise, or incorrect amount of levelling agents.

2. Faulty emulsification of carrier or insufficient circulation or distribution of carrier in the dye bath and through fibre or fabric.

3. Poor dispersion of dye before adding it to the dyebath.

4. Incomplete solution of the assistants before adding them to the dyebath.

5. Uneven heat-setting before dyeing.

6. Incorrect preparation of the fabric before dyeing.

Three approaches should be considered to problems of this type:

1. The shade, in some cases, may be levelled in a bath containing the original amount of carrier with the additions of dye to correct the shade. When a fabric is shaded, it is advantageous to reheat-set before redyeing.

2. The dyeing may also be stripped as much as 50 per cent by treatment in a bath containing non-ionic chlorinated solvents and non-ionic detergent. After rinsing, the goods may be redyed in a new bath.

3. If neither of the above methods is successful, nearly all the dye can be stripped by treatment for 1 hour at 82–93°C. in a bath containing:

| | |
|---|---|
| Sodium chlorite | 1–4 g./l. |
| Formic or oxalic acid | 1–4 g./l. |
| Non-ionic chlorinated solvents | |
| (e.g. 'Tanalon Special') | 1–4 g./l. |

After treatment, the goods should be rinsed thoroughly and neutralized.

## STRUCTURE AND PROPERTIES

PET polyester fibres are supplied in the form of filament yarns, staple fibre and tow. The properties of the fibre can best be considered by dividing the various types of fibre into three main groups, (1) high tenacity filament yarn, (2) medium tenacity filament yarn, and (3) staple fibre. These groups differ considerably from each other in respect of certain physical properties, but within any one group differences are generally only slight, although alteration in denier may affect some characteristics.

The values for the physical properties described below are those which may be regarded as typical of fibre in the state in which it is manufactured and supplied, either as filament or staple. The properties of material which has been processed in some manner, e.g. heat-set or spun into staple yarn will usually differ from the properties of fibre as supplied by the producer.

The information in this section is based upon the properties of PET polyester fibres supplied by I.C.I. Fibres Ltd., under the trade name 'Terylene'.

· 'Terylene' filament yarns are produced in a wide range of types and counts. Many are intermingled or producer textured.

'Terylene' filament yarn is supplied with a nominal 'S' twist of ¼ turn/inch (30 turns/metre). This is known as 'producer twist', and the actual level varies with the denier of the yarn concerned, as shown in the following table:

*Producer Twist of 'Terylene' Filament Yarn*

| Yarn dtex(den) | Average 'S' Twist | |
|---|---|---|
| | *Turns/inch* | *Turns/metre* |
| 28 (25) | 0.3 | 12 |
| 44 (40) | 0.3 | 12 |
| 56 (50) | 0.3 | 12 |
| 84 (75) | 0.4 | 16 |
| 110 (100) | 0.5 | 20 |
| 140 (125) | 0.5 | 20 |
| 167 (150) | 0.8 | 32 |
| 276 (250) | 1.0 | 40 |

Medium tenacity filament yarn is produced in a range of lustres classified as bright, dull and extra dull, and in a range of counts and numbers of filaments per yarn. In the widely used counts of 56, 84, 110 and 167 dtex (50, 75, 100, 150 den) the individual filaments are each of 2.2 dtex (2 den) approx.

352

Fibre suitable for normal textile purposes or for bulking by standard processes, e.g. false-twisting, is produced. In addition, a producer-bulked yarn, known as type 500, is available. In some cases, trilobal cross-section yarn is provided to give variation in handle and lustre.

Stabilized bulked yarns are also produced from 'Terylene' polyester fibre. They differ from certain other bulked yarns, e.g. false-twist yarns, in having low shrinkage, low extensibility and freedom from torque in singles form. Garments can be knitted to size more easily, and finishing is facilitated; prestabilization before dyeing is not required.

High tenacity filament is of a bright lustre in a range of yarn counts. The individual filaments are, with a few exceptions, each of 5.6 dtex (5 den) approx. Several types of high tenacity yarn vary in tenacity and also in extension characteristics.

Staple fibre is made in a range of counts from 1.7–11 dtex (1.5–10 den) in dull lustres. Various types, which may differ appreciably in properties, are designed specifically for use on the various spinning systems, such as the worsted, woollen, cotton or flax systems. Some staple for particular applications is also produced in bright lustre, and mass-coloured fibre is available in certain types.

Crimped tow is supplied for processing on the converter systems.

In addition to 'standard' staple fibre, two varieties with lower pilling tendencies are made for use in woollen and worsted blends with wool. Since this characteristic is gained at the expense of tenacity and durability, the particular type to be used in any given end-use must be chosen with care. For example, the lowest pilling variety can be used in loosely-woven ladies' wear, where a soft handle is required and where extreme durability is not of prime importance.

High tenacity, high modulus fibre is produced for use on the cotton system. This is particularly designed for blending with cotton where yarns and fabrics of increased strength are required. Medium tenacity 'Terylene', although a strong fibre, has a comparatively high extensibility, the extension at break being 30–50 per cent. Large quantities of this fibre are blended successfully with cotton but the much lower extensibility (modulus) of cotton, whose extension at break is only 5–10 per cent, prevents the attainment of maximum strength in the blend. As a result, 67/33 medium tenacity 'Terylene'/cotton yarns and fabrics have strengths little greater than those of 100 per cent cotton. The increased modulus of the high-tenacity 'Terylene' staple fibre makes it more compatible with cotton, and yarns and fabrics of greatly increased tensile and tear strengths can be obtained.

'Terylene' staple fibre differs from 'Terylene' filament yarn in two important respects. Firstly, staple fibre (with the exception of some for processing on the flax system) possesses a heat-set crimp; secondly, all staple fibre is heat-stabilized during manufacture. The latter results in staple and filament having differing shrinkage characteristics.

### Fine Structure and Appearance

'Terylene' fibres are smooth and rod-like, generally with a circular cross-section, but some types have a trilobal cross-section. Staple fibre is crimped. The fibres may or may not contain pigment.

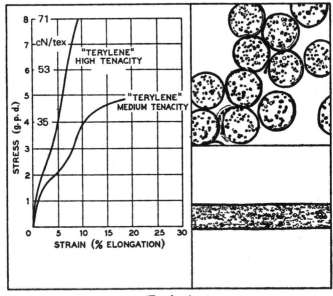

'Terylene'

**Tenacity** (cN/Tex)

| | Filament | | Staple | | |
| | High ten. | Med. ten. | High ten. | Med. ten. | Low-pilling |
|---|---|---|---|---|---|
| Std. (dry) | 56.5–70.6 | 35.3–44.1 | 48.6–57.4 | 35.3–44.1 | 22.1–30.9 |
| Std. (wet) | ” | ” | ” | ” | ” |
| Std. (loop) | About 80% | About 95% | – | – | – |
| Std. (knot) | About 70% | About 85% | – | – | – |

## Tensile Strength

High tenacity filament: 7,350–8,750 kg/cm² (105,000–125,000 lb/in²)

Medium tenacity filament: 4,900–5,950 kg/cm² (70,000–85,000 lb/in²)

High tenacity staple: 5,250–7,350 kg/cm² (75,000–105,000 lb/in²)

Medium tenacity staple: 4,900–5,950 kg/cm² (70,000–85,000 lb/in²).

## Elongation

High tenacity filament: 8–11 per cent (depending on type)
Medium tenacity filament: 15–30 per cent
High tenacity staple: 20–30 per cent
Medium tenacity staple: 30–50 per cent
   The extension at break is virtually unaffected by moisture.

## Elastic Recovery

'Terylene' fibres have a good recovery from stretch, and from compression, bending and shear.

## Initial Modulus

'Terylene' has a high resistance to tensile deformation in the region of extension to which fibres are most subject during use (0–5 per cent). The initial moduli of elasticity for the four types of fibre are as follows:

High tenacity filament: 971–1,148 cN/tex (110–130 g/den)
Medium tenacity filament: 883–1,015 cN/tex (100–115 g/den)
High tenacity staple: 706 cN/tex (80 g/den) approx.
Medium tenacity staple: 265–530 cN/tex (30–60 g./den.approx.)

## Average Toughness

High tenacity filament: 0.325 g.-cm./denier-cm.
Medium tenacity filament: 0.50 g.-cm./denier-cm.
Medium tenacity staple: 0.61 g.-cm./denier-cm.

## Creep Characteristics

At the low extensions to which, partly because of their high modulus, 'Terylene' yarns and fibres are most subject in use, they exhibit negligible creep. 'Terylene' filament yarn, for example, recovers completely from an extension of 1 per cent, and recovery is more than 90 per cent complete after an extension of 3 per cent. The recovery figures for staple fibres are lower than those for filament yarn, due to the partially non-recoverable removal of crimp from staple fibre during loading.

## Abrasion Resistance

For identically constructed materials, the abrasion resistance of standard 'Terylene' is of a high order compared with that of most textile fibres, both natural and man-made, being exceeded only by nylon among the commoner fibres. The abrasion resistances of the lower-pilling varieties are less than that of the standard type, being generally of the same order as wool.

## Specific Gravity

1.38.

## Effect of Moisture

'Terylene' absorbs only a very small amount of moisture, and the tenacity and elongation are unaffected by moisture. The moisture regain is approximately 0.4 per cent at 65 per cent r.h. and 20°C.

Prolonged exposure of 'Terylene' to moisture at high temperatures, e.g. boiling water or steam, will bring about a slow deterioration in the physical properties of the fibre due to hydrolysis of the polyester polymer.

## Thermal Properties

*Softening point*: 260°C. approx.

*Sticking temperature*: 230–240°C.

*Effect of Low Temperature*: At −40°C., the tenacity of 'Terylene' increases by 6 per cent and its extensibility decreases by 30 per cent relative to the tenacity at 20°C. At −100°C., the tenacity increases by about 50 per cent and the extensibility decreases by about 35 per cent.

*Effect of High Temperature*: At 180°C., 'Terylene' retains about half its tenacity at normal temperature. The effect on the tenacity of 'Terylene' of prolonged exposure to high temperatures is small in comparison with the effect on many other fibres, both man-made and natural.

Heated in air at 150°C., 'Terylene' is only slightly discoloured after one month; it retains about 85 per cent of its original strength after 1 month and 55 per cent after 6 months.

### Flammability

'Terylene' fibre fuses and forms a hard bead when heated at a temperature below its ignition point. A flame applied to the fibre causes it to melt, with the formation of a molten bead which ignites with difficulty and burns with a sooty flame. Generally, the molten bead drops from the burning material which is thereby extinguished, leaving no afterglow. The fire risk of 'Terylene' fibre is as low as that of loose wool. The majority of 100 per cent 'Terylene' fabrics, when tested according to B.S.2963, can be described as being of low flammability (B.S.3121). Construction, additives, finishes and the presence of other fibres will, of course, influence considerably the burning characteristics of any particular fabric.

*Specific Heat*: 0.27 cal./g. deg. C. at 20°C.

*Latent Heat of Fusion.* 11–16 cal./g.

*Thermal Conductivity.* $2 \times 10^{-4}$ cal./g. cm. deg. C.

*Coefficient of Cubic Expansion*
  Below 60°C.: $1.6 \times 10^{-4}$
  90–190°C.: $3.7 \times 10^{-4}$

357

## Shrinkage Properties

'Terylene' filament yarn as supplied shrinks approximately 3 per cent in air at a temperature of 100°C. and 10 per cent in air at a temperature of 150°C. High tenacity 'Terylene' yarn has a higher degree of shrinkage than medium tenacity yarn, particularly at higher temperatures.

The shrinkage of 'Terylene' filament yarns in boiling water is approximately 6 per cent, and this is higher than the hot-air shrinkage at 100°C. Again, high tenacity yarn shows a slightly higher shrinkage than medium tenacity yarn.

The shrinkage characteristics of 'Terylene' filament yarn differ from those of nylon, particularly at temperatures above 120°C. Thus, whereas the boiling water shrinkages of 'Terylene' and nylon medium tenacity yarns are about 6 and 10 per cent respectively, in hot air at 160°C. they are 10 and 7 per cent, and at 200°C., 14 and 9 per cent.

'Terylene' staple fibre differs from 'Terylene' filament yarn in being heat stabilized during manufacture, and staple fibre shrinks less than 1 per cent in boiling water. Shrinkages in excess of this amount are, however, experienced in dry air above 120°C., and at 150°C. the shrinkage of staple fibre may exceed 2 per cent, increasing as the temperature is raised still further. No felting shrinkage occurs under moist conditions with fabrics constructed wholly of 100 per cent 'Terylene' yarns.

## Effect of Sunlight

'Terylene' has a high resistance to degradation by light. After prolonged exposure, it suffers a gradual loss of strength, but does not discolour. When exposed behind glass, 'Terylene' shows a considerable increase in resistance to sunlight, and shows a marked superiority to most other fibres under these conditions.

## Chemical Properties

### Oxidizing and Reducing Agents

'Terylene' polyester fibre has an excellent resistance to oxidizing and reducing agents, and the fibre will withstand bleaching processes more severe than those normally employed for textile fibres. 'Terylene' materials may be subjected without damage to any of the usual bleaching agents, including those based on hypochlorite, chlorite, hydrogen peroxide, per-salts and the reducing sulphur compounds.

## Miscellaneous Agents

'Terylene' has an excellent resistance to the chemicals encountered in normal textile use. The only chemicals which, as a class, will dissolve 'Terylene' fibre at normal or moderate temperatures are the phenols. Most phenols will swell or dissolve 'Terylene', depending on the temperature and concentration employed.

Resistance to dilute forms of phenol, such as creosote from wood tar, is good at normal temperatures. Creosote preservatives are not liable to cause serious damage to 'Terylene'.

Cyclohexylamine, which is commonly added to boiler water to minimize corrosion, will accelerate the degradation caused by steam in contact with 'Terylene'.

Perspiration does not have any deleterious effect on 'Terylene'.

## Acids

'Terylene' is a polyester, and is therefore subject to hydrolysis. Acids, alkalis or water alone can all promote the hydrolysis of the polyethylene terephthalate, but the effects are not the same in each case.

Under the influence of acid, gradual degradation of the fibre will occur to an extent dependent on the severity of the conditions. Except under extreme conditions, the rate of acid hydrolysis of a polyester is unexpectedly slow, and consequently good resistance to the majority of mineral and organic acids is a characteristic property of 'Terylene' fibre.

## Alkalis

The action of aqueous alkalis (with the exception of ammonia and its derivatives) is fundamentally different from that of water, acids and ammonia and its derivatives. In the former case, progressive solution of the fibre occurs, where as in the latter case, degradation occurs without solution taking place.

The resistance of 'Terylene' fibre to alkalis is fully satisfactory for the purposes of a textile fibre. It will, for example, withstand the conditions encountered in mercerizing and in dyeing with vat colours. On the other hand, 'Terylene' should not be pressure kier boiled in the presence of alkalis, as these conditions favour an accelerated attack on the fibre.

### Effect of Organic Solvents

'Terylene' polyester fibre shows high resistance to most of the common organic solvents. Examples of these, which include the agents normally used for dry cleaning, are as follows: acetone, dioxane, ether, methyl and ethyl alcohol, benzene, toluene, xylene, light petroleum, methylene chloride, chloroform, carbon tetra-chloride, perchloroethylene and trichloroethylene. At room temperature, these have little effect on the strength of 'Terylene'. Continuous immersion for 6 months in methyl alcohol at 30°C. causes negligible loss of tenacity, while at 50°C. reduction is of the order of 15 per cent.

The treatment of unset forms of 'Terylene', such as unfinished filament yarn, in solvents may cause appreciable shrinkage. With methylene chloride and chloroform, this will occur even at room temperature, and with many of the other solvents mentioned above, shrinkage is considerable at their individual boiling points. Solvents causing the greatest shrinkage are benzene, toluene, xylene, dioxane and the chlorinated solvents with the exception of carbon tetrachloride, which causes no shrinkage. Care must be taken, therefore, to ensure that any such 'Terylene' yarns or fabrics which are to be exposed to hot solvents have previously been heat-set.

The range of chemicals which will dissolve 'Terylene' polyester fibre at normal or moderate temperatures is limited, and the only chemicals which as a class will do this are phenols. Most phenols will swell or dissolve 'Terylene', depending on the temperature and concentration used. Solution is not accompanied by immediate degradation, and the viscosity of a solution of 'Terylene' in ortho-chlorophenol may be used as a measure of the molecular weight of the polyester.

'Terylene' will also dissolve in mono-, di-, and tri-chloroacetic acid. These compounds dissolve 'Terylene' at temperatures above their melting points, which are respectively 63°C., 10°C. and 55°C.

Some other substances will dissolve 'Terylene' at higher temperatures, solution being rapid in the boiling solvent. Among such solvents, with their boiling points, are tetrachloroethane (146°C.), cyclohexanone (155°C.), benzyl alcohol (205°C.), nitrobenzene (210°C.), naphthalene (218°C.), diphenyl (254°C.), and dimethyl phthalate (282°C.).

'Terylene' is virtually unaffected by dimethyl phthalate at normal temperatures. Dichlorodifluoromethane ('Arcton' 6 or 'Freon' 12) and monochlorotrifluoromethane ('Arcton' 4 or 'Freon' 22), as commonly used in refrigeration units do not affect 'Terylene' to any noticeable degree at temperatures between $-20°C.$ and $+20°C.$

Resistance of 'Terylene' to hydrocarbon oils is good.

### Insects

PET polyester is not a source of nourishment to living creatures, and 'Terylene' has an excellent resistance to white ants, dermestid beetles, silver fish, moth larvae and similar pests.

### Micro-organisms

Fungi and bacteria do not attack 'Terylene' itself, but it must be remembered that some fungi and bacteria are capable of growing even on the very small amount of contaminants which can be present on the surface of fibres comprising yarns or fabrics. Such growth will have no effect on the tensile properties of the material, but products generated by the organisms can cause serious discolouration. It is desirable, therefore, to take normal precautions against fungal attack, such as the avoidance of damp storage.

### Electrical Properties

In common with other synthetic fibres having a low moisture regain, 'Terylene' is a very good insulator.

*Dielectric constant*:
  3.17 at 20°C. and 1 kc./sec.
  2.98 at 20°C. and 1 Mc./sec.

*Volume resistivity*:
$1.2 \times 10^{19}$ ohm-cm. at 25°C. and 65 per cent r.h., measured on a 1 mil. film. This is some $10^9$–$10^{12}$ times the resistivity of silk, nylon, cotton or rayon.

Due to the low moisture regain of 'Terylene', the exceedingly high resistivity is maintained at high humidities.

At high temperatures, too, the value for volume resistivity remains high. At 180°C., for example, it is about $3 \times 10^{12}$ ohm-cm.

*Breakdown voltage*:

2.5 kV./mil., measured on 1 mil. thick film.

'Terylene' is non-tracking, so that accidental flash-over cannot produce a conductive track.

### Allergenic Properties

'Terylene' is physiologically inert, and does not cause any irritation to the skin.

### Coefficient of Friction

'Terylene' has a relatively high coefficient of friction against smooth surfaces, especially of metal or porcelain. The coefficient of friction of a yarn against, for example, a yarn guide decreases with increasing tension.

The coefficient of friction against a matt guide, i.e. a guide with a discontinuous surface, is much less than against a polished guide, owing to the smaller area of actual contact between the yarn and the matt surface. The dynamic coefficient of friction of 'Terylene' filament yarn against a matt steel surface, for example, is much lower than that against bright steel.

With staple fibre, however, high frictional co-efficients between spun yarns and guides are not obtained. This is due to the arrangement of the individual fibres in the yarn being such that intimate contact between the yarn and the guide surface does not occur.

### Refractive Index (20°C.)

Parallel to fibre axis: 1.72
Perpendicular to fibre axis: 1.54

## PET POLYESTER FIBRES IN USE

### General Characteristics

*Mechanical Properties*

Considered superficially, PET polyester fibres are generally similar in their mechanical properties to nylon fibres. They are high-strength fibres which display an extension at break that reflects great toughness. But on looking more closely at the mechanical properties of the two types of fibre, important differences may be seen.

Most significant of the differences between PET polyester and nylon fibres lies in the initial moduli. PET polyester fibres have a higher initial modulus than nylon fibres. This means that PET polyester fibres have a greater resistance to stretching in response to a tensile force. The application of a small load will cause 2 or 3 times as great an extension in nylon as in PET polyester fibre.

This high resistance to deformation displayed by PET polyester fibres is shown also in their high resistance to bending. PET polyester fibres are much stiffer than nylon fibres.

The high modulus and stiffness of PET polyester fibres confer great dimensional stability on fabrics made from them. Goods containing PET polyester fibres are not readily deformed. There is a characteristic firmness to PET polyester fabrics, which display a crisper handle than those of nylon.

The stiffness of PET polyester fibres may be demonstrated by comparing PET polyester fibres with wool of comparable diameter. A 4.4 dtex (4 den) fibre, for example, of diameter $20.3\mu$ is comparable in diameter with a 70s wool ($20\mu$). Its handle, however, is more like that of a 64s wool (diameter $22\mu$).

On the other hand, a fine cotton will provide a crisper handle than a PET polyester fibre of comparable diameter; the initial modulus of PET polyester staple fibre is lower than that of cotton.

These characteristics of high strength, high resistance to stretch, toughness and stiffness are of inestimable value in many fields of textile application. They confer great dimensional stability and contribute to the crease-resistance and shape-retention which are such important features of PET polyester goods. But they militate against the use of PET polyester fibres in some applications. In ladies' stockings, for example, it is essential that the fibre should have sufficient 'give' to accommodate the stretching that takes place at the knee. The fibre must also have sufficient elastic recovery from such stretching to return to its original shape when the knee is straightened.

Nylon, with its low initial modulus and high elastic recovery, is admirable for this end-use. But PET polyester fibre, with its high resistance to stretch and lower elastic recovery, is less satisfactory.

At the low extensions to which, partially because of their high modulus, they are most subject in use, PET polyester yarns and fibres exhibit negligible creep. Filament yarns recover completely, for example, from an extension of 1 per cent, and recovery is more than 90 per cent complete after a 3 per cent extension. The recovery figures for staple fibre are lower than those for filament yarns, due to the partially non-recoverable removal of crimp from staple fibre during loading.

PET polyester fibre possesses an ample amount of delayed elastic deformation to provide 'give' under shock-loading conditions. Nevertheless it also displays a high degree of springiness under less rapid loading conditions. This high immediate elastic recovery of PET polyester fibre explains the good wrinkle performance in wear of fabrics containing high proportions of PET polyester fibre. After having been subjected to the bending deformations which fabrics undergo in use, PET polyester fibres recover quickly to their original configuration.

In common with other textile fibres, PET polyester fibre has low shear strength because of its non-isotropic nature. As PET polyester is not, however, a brittle fibre, high shear-stress concentrations cannot be set up because the fibre or yarn will deform and convert these to tensile forces. For this reason, the loop strength of PET polyester fibre is not appreciably less than the straight tensile strength.

*Moisture*

PET polyester fibres have a very low moisture absorbency, and this affects the practical use of the fibre in a number of ways. The mechanical properties of the fibre, for example, are virtually unaffected by moisture, the tensile strength and elongation remaining unchanged. PET polyester fabrics have good dimensional stability during wet processing and washing.

The low moisture absorbency makes for rapid drying, and PET polyester fabrics have excellent ease-of-care properties.

In common with other hydrophobic fibres, PET polyester tends to accumulate charges of static electricity, and these may prove troublesome in processing and during fabric use. Garments may attract dust and dirt, becoming soiled more readily than a fibre that does not so readily acquire electrostatic charges. This difficulty may be overcome by the application of antistatic finishes, of which a number of washable types are now available.

Fibres of low moisture absorption tend to be unsatisfactory for use in underwear and other garments in contact with the skin. They do not absorb the perspiration as wool, for example, does, and they may feel damp and clammy. This is mitigated to some extent in the case of PET polyester fabrics, as the fibres tend to 'wick' readily. Moisture is carried rapidly through the fabrics, so that it can evaporate quickly as it reaches the outer surface.

The low moisture penetrability of PET polyester fibres caused considerable difficulty in the development of dyeing techniques for this fibre. PET polyester fibres are now dyed commonly at high temperatures, and often under pressure.

*Thermal Properties*

In common with other thermoplastic fibres, PET polyester fibres will soften when heated. The sticking and melting points of these fibres are, however, high enough for all normal textile uses, and no difficulties are experienced in this respect.

PET polyester fibres have an excellent resistance to the effect of prolonged exposure at elevated temperatures below the softening point, and this characteristic is proving important in many industrial applications.

PET polyester yarns as supplied by the manufacturer are usually heat sensitive in that they will shrink when heated in water or air. If necessary, PET polyester materials may be stabilized to heat. One method is to shrink to stable dimensions either by steaming or by subjecting to dry heat under conditions which allow free relaxation.

The method of heat relaxation, however, confers increased extensibility, and an alternative method is to heat-set to fixed dimensions, allowing very little or no shrinkage. Under these conditions less change takes place in the physical properties of the fibre, the values for denier, tenacity, modulus and extensibility remaining almost unaltered.

*Heat-Setting*

One of the most important and useful properties of PET polyester fibres is their ability to take on a 'permanent set' when shaped at high temperatures. For example, pleats may be inserted in a PET polyester fabric in such a way that they are highly durable to wearing and laundering. Provided it is present in sufficient proportion, PET polyester fibre also confers this property on blended fabrics containing it.

The ability of PET polyester fibre to accept a 'permanent set' is not influenced by the temperature of any prior heat-setting treatment carried out to confer dimensional stability. In fact, it is essential that any fabric to be pleated or embossed be previously heat-set at a substantially higher temperature than that required in the shaping treatment in order to eliminate the possibility of uncontrolled shrinkage during the pleating process.

Pleats set in a garment in this way will be permanent so long as the setting temperature is not subsequently exceeded. The temperature used in setting is selected with this in view, especially when the garment must withstand ironing. An apparel fabric, for example, will commonly be heat-set at temperatures in the region of 200–220°C.

In a properly heat-set fabric, the fibres will tend always to revert to their set position. Heat-setting contributes greatly, therefore, to the dimensional stability and wrinkle resistance of PET polyester fabrics.

During heat-setting, further crystallization of the polyester takes place, resulting in an increase in the stiffness of the fibre. This may affect the handle of the fabric, which becomes crisper. This effect is alleviated by subsequent mechanical treatment of the fabric, and the sequence of finishing operations is often selected with this in mind. If heat-setting is carried out before scouring and/or dyeing, for example, the increased stiffness of the fabric will be largely worked out during the latter operations. It is essential that heat-setting should be carried out under accurately controlled conditions, as a 5°C. variation in temperature during heat-setting is sufficient to cause a change in shade on subsequent dyeing.

If heat-setting, for one reason or another, is carried out after dyeing, it is necessary to select the dyestuffs with particular care in order to minimize the effect of dyestuff volatilization. It may be necessary under these circumstances to soften the heat-set fabric by calendering or other mechanical treatment.

The development of 'permanent press' techniques during recent years has stimulated an increased demand for PET polyester fibres. Permanent press is a logical step forward in the creation of ease-of-care garments, in which the heat-setting of the fibre takes place after the garment has been made up. The garment itself is set in the desired shape.

Polyester fibres generally have proved most successful in this respect, commonly in blends with cotton, and the rapid advance of permanent press has been a major factor in the polyester field.

*Sunlight*

All fibres are affected by the radiations in sunlight, which bring about deterioration on prolonged exposure. PET polyester is better than nylon in its resistance to sunlight, and this superiority is most pronounced when the fibre is behind glass. PET polyester fibres have established a useful market in the curtain field, largely through this increased resistance to the effects of sunlight.

*Chemical Properties*

PET polyester fibres have an excellent all-round resistance to chemicals. They have a surprisingly good resistance to acids, but are less resistant to alkali, and this must be borne in mind in the selection of fibres for industrial applications.

PET polyester fibres are not affected appreciably by any of the normal bleaching techniques.

*Electrical Properties*

The excellent electrical properties of PET polyester fibres are allied with good resistance to the effect of high temperatures, e.g. to 180°C. As the modern trend is to run electrical motors at higher temperatures, there is a demand for good insulators which will withstand elevated temperatures. PET polyester fabrics, heat-set at temperatures above those encountered in practical use, have found many applications in this field.

*Pilling*

The phenomenon of pilling is a familiar one in the wearing of fabrics from all types of staple fibre. Small bundles of fibres collect on the surface of the fabric during use, forming 'pills' which are held on to the fabric surface by fibres gripped at one end by the yarn and at the other end by entanglement in the pill bundle.

In the case of fibres such as wool, these pills may be removed fairly easily, e.g. by brushing. But in the case of a strong fibre such as PET polyester, the fibres holding the pills do not break easily, and the pills remain attached to the garment surface.

Pilling may be reduced by using tightly constructed yarns and weaves, and the tendency to pill can be removed effectively from PET polyester cloth by careful singeing of the loose fibre ends. This is carried out preferably after dyeing, as fused polyester fibres tend to dye to a darker colour.

Many manufacturers are now producing polyester fibres of lower tenacity, which have a reduced pilling tendency.

## Textured Yarns

PET polyester filament yarns may be subjected to the processes used in producing textured yarns. Some of these yarns show advantages over textured nylon yarns, for example in maintaining superior wash/wear characteristics, wrinkle resistance and recovery from stretch.

Producer-textured yarns have had a notable success, such as the false-twist bulked yarns which are characterized by low stretch and high length stability. These yarns are used, for example, in the manufacture of double jersey and other knitted fabrics; these are more stretch resistant and less prone to shrinkage during processing than comparable fabrics made from bulked nylon.

## Blends

A high proportion of the output of PET polyester fibre is in the form of staple, which is widely used in blends with wool, cotton, viscose and flax.

Blended with cotton, PET polyester fibre increases the wear and abrasion resistance, especially as the proportion of polyester fibre reaches 50 per cent and more. Crease recovery improves, and a fabric containing 67 per cent PET polyester/33 per cent cotton has proved particularly successful for shirts, rainwear, jackets and suits. The fabric has the hard wear, ease-of-care properties associated with the PET polyester fibre, and the handle, opacity and absorbency of cotton.

368

Blended with wool, PET polyester adds crispness of handle, wear resistance, toughness and strength. The effect of the PET polyester is most noticeable in fabrics subjected to damp conditions, in which the wool becomes limp and easily creased. The presence of PET polyester provides crease resistance and dimensional stability, for example in those regions of the body where high humidity is liable to rumple and crease an all-wool fabric.

### Washing

Most fabrics containing PET polyester fibre are washable, but in blend fabrics the nature of the other component will generally control the behaviour of the fabric under different washing conditions. Thus, wool, silk, regenerated celluloses and cellulose acetate fibres are relatively delicate fibres which could be damaged if too severe washing conditions were used. Blends of PET polyester fibres with cotton or flax, on the other hand, are generally more robust. The conditions used in laundering blends of PET polyester fibre with other fibres are therefore controlled more by the other fibres than by the PET polyester.

PET polyester fibre has the valuable characteristic of conferring 'minimum care' properties on fabrics containing it. The degree of 'minimum care' achieved depends largely on the proportion of PET polyester fibre present, but most articles containing PET polyester fibre, if washed correctly, require at most only light ironing.

PET polyester fibres have high strength and abrasion resistance, and they will withstand vigorous mechanical washing treatments. It is not necessary to treat 100 per cent PET polyester goods as if they were made from a delicate fragile material, even though lightly constructed fabrics may give this appearance.

PET polyester fibres have a high resistance to chemical attack, and none of the detergents or chemicals commonly used in washing, including peroxides, per-salts, hypochlorites, acids and alkalis, will bring about any significant deterioration of the PET polyester fibre under conditions likely to be encountered in the laundering of textiles. Vigorous techniques may thus be used, enabling soil to be removed effectively.

Fabrics may behave perfectly satisfactorily during laundering, but the performance of made-up garments will depend on correct trimmings and on the use of suitable making-up techniques. Thus, many garments made from heavier-weight PET polyester/

wool uniform fabrics, suitings and trouserings are suitable only for dry-cleaning, as they have not been designed to be washable.

Most durably-pleated garments containing PET polyester fibre may be washed satisfactorily in a machine or by hand, but the use of excessive washing temperatures may result in disturbance of the pleats; hand or mild machine washing is generally to be recommended for pleated garments.

All properly constructed and finished PET polyester fabrics will possess satisfactory dimensional stability to the washing, dry-cleaning and ironing conditions they are expected to encounter in use.

### Washing Assistants

Normal liquid detergents are good grease and oil emulsifiers, but have poor soil-suspending properties. They are inadequate for washing heavily soiled articles, as the soil may be re-deposited on the fibre. Synthetic fibres in particular are prone to the deposition of soil from such wash liquors, and a better result is obtained using either soap or a 'built' detergent, both of which have superior suspending powers.

Soap is suitable for soft-water areas, but it cannot be used alone in hard water. In hard water districts, the addition of a sequestering agent will prevent the deposition of a lime scum. The use of 'built' detergent-based washing powders of the type used for white cotton articles is also recommended.

### Washing Temperature

The washing temperature should be kept as low as possible, and in general should not exceed 40–50°C. in domestic washing, although higher temperatures can be used in commercial laundry processes.

PET polyester is a thermoplastic fibre, and it is liable to crease at temperatures above 50–60°C. If cooled in this condition, the creases persist and can be removed only by ironing. Creasing tendencies can be greatly reduced by progressive cooling of the rinse liquor from washing temperature down to cold-water temperature while continuing to agitate the garments.

Progressive cooling of this nature is not as practicable in domestic washing as in commercial laundering or fabric finishing, but it is possible to follow a hot wash with a warm rinse and achieve a reasonable compromise.

## Fog Marking

The deposition of dust and smoke particles may cause 'fog marking' of textiles, especially in dry atmospheres and on fabrics which accumulate an electrostatic charge. Fog marks are noticeable as dark bars in folded cloth, or as a 'greying' round the lower edges of slips or similar garments. The effect is of secondary importance in shirts and other articles which soil primarily by physical contact.

Vigorous laundering will generally remove fog marking, and the addition of a non-ionic detergent or cationic softener to the final rinsing water will prevent recurrence.

## Localized Soiling of Garments

Soiling at the collars and cuffs of shirts, blouses and similar fitted garments is due primarily to contact with the skin rather than to static charges. Such soil is composed partly of greasy matter, partly of the skin pigment melanin, and partly of particles of solid matter which have worked into the yarns.

If the stains are heavy, they are best removed by spotting with neat soap or detergent and flexing the fabric by hand. The mechanical action helps to work the solid particles out of the yarns; at the same time the greasy matter is emulsified, both types of soil being removed by subsequent washing. In difficult cases, a metasilicate treatment is desirable.

## Grease and Oil Stains

Ingrained grease stains are difficult to remove from PET polyester fabrics, and are particularly obvious on plain, self-coloured materials. The best method of removal is to spot the stain with neat liquid detergent, work this into the fabric, and leave for 2–3 hours prior to laundering. Alternatively, dry cleaning may be necessary.

For garments that are extensively soiled with oil, e.g. engineers' overalls, detergents rather than soaps are recommended for laundering. Soap should not be used in the early stages of laundering articles which are heavily contaminated with oily soil, because of the danger of the soap/oil emulsion cracking.

Overalls which are consistently used under conditions where they become heavily soiled with oil-bound dirt will benefit from dry cleaning at intervals of 15–20 washes. The most suitable frequency will be determined by experience in each particular case.

### Drying

PET polyester fabrics absorb very little water, and most of the water in a wet article is in the interstices of the cloth. Garments dry quickly and easily when hung in the air.

The spin-drying of hot wet articles, followed by rinsing in cold water, is particularly to be avoided, as is prolonged spinning, which has a strong cooling effect. Bad practice of this nature is most likely to occur when using 'twin-tub' washing machines.

Tumbler drying in hot air helps to remove washing creases. After the garments are dry, tumbling should be continued for 5–10 minutes without heat to allow progressive cooling.

### Ironing

PET polyester articles which have been washed correctly will need little or no ironing. If necessary, however, ironing may be carried out effectively by steam ironing or dry ironing at 'synthetic' setting on the reverse side, or with the help of a press cloth. Ironing temperatures should not exceed about 135°C. Sticking occurs at about 250°C.

### Dry Cleaning

PET polyester fibres are highly resistant to all the chemicals likely to be encountered in dry cleaning, and no special precautions are necessary.

### End Uses

The important properties of PET polyester fibres, filament and staple, may be summarized as follows:

1. Low moisture regain.
2. High dry and wet strengths.
3. High initial modulus.
4. High resistance to and recovery from bending.
5. Low creep.
6. Ability to be heat-set.
7. High abrasion resistance.
8. Good electrical insulation properties.
9. Good resistance to exposure to elevated temperatures.

10. Good resistance to most common chemicals, including oxidizing and reducing agents. Good resistance to acids. Good resistance to dilute alkalis, but attacked by concentrated, hot alkalis.
11. Good resistance to common solvents.

The end-uses for PET polyester fibres have developed largely around these important characteristics, enabling PET polyester fibres to become one of the most versatile of all modern synthetic fibres.

### Apparel Fabrics

PET polyester fibres have made their way into virtually every type of apparel end use. Alone or in blends with wool, cotton, flax and other fibres, PET polyester fibres provide an extensive range of hard-wearing, comfortable, easily-looked-after garments with high dimensional stability and wrinkle resistance.

The strength and dimensional stability of PET polyester filament yarns enable the manufacturer to produce sheer, lightweight textiles. Tulle, voiles, taffetas, poults, satins, brocades, organdies and delicate dress materials are easily laundered and retain their shape and appearance. Continuous filament PET polyester fabrics have found important outlets in curtain nets, ties, shirts and lingerie.

Full-handling filament fabrics, such as double jersey fabrics, are made from producer-textured PET polyester yarns such as stabilized bulked yarn made from 100 per cent 'Terylene'. This type of bulked filament yarn differs from normal false-twist bulked yarns in that it is dimensionally stable, and shrinkage during finishing is therefore of a low order.

PET polyester staple fibre provides yarns and fabrics with the added fullness characteristic of staple fibre textiles. With this fullness goes the warmth resulting from the insulating effects of entrapped air.

Suiting fabrics made from PET polyester staple are comfortable, hard-wearing and keep their shape well. Suits made from PET polyester, or from mixtures of this fibre with other fibres (with an adequate proportion of PET polyester), retain their shape even after prolonged continuous wear.

### Curtains

The good resistance to sunlight which is a feature of PET polyester fibres, especially behind glass, has enabled these fibres to establish an important outlet in the curtain trade. The fast dyes used in dyeing PET polyester fabrics have contributed greatly to success in this field.

## Floor Coverings

PET polyester fibres have made good headway in the field of floor coverings, notably in the production of sliver knit rugs and tufted carpeting.

## Laundry Equipment

The heat resistance of PET polyester fibres, allied with strength, hard wear and abrasion resistance have proved of great value in laundry applications. Laundry bags, dye bags, polished head press covers, laundry blankets, packing flannel and calender sheeting are now commonly made from PET polyester fabrics, which outlast the fabrics previously used.

## Conveyor Belts

PET polyester filament yarns are used as reinforcement in rubber conveyor belts, and have proved particularly valuable in industries where belts are subjected to acid conditions. The PET polyester reinforcement stands up well to acid which penetrates into damaged belts.

## Fire Hose

The high strength and high initial modulus of PET polyester fibres have made possible the production of fire hoses which are strong and light. These hoses are absolutely rot-resistant, and may be left outside without suffering significant damage.

## Ropes, Nets, Sailcloth

The strength, high modulus, resistance to light, heat, microbes, chemicals, water and other influences have opened up an important field of application in ropes, twines, nets, sailcloth, awnings and other goods of this type which are used outdoors.

## Filling

PET polyester staple is used as filling in pillows, quilts, and the like. Polyester fibre filling is easy to wash and dry, non-allergenic, springy and crush-resisting.

## Sewing Thread

High strength and high modulus are useful characteristics in a sewing thread, and PET polyester yarns have come into widespread use in this application.

## Hosiery Dyeing

PET polyester dye-bags are used in the hosiery trade for dyeing nylon stockings. The PET polyester fibre resists dyes used with nylon.

## Papermaking

The felts used in paper manufacture must have a good resistance to moisture, heat and abrasion, and must be able to stand for long periods without rotting. PET polyester fibre has proved especially satisfactory in this application, having the added advantage of resisting the acid conditions which result when aluminium sulphate is used as filler.

## Electrical Insulation

The good dielectric properties and high strength of PET polyester fibres are maintained at temperatures above 100°C., and fabrics made from these fibres are used in the insulation of electric motors designed to operate at elevated temperatures.

## Tyres

PET polyester cords have made substantial headway in this application. This development has resulted to some degree from the increasing use of radial ply tyres, in which the reinforcement is required to have an optimum resistance to stretch. The high modulus of PET polyester fibres is an advantage which enables them to compete effectively against rayon and nylon in this field.

## (2) POLY–1, 4–CYCLOHEXYLENE–DIMETHYLENE TEREPHTHALATE FIBRES (PCDT POLYESTER FIBRES)

Fibres spun from poly–1,4–cyclohexylene–dimethylene terephthalate:

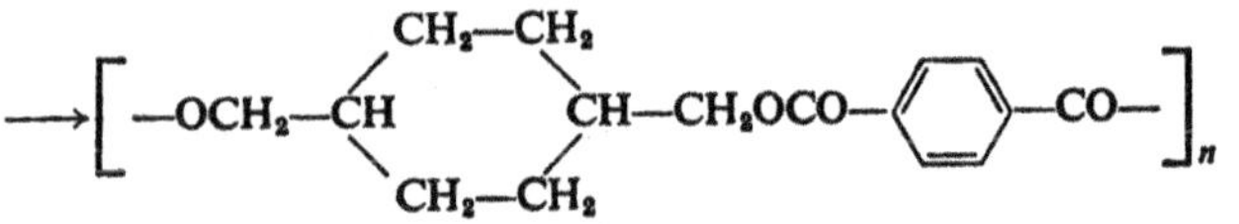

## INTRODUCTION

In 1958, a new type of polyester fibre was introduced to the textile trade by Eastman Chemical Products Inc., under the trade mark Kodel.*

Kodel fibre is spun from the polymer made by condensing terephthalic acid with 1,4–cyclohexanedimethanol (see page 377) and it is therefore of fundamentally different constitution from the polyethylene terephthalate polyesters which form the bulk of commerical polyester fibres. The polymer from which Kodel 211 is spun is poly–(1,4–cyclohexylene–dimethylene terephthalate), which may be shortened conveniently to PCDT polyester.

Since Kodel was introduced, Eastman Chemical Products Inc. have marketed another form of Kodel, which is based on polyethylene terephthalate (PET polyester). The PCDT polyester fibres are now designated Kodel 200—series, and the PET polyester fibres are Kodel 400—series fibres.

The information which follows is based upon data available for Kodel 211, which may be regarded as the original fibre of the PCDT polyester type.

## TYPES AND SIZES

PCDT polyester fibres are produced mainly in the form of staple, tow and top suitable for processing on all the usual systems.

---

* Registered trade mark of Eastman Kodak Company, Rochester, N.Y., U.S.A.

They are available in a range of counts, crimp and staple lengths to meet all requirements.

## PRODUCTION

PCDT polyester fibres are spun from poly–1,4–cyclohexylene–dimethylene terephthalate made by condensing terephthalic acid with 1,4–cyclohexanedimethanol:

$$HOCH_2-CH \underset{CH_2-CH_2}{\overset{CH_2-CH_2}{\diagdown\diagup}} CH-CH_2OH + HOOC-\langle\ \rangle-COOH$$

$$\longrightarrow \left[ -OCH_2-CH \underset{CH_2-CH_2}{\overset{CH_2-CH_2}{\diagdown\diagup}} CH-CH_2OCO-\langle\ \rangle-CO- \right]_n$$

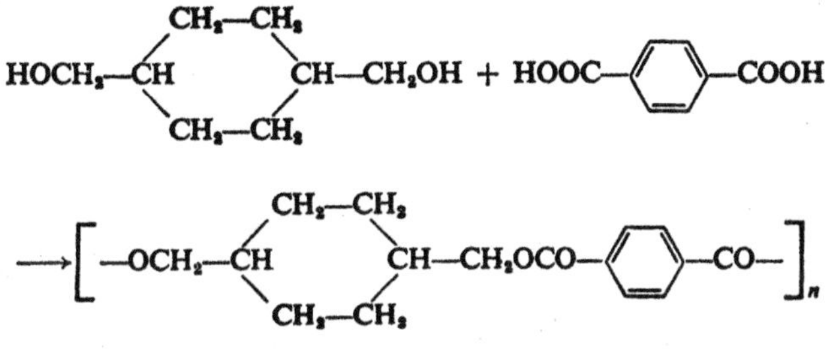

**Reactant Synthesis**

(a) *Terephthalic Acid*

This is produced from p.xylene as described on page 332.

(b) *1,4–Cyclohexane Dimethanol*

Two isomeric forms of this substance are possible, the *cis* and the *trans*. The material used in producing PCDT polymer is a mixture of the two isomers.

**Polymerization**

Dimethyl terephthalate, made by esterification of terephthalic acid, is mixed with 1,4–cyclohexane dimethanol, and the two reactants are heated to 200°C. in the presence of an ester interchange catalyst. As condensation proceeds, methyl alcohol distils off. The temperature is raised gradually to about 300°C., vacuum being applied in the later stages, and heating is continued until the polymer has attained the desired molecular weight. The polymer produced in this way from a mixture of *cis* and *trans* isomers of 1,4–cyclohexane dimethanol has a melting point of about 290°C.

### Spinning

The polymer is melt spun in the usual way, the filaments solidifying as they meet the cold air. They are drawn to 4½ to 5 times their original length at a temperature of about 120°C.

## PROCESSING

### Desizing

Knitted and woven fabrics will often contain sizing materials in addition to dirt, oil and grease. Desizing and cleaning must be carried out before dyeing.

Water-insoluble sizing materials may be removed with a proteolytic or amylolytic enzyme. These enzymes are usually active at temperatures up to 60°C., and 30–60 minutes at this temperature will usually be sufficient to solubilize the sizes. There are many rapid desizing agents available which will digest starch in 2–3 minutes. Sodium bromite desizing is also used.

### Scouring

Except for lubricants applied in manufacturing, PCDT polyester fibre is usually clean and free from foreign matter. The lubricants can be removed prior to dyeing by a warm water (71°C.) rinse.

Yarns and other goods may be contaminated with dirt or oil, and a mild scour with non-ionic detergent and alkali (tetrasodium pyrophosphate) should be used. If the material is extremely dirty or contains oil or grease stains, it is advisable to scour with a petroleum solvent that has been emulsified with a non-ionic emulsifying agent. Alkalis, such as tetrasodium pyrophosphate, trisodium phosphate, sodium carbonate or sodium hydroxide should be used with the petroleum solvent for best results.

If the PCDT polyester fibre is blended with other man-made or natural fibres, the usual scouring or bleaching processes used to prepare the other fibre will not normally affect the PCDT fibre.

A thorough rinsing should follow any preparatory treatment to ensure that residual chemicals or foreign matter have been completely removed.

It is advantageous to boil off or scour many fabrics in open width form before processing in rope form. This treatment partially stabilizes the goods and helps to prevent cracks or streaks. Care should be taken, however, to avoid fixation of any identification tint that might be present in the goods.

### Miscellaneous

Blend fabrics containing PCDT polyester fibres will withstand all the usual treatments encountered in mill processing. By contrast with PET polyester fibres, PCDT polyester fibres will withstand kier boiling, and this process may be used on blends of cotton and PCDT fibres. Mercerizing will not affect PCDT polyester fibres, so long as the temperature of the caustic solution is kept at 32.2°C. or lower.

Blends of PCDT polyester fibre and wool may be carbonized, but they should be neutralized as soon as possible after baking.

### Heat-Setting

Fabrics of PCDT polyester do not require heat-setting. Fabrics should be heat-treated to remove residual carrier after dyeing. The time and temperature required to remove the carrier varies with type of carrier used. Manufacturers' instructions should be followed closely.

### Bleaching

PCDT polyester fibre is produced as a white fibre which does not normally require bleaching. In blends with other fibres, however, bleaching may become necessary, and any of the techniques used to bleach the complementary fibre may be used.

PCDT polyester blends have been bleached by both batch and continuous methods with hydrogen peroxide. Sodium hypochlorite and sodium chlorite may also be used to bleach blended fabrics with no ill effect on the PCDT polyester fibre.

It is common practice to improve the whiteness of bleached fabrics with the aid of various blue and/or violet dyes or pigments, optical bleaches or whiteners or combinations of both. Optical brighteners that produce good whites on cellulosics have little or no effect on PCDT polyester fibre. Experience has shown that a combination of two optical brighteners, one for PCDT and the other for the cellulosic fibre, produce good whites.

### Dyeing

PCDT polyester fibres may be dyed either with disperse or azoic (developed) dyes.

Disperse dyes provide a complete range of shades from pastels to blacks. Azoic dyes are most commonly used to produce blacks.

In common with other types of polyester fibre, PCDT polyester fibres require the use of temperatures above the boil, or of carriers.

### *Carriers*

Because of a lack of pressurized equipment, most dyeing is done with the aid of a carrier. The proper selection and use of a carrier is important. The self-emulsifying butyl benzoate carrier is suggested for knitted and woven fabrics. This material has low toxicity and produces level, well-penetrated dyeings. If the product is properly formulated, it yields a stable emulsion merely by adding warm water while stirring vigorously.

Biphenyl carriers are widely used for dyeing carpets of PCDT polyester fibres. These carriers should be used according to manufacturers' instructions.

Emulsions of methyl salicylate have also given excellent results. A larger concentration of this carrier is necessary, however, to produce dye exhaustion comparable with that obtained with butyl benzoate. For this reason, methyl salicylate is more expensive to use.

### *High Temperature Dyeing*

A reduction in both time and chemical cost may be obtained by dyeing PCDT polyester fibre under pressure at elevated temperatures. Temperatures of up to 121°C. are recommended. The use of a carrier is necessary to obtain complete exhaustion of the dyebath, and the butyl benzoates are particularly effective.

The heat fixation process for dyeing polyesters is especially effective in the application of disperse dyes to PCDT polyester fibres and to blends of these fibres with cellulosic fibres.

### *PCDT Polyester/Cellulosic Blends*

Blend fabrics of PCDT polyester fibres with cellulosic fibres can be dyed in a one-bath procedure using disperse dyes on the polyester fibre and after-treated direct dyes on the cellulosic fibres. Both classes of dyes, the carrier and any necessary assistants should be added to the dyebath at the start of the dyeing cycle.

In cases where fastness requirements cannot be met with after-treated direct dyes, a two-bath exhaustion technique can be used to apply vats, naphthols, sulphurs, soluble vats or fibre reactive dyes to the cotton or rayon. The PCDT polyester fibre is dyed first with disperse dyes. The complementary fibre, either cotton or rayon, is then dyed in a new bath following routine methods for the specific dye used.

The dyer should use the sequence of method best suited to his own equipment when handling PCDT polyester/cellulosic blends. The polyester fibre may be dyed by the carrier method, the heat-fixation process or under pressure. The cotton or rayon can be dyed on a continuous range or it can be batch dyed at a later time. If the cotton or rayon is to be dyed with vats, naphthols, sulphurs, soluble vats or fibre reactive dyes after the polyester is dyed, an intermediate scour is not necessary, as scouring must be carried out after the second dyeing operation.

## PCDT Polyester/Wool Blends

PCDT polyester fibre in blends with wool can be dyed by either a one-batch or two-batch method, depending on fastness requirements. The PCDT polyester fibre is dyed with disperse dyes, the wool with neutral-dyeing acids, chrome or premetallized dyes.

Disperse dyes will stain the wool to various degrees. Careful dye selection is necessary to minimize the stain and not impair fastness qualities.

A one-bath procedure can be used if the polyester dyes do not stain the wool enough to affect the shade or fastness. If the one-bath method is used, the dyes for the wool should be selected from those that will dye at the slightly acid condition used for applying disperse dyes on the polyester fibre. The neutral dyeing premetallized dyes can be used to provide men's wear fastness properties. If colours brighter than those obtainable with this class of dyes are required, there are a number of weak acid dyeing milling dyes that may be used. Small additions of ammonium sulphate might be needed to aid dye exhaustion on the wool after the polyester fibre has been brought to shade.

381

The wool stain may be cleared by treatment at 60°C. for 20 minutes with the following:

Non-ionic detergent 2.0 per cent.
Ammonium hydroxide 1.0 per cent.

The reducing method of clearing is usually unnecessary, as the cleared wool will pick up dye from the polyester during the wool dyeing.

If it should be necessary to use the reducing method, the following formula may be used:

Ammonium hydroxide 2 g./l.
Sodium hydrosulphite 1 g./l.
Non-ionic detergent 2 g./l.

Treat at 54°C. for 30 minutes.

If optimum fastness is required, a two-bath procedure should be used. After applying the normal procedure for dyeing 100 per cent PCDT polyester, the fabric is scoured to clear the stain on the wool. Rinsing and then dyeing in a new bath will give the desired shade to the wool.

### Stripping

When imperfect dyeings are obtained with PCDT polyester fibres, the following approaches should be considered:

1. The shade, in some cases, can be levelled by placing the fabric in a bath with the original amount of carrier plus additional dye to correct the shade if necessary.

2. The problem of shading in the fabric can also be solved by reheat-setting the fabric at a higher temperature than that originally used prior to dyeing. Fabrics unevenly heat set in the beginning cannot be dyed level unless uniform heat is applied during reheat-setting. The reheat-setting temperature must be higher than the original heat-setting temperature. Streaks, spots or similar fabric defects not caused by the original heat setting should not, of course, be re-heat set.

3. The dyeing may also be stripped as much as 50 per cent by treatment in a bath containing 10–15 g./litre butyl benzoate carrier, plus 2–4 g./litre of non-ionic detergent. The goods should then be rinsed and redyed in a new bath.

4. If none of the above methods is successful, the dyeing may

be stripped almost completely by the following treatment for one hour at 82–93°C.:

Sodium chlorite 1–4 g./l.
Formic or oxalic acid 1–4 g./l.
'Tanalon Special' 1–4 g./l.

Rinse thoroughly and neutralize.

*Note.* This stripping formula must not be used on fabrics containing wool.

### Shearing

Shearing is commonly carried out on PCDT polyester/worsted, PCDT polyester/woollen and some PCDT polyester/cellulosic blend fabrics. Shearing has a significant effect on the appearance, performance and handle of these fabrics.

Fabrics with a fuzzy surface requiring a clean, smooth finished appearance should first be sheared and then singed. Shearing removes the longer fibres which would otherwise melt and form large beads during singeing, giving the fabric a harsh handle. Shearing also increases the pill resistance of the fabric.

### Singeing

Singeing may be carried out without difficulty on fabrics containing PCDT polyester fabrics, resulting in improved appearance due to a cleaner surface. As indicated above, it may be advantageous to shear the fabric before singeing in order to remove the longer fibres.

Gas-flame singeing or plate singeing may be used effectively, adjustments for optimum conditions being made on each individual machine. Low fabric speeds and/or high flames should be avoided, as a severe strength loss may result from oversingeing. Speeds of at least 67.5 m/min (75 yd/min) are recommended.

Melted fibre ends resulting from singeing will dye noticeably heavier by exhaust techniques than the normal fibre surface, and singeing should be carried out after dyeing in order to avoid a speckled effect. Fabrics dyed by heat-fixation methods, however, may be singed prior to dyeing.

### Chemical Finishing

Chemical finishes are used on blends of PCDT polyester fibres with other fibres, notably cellulosics, to modify handle and impart special characteristics. These finishes may be resins, reactants, antistatic agents, softeners, water repellents or combinations of such products.

**Mechanical Finishing**

The handle of PCDT polyester blended fabrics may be altered significantly by mechanical finishing, which may or may not be carried out in conjunction with chemical finishing.

*Calendering*

All types of calenders, e.g. polishing, Schreiner, embossing and silk calenders may be used to good advantage. Calendering may be carried out in any of the following stages of the chemical finishing operation: (1) before the resin finish is applied, (2) between the drying and curing operations, or (3) after the curing operation.

*Semi-decatizing*

This step may be carried out in the final stages of the finishing procedure, giving additional crispness and smoothness to the fabric.

*Compressive Shrinkage*

All fabrics of PCDT polyester/cotton blends should be compression shrunk to ensure dimensional stability, and to provide a luxurious handle which is not obtainable with other mechanical finishing devices.

## STRUCTURE AND PROPERTIES

The properties of PCDT polyester fibres are generally similar to those of PET polyester fibres, but there are important differences, for example in mechanical properties, specific gravity, etc. The information which follows in this section is based upon data for the PCDT polyester fibre Kodel 211.

**Fine Structure and Appearance**

Smooth-surfaced fibre of round cross-section.

**Tenacity**

22–26.5 cN/tex (2.5–3.0 g/den)

**Elongation**

24–34 per cent.

**Elastic Properties**

265 cN/tex (30 g/den) at yield point (cf. 353 cN/tex; 40 g/den for PET fibre).

**Average Stiffness**

97 cN/tex (11 g/den).

**Average Toughness**

0.5 grams per cm./den.cm. (cf. PET polyester: 1.04).

**Resilience**

Work recovery percentage at   2 per cent extension: 85–95
                              5 per cent extension: 50–60
                              10 per cent extension: 30–40

   (Corresponding figures for PET polyester are 75–85, 35–45, 15–25.)

**Specific Gravity**

1.23.

**Effect of Moisture**

Regain: 0.4 per cent.

**Thermal Properties**

*Melting point*: 290°C.

*Flammability*: Yarns and fabrics burn slowly, but the burning material melts and drops off when hanging free. In blends, other fibres must be considered.

**Effect of Sunlight**

Excellent resistance.

### Chemical Properties

Excellent resistance, comparable generally to PET polyester, including high resistance to acids and alkalis.

### Effect of Organic Solvents

Excellent resistance to solvents and cleaning agents commonly encountered in normal textile use. Trichloroethylene and methylene chloride may cause some shrinkage, and should be avoided. The fibre is swelled by phenols, toluene, ethyl acetate and acetone. It dissolves in a 60/40 mixture of phenol and tetrachloroethane at 100°C.

### Insects

Similar to PET polyester.

### Micro-organisms.

Similar to PET polyester.

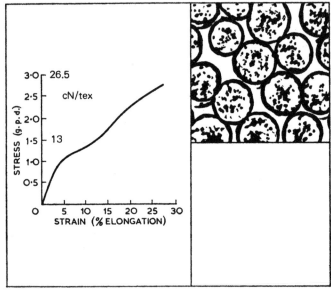

*PCDT Polyester Fibre ('Kodel')*

## PCDT POLYESTER FIBRE IN USE

### General Characteristics

PCDT polyester fibres have a general resemblance to PET polyester fibres, but the differences are such as to exert a significant effect on the uses of the fibres. PCDT polyester fibres have lower tenacity and elongation than the PET polyester fibres, but they have superior recovery from stretch. They serve in applications where resilience and bounce are of greater importance than high tenacity, e.g. in carpets, rugs, knitwear, etc.

The lower tenacity of these fibres contributes to their improved pilling properties. Fabrics will shed their pills more readily than fabrics made from the stronger polyester fibres.

The moisture absorption of PCDT polyester fibres is virtually identical with that of the PET polyester fibres, and this affects the behaviour of the PCDT polyester fibres in a similar way. The mechanical properties are not affected by moisture; the electrical properties are excellent; the accumulation of electrostatic charges may cause difficulties.

PCDT polyester fibres have a lower specific gravity than PET polyester fibres, giving them increased covering power. They provide lightweight fabrics of great warmth and comfort.

The high melting point of PCDT polyester fibres makes for a high safe-ironing temperature (about 218°C.).

Fabrics containing PCDT polyester fibre may be heat-set at a relatively low temperature, e.g. about 160°C., and this is particularly useful in the finishing of blended fabrics containing wool.

### Washing

Similar to PET polyester fibres.

### Drying

Similar to PET polyester fibres.

### Ironing

PCDT polyester fibres may be ironed safely at temperatures up to about 218°C.

**Dry Cleaning**

Trichloroethylene and methylene chloride should not be used in dry-cleaning PCDT polyester fibres, as they may cause shrinkage. Perchloroethylene should be used with care.

**End Uses**

*Apparel Fabrics*

The excellent recovery from stretch and high resistance to pilling have been important factors in the acceptance of PCDT polyester fibres in the apparel fabric field. Blends with wool and acrylic or modacrylic fibres are particularly suitable for knitted goods, providing good dimensional stability, ease-of-care, comfort and warmth, together with the characteristics associated with polyester fibres generally. Sweaters, for example, are machine washable and dryable.

*Floor Coverings*

The excellent resilience and recovery of PCDT polyester fibres has stood them in good stead in the floor covering field. Rugs, mats and broadloom carpets are made from 100 per cent PCDT polyester fibre. They are soft, luxurious and hard wearing, with good resistance to matting and clumping.

*Fiberfill*

The low specific gravity of PCDT polyester fibres is a useful characteristic in applications such as pillows, quilts, padded clothing and the like, where the fibre is used in the form of a fiberfill. Low specific gravity is not necessarily the most important property of a fiberfill material, however. PCDT polyester fibre has outstanding resiliance that permits it to support the voids within the fiberfill batt.

## (3) OTHER TYPES OF POLYESTER FIBRE

Since its introduction during the early 1950s, polyethylene terephthalate fibre (PET polyester fibre) has made rapid headway and attained a dominating position in the field of synthetic textile fibres. PET polyester fibres are now being produced by many firms throughout the world.

PET polyester fibres have reached this position, as in the case of nylon 6.6 and nylon 6, by providing the textile manufacturer with fibres that serve effectively over a wide range of textile and industrial applications. But, inevitably, there are fields of application in which the range of properties offered by PET polyester fibres are inadequate or unsatisfactory. And fibre manufacturers have sought ways of producing new types of polyester fibre which would retain the broad characteristics associated with polyesters, but provide fibres capable of serving where established polyester fibres are inadequate.

As in the case of polyamide fibres, attempts to provide new types of polyester fibre have followed three main routes:

(A) Physical modification of established types of polyester fibre.

(B) Chemical modification of established types of polyester fibre.

(C) New types of polyester.

Already, as we have seen, the third route has provided a new type of polyester fibre, PCDT polyester, which has become established on a commercial basis alongside PET polyester fibres.

## (A) PHYSICAL MODIFICATION OF ESTABLISHED POLYESTER TYPES

### Modification of Polymer

The mechanical properties of polyester fibres may be varied by controlling the physical characteristics of the polymer. Reduction of the molecular weight, for example, will affect the mechanical properties, and this is used in producing polyester fibres with reduced pilling tendencies.

The orientation and crystallinity of the polymer may be affected by adjustment of the conditions used in drawing the fibre. Two-stage drawing, for example, is used for the production of high modulus polyester staple which is suitable for blending with cotton.

Control of the drawing process also makes possible the production of filaments in which there are sections of drawn material alternating with undrawn material. This thick-and-thin type of filament gives attractive dye-fleck effects.

The shrinkage of filaments is influenced by the drawing process, and this is used in the production of yarns which provide special surface effects on fabrics.

### Modification of Fibre

#### *Multilobal Cross-Section*

As in the case of polyamide fibres, polyester fibres of special cross-section may be made by extruding the filaments through holes of appropriate shape.

#### *Bulk Staple*

Differential cooling of the filaments as they emerge from the spinneret produces a spiral effect in the filaments. This technique is used in producing bulked polyester yarns.

#### *Textured and Bulked Yarns*

Polyesters are thermoplastic, and the fibres may be subjected to all the texturing processes in common use.

#### *Bicomponent Fibres*

Bicomponent filaments are produced from polyesters, with effects similar to those obtained from bicomponent polyamide yarns.

### (B) CHEMICAL MODIFICATION OF ESTABLISHED POLYESTER TYPES

The polyesters from which PET and PCDT polyester fibres are spun are chemically inactive materials, and this has been an important factor in the difficulties associated with dyeing these fibres. There are no active groups in the polymer molecule to which dyestuffs are attracted.

Much of the research on polyester fibres during recent years has been concerned with this problem of improving the dyeing characteristics of the fibre, and progress has been made by modifying the chemical nature of the polymer molecule. Special types of PET polyester fibres are now produced, for example, in which sulphonic groups have been introduced into the polyester molecule, providing sites to which basic dyestuffs may become anchored. Halogens or phosphonate groups substituted in the polymer structure enhance the flame retardance, especially if antimony oxide is included in the polymer matrix.

Affinity for dyestuffs and other modifications to the fibre properties are also bestowed upon PET polyester fibres by introducing a small amount of another component into the polymerization. A fibre 'Grilene', for example, was made by Nippon Rayon Co. Ltd. by condensing terephthalic acid, ethylene glycol and a small amount of p.hydroxy benzoic acid to form a polyester/polyether copolymer.

## (C) NEW TYPES OF POLYESTER

The most radical way of producing polyester fibres with different ranges of properties is to spin the fibres from new types of polyester. Poly-p-ethyleneoxybenzoate, for example, has been spun into fibres which were marketed in Japan as A-Tell. Produced by reacting ethylene oxide with p-hydroxybenzoic acid, the polymer melts at $224°C$ and is melt spun. The fibres are similar in many respects to PET polyester fibres. Tenacity is 35.32–46.80 cN/tex (4.0–5.3 g/den), elongation at break 15–30%, with almost 100% recovery from low elongation. Moisture regain 0.4% and specific gravity 1.34 are very similar to those of PET polyester fibres. Resistance to acids and alkalies is greater than that of PET polyester fibres.

Poly-p-ethyleneoxybenzoate fibres, resembling closely the established polyester fibres, were unable to offer advantages that might have given them a worthwhile place in the textile market, and production was discontinued.

## 3. POLYVINYL DERIVATIVES

### Introduction

It has long been known that certain compounds containing a double bond would undergo spontaneous change, during which a liquid, for example, was converted gradually into a solid material. The low-boiling liquid vinyl chloride, more than a century ago, was shown to change in this way.

These changes are now recognized as polymerizations, during which the small molecules join together via the double bond to form long chain molecules of polymer. (*See below.*)

Compounds containing the vinyl group, i.e. $CH_2$=CH—, will commonly undergo polymerization in this way, and the polymers that are formed are called *polyvinyl compounds*. The linking together of the small molecules of a vinyl compound results in the formation of a long molecule consisting of carbon atoms, the remaining atoms that were present in the vinyl compound

## POLYMERIZATION OF VINYL COMPOUND

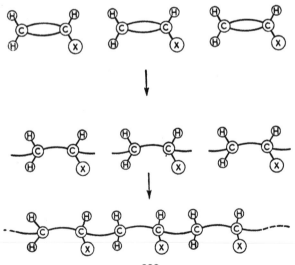

forming pendant groups of atoms attached to the long carbon backbone of the polymer molecule.

As the formation of polymer takes place by addition of one small molecule to another, without the elimination of water or other materials, this type of reaction is an *addition polymerization*.

Polyvinyl compounds were among the first polymers to be studied as potential sources of synthetic fibres. The early German fibre 'Pe Ce' was made by extruding solutions of a chlorinated polyvinyl chloride. It was not until the early years of World War II, however, that polyvinyl compounds began to show real promise in the field of genuine textile fibres. Fibres were spun from solutions of polyacrylonitrile (polyvinyl cyanide), and these have since developed into one of the most important of all classes of synthetic fibres.

## POLYACRYLONITRILE FIBRES

Fibres spun from polymers or copolymers of acrylonitrile:

$$CH_2=CH.CN \quad \rightarrow \quad -CH_2-CH-CH_2-CH-$$
$$\begin{array}{ccc} & | & | \\ & CN & CN \end{array}$$

Acrylonitrile           Polyacrylonitrile

### INTRODUCTION

Acrylonitrile was made in Germany by Moureu in 1893. From that time, up to almost the outbreak of World War II, acrylonitrile remained a laboratory curiosity. But during the late 1930s, it acquired a new status by becoming a constituent of one of the more important types of synthetic rubber under development in Germany and the U.S.A.

During World War II, the synthetic rubber industry underwent a mushroom growth in the U.S.A., and this precipitated a large-scale manufacture of acrylonitrile. By the end of the war, acrylonitrile was a relatively cheap industrial chemical, and was available in large quantities. Polymers of acrylonitrile came under intensive review in the rubber, plastics and synthetic fibre industries.

Acrylonitrile undergoes addition polymerization readily, and polyacrylonitrile had been examined as a potential fibre-forming polymer during the late 1930s. Practical difficulties had prevented any substantial progress being made in this direction, however, the main problem being to find a way of spinning a polymer which was virtually infusible and insoluble in any of the solvents then examined. At that time, attempts to spin polyacrylonitrile were directed largely at producing more soluble copolymers, e.g. containing vinyl chloride, which would dissolve in common solvents such as acetone. Fibres were spun from these copolymers by dry spinning techniques such as those used in spinning acetate fibres, and some of the copolymer fibres had properties of considerable interest.

Shortly before the war, solvents for 100 per cent polyacrylonitrile were found, such as dimethyl formamide, and the experimental production of polyacrylonitrile fibres continued throughout the war years in Germany by I.G. Farbenindustrie, and in U.S.A. by E. I. du Pont de Nemours and Co. Inc. By 1942, du Pont were producing a polyacrylonitrile fibre which was offered to the U.S. Government for military applications. The evaluation of this polymer was so promising that in 1944 du Pont announced their decision to begin pilot plant production. By 1945, du Pont were producing the world's first polyacrylonitrile fibre, which was provisionally named Fiber A.

These early polyacrylonitrile fibres were almost as strong as nylon, and had a high resistance to chemicals (notably acids) and to sunlight. They were very difficult to dye, and it seemed that their most promising commercial outlets must lie in industrial and outdoor applications. As dyeing techniques were evolved, however, it became apparent that polyacrylonitrile fibres could also become of importance in the apparel fabric field.

In 1948, du Pont made plans for the large-scale production of Fiber A, and in July 1950 a plant came into production at Camden, South Carolina. The fibre was produced as continuous filament yarn, and it was given the trade name 'Orlon'. This first fibre was a 100 per cent polymer of acrylonitrile, and it still presented problems in dyeing. In due course, however, the introduction of small proportions of a second monomer provided modified forms of fibre with improved dyeability.

In March 1952, a second 'Orlon' plant at Camden began producing staple fibre.

Meanwhile, in 1950, a second U.S. manufacturer entered the polyacrylonitrile fibre field, when Chemstrand Corp. constructed a pilot plant. This was followed by a commercial plant which came on stream at Decatur, Alabama in late 1952, producing a polyacrylonitrile fibre which was marketed as 'Acrilan'. After overcoming initial teething troubles, Chemstrand Corp. began large-scale production of 'Acrilan' in 1954. In 1955, Chemstrand Corp. established a British company, Chemstrand Ltd., and in 1958 a second 'Acrilan' plant was opened at Coleraine in Northern Ireland.

During the period 1955–60, polyacrylonitrile fibre plants began to spring up throughout the world. Germany, Japan, Italy, Belgium, Canada, France and Holland entered the polyacrylonitrile fibre production field. Since then, many other countries have followed suit.

## TYPES OF POLYACRYLONITRILE FIBRE

Polyacrylonitrile fibres are now produced by many manufacturers under a wide variety of trade names. More often than not, individual manufacturers will market a family of polyacrylonitrile fibres, each of which is designed to serve in a specific field of textile applications. The user of textile fibres, in consequence, is presented with a multitude of polyacrylonitrile fibres of embarrassing range and scope.

These polyacrylonitrile fibres all differ from one another to a greater or lesser degree, notably in certain characteristics such as dyeing properties. And in this respect, it is more difficult to consider polyacrylonitrile fibres as a general class than in the case of polyamide or polyester fibres.

In the early days of polyacrylonitrile fibres, there was a tendency to describe all these fibres simply as 'acrylic' fibres. It soon became apparent, however, that this term was describing a range of fibres of widely differing characteristics. It had become necessary to break down the polyacrylonitrile fibre class into smaller groupings in which fibres of similar characteristics could be brought together.

This problem was considered by the U.S. Federal Trade Commission in the formulation of the Rules and Regulations for Fibre Identification which came into force on 3 March 1960.

Polyacrylonitrile fibres were subdivided into two classes, depending upon the proportion of acrylonitrile in the polymer (see below). Those polyacrylonitrile fibres containing at least 85 per cent of acrylonitrile units are described as *acrylic* fibres, whereas those fibres containing between 35 and 85 per cent of acrylonitrile are described as *modacrylic* fibres.

This subdivision is helpful in that it separates two groups of polyacrylonitrile fibre which have little in common from the practical point of view. But it still leaves room for great variation in the characteristics of individual fibres within each group. The term 'acrylic', for example, does not distinguish between the fibre which is a copolymer of acrylonitrile with a small proportion of a second component, and a fibre spun from a graft copolymer of acrylonitrile.

A further source of confusion arises from the fact that the official F.T.C. term 'modacrylic' refers to polymers containing 35–84 per cent of acrylonitrile. It thus includes copolymers in which acrylonitrile may form the minor component; a fibre spun from a copolymer containing 60% of vinyl chloride and 40% of acrylonitrile, for example, is more properly considered from the chemical standpoint as a polyvinyl type fibre, despite its F.T.C. classification as a modacrylic fibre.

## NOMENCLATURE

### Federal Trade Commission Definitions

The generic names *acrylic* and *modacrylic* were adopted by the U.S. Federal Trade Commission for fibres of the polyacrylonitrile class. The definitions are as follows:

*Acrylic.* A manufactured fibre in which the fibre-forming substance is any long chain synthetic polymer composed of at least 85 per cent by weight of acrylonitrile units.

*Modacrylic.* A manufactured fibre in which the fibre-forming substance is any long chain synthetic polymer composed of less than 85 per cent but at least 35 per cent by weight of acrylonitrile units, except fibres qualifying under sub-paragraph (2) of paragraph (j) (rubber) of this section and fibres qualifying under paragraph (q) (glass) of this section.

**Note**

In the section which follows, polyacrylonitrile fibres are considered under two classes, depending upon the proportion of acrylonitrile units in the polymer:

(1) *Acrylic Fibres*, which are spun from polymers composed of at least 85 per cent by weight of acrylonitrile units (i.e. corresponding to the F.T.C. definition of *acrylic* fibre).

(2) *Modacrylic Fibres*, which are spun from polymers composed of less than 85 per cent but at least 50 per cent by weight of acrylonitrile units (i.e. corresponding to the F.T.C. definition of *modacrylic* fibre, but including only those modacrylic fibres in which acrylonitrile forms the major component of the copolymer).

## (1) ACRYLIC FIBRES

Fibres spun from polymers consisting of at least 85 per cent by weight of acrylonitrile units (—CH$_2$—CH(CN)—).

## TYPES AND SIZES

The early types of polyacrylonitrile fibre, e.g. 'Orlon' Types 41 and 81, were spun from 100 per cent polyacrylonitrile, but almost all modern types of acrylic fibre are spun from copolymers. These may be the 'normal' type of copolymer in which the second component is polymerized with the acrylonitrile, or they may be of the 'graft' copolymer type, in which the second component is incorporated by grafting on to the polyacrylonitrile.

The nature and proportion of the second component used in individual acrylic fibres are rarely disclosed, but a very large number of acrylonitrile copolymers has been described in the patent literature. Vinyl acetate, vinyl chloride, methyl acrylate and 2–vinyl–pyridine are among the monomers which are probably used commercially.

Acrylic fibres are used as filament yarn, tow for conversion, and as staple fibre. Staple is produced in counts and length suitable for all spinning systems.

### High Bulk Fibre

Acrylic fibres are unusual in their ability to attain a metastable state on hot stretching. When hot-stretched fibres are cooled, they will remain in their stretched state until subsequently heated, when they revert to their unstretched dimensions. High shrinkage fibres may be made in this way, with shrinkages of 30 per cent and higher, and by blending these high-shrink fibres with normal staple, followed by subsequent steaming, high bulk effects are obtained.

Many types of high-shrinkage acrylic fibre are now produced.

## PRODUCTION

### Monomer Synthesis

Acrylonitrile is available in more than adequate quantity to meet the demands of the acrylic fibre industry. It is made usually by any of four main routes.

### (a) *Ethylene Cyanhydrin Dehydration*

Ethylene cyanhydrin is made either by treatment of ethylene oxide with hydrogen cyanide (1) or by reaction of ethylene chlorhydrin with alkali cyanides (2).

The ethylene cyanhydrin is dehydrated (liquid phase) at 250–350°C. in the presence of alkaline catalyst, or (vapour phase) at 350°C. in the presence of alumina (3).

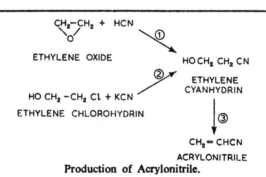

Production of Acrylonitrile.

**(b)** *Acetylene and HCN*

Acrylonitrile is made directly from acetylene by the addition of HCN.

$$CH{\equiv}CH + HCN \rightarrow CH_2{=}CHCN$$

**(c)** *Propylene Route*

Propylene is oxidized to acrolein (1) which is then reacted with ammonia to form a hydroxy amino compound (2). This is dehydrated and dehydrogenated to acrylonitrile (3).

$$CH_2{=}CH{-}CH_3 \overset{(1)}{\rightarrow} CH_2{=}CH{-}CHO \overset{(2)}{\rightarrow} CH_2{=}CH{-}CH{\Big\langle}{}^{NH_2}_{OH}$$

$$\overset{(3)}{CH_2{=}CH.CN}$$

**(d)** *Acetaldehyde Route*

Hydrogen cyanide is added to acetaldehyde to form the cyanhydrin (1). This is dehydrated to acrylonitrile (2).

$$CH_3CHO + HCN \overset{(1)}{\rightarrow} CH_3CH{\Big\langle}{}^{OH}_{CN} \overset{(2)}{\rightarrow} CH_2{=}CH.CN$$

**Polymerization**

The polymerization of acrylonitrile and its co-monomer is commonly carried out by stirring the monomers with water in the presence of catalyst and surfactants. Some of the acrylonitrile dissolves in the water to form a 7 per cent solution, the excess monomer forming an emulsion.

As polymerization proceeds, the polymer (which is insoluble in water) is precipitated to form a slurry. This is filtered, and the polymer is washed and dried.

Polymerization may be carried out as a batch process, or on a continuous basis. In the latter case, monomer, water and other materials are fed into a reaction vessel, and slurry is withdrawn continuously. Unreacted monomer is recovered and returned to the polymerization.

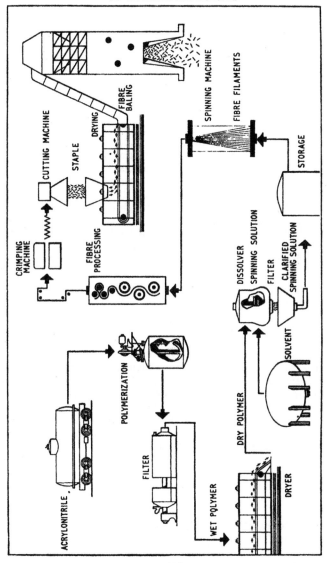

'Acrilan' Flow Chart

401

## Spinning

Polyacrylonitrile, 100 per cent or containing up to 15 per cent co-monomer, tends to decompose on melting, and fibres cannot be produced by melt spinning processes as used for polyamide and polyester fibres. Acrylic fibres are therefore produced from solutions, either by dry or wet spinning.

### Dry Spinning

The polymer is dissolved in an organic solvent, such as dimethyl formamide, to form a solution containing 25–40 per cent of polymer. This is degassed, filtered and heated almost to boiling point, and then extruded through spinnerets.

The fine jets of solution emerge into a vertical tube or spinning cell, through which air or other gas at high temperature (e.g. 400°C.) is flowing. As the jets fall through the tube, the solvent evaporates to leave solid filaments of polymer.

The filaments are brought together at the base of the spinning cell and stretched hot to 3–10 times their original length. If continuous filament yarn is being produced, the filaments are oiled, twisted and then wound on to bobbins.

If staple fibre is required, a number of yarns are brought together into a tow. This is then crimped and cut into staple of the desired length. Uncut tow is used for conversion.

### Wet Spinning

Polymer is dissolved in dimethyl formamide or other solvent, and the solution is degassed and filtered. It is then pumped through spinnerets into a coagulating bath containing a liquid in which the solvent is soluble but the polymer insoluble. The jets of solution coagulate into fine filaments, forming a tow which is washed after emerging from the bath. The tow is heated and drawn, dried, oiled and crimped. It may then be heated to relax the fibre before being cut into staple.

### PROCESSING

### Dyeing

The early forms of polyacrylonitrile fibre, such as 'Orlon' Type 41 staple and Type 81 filament yarn, were difficult to dye satisfactorily. Basic dyes were used, but light fastness was poor. A satisfactory range of shades was provided, however, by making use of the cuprous ion process developed in 1950 for the dyeing of 'Dynel'.

The preferred technique was to make use of acid dyes in the presence of cuprous ions. The process has been of great value in dyeing all fibres based on acrylonitrile; the copper acts as a bridge which links the nitrile group of the fibre to the dye molecule.

Using the cuprous ion technique, it was possible to obtain fast colours covering a wide range on 'Orlon' staple fibre, although the more highly orientated continuous filament 'Orlon' did not dye so readily. Good results were obtained, however, by dyeing under pressure at temperatures up to 120°C.

When manufacturers began producing copolymers of acrylonitrile, the position changed. The main reason for introducing a second component into the polymer was to establish increased dyestuff receptivity by providing anchor points for dyestuff molecules. Monomers used in the production of acrylonitrile copolymers were chosen for their effectiveness in this respect.

The monomers used today as second components in acrylonitrile copolymers are commonly of three general classes, depending on the type of site they offer to dyestuffs:

(1) Monomers which provide non-ionizing polar groups which are able to form co-ordination complexes with dyestuffs, e.g. vinyl acetate, methyl acrylate.

(2) Monomers containing basic (e.g. amino) groups capable of forming ionic bonds with acid and direct dyes.

(3) Monomers containing acid groups (e.g. sulphonic, phosphoric, carboxylic) capable of forming ionic bonds with basic dyes.

In addition to providing anchor points for dyestuff molecules, the monomers introduced into acrylic polymers create irregularities in the polymer molecule. The 'packability' of the molecules is reduced, and the degree of crystallinity is lowered. Fibres spun from these copolymers are more receptive to solvents, being more readily swelled and more easily penetrated by dyestuff and other molecules.

This latter effect is not apparent, however, in the case of graft copolymers, as used in the production of the nitrile alloy fibres.

403

The introduction of the graft does not affect the form of the polyacrylonitrile backbone, and the resistance to solvents in this case may actually be increased.

Today, acrylic fibres are produced in which receptivity to all types of dyestuff has been built in.

## STRUCTURE AND PROPERTIES

The properties of polyacrylonitrile fibres vary over a wide range, each manufacturer producing a fibre or fibres that will have the blend of properties that he feels will meet his particular requirements.

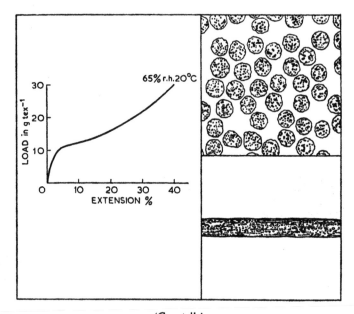

*'Courtelle'*

404

The acrylic fibres offer unusual scope in this respect. The characteristics of the fibre may be varied to almost an infinite degree by selection of a particular type of acrylonitrile copolymer, controlling the spinning conditions (wet or dry spinning) and the aftertreatment (e.g. drawing) to which the fibre is subjected.

## Fine Structure and Appearance

Acrylic fibres are produced in a variety of cross-sections, depending upon the conditions under which the fibres are spun. In general, wet spun fibres have a round or kidney-bean shaped cross-section. Dry spun fibres may be of dog-bone or flat cross-section. Many bicomponent acrylic fibres are produced, yielding high-bulk fibres on processing.

## Tenacity

The early types of polyacrylonitrile fibre, which consisted of 100 per cent acrylonitrile polymer, were highly orientated and crystal-

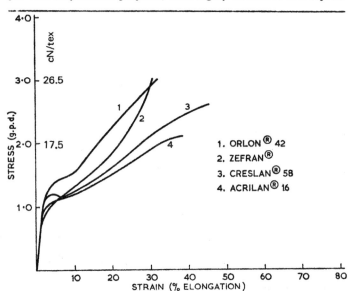

Stress-strain Curves for Some Commercial Acrylics. (Single fibres tested at standard conditions)—*Chemstrand Ltd.*

line fibres in which the molecules were held together tightly by the intermolecular forces acting through the nitrile groups.

The introduction of a second component into the polymer, with the primary object of increasing dyeability, reduced the packability of the long molecules and affected the mechanical properties of the fibre. The tenacity of copolymer acrylic fibres, for example, is lower than that of the 100 per cent acrylonitrile fibres, other things being equal.

It soon became apparent, however, that acrylic fibres were going to make their way in the textile market through attractive handle, resilience, ease-of-care and other properties, rather than by virtue of high tenacity. The reduction of tenacity which followed introduction of a second component into the polymer, therefore, was not of serious consequence. The tenacity of acrylic fibres remains high enough for the type of applications in which these fibres serve.

Typical tenacity ranges of modern acrylic fibres are as follows:

(1) *Staple and Tow*

| | |
|---|---|
| Dry: | 17.7–31.8 cN/tex (2.0–3.6 g/den) |
| Wet: | 14.1–23.9 cN/tex (1.6–2.7 g/den) |
| Std. Loop: | 15.9–20.3 cN/tex (1.8–2.3 g/den) |
| Std. Knot: | 15.0–20.3 cN/tex (1.7–2.3 g/den) |
| H.T.: | 29.1–37.1 cN/tex (3.3–4.2 g/den), dry |
| | 25.6–31.8 cN/tex (2.9–3.6 g/den), wet. |

(2) *Filament* ('Creslan')

Dry: 35.3–36.2 cN/tex (4.0–4.1 g/den)
Wet: 26.5–33.5 cN/tex (3.0–3.8 g/den).

**Tensile Strength**

Staple: 2,100–3,150 kg/cm$^2$ (30,000–45,000 lb/in$^2$)
Filament: 3,500–5,250 kg/cm$^2$ (50,000–75,000 lb/in$^2$).

**Elongation**

Staple: 20–55 per cent.
Filament: 30–36 per cent.

**Elastic Recovery**

Acrylic fibres have a high elastic recovery from small extensions, e.g. 90–95 per cent at 1 per cent extension. The recovery trom higher extensions is moderate, e.g. 50–60 per cent at 10 per cent extension. In general, the recovery characteristics resemble those of wool.

### Initial Modulus

Acrylic fibres have a high initial modulus, commonly in the region of 353–441 cN/tex (40–50 g/den).

### Average Stiffness

Staple fibre:  62–88 cN/tex (7–10 g/den).
Filament:     141–362 cN/tex (16–41 g/den).

### Average Toughness

Staple:    0.40–0.70
Filament:  0.22–0.49

### Specific Gravity

1.16–1.18

### Effect of Moisture

Regain: 1.0–3.0 per cent.
Water absorption at 20°C. and 95 per cent rh: 2.0–5.0 per cent.

The water absorption by acrylics is relatively low, but it is sufficient to reduce the difficulties associated with development of static charges (e.g. by comparison with polyester fibres). It is also a significant factor in the comparatively good dyeability of acrylic fibres.

The tensile properties of acrylic fibres are affected to some degree by water, the tenacity, for example, being reduced to 75–95 per cent of the dry tenacity.

### Thermal Properties

Acrylic fibres do not have true melting points, but tend to stick to metal surfaces at 215–255°C. when pressed against them.

### Effect of High Temperature

The mechanical strength of acrylic fibres is not seriously impaired by exposure to heat. The tenacity after 100 hours at 155°C. is typically about 96 per cent of the original tenacity.

After 15 minutes at 130°C., acrylic fibre may become cream-coloured. As heating becomes more severe, the fibre becomes progressively more discoloured. At 150°C. it is noticeably yellow after 1 hour. At 250°C. the fibre darkens through yellow and brown to black within 5 minutes and becomes insoluble in the usual solvents.

When acrylic fibres are heated, the modulus falls and they are easily stretched. This stretching, if carried out carefully, is a reversible process caused by increased orientation of the polymer molecules, without any appreciable plastic flow taking place. If the stretched fibre is cooled before tension is released, the fibre remains in its stretched form. If the stretched and cooled fibre is heated again, however, it will relax and return to its original length.

This ability of acrylic fibres to assume a metastable state after being stretched at high temperature is used in the production of high shrinkage fibres.

### Flammability

Acrylic fibres will burn, but they are not dangerously flammable fibres.

### Effect of Sunlight

Acrylic fibres have excellent resistance to the effects of sunlight. After 600 hours exposure, the tenacity of a fibre is typically 96 per cent of the original tenacity.

### Chemical Properties

### Acids

Acrylic fibres are unaffected by dilute solutions of strong mineral acids, but they tend to be attacked on prolonged immersion in concentrated solutions.

### Alkalis

Dilute solutions of caustic soda and all solutions of sodium carbonate and bicarbonate have no effect on the mechanical properties of acrylic fibres. Strong alkalis attack the fibre.

### General

Acrylic fibres are resistant to most common organic substances, the fibre strength being appreciably unaffected by concentrated carboxylic acids, phenols, alcohols, ketones, hydrocarbons, chlorinated hydrocarbons or detergents.

Some less common organic substances such as dimethyl formamide, α–butyrolactone, dimethylsulphoxide and ethylene carbonate are solvents for acrylic fibres.

Most salts are without effect, but very concentrated solutions of sodium and calcium thiocyanate, zinc chloride and certain other salts act as solvents.

### Effect of Organic Solvents

Acrylic fibres generally have a good resistance to common organic solvents, including those normally used in dry cleaning.

### Insects

Acrylic fibres are not attacked by moth larvae or other insects.

### Micro-organisms

Acrylic fibres are not attacked by micro-organisms. Fabric buried in soil containing a variety of micro-organisms retained its original bursting strength after 6 months. Cotton, under similar conditions, had undergone a complete loss of strength after 2 weeks' incubation.

### Electrical Properties

The electrical resistance of acrylic fibres is of the same order as that of other man-made fibres. The following table, for example, shows the electrical resistance of 'Courtelle' in comparison with some other fibres. The measurements were taken on scoured samples, after conditioning at 50 per cent r.h. and 20°C. For each measurement 1.5 grams of the sample were placed between two circular electrodes, and the resistance measured on the B.P.L. Megohmeter at a test pressure of 500 volts while the sample was under a load of 30 g./sq.cm.

|  | Resistance ($\times$ $10^{12}$ ohms) |
|---|---|
| Acetate | 5.0 |
| Cotton · | 0.006 |
| 'Courtelle' | 5.0 |
| Nylon 6.6 | 3.0 |
| PET Polyester | 5.0 |
| Rayon staple ('Fibro') | 0.014 |

**Allergenic Properties**

Acrylic fibres applied to the human skin for long periods have shown no dermatological or other ill effects. The fibres have no known toxicological effects.

**Refractive Index**

1.52 ('Courtelle').

## ACRYLIC FIBRES IN USE

**General Characteristics**

*Cross-Section*

Acrylic fibres are made in a variety of cross-sectional shapes, and this has an important effect on the nature of the fabrics produced from them. Wet spinning, in general, yields fibres of round or bean-shaped cross-section. Dry spun fibres are generally of dog-bone or flat cross-section.

The bending stiffness of a flattened cross-section is less than that of a round cross-section fibre of equal cross-sectional area; the

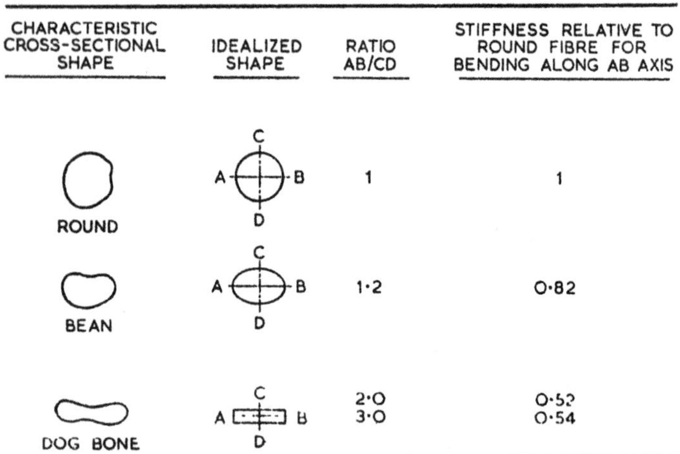

| CHARACTERISTIC CROSS-SECTIONAL SHAPE | IDEALIZED SHAPE | RATIO AB/CD | STIFFNESS RELATIVE TO ROUND FIBRE FOR BENDING ALONG AB AXIS |
|---|---|---|---|
| ROUND | | 1 | 1 |
| BEAN | | 1·2 | 0·82 |
| DOG BONE | | 2·0 3·0 | 0·52 0·54 |

Cross-sectional shape. Effect on Bending Stiffness—*Chemstrand Ltd.*

bending stiffness of a flattened cross-section with a three-to-one ratio of principal axes is approximately one-third that of a round fibre of equal orientation. Many dry-spun fibres have about a three-to-one ratio of principal axes.

Bending stiffness is also directly related to modulus, and it follows, therefore, that orientation (which strongly affects modulus) will have a significant effect on bending stiffness. Changes in orientation may thus tend to mask or reinforce the effects of cross-section on bending stiffness, depending on the way the two factors are combined.

The high bending stiffness of the round or bean-shaped cross-section acrylic fibres is particularly advantageous in carpet fibres, contributing to resilience or spring-back. A flattened or dog-bone cross-section, on the other hand, is conducive to a softness of touch in fabrics made from this type of fibre. The dog-bone type of acrylic fibre also has a distinctive effect on the reflection of light, and this is often noticeable as a sheen or lustre in certain fabric constructions.

*Mechanical Properties*

The stress-strain relationships of different acrylic fibres vary over a wide range (see page 405), each fibre being the product of its particular chemical and physical structures.

The degree of orientation resulting from drawing is a most important factor in determining the mechanical properties of the fibre. This may be seen from the diagram on page 405, showing the stress-strain characteristics of a spectrum of acrylic fibre types. Continuous filament, as exemplified by 'Creslan' 3, is usually more highly oriented. 'Orlon' 29, on the other hand, has a very low degree of orientation. All possible variations between, and even beyond, these limits are possible.

In general, the tenacities of acrylic fibres may be regarded as lying in a range between that of standard rayon and nylon. Acrylics are strong enough for all the normal apparel applications but are not usually considered for high-strength applications such as those in which nylon and polyester fibres serve.

Allied with good tenacity and extension is an excellent elastic recovery and high initial modulus. The mechanical properties as a whole, therefore, are such as to make for the production of fabrics of good dimensional stability. Most important of all, perhaps, is the fact that these mechanical properties are retained in conditions which would tend to bring about deterioration in many other fibres. Acrylics are not significantly affected, for example, by exposure to sunlight, moisture, micro-organisms, chemicals or solvents.

The mechanical properties of acrylic fibres combine to provide the attractive handle which is so important a characteristic of fabrics constructed from these fibres.

## Specific Gravity

Acrylic fibres have a low specific gravity, slightly higher than that of nylon. They provide lightweight fabrics which bulk well.

## Moisture

The water absorbency of acrylic fibres is generally low, but it is higher than that of polyester fibres. Acrylics do not, as a rule, present the difficulties caused by accumulation of static electricity to the extent that polyester fibres do. Also, the small water absorption contributes significantly to the dyeability of acrylic fibres.

Water brings about only a negligible amount of swelling of acrylic fibres, and it has little effect on mechanical properties. This contributes greatly to the dimensional stability, wrinkle resistance and ease-of-care characteristics of fabrics made from acrylic fibres.

Despite the low moisture absorption, acrylic fibres are comfortable when worn next to the skin. This is due very largely to the readiness with which these fibres will remove water by wicking.

## Thermal Properties

Acrylic fibres tend to be heat-sensitive, and heat-setting is necessary to achieve good stability. Heated to high temperatures, the fibres decompose rather than melt, but the temperature at which this takes place is sufficiently high to cause no practical difficulties in normal textile applications.

In common with other thermoplastic fibres, acrylics may be heat-set. Fabrics made from 100 per cent acrylic fibre or from blends containing 50 per cent or more acrylic fibre may be durably pleated.

Acrylic goods are moderately flammable.

## High Bulk Fibres

The long polymer molecules in a polyacrylonitrile fibre behave as though they are composed of a mixture of reasonably well-ordered crystalline areas and less well-ordered or amorphous areas. Even fibres spun from pure acrylonitrile do not show the same degree of order as is possible in nylon and polyester fibres.

Furthermore, acrylic fibres do not develop much additional crystallinity on orientation or heat-setting. They differ in this respect from polyamide and polyester fibres, and acrylic fibres are in consequence more difficult to heat-set.

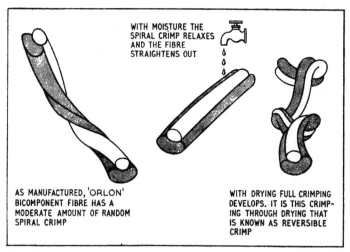

WITH MOISTURE THE SPIRAL CRIMP RELAXES AND THE FIBRE STRAIGHTENS OUT

AS MANUFACTURED, 'ORLON' BICOMPONENT FIBRE HAS A MODERATE AMOUNT OF RANDOM SPIRAL CRIMP

WITH DRYING FULL CRIMPING DEVELOPS. IT IS THIS CRIMP-ING THROUGH DRYING THAT IS KNOWN AS REVERSIBLE CRIMP

*'Orlon' Bicomponent Fibre.* The crimp in this fibre is an inherent property resulting from the bicomponent structure of the fibre. Two filaments of differing chemical composition are fused during spinning, forming a double filament in which each component behaves differently with respect to its swelling characteristics. This results in a three-dimensional crimping effect.

The effect is similar to that shown by wool, in which the fibre crimp results from the bicomponent nature of the protein forming the cortex of the fibre.

413

When amorphous polymers are stretched, they undergo some molecular orientation. With increasing stretch and temperature, the molecules begin to slide past one another, and plastic flow takes place. If orientation is allowed to occur without plastic flow, and the stretched fibres are cooled, they will remain in their stretched form until such time as they are again heated to the point at which the molecules are free to move. The fibres will then retract to their unstretched form.

Acrylic fibres behave in this manner when heated and stretched, but they also have a proportion of well-ordered crystalline regions. These act as anchor points for the molecules, preventing plastic flow from taking place and ensuring that the molecules return to their original positions when the stretched fibre is heated and allowed to relax.

This shrinkage memory of acrylic fibres may be lost, of course, if excessive plastic flow takes place by persistent stretching at high temperatures, or by annealing the fibre at constant length. These techniques are, in fact, used by fibre producers to make fibres of higher modulus.

Precisely controlled, shrinkable fibres are made by most acrylic fibre manufacturers for use especially in sweaters and knit goods. These fibres have built-in shrinkages of 20–23 per cent for optimum bulking effects, and are referred to as regular shrinkage, high-bulk or hibulk acrylic fibres.

These high-bulk acrylic fibres are used for creating lightweight yarns and fabrics which are bulky and warm. High-bulk yarns are made, for example, by spinning together stretched and regular fibres. After the yarn has been spun, it is heated under conditions which bring about relaxation of the stretched fibre. This shrinks to its pre-stretched length, forming a 'core' in the yarn. The unstretched fibres buckle and crimp, and are forced to the outside of the yarn as the stretched fibres shrink.

Fabrics made from high-bulk acrylic yarn are lofty and warm, and yet possess the stability to wear and washing that is characteristic of acrylic fibres generally.

Acrylic fibres of even higher controllable shrinkages are also produced. Fibres of 30 to 45 per cent shrinkage, for example, are used in pile fabrics, where they are combined with non-shrinking, heavy-denier 'guard hair' acrylic fibres. The shrinkage fibres become the dense, soft inner layer of simulated fur, while the guard hair fibre provides the soft aesthetics and appearance of natural fur.

Heavy denier acrylic fibres capable of shrinking 15 per cent are used for special styling effects in carpets, or for achieving especially dense pile for heavy wear applications. Medium shrink fibres of this type are also used where less shrinking and therefore less bulking is desirable in both knitted and woven applications.

Processing techniques have been developed for making use of the high-shrinkage characteristics of acrylic fibres. The $A$ to $Z$ process developed by the Linen Industries Research Association in Northern Ireland is based on stretching acrylic tow or other synthetics to approximately 20 per cent and then feeding the tow through a relaxing zone with constant take-off and varying the speed of the input. The tow is then cut to staple and processed in the conventional worsted manner, enabling a whole range of fibre shrinkages from 0 to 20 to be produced. If required, higher shrinkages may be achieved by this process.

Acrylic knitting yarns processed by the $A$ to $Z$ technique have been used successfully in the field of children's schoolwear. A high bulk-to-weight ratio makes this yarn very economical for hand-knitting, and the garments are machine-washable.

Another process which makes use of the high-shrinkage capability of acrylic fibres is the Perbul process developed by Mitsubishi Rayon and Daido Worsted Mills of Japan. This process develops alternating segments with shrink, no-shrink properties within individual fibres; it is designed for the woollen system.

### Environmental Conditions

Acrylic fibres have outstanding resistance to sunlight, micro-organisms, insects and ageing, and have excellent outdoor weathering resistance.

### Chemical Resistance

Acrylic fibres have a good resistance to all the chemicals likely to be encountered in normal textile use, including bleaches, dilute acids and alkalis, dry-cleaning solvents, etc.

*Electrical Properties*

The low moisture absorption tends to encourage accumulation of static electricity, but this may be overcome by the use of suitable antistatic finishes.

*Helical Crimp*

Fibres with a helical crimp develop more bulk than fibres with a planar, zigzag crimp. Also the helically-shaped fibre has more resilience than a similar fibre of zigzag form. Such fibres resist matting and inter-fibre slippage better than fibres with planar crimp.

Crimped natural fibres, like wool, tend to have a helical crimp, whereas man-made fibres commonly have a planar, zigzag crimp resulting from a gear or stuffing-box process operated with the help of heat and pressure.

Mechanically-induced crimp tends to be pulled out rather readily from acrylic fibres during normal textile processing. This can be prevented to some extent by heat setting.

Helical crimp may be introduced into acrylic fibres by mechanical means, e.g. by a false twisting and simultaneous heat setting operation like that used for nylon and polyester fibres. It may also be introduced by spinning the fibre in bicomponent form.

In 1959, du Pont introduced their 'Orlon' Type 21 fibre, which subsequently became known as 'Sayelle'. This fibre combines two filaments in a single strand, the filaments being fused lengthwise to form an acorn-shaped cross-section. A difference in shrinkage characteristics produces a permanent and reversible crimp, providing a fibre with excellent cover and compressional resilience, improved bulk and loft, good wrinkle resistance and ease-of-care properties. Helically crimped fibres were also developed in which the helical crimp was unaffected by moisture

This type of bicomponent acrylic staple has proved particularly useful as a fiberfill for pillows, quilting and the like. The fibres tend to resist matting by remaining interlocked, even through washing and drying.

Helically crimped acrylic continuous filament yarns were subsequently introduced for sweaters and knit goods, the crimp being fully developed after boiling off. Goods made from these yarns have a crisp, wool-like handle and excellent elasticity and dimensional stability.

## Continuous Filament Yarns

The original uses for acrylic fibres were primarily in industrial and outdoor applications. This was largely because of the good solvent and chemical resistance and the outdoor ageing properties. It soon became apparent that volume would be limited for the relatively high-priced continuous filament fibre. Early techniques for conferring dyeability and level dyeing in continuous filament were not sufficiently successful to be economically attractive, however, and most of the development of acrylic fibres centred upon staple fibre.

Despite this concentration on staple fibre, a few firms persevered with continuous filament acrylic fibre and began producing yarns. Fabrics made from continuous filament acrylic fibres have a dry, silk-like hand, and are smooth and lustrous in appearance Fleece and textured knit fabrics are naturally dry and soft to the touch.

## Producer Coloured Fibre

The dispersion of pigments in the spinning solution provides producer coloured or spun-dyed yarns. Alternatively, it is possible to use substantive dyes to colour the fibre while it is in the swollen state after coagulation and solvent removal.

Pigments or dyes are selected with excellent fastness properties and this type of fibre is ideal for outdoor uses, such as tarpaulins, awnings and tents.

A number of producer coloured acrylic fibres have appeared on the market in recent years.

## Modified Surfaces

The surface characteristics of acrylic fibres may be changed to affect important properties such as surface friction, antistatic behaviour and wettability.

Fibres which combine dog-bone cross-section, low crimp and lower surface friction provide a smooth hand and lustrous appearance to fabrics. Fibres of flat cross-section and lower-than-normal orientation result in a fairly soft hand, yet such fibres are sufficiently resilient and stiff to form loops for fabric surface effect.

Acrylic fibres with a durable antistatic surface have been developed especially for blankets. These fibres do not build up static charges as readily as unmodified acrylic fibres, and dissipate static electricity quickly. The antistatic property is permanent, at least through ten washings in a home laundry.

Silicone-treated acrylic fibres are produced for the U.S. Navy. These fibres are made into jackets which are water-repellent and have excellent life-saving flotation properties. The jackets have high thermal insulation value and offer protection also against shotgun blasts, fragmentation and even bullets.

### Modified Heavy Denier for Carpets

Special types of heavy denier acrylic fibre are produced for use in carpets.

### Washing

Knitted and lightweight garments of 100 per cent acrylic fibre should be washed in warm water (40°C., 104°F.) using a detergent or soapflakes, while more robust garments such as shirts may be washed in hand-hot water (48°C., 118°F.). Thorough rinsing in lukewarm water, followed by cold water, is recommended to preserve the softness for which acrylic fibre fabrics are outstanding. The use of a proprietary softening agent in the final rinse will also help to retain softness and colour.

Pleated garments should be washed by hand and given a warm rinse followed by a hand-hot (48°C., 118°F.) rinse and drip-dried.

Acrylic sweaters and lightly constructed garments are preferably washed by hand. Machines may be used for garments which are labelled as suitable for such treatment.

### Drying

Acrylic fibres absorb very little water, and garments made from them dry easily. A spin drier may be used for acrylic garments which are not pleated, provided that the garments are quite cold when put into the drier. In this way, garments should be dried very quickly, although slightly more ironing may be necessary.

Acrylic garments require the minimum spinning time (approximately 15 seconds). It is recommended that this period is not exceeded, and that the machine is not packed too tightly with garments.

The majority of woven and firmly knitted garments may be drip-dried, but heavy knits should be dried flat. Tumbler drying is not recommended unless the temperature can be controlled to keep below 60°C. (140°F.) followed by cold tumbling.

### Ironing

Knitted garments are usually ready to wear as soon as they have dried. Some garments and woven goods may require a minimum of ironing, which should be carried out with a cool iron (HLCC Setting 1) on the reverse side of the fabric when the garment is dry.

### Dry Cleaning

Acrylic fibres are not affected by the solvents usually used in dry cleaning, and garments may be dry cleaned without difficulty.

The tumbling temperature must not exceed 60°C., (140°F.). Any pressing treatment given should be light to avoid glazing.

### End Uses

*Cheap Raw Material*

There were a number of reasons for the rapid increase in the production of polyacrylonitrile fibres. Acrylonitrile, the basic raw material, was available in plentiful supply and at a very low price compared with raw materials for other synthetic fibres. Also, the production of polyacrylonitrile fibres is relatively simple, and the patent position makes it possible for a producer to manipulate the basic principles and still evolve a polyacrylonitrile fibre with traditional characteristics of this type of fibre.

Despite the astonishing progress of polyacrylonitrile fibres, and their ready acceptance into a wide range of applications, the early years were beset with many troubles. High fibre price and serious dyeing problems held up progress during the early years.

At first, disperse dyestuffs were virtually the only class that could be used on polyacrylonitrile fibres. These dyes gave satisfactory fastness in pale shades, and they are still used mainly for this purpose. But they do not build up to medium and deep shades, and for the production of heavy shades, acid dyestuffs were applied, using the 'cuprous ion' technique. This was not adopted widely, however, owing to difficulties in obtaining level dyeings and the restricted range of acid dyestuffs having good dyeing properties by the cuprous ion technique.

The basic (cationic) dyestuffs were found to be complementary to the disperse dyestuffs in that they would provide deep shades of very good fastness to washing, and moderate fastness to light. They were not entirely satisfactory for pale shades, however, because of a very high initial rate of dyeing. The main defect of the old type basic dyes on polyacrylonitrile fibres was their moderate fastness to light. The newer types of cationic dyes have a very high fastness to light and improved dyeing properties.

Improvement in the dyestuffs and dyeing techniques has been accompanied by the development of polyacrylonitrile fibres with modified properties. The incorporation of small amounts of second component monomers has provided acrylonitrile copolymers with increased affinity for anionic dyestuffs applied by conventional dyeing methods. These acid dyeable fibres have found ready acceptance for blending with basic dyeable fibres to produce cross-dyeable fabrics.

## Knitted Outerwear

As the dyeing problems were overcome and prices reduced gradually, polyacrylonitrile fibres found their first substantial market in Turbo-processed yarns for knitted outerwear. Tow-to-top conversion via the Turbo stapler provided high-bulk knitting yarns with a soft luxurious hand and excellent bulk for weight. Garments made from these yarns were machine washable, whereas wool sweaters had to be hand-washed and block dried.

## Carpets

While the fine denier polyacrylonitrile fibres were making headway in knitwear, development of heavier fibre – e.g. 17 dtex (15 den) – was undertaken with a view to penetration of the carpet market, which had traditionally been dominated by wool. An 'Acrilan' carpet fibre, for example, was introduced in 1956, and carpets made from this fibre were equivalent in performance to those of wool in soilability, abrasion resistance and crush resistance, and were superior in cleanability and shampooability.

Brighter, cleaner shades could be obtained with 'Acrilan', and yarn yield factors of 95–98 per cent were realized with 'Acrilan', compared with about 85 per cent with wool. Despite this difference, however, polyacrylonitrile fibres were initially more expensive than wool, and they remained a luxury carpet fibre for several years, until eventually the price was reduced to a level comparable with wool. The excellent characteristics of the polyacrylonitrile fibres, coupled with uniformity and steady price, enabled the new fibres to make rapid headway in the carpet trade. Stock dyeing or conventional hank dyeing are conventional practices with polyacrylonitrile carpet staple, and either woollen system or long staple worsted spinning is used in yarn manufacture.

## Furnishing Fabrics

Acrylic fibres have made substantial progress in the field of furnishing fabrics, providing materials that combine luxurious appearance with first-rate practical performance. Curtains are made in a wide variety of weaves, ranging from sheer fabrics that have enough body to hang well as perfect summer curtains, to heavy velvets that provide richness and warmth allied with easy-clean properties that are a feature of acrylic fibres.

These same characteristics of acrylic fibres serve them well in blankets and bedspreads, upholstery fabrics, tablecloths and other furnishing and household applications.

## Apparel Fabrics

Following the successful introduction of acrylic fibres into the knitwear and carpet trades, the acrylic manufacturers explored other markets for this type of fibre. They had a fibre with many properties similar to those of wool, but with characteristics which made it preferable to wool in many instances. Resistance to micro-organisms and insects, dimensional stability to washing, excellent resistance to outdoor exposure, sunlight resistance, bulk without weight, good covering power and a steady price are some of the attractive features offered by acrylic fibres which prompted manufacturers to use the fibre in blankets, blends with viscose for men's and boys' slacks, blends with wool for ladies' dresses and skirts, single jersey fabrics for men's and boys' shirts, ladies' blouses and dresses, and blends with wool for men's slacks and suits.

## Single Jersey Fabrics

Single jersey fabric made from a blend of regular and high-bulk staple spun on the cotton system to 21s count was an immediate success. The piece-dyed fabric is machine washable and very stable. Acrylic jersey sports shirts are now established in a field that was previously dominated by cotton.

## Tufted Pile Liners

In tufted pile liners, the use of acid dyeable and basic dyeable acrylic fibres in combinations or blends has enabled patterned liners to be made. This has become an important outlet for acrylic fibres.

## Circular Sliver Knit Fabrics

Acrylics have become established in the Wildman circular sliver knitting trade. Liner fabrics and certain types of outerwear fabrics simulating natural animal fur fabrics were first produced from stock-dyed fibre and later with solution-dyed or spun-dyed fibres. In this end use, the acrylics offered advantages that could not be matched by other fibres, enabling manufacturers to make reasonably priced, high quality fabrics for the mass market.

In addition to the liner and outerwear fabrics produced using Wildman circular sliver knitting machines, another interesting use has developed in the medical field. Tests conducted in hospitals with acrylic Debicure pads made from sliver knit fabrics proved that these could be a remedy for bedridden patients suffering from bed sores. Moreover, fabrics of this type do not support bacteria, and can be cleaned easily, dried quickly and put back into service.

## Outdoor Fabrics

Acrylic fibres have excellent resistance to sunlight, insects and micro-organisms, and acrylic fabrics have always found a ready outlet in outdoor applications. With the advent of solution-dyed or spun-dyed acrylics, new opportunities were created in this field. Awning fabrics moved into the acrylic painted cotton and vinyl-coated cotton areas and made good headway against established cotton fabrics.

Pigments with excellent fastness properties are available for this type of use, capable of withstanding thousands of hours of sunlight without noticeable change of shade. This, coupled with the good fabric strength retention of acrylics after hours of sunlight and weather exposure, has made other outdoor uses such as flags, ski-jackets, hunting jackets, boat covers and swimming pool covers natural outlets for the fibre.

## Flocking

The process of flocking has become an important textile manufacturing process in recent years. This has been due partly to the introduction of synthetic fibres generally, and also to the considerable advantages offered by the acrylics and nylons over the natural fibres. Acrylics can be shipped in tow form which is ideal for cutting into flock. The availability of various types of acrylic in low counts, its luxurious hand and good recovery from crushing make it admirable for apparel flocking use.

Acrylic flocked fabrics are now established in various markets. Pile flocked fur-like fabrics may be made from acrylic fibre, for example, and then printed, brushed and finished. Other applications include velvet and corduroy apparel fabrics, suede, toy animal furs, souvenir pennants, wall panels, paint rollers and record player turn-tables.

Flocked floor coverings are now being produced from acrylic fibre, and acceptance has been good in many applications. The motor car industry, for example, is using flocked floor coverings in appreciable amounts.

*Tufting*

Fine gauge tufting has assumed a position of some importance in the tufting trade, especially in the blanket section of the industry. Tufted blankets and liners have added to existing available products and extended consumer choice in this field.

By using various combinations of bright and dull fibre a wide range of appearances and textures may be obtained, and by polishing and tigering such fabric a variety of cloths has been produced for a range of markets.

*Non-woven Fabrics*

In the non-woven fabrics trade, the Chatham Fibrewoven process for producing needle-punched blankets was the first significant technical development in which acrylics played an important role.

The Fibrewoven process is a sophisticated improvement over ordinary needle-punching techniques. The usual needle loom employs needles operating reciprocally in a vertical position normal to the plane of the web. The Fibrewoven machine employs needles operating at an angular displacement with respect to the line of travel of the web. By suitable design of needle penetration and web advance, a mechanical interlocking of fibres in the machine direction is effected, resulting in improved fabric properties.

Blankets represent the initial product of this process, but many other products are of great potential importance. The fibre spends only 8 minutes in transit from the feed hopper of the card to the roll of grey-state fabric. If this is compared with the time associated normally with fibres in process in woven blankets or other fabrics, it is easy to see the advantages offered by the Fibrewoven process.

Acrylic fibres, already well established in the blanket trade, were the fibres selected for the development of blankets by the Fibrewoven principle, and as a result the acrylics are now playing an even greater role in the blanket industry as a whole.

Felts of needle-punched acrylics offer significant strength retention properties and resistance to heat, chemicals and abrasion. They are unaffected by mildew, rot and fungi, are water repellent, easily washed and cleaned. Acrylics give unique fabric cohesion due to the action of the needles catching random fibres and entangling them. Steam or heat treatment draws the fibres even closer together so that the felts can be produced in almost any desired weight, thickness and density – to rigid specification. They have numerous applications in many industries including filtration, padding, thermal insulation (weather stripping for windows), shock absorption, interlinings and packaging. Such non-woven fabrics are gaining slowly but steadily a hold on the market in place of wool, hair, cotton and cellulosic fibres.

## Paper

Certain types of acrylic fibre may be beaten into paper. These papers are expensive, but for specialized applications they serve a useful function. Acrylic papers are used, for example, in geological survey maps.

## Core Yarns

Elastomeric yarns of the polyurethane type have brought revolutionary changes in the textile industry. These yarns have low tenacity, elongation in the region of 500 per cent and a rubber-like elasticity. They were developed primarily to take the place of rubber threads in the foundation garment industry, but they are now used in the weaving and knitting industries as well.

Core-spinning, an old technique, has been revived with elastomeric yarns, and acrylics in 100 per cent form or in blends with other fibres are being used as the sheath fibres around the elastomeric core. This type of yarn is being used in cuffs, neck bands and waist bands of sweaters, and in some instances throughout the sweater, providing comfort, good shape retention and improved elastic properties. Bathing suits, socks and double jersey fabrics are applications in which acrylic/elastomer core-spun yarns are gaining popularity.

In the weaving industry, core-spun yarns are assuming an increasing importance, with stretch woven fabrics finding immediate markets in ladies' slacks, men's trousers and suits, shirtings and upholstery fabrics. Elastomeric core-spun yarns are being produced on cotton, woollen and worsted spinning systems, providing another outlet for acrylic fibres in the years ahead.

## (2) MODACRYLIC FIBRES

This category of polyacrylonitrile fibres includes modacrylic fibres spun from polymers consisting of less than 85 per cent by weight of acrylonitrile units ($-CH_2-CH(CN)-$), but excluding those copolymers in which acrylonitrile is not the major component.

### TYPES OF FIBRE

The fibres in this category are spun from an extensive range of copolymers of acrylonitrile, in which the nature and proportion of the second (and possibly other) components may vary within wide limits. The types of modacrylic fibre that can be produced within this broad category are capable of wide variations in properties, depending on their composition and method of manufacture.

As in the case of acrylic fibres, details of the chemical structure of individual modacrylic fibres are not always available. The second component of the polymer is, however, commonly chosen from vinyl chloride, vinylidene chloride or vinylidene dicyanide.

The information which follows is based on data for VEREL* fibre, which may be regarded as a representative fibre of this type.

---

* Registered trade mark of Eastman Kodak Company, Rochester, N.Y., U.S.A.

## VEREL

VEREL fibre is of the polyacrylonitrile type; it is produced by Tennessee Eastman Company, U.S.A. It is spun from a copolymer of undisclosed composition, in which acrylonitrile is probably the major component, being present to the extent of some 60 per cent.

By F.T.C. definition, VEREL is a modacrylic fibre.

### TYPES OF FIBRE

VEREL fibre is manufactured only in staple form. A range of staple lengths is available to suit all processing systems, in the following deniers: 3, 5, 8, 12, 16, 24 and 40. The fibre is produced in bright or dull lustre, and various crimp levels and different degrees of crimp permanence are available.

VEREL fibre is produced in two basic cross-sections, peanut and ribbon.

### PRODUCTION

#### Monomer Synthesis

*Acrylonitrile.* See page 399.

#### Polymerization

No information available.

#### Spinning

The polymer is dissolved in solvent (probably acetone) and the spinning solution is pumped through a spinneret. The fine jets of solution emerge probably into a coagulation bath (wet spinning), in which the solvent is removed to leave solid filaments. These are gathered into a bundle or tow which is processed to stabilize it. Lubricating oils are added to aid in the subsequent spinning of yarns, and a crimp is added.

The tow moves to a cutting machine where the continuous strands are cut into staple fibre.

## PROCESSING

### Scouring

Scouring of fabrics of VEREL fibre follows the general pattern established for other man-made fibres. A neutral scouring with detergent will remove dirt acquired from previous processing.

Alkaline scouring should be carried out when necessary with a mild alkali such as disodium phosphate or tetrasodium pyrophosphate at 60°C. A stronger alkaline scour will discolour VEREL.

### Bleaching

VEREL fibre does not as a rule require bleaching, as it is white enough for most end-uses. Sodium chlorite and formic acid may be used to slightly bleach VEREL if necessary. The use of optical brighteners plus small amounts of cationic dye is the most practical and effective way to brighten VEREL.

### Dyeing

(a) *100 per cent VEREL Fibre*

VEREL dyes readily and no special equipment is needed. In dyeing most shades a dyeing assistant is used to ensure complete exhaustion of the dye and to provide best fastness properties.

VEREL can be dyed as raw stock, skein or piece goods at an operating temperature of 71°C for best results.

Three classes of dyes may be used successfully in dyeing VEREL: (1) basic, (2) disperse, and (3) neutral premetallized.
*Basic Dyes*

Basic or cationic dye combinations are available with excellent fastness properties on VEREL fibre. When used in combination with the proper dyeing'assistants, basic dyes level well and exhaust almost completely. In addition, a complete range of shades can be obtained with these dyes.

Proper selection of dyes is important. Many basic dyes have poor fastness properties and certain basic blues are sensitive to heat; they tend to discolour or reduce if exposed to high temperatures for long periods of time. The use of these dyes should obviously be avoided.

A disadvantage of basic dyes is that they are difficult to strip or level once an off-shade or unlevel dyeing has occurred.

Stripping or levelling of basic dyes is difficult, but it can be done. Most dyes can be stripped with a combination of a dyeing assistant, a levelling agent, and an oxidizing or reducing agent or both, depending on which dyes are used.

Because of the excellent affinity of VEREL for basic dyes, retarding agents are sometimes needed to prevent unlevel dyeings.

A retarding agent such as Ahcol 177 (ICI United States) or Cordon AES—65—SY (Finetex Inc.) should be used when-dyeing VEREL with basic dyes. Level dyeings can be obtained by using a slow rate of temperature rise (1°C/min).

Basic dyes are commonly used on 100% VEREL fibre and in VEREL blends. Disperse and neutral premetallized dyes are rarely used.

### Disperse Dyes

Selected disperse dyes have good colour fastness and levelling qualities. The range of shades obtainable with these dyes is limited. Colour-fastness to washing in medium and heavy shades is generally poor. Blacks formulated with disperse dyes are economically impractical. Care should be used in the selection of dyes as many also have poor fastness to light and crocking.

### Neutral Premetallized Dyes

These dyes have been used very successfully, especially for dyeing carpets of VEREL fibre. The shade range is limited in that very bright shades are impossible to obtain.

### Notes

#### 1. VEREL Dyeing Assistant

Eastman MDA—1 modacrylic dyeing assistant is a self-emulsifying organic ester used to promote exhaustion of dyes on modacrylic fibre. Eastman MDA—1 is prepared with water at 24-38°C to provide a stable emulsion. The emulsion is added slowly to the dyebath before the dyes are introduced.

Eastman MDA—1 is used in concentrations of 0.5—6.0% (based on the weight of the fibre) depending on the depth of shade.

## 2. *Low Temperature Dyeing*

Although modacrylic fibres can be dyed at temperatures above 88°C, Eastman has developed a method of dyeing VEREL fibre at 71°C that prevents packed dyed cakes, yarn and fabric distortion, and severe delustring. Shades with good penetration and fastness can also be obtained at temperatures as low as 60°C. Low temperature dyeing, however, requires a good balance of dyeing assistants and careful selection of dyes.

All types of VEREL fibre can generally be dyed at low temperatures, but basic dye may not adequately penetrate tightly woven fabrics made with high-twist yarns. Good penetration will result from the use of the proper retarding agent, to prevent fast strike and permit slow exhaustion. If it is necessary to exceed 71°C, 50% common salt should be included for the last 30 minutes of the dye cycle. VEREL fibre dyed in this manner can be dried immediately at 82–138°C and restored to full lustre.

## 3. *Delustring*

After fabrics of VEREL fibre have been dried, they can exhibit delustring which is easily recognised by weak shade, dull appearance and harsh hand. Delustring of VEREL fibre can occur at wet-processing temperatures as low as 60°C, and it becomes increasingly severe between 60°C and the boil. Additionally, air drying can increase the severity of delustring. If delustring occurs, fabrics of VEREL fibre can be restored to normal appearance and character in any of several ways; for example, by exposure to dry steam or to hot (71°C) aqueous salt solutions, followed by proper drying. The use of dry heat in conventional stock, package, skein, loop or tenter dryers, however, is usually the most practical way of relustring mildly delustred VEREL fibre. Drying the fibre at 82–138°C should completely restore dyed VEREL fibre to its original lustrous condition. The fibre should be kept moist before being dried. Exposure to hot aqueous salt solutions (20–50%), followed by drying at the maximum drying temperature, is the most effective way of relustring severely delustred VEREL fibre. The larger denier VEREL fibres are more difficult to relustre than the smaller denier fibres.

## (b) *Blends of VEREL and Acetate*

VEREL can be dyed and acetate left reasonably clear through the use of selected neutral premetallized dyes. It is not possible to dye the acetate and leave VEREL white. Any dye that will colour acetate will also dye VEREL. By properly selecting disperse dyes, unions may be obtained on these two fibres. If necessary, the VEREL can be dyed to shade by adding neutral premetallized dyes which will not affect the acetate. Cross-dyed effects on this blend are limited. It is possible to select colours that will dye acetate heavier than VEREL. The VEREL can then be shaded with basic or neutral premetallized dyes for two-colour effects. Such dyeings are, however, difficult to control.

## (c) *Blends of VEREL and Acrylic*

Because of the wide variation in dyeing behaviour of different acrylic fibres, it is not possible to make general statements about colour effects on VEREL/acrylic blends.

## (d) *Blends of VEREL and Cotton, and VEREL and Rayon*

VEREL blended with cellulosic fibres permits a wide range of colour effects. Such blends may be union dyed, cross-dyed, or either fibre can be coloured and the other left white. Combinations of direct and neutral premetallized dyes can be applied from a single bath for unions or cross-dyes with good fastness properties. Disperse dyes may also be used in the same bath with direct dyes. Basic dyes can be applied in the first of two baths followed by direct dyes on the cotton or rayon. Successful plant runs have been made by dyeing with basic and direct dyes in one bath. The compatibility of basic and direct dyes should be verified.

Bright shades of outstanding wet fastness properties can be produced by first dyeing the cellulosic fibre with napthols and then topping the VEREL in a second bath with basic dyes. Selected disperse or neutral premetallized dyes will leave cotton or rayon white or slightly stained. Basic dyes can also be used for colour and white effects when bright shades are required. Selected direct dyes and naphthols, in the presence of cotton or rayon, will leave VEREL white. Many vat and sulphur dyes will stain VEREL in the presence of cotton or rayon. The high concentration of sodium hydroxide used in vat dyeing may cause VEREL to discolour.

### (e) *Blends of VEREL and Nylon*

VEREL can be dyed with selected basic dyes leaving nylon white or only slightly stained. Nylon can be dyed with selected acid or acid premetallized dyes leaving VEREL white or only slightly stained. Unions are possible by any one of three methods:

1. Selected neutral premetallized dyes with a nylon retarding agent to achieve balance.
2. Selected disperse dyes using a retarding agent for nylon to achieve balance.
3. Combination of acid and basic dyes applied from either a one- or two-bath method. Unions or cross-dyes can be obtained.

### (f) *Blends of VEREL and Polyester*

This blend is used in scatter rugs and knit goods. It is possible to obtain solid colours by dyeing the VEREL and polyester fibres separately, and then blending the fibres.

In yarn or piece dyeing, selected basic dyes may be used to dye the VEREL only and leave the polyester relatively unstained. Basic dyes will produce a wide range of colours with excellent light and wash fastness properties. A limited range of colours may be obtained by using either disperse or neutral premetallized dyes.

### (g) *Blends of VEREL and Wool*

Several methods may be used for dyeing this blend. By using selected acid, chrome, or acid premetallized dyes, the wool can be dyed leaving the VEREL fibre white. Basic dyes may be used to dye the VEREL and leave the wool relatively unstained. Stain may be removed from the wool by means of zinc sulphoxylate formaldehyde reducing agents, without affecting the shade of the VEREL. Unions may be dyed with selected neutral premetallized dyes. Retarding agents of various types have been used to control unions. Two-bath processes using acid and basic dyes may also be used for desirable unions.

## Printing

The inherent flame resistance of VEREL fibres makes them useful for many kinds of printed home furnishings. Twills, velours and basket and plain weaves are a few of the fabric types in which VEREL modacrylic fibre can be used for making finished goods such as draperies, bedspreads, sleepwear, upholstery and mattress ticking. Other qualities which have added to the widespread acceptance of VEREL modacrylic fibre are its warm, luxurious and soft hand, and its resilient texture and excellent dyeability.

### *Printing Assistants and Methods*

VEREL modacrylic fibre is easily printed with cationic, disperse, and neutral-premetallized dyes. Also, resin-bonded pigments can be applied to fabrics made from VEREL fibre. Conventional printing equipment and methods can be used to print a wide variety of fabrics made with VEREL fibres.

*Viscosity Control.* Thickeners are used to produce optimum print-paste viscosity. Strongly anionic thickeners, however, should not be used on fabrics made from VEREL when printing with cationic dyes.

*pH Adjustment.* In addition to the usual pH-adjustment additives, a hygroscopic agent such as glycerine should be added to the print paste.

*Drying and Developing Dyes.* After the fabric has been printed, it may be dried. This step, however, is not essential.

The dyes or pigments used to print the fabric are fixed by atmospheric steaming; the use of carriers or other dyeing assistants is not necessary. When the dyes have been developed, the printed fabrics may be scoured and dried. During scouring, staining of the unprinted background by cationic dyes can be minimized by the use of a strongly anionic retarder. Such retarders have no effect on disperse or neutral-premetallized dyes.

*Pigment Printing.* As pigment printing involves the use of a binding agent to bond the pigment to the fabric, the affinity of individual fibres for the pigment is not a factor in printing with such colorants. The main consideration is that curing temperatures and times must be kept below those which can cause the fibre to yellow.

## Dyes for Printing VEREL

Each class of colorant has some advantages over others.

*Cationic Dyes.* These dyes produce vivid colours of excellent tinctorial power and good fastness. They are generally the best dyes for printing VEREL fibre.

*Disperse Dyes.* These are less expensive than cationic dyes, but the wash and sublimation fastness and the colour brilliance are not, in general, as great as those of cationic dyes.

*Neutral-Premetallized Dyes.* These are usually colourfast and easy to apply, but are not as bright or as fast to washing and sublimation as cationic dyes.

*Resin-Bonded Pigments.* Although easy to apply and economical in light to medium shades, resin-bonded pigments can adversely affect the hand, dry-cleanability and flammability of fabrics. With careful selection, however, these colorants give acceptable colour-fastness. By limiting the amount of coverage, most flammability requirements can be met.

## Delustring During Steaming

Delustring during steaming of VEREL printed fabrics is a potential problem only if the printed fabric is wet (other than with print paste) or if saturated steam is used to fix the dyes. If delustring is encountered during the steaming operation, the fabric should be exposed to a hot solution (10–20 g/1) in a wash box at 94°C and cooled slowly through subsequent wash-boxes. The fabric should then be dried immediately at 104–121°C.

A more desirable solution to this problem would be to eliminate the cause of delustring by drying the printed fabric prior to steaming and by using well-trapped steam to develop the dyes. Objectionable delustred spots may also occur if condensate drips onto the fabric during steaming.

## Stripping

### *Basic Dyes*

Stripping or levelling of basic dyes can present problems. Most dyes can be stripped from VEREL, however, by using a combination of a dyeing assistant (Tanadel V − Tanatex Chemical), a levelling agent (Migrassist AC − Tanatex Chemical), and an oxidising agent or reducing agent or both, depending on which dyes are used.

### *Disperse Dyes*

Most disperse dyes can be partially stripped with 5−10 per cent soap chips at 60°C. for 30 minutes. Sodium chlorite will completely strip many disperse dyes, and zinc sulphoxylate formaldehyde will strip the dischargeable types. Various non-ionic materials have been used successfully to level shaded dyeings.

### *Neutral Premetallized Dyes*

Sodium chlorite and sodium hypochlorite have both been used to strip these dyes from VEREL fibres with good success. Many of the levelling agents suggested for neutral premetallized dyes on wool and nylon also work well on VEREL.

### Finishing

VEREL fibre is used in a wide variety of blends and fabric constructions, each having different end-use requirements. Finishing techniques are selected to suit particular needs, and there is no single finishing process which applies to all fabrics.

### *Resin Treatment*

Resins behave on VEREL fibre as they do on other hydrophobic fibres, i.e. they simply form a coat on the surface of the fibre or fabric. They do not cross-link or otherwise react with the fibre.

A blend of VEREL and cellulosics may be successfully treated with resin if curing times and temperatures are closely controlled.

435

At a curing temperature of 135°C., for example, 5 minutes might be the minimum time to effect a cure. A curing temperature of 149°C. might require only 1½ minutes for curing. Temperatures exceeding 150⁰C will cause VEREL to yellow, the extent of the discolouration depending on the length of time of exposure to elevated temperatures.

The use of resins or other finishes on fabrics designed for maximum flame resistance may sometimes cause the fabric to lose some of its inherent resistance to burning.

### Shrinkage Control

Fabrics containing high percentages of VEREL fibre may be finished with good dimensional stability if they are relaxed completely by heating in the finishing routine. This may be accomplished in a loop or airlay dryer, or similar equipment, at temperatures up to 127°C. Stretching of the fabric during any phase of processing should be avoided.

### Shearing

Fabrics of VEREL fibre are easily sheared on conventional shearing equipment. The fusing of fibre tips may be a problem on dense cut pile fabric, but this can be eliminated by passing the goods through the shear at a slower speed. It is important that the shear blades are kept sharp.

### Singeing

VEREL fibre forms charred black beads on the fabric surface during singeing. Fabrics containing a low percentage of VEREL, however, may be singed successfully both in greige and finished forms.

## STRUCTURE AND PROPERTIES

The information which follows refers to VEREL Type 163, 3D/F lustre fibre.

### Fine Structure and Appearance

A white fibre, available as bright or dull lustre. Some types are available in spun dyed black. Peanut-shaped cross-section. X-ray diffraction patterns show VEREL to be a well-orientated, slightly crystalline fibre.

### Tenacity

15.9– 22.1 cN/tex (1.8--2.5 g/den), dry;
15.0–21.2 cN/tex (1.7–2.4 g/den), wet.

**Tensile Strength**

2,940–3,290 kg/cm$^2$ (42,000–47,000 lb/in$^2$).

**Elongation**

35–40 per cent, dry or wet.

**Elastic Recovery**

88 per cent at 4 per cent; 55 per cent at 10 per cent.

**Initial Modulus**

247.2 cN/tex (28 g/den)

**Average Stiffness**

8.0.

**Average Toughness**

4.33 cN/tex (0.49 g/den)

**Yield Stress**

5.83 cN/tex (0.66 g/den)

**Yield Strain**

3.6 per cent E.

**Compliance Ratio**

1.18.

**Specific Gravity**

1.37.

**Effect of Moisture**

Regain: 3.0–3.5 per cent.

**Thermal Properties**

*Sticking Temperature:* 180–185°C. Maximum safe ironing temperature: 150°C.

*Effect of Low Temperature*

Physical characteristics maintained to extremely low temperatures.

## *Effect of High Temperature*

Regular fibre: shrinkage in boiling water, 0.2 per cent; hot air (140°C), shrinkage 4—6 per cent. A rapid change of properties takes place, with fall in tenacity and modulus, and increase in elongation.

Temperatures exceeding 150°C will cause VEREL fibre to yellow, the extent of discolouration depending on the length of time of exposure to elevated temperatures.

## *Flammability*

VEREL fibre has a very good flame resistance. It is very difficult to ignite, and is self-extinguishing. It leaves a hard black char, and does not drip.

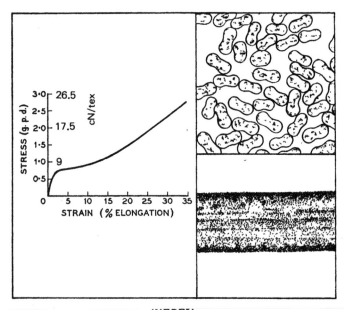

'*VEREL*'

**Effect of Age**

There is virtually no change in the properties of VEREL fibre over an extended period of time.

**Effect of Sunlight**

VEREL has good weathering characteristics; it is better in this respect than any natural fibre and most of the synthetic fibres, including viscose, acetate, polyamides and polyesters.

After 50 weeks' outdoor exposure, VEREL retains good strength and elongation under conditions which destroy acetate, wool and cotton.

**Chemical Properties**

*Acids*

Excellent resistance even at high concentrations.

*Alkalis*

Good general resistance. Alkalis under moderate conditions have no effect on tenacity, but cause some discolouration.

*General*

VEREL has a high degree of resistance to a wide range of chemicals.

**Effect of Organic Solvents**

VEREL fibre resists all dry cleaning solvents and most common organic solvents. It dissolves in warm acetone.

**Insects**

Not attacked

**Micro-organisms**

VEREL fibre is not attacked by mildew and other micro-organisms. It is unaffected after being buried for 12 weeks in moist, biologically active river loam at 25°C. Cotton decomposes after 6–7 days in the same test.

**Electrical Properties**

Dielectric strength exceeds 1,500 volts/mil (film); dielectric constant 410 at 60 cycles.

**Allergenic Properties**

VEREL fibre is non-allergenic.

**Refractive Index**

1.538. VEREL fibre exhibits practically no birefringence.


## VEREL IN USE

**General Characteristics**

VEREL fibre possesses many attractive characteristics from the point of view of general textile use. It combines a soft, warm handle with the capacity for absorbing moisture that makes for comfortable wear. It has a good whiteness and dyes easily and effectively, and may be electropolished to a high lustre.

*Mechanical Properties*

VEREL is a strong, tough fibre. Its mechanical properties are of the same order as the average type of acrylic fibre. Tenacity, elongation and elastic recovery are adequate for normal textile applications, and are not affected significantly by water.

*Specific Gravity*

The specific gravity of VEREL fibre is relatively high (1.37).

*Moisture*

The moisture regain of VEREL fibre is unusually high by comparison with other polyacrylonitrile fibres, and this is an important feature of the fibre. It makes for greater comfort in wear, especially in fabrics worn close to the skin. Despite this characteristic, VEREL is not adversely affected by moisture, its mechanical properties being only slightly affected. Fabrics retain good dimensional stability and wrinkle resistance.

*Thermal Properties*

VEREL is a thermoplastic fibre, with a fairly low softening temperature (120–125°C) by comparison with acrylic fibres, and care must be taken in all processing which involves elevated temperatures.

The non-flammability of VEREL is a great advantage, putting this fibre into almost the same class as polyvinyl chloride type fibres. Its flame resistance is greatly superior to that of the acrylic fibres.

## Environmental Conditions

VEREL fibre has excellent resistance to sunlight, ageing and general weathering conditions. It is completely resistant to attack by insects and micro-organisms.

## Chemical Resistance

The resistance of VEREL fibre to most types of chemical, including acid and alkali, is excellent. It will withstand bleaching conditions and is not affected by the chemicals used in normal textile processing. None of the common dry cleaning solvents affect the fibre.

## Pilling Resistance

VEREL fibre exhibits little tendency towards pilling, even in fabrics made with low twist yarns.

### Washing

Fabrics of VEREL fibre may be washed by hand or by machine, at a temperature not exceeding 50°C., using a neutral soap or detergent. Gentle agitation should be used.

### Drying

Fabrics may be tumble dried or drip dried. Elevated temperatures should be avoided.

### Ironing

Fabrics may be ironed, preferably using a damp cloth, with the iron at 'synthetic' setting. Maximum safe ironing temperature is 150°C. Care should be taken to avoid distorting the fibres to produce a glazed effect.

**Dry Cleaning**

Fabrics of VEREL fibre may be dry cleaned by the normal methods. The fibre is not affected by the usual dry cleaning solvents.

**End Uses**

*Pile Fabrics*

The softness of VEREL fibre, combined with whiteness, excellent flame-resistance and controlled shrinkage have proved advantageous in the production of pile fabrics. These include both the woven and knitted types of construction for a wide variety of end uses ranging from coatings, liner fabrics and floor coverings, to trimmings for collars, cuffs, boots and shoes.

The availability of VEREL in both regular cross-section and ribbon cross-section forms has been put to good use in the manufacture of simulated fur fabrics.

Pile fabrics made from VEREL provide warmth without weight, associated with soft, attractive handle and great comfort.

*Knit Goods*

The ability of fabrics of VEREL fibre to retain a soft hand after repeated washing and drying is advantageous in knitwear such as sports shirts, underwear and children's garments. A blend containing some 25 per cent of VEREL with cotton realizes the attractive properties of VEREL at an economic price.

*Three-dimensional Fabrics*

The use of VEREL shrinkable fibre makes possible the production of three-dimensional fabrics with a wide range of blister and pucker effects.

*Industrial Applications*

The resistance of VEREL fibre to many types of chemical, including acids and alkalis, has brought it many applications in the industrial field. It is used for filter cloths, protective clothing, etc.

*Drapery; Upholstery*

VEREL fibre provides excellent drapery and upholstery fabrics which have the special advantage of fire resistance.

*Carpets*

VEREL fibre has many valuable characteristics to offer to the carpet trade, including good abrasion resistance, good covering power, a wide range of dyeability, high resistance to soiling and ease of cleaning.

Carpets retain their appearance and texture over long periods of service, and tests have shown VEREL to be better than wool in this respect.

The soil resistance of VEREL is excellent, and it is cleaned easily and effectively with the aid of neutral detergents. In wear tests, VEREL carpets gave 20 to 50 per cent greater wear life than acrylic fibres.

Carpets of VEREL fibre have a crush resistance similar to wool. They display excellent stain resistance and flame resistance, and are inherently mothproof and rot-resisting.

*Note*

*Lastrile. (F.T.C. Definition)*

A manufactured fibre in which the fibre-forming substance is a copolymer of acrylonitrile and a diene (such as butadiene) composed of not more than 50% but at least 10% by weight of acrylonitrile units.

## POLYVINYL CHLORIDE FIBRES

Fibres spun from polymers or copolymers of vinyl chloride:

$$CH_2 = CHCl \longrightarrow --CH_2-CH-CH_2-CH---$$
$$| \qquad | $$
$$Cl \qquad Cl$$

**VINYL CHLORIDE**          **POLYVINYL CHLORIDE**

### INTRODUCTION

Polyvinyl chloride (P.V.C.) was first made by the French chemist Henri Regnault, who prepared and polymerized vinyl chloride in 1838. When all known polymers came under examination as potential sources of synthetic fibres, during the early part of the present century, it was natural that polyvinyl chloride should be among those considered.

Many attempts were made to dissolve polyvinyl chloride and produce filaments by extrusion of solutions through spinnerets. But the polymer was difficult to dissolve, and solvents suitable for use on a commercial scale were not discovered.

An experimental P.V.C. fibre was spun in Germany in 1913, but did not come into large-scale production. During the years following World War I, attempts were made to modify the polymer in order to make it more readily soluble. It was found that this could be achieved by introducing a small proportion of another vinyl compound into the polymerization, forming a copolymer in which the molecular chains did not pack together so closely as in the case of P.V.C. itself. Experimental fibres were made from a copolymer containing 85 per cent vinyl chloride and 15 per cent vinyl acetate in Germany in 1928.

In 1934, it was found that the solubility of polyvinyl chloride could also be increased by chlorinating the polymer. The large chlorine atoms introduced at intervals along the P.V.C. molecules had an effect similar to that of the acetate groups in the copolymer, reducing the degree of attraction exerted between the long molecules and increasing solubility.

'Pe Ce' fibres, made from chlorinated P.V.C., were introduced in Germany in 1936. They achieved some limited success, but have never made real progress in the general textile field. In 1937,

**444**

fibres spun from a copolymer of vinyl chloride and vinyl acetate, called 'Vinyon', were produced in the U.S.

In 1940, French chemists discovered that P.V.C. itself could be dissolved in a mixture of acetone and carbon disulphide, and this solution was suitable for the production of P.V.C. fibres on a commercial scale.

P.V.C. fibres are now manufactured in a variety of forms and modifications. They have achieved a limited success in the textile trade, but their range of applications is restricted by their low softening point. Many P.V.C. fibres will soften at temperatures as low as 70°C.

On the other hand, P.V.C. fibres have characteristics which have created a demand for them in certain applications. They do not burn, and have a high resistance to many chemicals. Their tendency to shrink at relatively low temperatures is made use of in the production of high bulk yarns.

## TYPES OF POLYVINYL CHLORIDE FIBRE

Fibres are produced from 100 per cent polyvinyl chloride, from copolymers containing small proportions of various second components, and from polyvinyl chloride which has been chemically modified, e.g. by chlorination.

These fibres are available in a variety of forms, e.g. of controlled shrinkage, to suit particular applications. They are produced as continuous filament yarns, staple fibre and tow.

*Note*

In the section which follows, polyvinyl chloride fibres are considered under three type classifications, as follows:

(1) Polyvinyl Chloride Fibres (100 per cent P.V.C.)
(2) Vinyl Chloride Copolymer Fibres
(3) Chemically-Modified Polyvinyl Chloride Fibres.

## NOMENCLATURE

The name 'Vinyon' was registered as a trade mark by Union Carbide Corporation for the fibre spun from a copolymer of vinyl chloride and acrylonitrile. This particular filament yarn was designated 'Vinyon' N.

'Vinyon' N was subsequently followed by another type of polyvinyl chloride fibre, called 'Vinyon' HH, which was spun from a copolymer of 86 per cent vinyl chloride and 14 per cent vinyl acetate.

The 'Vinyon' trade mark was never enforced by Union Carbide, and it was released for generic use. The term was adopted by the U.S. Federal Trade Commission as an official definition for fibres of the polyvinyl chloride type.

### Federal Trade Commission Definition

The generic term *vinyon* was established by the U.S. Federal Trade Commission for fibres of the polyvinyl choride type, the official definition being as follows:

*Vinyon*. A manufactured fibre in which the fibre-forming substance is any long-chain synthetic polymer composed of at least 85 per cent by weight of vinyl chloride units ($-CH_2-CHCl-$).

### Chlorofibre

The term *chlorofibre* is also widely used to denote polyvinyl chloride fibres as defined by the Federal Trade Commission. This term has the advantage of avoiding any possibility of confusion with the trade name 'Vinyon'.

## (1) POLYVINYL CHLORIDE FIBRES
## (100 PER CENT P.V.C.)

### INTRODUCTION

The production of 100 per cent P.V.C. fibres has made steady progress in a few countries. The French firm of Rhone Poulenc Textile has, in particular, persevered with the development of these fibres. The first P.V.C. fibres were spun at the Tronville-en-Barrois plant in 1949, and by 1976 some 9 million kg of fibre were being produced annually.

In Italy, by 1967, Societa Polymer were producing 'Movil' P.V.C. fibres in a plant with a capacity of 4 million kg per annum, and Applicazioni Chimiche Societa per Azioni (ACSA) began producing 'Leavin' P.V.C. fibres in a plant with a capacity of 5 million kg per annum. P.V.C. fibres were also produced in Japan and in West Germany. Production in Italy and West Germany subsequently ceased.

The progress of P.V.C. fibres since 1949 has been steady rather than spectacular, and the use of the fibre for large-volume textile applications has been restricted by a lack of stability to heat and to dry cleaning solvents. The attractive characteristics of P.V.C. fibres (see page 455) coupled with the low price of the polymer have, however, encouraged producers to seek ways of overcoming the fibre's deficiencies in order to extend the range of its textile applications.

By 1964 an important step forward had been taken by the firm Rhone Poulenc Textile, who developed a blend of polymers consisting of standard P.V.C. and chlorinated P.V.C. to obtain better dimensional stability at higher temperatures, e.g. boiling water.

## PRODUCTION

### Monomer Synthesis

Vinyl chloride is produced in very large quantity for the production of polyvinyl chloride plastics, and fibre manufacture makes use of only a small part of the output. Two routes are commonly used.

### (a) *Acetylene and Hydrogen Chloride*

Acetylene is reacted with hydrogen chloride in the presence of a mercuric chloride catalyst

$$CH{\equiv}CH + HCl \rightarrow CH_2{=}CHCl$$

## (b) *Ethylene and Chlorine*

Ethylene dichloride is produced by reaction of ethylene with chlorine (1). This is then heated under pressure when hydrogen chloride is released, leaving vinyl chloride (2).

$$\overset{(1)}{\phantom{CH_2=CH_2+Cl_2\rightarrow}}\overset{(2)}{\phantom{CH_2Cl.CH_2Cl\rightarrow}}$$

$$CH_2 = CH_2 + Cl_2 \rightarrow CH_2Cl.CH_2Cl \rightarrow CH_2 = CHCl$$

### Polymerization

Vinyl chloride is polymerized typically as an aqueous emulsion in autoclaves, under pressures of 45–50 atmospheres and a temperature of about 65°C. The polymer forms a suspension in the water, and is recovered by spray drying.

### Spinning

P.V.C. fibres may be spun by dry or melt spinning processes.

### *Dry Spinning*

P.V.C. is dissolved in a solvent, e.g. acetone/carbon disulphide mixture, and the solution at 70–100°C. is pumped through spinnerets. The fine jets emerge into a stream of hot air, the solvents evaporating to leave solid filaments of P.V.C. The solvents are recovered and re-used.

This process is used almost exclusively for the commercial spinning of textile-grade P.V.C. fibres.

### *Melt Spinning*

Molten P.V.C. may be spun by extruding it through spinnerets, but the spinning temperatures cannot be raised high enough to permit of the production of fine deniers needed for staple and tow. P.V.C. begins to decompose at about 200°C., and below this temperature the viscosity of the molten P.V.C. restricts the fineness of the extruded filaments to a diameter of about 0.2 mm.

## PROCESSING

The information which follows is based upon the P.V.C. fibres produced by Rhone Poulenc Textile, France, under the following designations:

LX: Standard P.V.C.

ZC: Standard P.V.C. + chlorinated P.V.C.

LX or ZC: Continuous filament.

'Fibravyl' LX: Staple with high shrinkage in boiling water (55%).

'Thermovyl': Heat stabilized form of staple.

'Retractyl': Staple fibre with shrinkage characteristics intermediate between those of 'Thermovyl' and 'Fibravyl'.

'Thermovyl' ZC: Staple of increased tenacity, dyeability and thermal stability.

'Fibravyl' ZC: Shrinkage in boiling water 25–30%.

### Dyeing

(1) *Shrinkable Fibres (100 per cent or blends).*
    *No-Shrinkage Conditions*

The standard procedure is to dye with disperse dyestuffs at temperatures below 70°C., in conjunction with assistants or swelling agents.

(2) *'Thermovyl'* (*100 per cent or blends*)

The techniques described above are equally suitable for dyeing 'Thermovyl' LX, but for dyeing 'Thermovyl' ZC it is possible to use higher dyeing temperatures (up to 100°C). This permits dyeing to be carried out without the help of a swelling agent. Cationic dyes have to be used under a specific process.

'Thermovyl' ZC may be dyed in any form, but it is commonly dyed as stock, combed top, yarn or in the piece.

(3) *'Fibravyl' LX or ZC (100 per cent or blends).*
    *Shrinkage Conditions.*

The temperature conditions used for shrinking these fibres correspond with those temperatures at which disperse dyes may be used effectively without assistants (generally 98–100°C).

*Resist Dyeing; Cross Dyeing*

P.V.C. fibres resist most of the types of dyestuff used for dyeing natural fibres, and resist or cross dyeing effects are obtainable in blends.

### Printing

'Thermovyl' ZC may be printed effectively with disperse dyes, conventional steaming being all that is required to set the dyes.

### Showerproofing

P.V.C. fibres are hydrophobic, but water will nevertheless penetrate through the interstices of a cloth made from them. P.V.C. fabrics may be showerproofed by means of the standard paraffin wax type agents, with or without the addition of aluminium salts.

### Waterproofing

P.V.C. fabrics may be rendered completely impermeable to water by the application of a suitable coating, and all forms of standard coating agents may be used. There are advantages in using a coating which will have the same non-flammability and resistance to degradation of the P.V.C. fibres, and special materials for this purpose are available.

If rubber coatings are required, it is best to use a self-vulcanizing latex at low temperature, rather than a process which requires vulcanization at temperatures above $70^{\circ}$C. for continuous filament or 'Fibravyl', or $100^{\circ}$C. for 'Thermovyl' ZC.

### Bonding

Special adhesives are available for the bonding of P.V.C. fabrics.

## High-bulk and Shrinkage Treatments

P.V.C. fibres of the shrinkable type will undergo a high degree of shrinkage when heated to appropriate temperatures, and this may be used for creating bulk and related effects, as in the case of acrylic fibres.

Shrinkage may be brought about by dry, wet or steaming treatment, the choice depending upon the nature of the goods. In all cases, it is essential to carry out the treatment as uniformly as possible.

Dry treatment may be carried out by subjecting the goods to hot air in a stenter. Steaming may be achieved on a cylinder drier or in an enclosed chamber.

Wet treatments are carried out on the jig, etc., the technique used depending on the nature of the goods. Yarns are treated in hank form, and piece goods are in open width, sewn end to end. After treatment, piece goods may be passed through the stenter to smooth out the fabric and set the required shrinkage.

Shrinkage carried out on mixed yarns, doubled yarns or yarns spun from blends of P.V.C. with other fibres creates a high degree of bulking. Shrinkage carried out on fabrics and knitted goods brings about a tightening of the yarns, and this may be used to advantage to obtain various effects depending upon the construction of the fabric.

### 1. Fabric with Warp of One Material and Weft of Another

When fabrics are made with P.V.C. weft, for example, and cotton warp, the shrinkage of the P.V.C. produces a fabric which is very tightly packed and very strong. Fabrics of this type are used for motor-car hoods, sports clothing and rainwear; they are very resistant to tearing, are showerproof and dimensionally stable.

Very dense velvets (collar velvets) may be made by shrinking a velvet of normal construction, but woven with a ground of P.V.C.

New types of jersey have been created by knitting P.V.C. on Interlock machines together with other types of yarn (cotton, wool, rayon, nylon), and shrinking the knitted fabric during finishing.

### 2. Homogeneous Fabric made with Mixed Doubled Yarns

Shrinkage of the P.V.C. yarns in this case creates a bulky fabric.

### 3. Homogeneous Fabric made with Blended Yarns

The effects of shrinkage in this case are similar to (2), but the general appearance is that of a tightly packed plain cloth woven with bulky yarns. Additional effects may be obtained by raising this type of fabric.

### 4. *Fancy and Cloque Effects*

Shrinkage effects may be localized by using the above techniques as patterns in the fabric. Using a plain fabric as a basis, fabrics with designs in relief, cloques, plaited weave, knop effects, etc., may be produced.

### Shaping; Moulding; Embossing

The thermoplastic nature of P.V.C. fibre, and the low temperature at which softening takes place, are put to good use in shaping techniques which are used on fabrics containing P.V.C. fibres.

Fabrics consisting totally or in part of P.V.C. fibres such as 'Fibravyl' LX may be shaped by heating them in such a way that they shrink on to a former.

Fabrics which have already been shrunk in the piece, or fabrics made from thermally stabilized fibre such as 'Thermovyl' LX may be embossed or moulded under pressure, with or without the help of heat.

Using the 'shrinking-on' method, it is possible to cover flexible metal or rubber tubes, electrical conductors, etc., with sleeves of P.V.C. tubular fabric, and then to shrink the cloth on to the tube or wire it covers

On the other hand, by moulding fabrics which have already been shrunk, it is possible to make a range of useful articles which are unaffected by humidity, such as loudspeaker grilles, upholstery and car fabrics.

### STRUCTURE AND PROPERTIES

(Based on continuous filament)

### Fine Structure and Appearance

Smooth, rod-like fibres, of near-circular cross-section.

### Tenacity

24–27 cN/tex (2.7–3.0 g/den), wet or dry.

### Tensile Strength

32–36 kg/mm$^2$.

**Elongation**

12–20 per cent, wet or dry.

**Specific Gravity**

1.4 ('Thermovyl' ZC: 1.38).

**Effect of Moisture**

The water absorption is virtually nil, and the fibre does not swell in water. Moisture has no effect on mechanical properties.

**Thermal Properties**

Shrinks on heating above 70°C. ('Thermovyl' ZC: above 100°C.) as the extended molecules tend to return to their pre-stretch configuration.

At higher temperatures, softening continues, and decomposition begins at about 180°C.

*Flammability*

P.V.C. fibres are inherently non-flammable. They will not burn, nor will they emit flames or release molten incandescent drops capable of spreading a fire on combustible materials.

When subjected to an intense flame, fabrics made from pure P.V.C. fibres will disintegrate, but the residues may be touched by hand for they are not hot. There is, therefore, no risk of burning.

*Effect of Low Temperature*

P.V.C. fibres retain their flexibility and strength at temperatures as low as $-80$°C.

*Thermal Conductivity*

$74 \times 10^{-2}$ (W.m$^{-1}$.K$^{-1}$)

**Effect of Sunlight**

Excellent resistance. Yarns exposed to direct sunlight lost 10 per cent of their strength after 5 months' exposure, 15 per cent after 10 months' exposure, and 25 per cent after 18 months' exposure.

**Chemical Properties**

*Acids*

No effect. The fibres remained unharmed, for example, after 4 years' steeping in concentrated nitric acid, sulphuric acid or aqua regia.

*Alkalis*

No effect. The fibres remained unharmed, for example, after 4 years' steeping in caustic soda or caustic potash (50 per cent).

*General*

P.V.C. fibres have outstanding resistance to a wide range of chemicals, including bleaches, urine, perspiration, reducing agents, oxidizing agents.

**Effect of Organic Solvents**

Alcohols, ether and petroleum hydrocarbons do not affect P.V.C. fibres, but the fibres are swelled by toluene, trichlorethylene, benzene, carbon disulphide, ethyl acetate, acetone, chloroform, methylene chloride and nitrobenzene. They are also attacked by phenols.

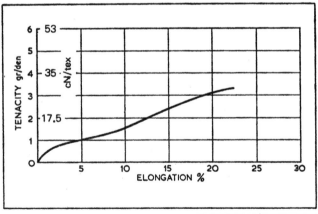

*Polyvinyl Chloride Fibre; Continuous Filament*

### Insects

Completely resistant.

### Micro-organisms

Completely resistant.

### Electrical Properties

P.V.C. fibres have a high dielectric constant.

### Allergenic Properties

P.V.C. fibres do not cause any irritation when in contact with the skin.

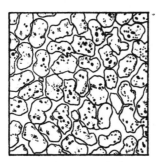

LX Continuous Filament

## P.V.C. FIBRES IN USE

### General Characteristics

#### *Fibre Structure*

P.V.C. fibres are commonly rather featureless and rod-like in structure, and in this respect contribute nothing unusual to potential applications.

#### *Mechanical Properties*

The regular types of P.V.C. fibre have a useful combination of tenacity and elongation, well suited to general textile applications. The mechanical properties are unaffected by moisture.

Fabrics made from P.V.C. fibres have an attractive handle, and are warm and comfortable against the skin. They have a high abrasion resistance and good wearing qualities.

## Specific Gravity

The specific gravity of 1.4 is fairly high, in the same region as the modacrylics.

## Moisture Relationships

P.V.C. fibres are absolutely non-absorbent, and the fibres do not swell in water. The mechanical properties of the fibre are unaffected, and fabrics made from P.V.C. are dimensionally stable and crease resistant when subjected to any form of wet processing. P.V.C. fabrics wash and dry easily and quickly.

## Thermal Properties

P.V.C. fibres have excellent thermal insulation properties, and this contributes to the warmth of properly-constructed P.V.C. garments. The good low-temperature properties are an advantage in certain applications.

The low softening temperature of P.V.C. fibre, on the other hand, has always been a drawback to its use as a general purpose textile fibre. The introduction of the newer types of P.V.C. fibre spun from standard and chlorinated P.V.C. with increased heat resistance, has increased the versatility of the fibre.

The high degree of stretch that may be locked into P.V.C. fibres by stretching when warm, and subsequently cooling, bestows upon these fibres the high bulking characteristics that have been so useful in acrylics.

P.V.C. fibres are absolutely flameproof, and this has proved one of their most important features. Mixed with other fibres, they will suppress the flammability of these fibres to a remarkable degree; 25 per cent of P.V.C. fibre blended with cotton or other cellulosic fibre, for example, will prevent the fabric or yarn from supporting combustion when ignited; 75 per cent of P.V.C. fibre in a blend of this sort will render the mixture almost non-flammable.

### Environmental Conditions

P.V.C. fibres have an excellent resistance to the degradative influences encountered in outdoor applications, including sunlight, insects and micro-organisms.

### Chemical Resistance

The high resistance of P.V.C. fibres to chemicals, including acids and alkalis, makes these fibres the first choice for many industrial applications.

The poor resistance to certain common solvents, on the other hand, has created problems in certain textile applications. P.V.C. goods cannot be dry cleaned with trichloroethylene and care must therefore be taken in dry cleaning to ensure that the correct solvents are used.

### Electricity

P.V.C. fibres generate negative electricity by friction with the skin, whereas most other fibres develop positive electric charges. It is claimed that this negative electricity has therapeutic effects, and is of value in the treatment of rheumatism and similar complaints.

When P.V.C. fibre is blended with wool the positive electricity generated on the wool neutralizes the negative electricity on the P.V.C., so that static is reduced or eliminated.

In blends with cotton or rayon, the electricity generated by the P.V.C. fibres is dissipated by the cellulosic fibre. Nylon, acrylic, acetate or silk fibres do not dissipate the charge to any significant extent.

If static electricity is causing trouble in the processing or use of P.V.C. fibres, it may be prevented by using an appropriate antistatic finish.

### Washing

Fabrics containing P.V.C. fibres may be washed easily and effectively, so long as care is taken to ensure that lukewarm water only is used. The water temperature must not be higher than $60^{\circ}$C and preferably between 30 and $40^{\circ}$C.

### Drying

P.V.C. fibre does not absorb water, and fabrics dry quickly and easily. Moderate temperatures only (below 65°C.) should be used.

### Ironing

Fabrics made from 100 per cent P.V.C. fibre, or from blends containing a high proportion of this fibre, are dimensionally stable and wrinkle resistant. They do not usually need ironing, but if it is considered necessary a cool iron should be used, with a damp linen cloth between iron and fabric.

Fabrics made from blends containing 25 per cent or less of P.V.C. fibre may be ironed usually without difficulty, using a low temperature setting.

### Dry Cleaning

Fabrics containing P.V.C. fibres may be dry cleaned with petrol (gasoline) or white spirit, but benzene and trichlorethylene should be avoided. Perchloroethylene should be used cautiously.

### End-Uses

P.V.C. fibres can be used in many applications. The main uses for Rhone Poulenc Textile P.V.C. fibres are hosiery and furnishing fabrics.

### *Industrial Applications*

The chemical resistance and non-flammability of P.V.C. fibres have enabled them to find important uses in the industrial field. Typical applications include:

Waddings, filter cloths, braiding, piping and other uses in the chemical industry;

Battery fabrics;

Protective clothing;

Tarpaulins, awnings, curtains, fishing nets, etc.;

Fairings and canvas awnings for aircraft, gliders, boats, buoys, etc.;

Orthopaedic materials, artificial limbs, saddlery, etc.;

Accessories for textile machinery, billiard cloths.

*Filtration.* One of the most important of these industrial uses is in the filtration of corrosive liquids, including strong acids, alkalis and oxidizing agents. Inorganic chemicals, as a general rule, do not attack P.V.C. fibres, but care must be taken to ensure that organic materials do not swell the fibres.

In the filtration of gases, the humidity of the gas does not influence the pore size of the fabric. Also, the accumulation of static electricity by the fibres may help in filtration by attracting particles of dust and dirt to the filter fabric.

Controlled shrinkage of the filter fabric may be used to create fabrics capable of filtering ultra-fine particles from liquid media.

P.V.C. fibres may be used in the form of linters, wadding or flock, in addition to the normal fabric form.

The choice of fabric to be used depends upon the temperature of the material to be filtered. If the temperature is below about 70°C., fabrics made from regular P.V.C. fibre ('Fibravyl') may be used. At temperatures above this, it is preferable to use a heat-stabilized type of P.V.C. such as 'Thermovyl'.

## Upholstery; Furnishings

The resistance of P.V.C. fibres to deterioration has brought many applications in upholstery and furnishing, especially in tropical countries. Mosquito netting, furnishing fabrics, awnings, tents, etc., are not affected by moulds even under the most humid conditions. The resistance to insects and other animal pests is complete.

These same properties, allied especially with the non-flammability of P.V.C., have opened up important fields of application in the furnishing of cinemas, theatres, ships and aircraft. P.V.C. fibres provide fabrics for curtains, awnings, upholstery, carpets, netting, hammocks and the like. P.V.C. waddings and felts are used as non-flammable insulation materials.

The motor car trade is an important outlet for P.V.C. fabrics, providing hoods, seat coverings and other fabrics.

The large variety of cloque, knop and other effects which are obtained by using the thermal shrinkage of P.V.C. fibres provides novelty effects for furnishing fabrics, which may be obtained simply and cheaply.

*Apparel*

The sensitivity of P.V.C. fibres to heat has tended to restrict their use in the field of apparel fabrics. Unstabilized P.V.C. fibres begin to shrink at temperatures which may be as low as 70°C., and great care must be exercised in the washing and ironing of fabrics containing these fibres. Also, the fact that certain commonly-used dry cleaning agents will soften P.V.C. fibres has proved a drawback to their use in clothing applications.

The development of the newer types of P.V.C. fibre (see page 447) has greatly improved the prospects of P.V.C. fibres in this field.

Despite the difficulties inherent in P.V.C. fibres, they have made considerable progress in some apparel applications. P.V.C. fabrics have an attractive handle, and are very pleasant to wear next to the skin. Used alone or in blends with other fibres, P.V.C. fibres are used for hosiery, sports and travel shirts, and baby clothes. These garments may be washed easily in lukewarm water, dried overnight, and used again without ironing.

Special effects are obtained by making use of the shrinkage of P.V.C. when boiled in water. Cloque fabrics, for example, are made by weaving alternate bands of P.V.C. with other yarns, followed by heat treatment which brings about shrinkage of the P.V.C.

The use of a P.V.C. weft and a warp of a different fibre provides a fabric which may be weft-shrunk to produce a tightly-packed warp. Velvets of great pile density, for example, may be made by using this technique. Water-resisting fabrics, likewise, may be made for use in rainwear, sports jackets, uniforms, etc.

*Blends with Wool*

P.V.C. staple fibre is blended with wool for the production of woollen goods, providing fabrics of increased strength. The milling capacity is increased by shrinkage at elevated temperature, yielding a cloth which would not otherwise felt. Shrinkage capacity is likewise added to fabrics which are normally incapable of being milled.

In worsted goods, the addition of P.V.C. fibres brings about an increase in strength, and makes possible the production of bouclé effects.

The addition of P.V.C. to wool is also a way of lowering the cost of a fabric without detracting significantly from the characteristic properties of wool, and at the same time adding desirable properties, such as non-flammability, which come from the P.V.C.

## Blends with Cotton and Rayon

P.V.C. fibres are blended with cotton and other cellulosic fibres to bring about increase in strength, and to make possible the effects obtainable from controlled shrinkage, e.g. bouclé effects. The handle, warmth and crease-resistance of rayon fabrics are increased.

## Blends with Nylon

P.V.C. fibre blended with nylon provides a fabric in which many of the essential characteristics of the nylon − such as rot and mildew resistance − are retained.

## (2) VINYL CHLORIDE COPOLYMER FIBRES

Fibres spun from copolymers of vinyl chloride with a smaller proportion of a second monomer.

The fibre 'Vinyon' HH, for example, is spun from a copolymer of vinyl chloride and vinyl acetate, which contains 85 to 86.5 per cent of vinyl chloride by weight.

$$CH_2=CHCl \quad + \quad CH_2=CHOCOCH_3 \quad \longrightarrow \quad -CH_2-CH-CH_2-CH-$$

$$\underset{Cl}{\big|} \qquad \underset{OCOCH_3}{\big|}$$

VINYL CHLORIDE     VINYL ACETATE     'VINYON' HH

The fibre 'Dynel' is spun from a copolymer of vinyl chloride and acrylonitrile, which are in the ratio of 60 parts to 40 parts by weight respectively. (Production suspended).

$$CH_2=CHCl \quad + \quad CH_2=CHCN \quad \longrightarrow \quad -CH_2-CH-CH_2-CH-$$

$$\underset{Cl}{\big|} \qquad \underset{CN}{\big|}$$

VINYL CHLORIDE     ACRYLONITRILE     'DYNEL'

461

## INTRODUCTION

Difficulties experienced in spinning 100 per cent polyvinyl chloride during the early 1930s led to many attempts to increase the solubility of the polymer. One technique was to introduce a small proportion of another monomer into the polymerization, to form a copolymer which would be expected to have increased solubility.

In 1933, Carbide and Carbon Chemicals in the U.S. developed a copolymer of vinyl chloride which was capable of being dissolved in a solvent and dry spun into fibres of useful properties. These fibres were given the name 'Vinyon'.

Commercial development of 'Vinyon' fibres did not take place until 1938, when American Viscose Corporation began producing the fibres for the first time.

'Vinyon' was marketed originally as a continuous filament yarn, 'Vinyon' CF. Production of this has since been discontinued, and 'Vinyon' is now available as a staple fibre, 'Vinyon' HH.

Today, a number of fibres are spun from copolymers in which vinyl chloride forms the major component.

## NOMENCLATURE

The system of classification used in the *Handbook of Textile Fibres* is based on the chemical constitution of the fibre, the polymeric unit which is present in greatest proportion being regarded as the chemical type of the fibre. Using this system of classification, all copolymer fibres in which vinyl chloride provides the major proportion of polymeric units are included under the heading of Polyvinyl Chloride Fibres. Thus, any copolymer made from two monomers, of which vinyl chloride forms more than 50 per cent by weight, is regarded as a polyvinyl chloride fibre.

### Federal Trade Commission Definitions

The definitions adopted by the U.S. Federal Trade Commission do not, unfortunately, follow a straightforward chemical system of classification. This leads to some confusion, for example, in the consideration of vinyl chloride copolymers.

According to the F.T.C. definition, vinyl chloride copolymers in which vinyl chloride forms at least 85 per cent by weight are known as *vinyon* fibres. The term *modacrylic*, on the other hand.

462

is used for those fibres in which the polymer is composed of less than 85·per cent and at least 35 per cent of acrylonitrile. Thus, a fibre based on a polymer containing, say, 60 per cent of vinyl chloride units and 40 per cent of acrylonitrile units is by definition a modacrylic fibre. Its name associates it with acrylonitrile, whereas the bulk of the polymer is polyvinyl chloride.

Fibres in this category have become of commercial importance, and are properly considered as fibres of the polyvinyl chloride type.

As in the case of any fibres based on copolymers, the properties vary greatly, even though the major component is the same. In the section which follows, two fibres are taken as examples of vinyl chloride copolymers:

(a) 'Vinyon' HH, which contains more than 85 per cent vinyl chloride, and thus comes under the F.T.C. definition of *vinyon*, and

(b) 'Dynel', which contains 60 per cent vinyl chloride, and comes under the F.T.C. definition of *modacrylic* because the minor component is acrylonitrile in more than 35 per cent proportion.

## (a) 'VINYON' HH

'Vinyon' HH is a fibre of the polyvinyl chloride type which is produced by Avtex Fibers Inc., U.S.A. It is spun from a copolymer of vinyl chloride (85–86.5 per cent) and vinyl acetate (15–13.5 per cent).

By F.T.C. definition, 'Vinyon' HH is a *vinyon* fibre.

## PRODUCTION

### Monomer Synthesis

(1) *Vinyl Chloride,* see page 447.

(2) *Vinyl Acetate*

Vinyl acetate is produced by the reaction of acetylene with acetic acid:

$$CH{\equiv}CH + CH_3COOH \rightarrow CH_2{=}CHOCOCH_3$$

The reaction may be carried out in one of two ways:

(a) acetylene is reacted with liquid acetic acid at temperatures up to 100°C., using mercuric salts as catalyst. Ethylidene diacetate is produced as a byproduct:

$$CH\equiv CH + 2\ CH_3COOH \rightarrow (CH_3COO)_2CHCH_3$$

This is pyrolyzed to split it into vinyl acetate and acetic acid.

(b) Acetylene is bubbled through acetic acid to provide a mixture of the two reactants in vapour form. The gas mixture is passed over a zinc or cadmium salt catalyst at 200–250°C.

### Polymerization

Vinyl chloride and vinyl acetate are copolymerized by emulsion polymerization or in a solvent for the copolymer, the process being similar to that used in making polyvinyl chloride (see page **448**). The polymerization is continued until a polymer of molecular weight in the range 12,000 to 27,000 has been reached.

### Spinning

The polymer is dissolved in acetone, and the solution is then filtered, de-aerated and stored in heated tanks. From the storage tank, the solution is pumped to the spinneret. As the fine jets emerge from the spinneret holes, they fall through a spinning tube through which hot air is passing. The acetone is evaporated to leave solid filaments of vinyl chloride/vinyl acetate copolymer.

The filaments produced in this way are brought together into a tow which is wound on to a spindle. The early types of 'Vinyon', e.g. 'Vinyon' CF, were stretched after spinning in order to orientate the molecules and increase the strength of the fibre. Modern 'Vinyon' HH is seldom used in applications requiring great tensile strength, however, and it is produced in the unstretched form.

The tow is lubricated and cut into staple.

*Note.* Titanium dioxide may be mixed into the spinning solution to produce a dull yarn, and coloured pigments may be introduced to provide a solution-dyed or spun-dyed yarn.

## PROCESSING

### Dyeing

Disperse dyes are used in conjunction with swelling agents such as dibutyl phthalate or o-hydroxydiphenyl. Dyebath temperature should not exceed 55°C.

## STRUCTURE AND PROPERTIES

### Fine Structure and Appearance

Smooth surfaced fibres of round or dog-bone cross-section.

### Tenacity

6.2–8.8 cN/tex (0.7–1.0 g/den), wet or dry

### Tensile Strength

840–1,190 kg/cm$^2$ (12,000–17,000 lb/in$^2$).

### Elongation

100–125 per cent, wet or dry.

### Specific Gravity

1.33–1.35.

### Effect of Moisture

'Vinyon' HH absorbs very little moisture. It has a regain of 0.1 per cent. It does not swell in water, and the tensile properties remain unaffected by water.

### Thermal Properties

'Vinyon' softens at 52°C. and shrinks at 60°C. It becomes sticky at 85°C. and melts at 135°C.

*Flammability.* Chars, but will not burn.

### Effect of Age

Negligible.

**Effect of Sunlight**

Negligible.

**Chemical Properties**

*Acids*

Unaffected at normal temperatures by mineral acids. Hot acids cause decomposition and embrittlement.

*Alkalis*

Unaffected by 30 per cent caustic soda or caustic potash.

*General*

Good resistance to most common chemicals, including perspiration.

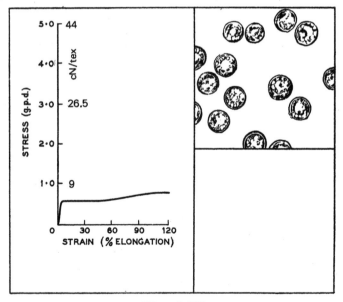

*'Vinyon' HH*

**Effect of Organic Solvents**

Resists alcohols, petrol (gasoline), paraffin and mineral oils. It is softened by aromatic hydrocarbons, esters and ethers. Dissolves in ketones and to some extent in chlorinated solvents.

**Insects**

Not attacked by moth grubs or beetles.

**Micro-organisms**

Not attacked by moulds or bacteria.

**Electrical Properties**

High dielectric strength of 650 volts per mil at 60 cycles.

**Allergenic Properties**

Not toxic to skin surfaces.

## 'VINYON' HH IN USE

**General Characteristics**

The low softening temperature of 'Vinyon' HH, quite apart from any other factor, denies it anything but very specialized textile applications. This being so, there is little inducement for the manufacturer to improve tensile and elongation properties in order to make the fibre more suitable for normal processing techniques. In this connection, it is interesting to note that the tenacity of the earlier Vinyon' CF was in the region of 30 cN/tex (3.4 g/den), and the elongation 18 per cent, so that it would be possible to effect considerable improvement in mechanical properties of 'Vinyon' HH if the need should arise.

The chemical and biological properties of 'Vinyon' HH are generally similar to those of 100 per cent P.V.C. fibres. The fibre is completely non-flammable, and is resistant to many chemicals. It is attacked, however, by a fairly wide range of common solvents.

**End-Uses**

Vinyon' HH is used as a bonding fibre in non-woven applications. Mixed with other fibres it becomes tacky when heated and bonds the mass of fibres together. It is used in this way for making felts, bonded fabrics and heat-sealable papers. Important applications include filter fabrics, tea-bags, door panels and webbings.

'Vinyon' is used for industrial and outdoor fabrics including garden furniture, tarpaulins, awnings, filter cloths and protective clothing.

## (b) 'DYNEL'

'Dynel' was a fibre of the polyvinyl chloride type produced by the Fibers and Fabrics Division of Union Carbide Corporation, in the U.S.A. It was spun from a copolymer of vinyl chloride (60 per cent) and acrylonitrile (40 per cent).

By F.T.C. definition, 'Dynel' was a *modacrylic* fibre.

### INTRODUCTION

The most serious disadvantage to the use of polyvinyl chloride type fibres for general textile applications lies in their low softening point. Many attempts were made to raise the softening point of vinyl chloride/vinyl acetate copolymers, as used in making 'Vinyon' HH, by adjustment of the relative proportions of the two components, but without success.

During the work on vinyl chloride copolymers, however, it was found that fibres of satisfactory softening point could be spun from copolymers of vinyl chloride and acrylonitrile, in which the vinyl chloride remained the major component.

An early fibre produced from copolymers of this type was 'Vinyon' N. This was a continuous filament fibre which was dry-spun from a solution of vinyl chloride/acrylonitrile copolymer. It is no longer produced.

In 1951, Union Carbide Corporation introduced on to the textile market a new fibre called 'Dynel'. This fibre was similar to 'Vinyon' N in that it was spun from a copolymer of vinyl chloride and acrylonitrile, with the vinyl chloride forming the major component. But it was wet-spun instead of being dry-spun, and it was available as staple and tow, instead of continuous filament.

Production of 'Dynel' was later discontinued, but information about it has been retained for its historic and technical value in the following section. 'Dynel' was of importance as an example of a fibre spun commercially from a copolymer of vinyl chloride and acrylonitrile.

## TYPES OF FIBRE

'Dynel' was marketed as staple and tow, bright, semi-dull and dull, and in a wide range of spun dyed colours.

There were three main types of 'Dynel':

(1) *Regular*. Standard grade staple and tow (Type 180).

(2) *Controlled High-Shrinkage*. Staple and tow with 30 per cent shrinkage in boiling water (Type 183). For high bulk fabrics or pile pull-down.

(3) *Carpet Fibre*. A special fibre for carpet use, with lower tenacity, higher elongation, greater toughness and abrasion resistance, high resilience and resistance to soiling.

## PRODUCTION

### Monomer Synthesis

(a) Vinyl Chloride. See page 447.

(b) Acrylonitrile. See page 399.

### Polymerization

Vinyl chloride (60 parts) and acrylonitrile (40 parts) are polymerized by an undisclosed technique.

### Spinning

The polymer is dissolved in acetone, and the solution de-aerated and filtered. It is then pumped through spinnerets, the jets emerging into a bath of water. The acetone is dissolved out by the water, and solid filaments of polymer are formed. The continuous filaments are stretched hot and then heated to anneal and stabilize them in their stretched form. They are then crimped and cut into staple fibre suitable for use on cotton or wool machinery.

## PROCESSING

### Desizing

Desizing with enzyme agents is recommended, especially in blends of 'Dynel' with cellulosic fibres.

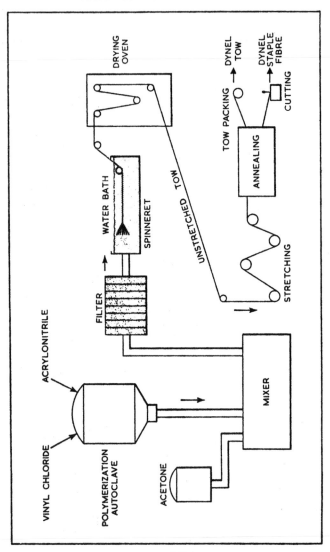

'Dynel' Flow Chart

### Scouring

Goods which have been soiled during processing may be scoured by treatment for up to 1 hour at 70°C. in a bath containing non-ionic surface active agent.

### Bleaching

'Dynel' is a white fibre which does not normally need bleaching. If bleaching should be considered necessary, sodium chlorite may be used in a liquor acidified with nitric acid.

### Dyeing

(a) *100 per cent 'Dynel'*

'Dynel' fibre, stock, tow, yarn or piece goods may be dyed effectively by conventional methods, on conventional equipment. A full range of colours may be obtained, from lights through darks, with excellent fastness to perspiration, crocking, washing and light, and resistance to gas fading.

Dyestuffs used include the entire range of disperse dyes, neutral premetallized dyes, most of the cationic dyes, and certain vat dyes. From this wide range it is possible to select dyestuffs to match any requirement, and to withstand virtually every type of service required of the finished fabric.

'Dynel' may be dyed at temperatures below the boil by using a carrier. This adds to dyeing costs, however, and substantial quantities of 'Dynel' are dyed at temperatures of 96°C. and above without a carrier. At temperatures of 96°C. with a carrier, the fibre's moisture absorption becomes high enough to be of practical value in dyeing. When the colour has been fixed in the fibre by the dyeing process, it is unavailable to the action of water and the various chemical agents that might be expected to discharge or alter the colour. This results in excellent fastness properties.

*Restoring Lustre*

'Dynel' absorbs water at dyeing temperatures, and this may be trapped in the hot fibre, causing a loss of lustre. This has a significant effect on the shade attained. As light fastness is markedly better on lustrous than on dull fibre, it is essential to bring the dyed material up to its full colour value by restoring lustre. This may be done by application of dry heat at 121°C., by semi-decatizing with steam for short cycles, or in the case of package dyeing by using a boiling sodium sulphate solution.

### Thermoplasticity

'Dynel' is a thermoplastic fibre, and this must be considered in all processes involving elevated temperatures, including dyeing. It must be given special attention during piece dyeing. Cooling at the end of dyeing must be done very slowly to avoid setting creases in the fabric. Tightly woven fabrics in particular must be given this care to avoid creasing.

### (b) Blends

In addition to 100 per cent applications, 'Dynel' is used in blends with natural and man-made fibres, and may be commercially union- and cross-dyed printed in a wide range of shades. Simple procedures are used for dyeing blends with rayon, rayon and acetate, rayon and wool, mohair, wool, cotton, acrylic and polyamide fibres.

## STRUCTURE AND PROPERTIES

### Fine Structure and Appearance

White (or spun-dyed) fibre. Cross-section of regular and high shrink types: irregular ribbon; carpet fibre: round to elliptical.

|  |  | Regular | High-Shrink | Carpet |
|---|---|---|---|---|
| **Tenacity** | cN/tex | 18–31 | 37 | 8.8 |
|  | (g./den.) | (2.0–3.5) | (4.2) | (1.0) |
|  |  |  | Same wet or dry | |
| **Tensile Strength** | kg/cm² | 4,060 | 4,900 | |
|  | (p.s.i.) | (58,000) | (70,000 | |
| **Elongation** | (per cent) | 30–42 | 14–17 | 90–120 |
|  |  |  | Same wet or dry | |

472

**Elastic Recovery**

| | (per cent) | 100 @ 2% |
|---|---|---|
| | | 98 @ 5% |
| | | 95 @ 10% |

**Average Stiffness**

| | cN/tex | 90.9 | 218 | 8.8 |
|---|---|---|---|---|
| | (g./den.) | (10.3) | (24.7) | (1.0) |

**Average Toughness**

| | 0.58 | 0.33 | 0.60 |
|---|---|---|---|

**Specific Gravity**

1.3

**Effect of Moisture**

Regain: 0.4
Water has virtually no effect on the mechanical properties of the fibre. There is no shrinkage or felting at temperatures below boiling point.

**Thermal Properties**

Regular fibre: strain release starts at 120°C.; fibre becomes mouldable in the range 135–163°C. (dry heat) or in boiling water. High-shrink fibre: shrinks 30 per cent in boiling water, or at 127°C. in dry heat.
Carpet fibre: strain release begins at 130°C.

*Flammability*

'Dynel' will burn if held in a flame, but it goes out if the flame is removed; it will not support combustion. It does not drip beads of molten material.

**Effect of Sunlight**

Prolonged exposure causes some loss of tensile strength. The fibre darkens gradually.

**Chemical Properties**

*Acids*

Excellent resistance to acids of all concentrations.

*Alkalis*

Excellent resistance to alkalis of all concentrations.

*General*

Not affected by strong detergents, soaps, etc. Resistant to a wide range of inorganic chemicals.

**Effect of Organic Solvents**

Unaffected by hydrocarbons, dry-cleaning solvents and most other organic solvents. Acetone, cyclohexanone and dimethyl formamide are solvents in varying degrees. Certain cyclic ketones and some amines exert solvent or swelling action at higher temperatures.

**Insects**

Completely resistant.

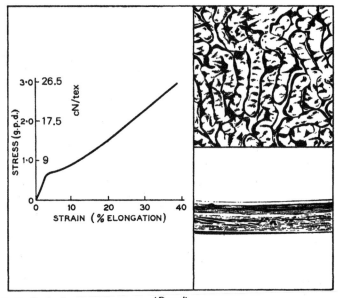

'Dynel'

474

**Micro-organisms**

Completely resistant.

**Electrical Properties**

Dielectric constant: 4.86 at 60 cycles; 4.29 at 1,000 cycles. Low moisture absorption encourages accumulation of static electricity.

## 'DYNEL' IN USE

### General Characteristics

'Dynel' fibre is of flat ribbon-like cross-section, and this contributes to the soft, attractive handle of 'Dynel' fabrics. Like wool, they feel warm and 'friendly' to the skin.

Some 'Dynel' fabrics have a texture, warmth and handle like that of fine vicuna. In general, 'Dynel' rates with the best natural fibres in respect to both warmth and compressional resilience.

The irregular, ribbon-shaped cross-section filaments contribute good covering power, and the high resilience of the fibre gives an unusual and permanent loft to fabrics.

### Mechanical Properties

'Dynel' has a combination of mechanical properties which are generally comparable with those of acrylic fibres. A moderate tenacity, good elasticity, good abrasion resistance and high resilience all contribute to good wearing qualities of 'Dynel' fabrics.

In blends with rayon and acetate, 25–30 per cent 'Dynel' result in a marked increase in fabric wear life. In blends with wool, up to 35 per cent 'Dynel' contributes greater strength and resistance to wear and tear. The stress-strain curve of 'Dynel' almost coincides with that of wool, and extends beyond to a tenacity of 26.5 cN/tex (3 g/den), (wool: 16–18 cN/tex; 1.8–2 g/den). This higher strength adds abrasion resistance to wool and 'Dynel' fabrics. Because these blended fabrics do not pill, fabrics containing 'Dynel' look better after being worn than wools blended with some stronger fibres that may cause pilling. In this respect, wool/'Dynel' blends are often more satisfactory than blends of wool with stronger fibres, as the 'Dynel' blends have a longer useful wear life.

*Dimensional Stability*

'Dynel' fibre is not a highly crystalline material and it does not, therefore, have a sharp melting point. When heated, the fibre will shrink slightly as it reaches a strain-release temperature. If the temperature is increased further, a greater shrinkage will occur until finally all the strains imparted to the fibre have been relieved.

When a 'Dynel' fibre has had strains relieved at a given temperature and time, it becomes dimensionally stable thereafter up to this temperature for that time. Fibres held under tension may be heated considerably above their strain release temperatures with only minor changes in molecular structure. When 'Dynel' fabrics are stabilized by dyeing at the boil, or by boiling off, they are dimensionally stable in boiling water and dry-heat temperatures up to 120°C.

If 'Dynel' is heated in air at high temperatures, it darkens gradually and loses weight. Its mechanical properties, however, are preserved to a remarkable degree. When 'Dynel' is heated for prolonged periods under tension, the tenacity at elevated temperatures increases.

In order to take full advantage of the dimensional stability of 'Dynel' in blended fabrics for apparel, it is preferable for the loom width of the goods to be greater than the desired width of the finished fabric. Finishing, then, should include one stage at which the fabric is allowed to relax completely, under heat and *no* tension, to its natural length and width.

*Novel Fabric Constructions*

The controlled shrinkage of 'Dynel' may be used to obtain selective differences between parts of a fabric in which 'Dynel' is used in conjunction with other fibres. Puckered and bouclé effects, for example, are obtained in this way.

Fabrics may be knit or woven and stabilized by heat shrinkage to give extremely tight constructions. The selection of other synthetic and natural fibres for blending extends the range of novel fabrics of this type.

*Heat Setting; Moulding*

The application of heat to 'Dynel' fabrics at temperatures above their strain-release temperatures is used in the imparting of permanent pleats, and in moulding fabrics into desired shapes.

These shaped fabrics are dimensionally stable, and remain so unless the shaping temperature is again equalled or exceeded.

The use of both the shrinkage and the thermoplastic properties of 'Dynel' are well exemplified in the moulding of hats, where the fabric is shrunk on to a former and then set in this shape.

For radio and high-fidelity grilles and the like, high temperatures are used to provide fabric stiffness and rigidity to the finished article.

### Fire Resistance

'Dynel' will not support combustion. The fibre will burn if held in contact with an open flame, but it stops burning when the flame is removed. Fabrics made entirely of 'Dynel' pass the ASTM Test D626–41 for flame-retarded textiles, but some dye-stuffs and finishes contribute to flammability, and fabrics should be tested before claims are made as to their fire resistance.

### Outdoor Exposure

'Dynel' has outstanding resistance to outdoor exposure under all types of conditions. When 'Dynel' fabrics were buried in soil and held under tropical conditions of 31°C. and 97 per cent relative humidity, no deterioration of the cloth was detected after 6 months. Eight-ounce cotton duck disintegrated completely in 10 days under the same conditions.

In other tests, such as the mineral-base-agar and free-hanging tests, no fungus attack was observed. 'Dynel' is equally resistant to attack by insects.

After uncovered exposures in Florida for 250 sun-hours (approximately 60 days) on yarns and tapes of low denier fibres, 'Dynel' retained more than 50 per cent of its original tenacity and elongation.

### Blends

'Dynel', by itself, combines many of the characteristics of the more expensive natural fibres with a number of man-made features not available in any natural fibre.

In blends with wool, 'Dynel' makes stronger fabrics that hold their shape and press. With rayon, 'Dynel' provides a variety of textures with improved shape and press retention and longer useful wear life. In blends with cotton, 'Dynel' helps the fabric to keep its loft, and therefore its softness and warmth, even after long use and repeated laundering. From 25 to 40 per cent 'Dynel', stock blended, is generally most advantageous.

### Washing

'Dynel' fibres are non-felting and non-shrinking. These properties, coupled with high resistance to chemical attack and high wet strength, make possible repeated laundering even with strong detergents under vigorous conditions. Fabrics may be disinfected with sodium hypochlorite solutions without affecting tensile properties or handle.

'Dynel' fabrics may be washed either by hand or by machine, the temperature of the wash water being as low as possible, preferably below 50°C. (lukewarm).

As with all textile fibres, napped fabrics may pill if subjected to extreme agitation in automatic 'home laundries'. With proper fabric construction, however, napped 'Dynel' fabrics can be washed with confidence in automatic machines.

### Drying

The low moisture regain makes 'Dynel' one of the fastest drying fibres. In certain constructions, water is held mechanically, and this may be removed immediately by centrifuging. Heavy napped fabrics that are whirl-dried and hung up at room temperatures will dry with remarkable rapidity.

Drip drying is preferred if possible; tumble drying in home driers may cause some shrinkage. Drying temperatures should not exceed 60°C.

### Ironing

Wrinkles are removed very easily from 'Dynel' fabrics by ironing at low temperatures. If the ironing temperatures are too high, the 'Dynel' may stiffen and shrink. To preserve the beauty and luxurious hand of 'Dynel' fabrics, the following pressing and ironing instructions must be closely observed.

When ironing all 'Dynel' fabrics, the lowest iron setting and a dry cover cloth of cotton or other fabric should be used. If no cover cloth is used, an iron with a lower than 'rayon' setting is necessary. All 'Dynel' fabrics can be steam-pressed at reduced pressures, and wrinkles can be removed by jet steaming, but steam irons, mangles, or hot-head presses should not be used.

The resistance of 'Dynel'-containing fabrics to shrinkage by heat is increased markedly by stock-blending with more heat-resistant fibres. When 25–40 per cent 'Dynel' is present, the fabric will be stabilized and will retain creases set by an iron. In general, such fabrics can be ironed at the 'rayon' setting or with a steam iron.

### Dry Cleaning

Fabrics of 'Dynel' may be dry cleaned effectively, being resistant to the solvents commonly used. As always, elevated temperatures must be avoided.

### End Uses

#### Pile Fabrics

One of the most successful outlets for 'Dynel' is in the production of fur-like pile fabrics. 'Dynel' provides pile that is soft and lustrous, yet stable to stretching and shrinking. Fabrics made from 100 per cent 'Dynel' and from 'Dynel' in combination with 'Orlon' and other fibres have proved extremely successful.

#### Suits; Dresses, etc.

Dresses, suits, skirts, slacks, jackets and rainwear are some of the applications for 'Dynel', commonly in blends with other fibres. In blends with rayon and acetate, 25–35 per cent of 'Dynel' markedly increases fabric wear life, without pilling. In blends with wool, up to 35 per cent 'Dynel' contributes strength and resistance to wear and tear.

To woven goods generally, 'Dynel' imparts warmth, washability, good draping qualities, crease retention even when wet, and a long useful wear life.

#### Underwear; Knitwear, etc.

In knitted goods, 'Dynel' imparts resistance to shrinkage and stretching, improved retention of loft, a warm luxurious hand, and resilience. Blends of 'Dynel' with other fibres are used in men's, women's and children's underwear, sleepwear and socks.

A blend of 25–50 per cent of 'Dynel' with cotton has proved particularly successful in this field, the 'Dynel' contributing lasting softness, shape and size retention through many washings and long wear. The cotton contributes high absorbency and assures that garments stay comfortable next to the skin.

### Furnishing Fabrics; Curtains, etc.

'Dynel's flame-resistance has proved an important property in the use of the fibre for curtains and draperies. 'Dynel' draperies, for example, are used on the luxury liners *United States* and *America*.

### Carpets

The special grade of 'Dynel' produced as carpet fibre has properties different from those of regular 'Dynel'. These properties have been built into the fibre to make it particularly suitable for carpet use. Carpets made from 'Dynel' combine strength and ease of maintenance characteristics with excellent soil resistance, high resilience, good appearance and resistance to moth and mildew.

### Blankets

Blankets made from 'Dynel' are shrinkproof and mothproof, and will withstand repeated laundering and cleaning. They dry quickly and retain their warmth and handle over a long period of hard wear. In this application, again, flame resistance is an important asset.

### Industrial Applications

Chemical resistance and flame resistance are important characteristics in a fibre used for protective clothing, and 'Dynel' excels in both respects. It is used for shirts, trousers, uniforms and other clothing worn by people exposed to corrosive chemicals.

'Dynel' is used also in laundry nets, filter fabrics, paint rollers and overlays for boats and industrial equipment.

## (3) CHEMICALLY MODIFIED POLYVINYL CHLORIDE FIBRES

Fibres spun from polyvinyl chloride (or a copolymer of vinyl chloride) which has been subjected to chemical modification.

Many modifications of polyvinyl chloride have been made on an experimental basis, but the only commercially-important process at the present time is chlorination. Fibres are spun from chlorinated polyvinyl chloride in which the chlorine content may have been increased from 57 per cent to as high as 79 per cent.

## INTRODUCTION

In 1934, German workers discovered that the solubility of polyvinyl chloride could be increased by chlorination of the polymer. The introduction of additional chlorine molecules into the polymer forced the long molecules apart, enabling solvent molecules to penetrate more easily between them.

In 1936, chlorinated polyvinyl chloride fibres were marketed in Germany under the trade name of 'Pe Ce'. They were spun from acetone solutions of the chlorinated polymer.

'Pe Ce' melted at too low a temperature (below 100°C.) to be of real value as a textile fibre, but it had a number of unique characteristics which served it well in specialized applications. In particular, it was non-flammable.

The production of 'Pe Ce' fibre continued in Germany until and during World War II. It is produced in East Germany as 'Piviacid', and in the U.S.S.R.

## PRODUCTION

### Monomer Synthesis

Vinyl chloride. See page 447.

### Polymerization

Vinyl chloride is polymerized in emulsion as described on page 448.

### Chlorination

Polyvinyl chloride is dissolved in tetrachloroethane to form an 8 per cent solution, and chlorinated by treatment with chlorine at 80–90°C. After 24–36 hours, the chlorine content of the polymer has increased from 57 per cent to about 62–65 per cent.

The hydrochloric acid produced during the reaction, together with excess chlorine, is removed under vacuum, and the polymer may be isolated either by precipitation with methanol or by spray-drying.

### Spinning

Chlorinated P.V.C. is dissolved in acetone to form a 28 per cent solution. After filtration, this is pumped through a spinneret, the jets of solution emerging into a water bath. The acetone dissolves in the water, leaving solid filaments of chlorinated P.V.C. which are stretched and dried.

The filaments are brought together into a tow which may be crimped and cut into staple.

## STRUCTURE AND PROPERTIES

### Fine Structure and Appearance

Smooth-surfaced fibre, of bean-shaped cross-section.

### Tenacity

15.9–17.7 cN/tex (1.8–2.0 g/den), dry or wet.

### Elongation

24–40 per cent, dry or wet.

### Specific Gravity

1.44.

### Effect of Moisture

Regain: 0.2 per cent. Moisture has virtually no effect on mechanical properties.

### Thermal Properties

Chlorinated P.V.C. fibres shrink usually at about 70°C., and soften at 100°C.

*Flammability.* Non-flammable. Do not support combustion.

**Effect of Sunlight**

30 per cent loss in strength after 1 year's exposure.

**Chemical Properties**

Good resistance to most chemicals, including acids and alkalis.

**Effect of Organic Solvents**

Soluble in methylene chloride, butyl acetate, acetone, xylene, o-dichlorobenzene.

**Insects**

Not attacked.

**Micro-organisms**

Not attacked.

**Electrical Properties**

The fibre softens at too low a temperature to be of real value as an electrical insulator.

## CHLORINATED P.V.C. FIBRE IN USE

Chlorinated P.V.C. fibres are of limited value as textile fibres, largely as a result of their low softening point. They have found a number of specialized uses, however, which derive mainly from the flame-resistance and resistance to chemicals. They are used, for example, in flame-resistant clothing, protective clothing in industry, tarpaulins, tents, filter fabrics and the like.

## POLYVINYLIDENE CHLORIDE FIBRES

Fibres spun from polymers or copolymers of vinylidene chloride:

$$CH_2 = CCl_2 \longrightarrow$$

$$---CH_2---\underset{\underset{Cl}{|}}{\overset{\overset{Cl}{|}}{C}}---CH_2---\underset{\underset{Cl}{|}}{\overset{\overset{Cl}{|}}{C}}---CH_2---$$

**VINYLIDENE CHLORIDE**           **POLYVINYLIDENE CHLORIDE**

### INTRODUCTION

In 1940, the Dow Chemical Co. of America introduced a new
type of synthetic fibre consisting of a copolymer of vinylidene
chloride and vinyl chloride. It was given the generic name *saran*.

Vinylidene chloride, the chief component of saran, is a colourless liquid that was made as long ago as 1838. In common with
other vinyl-type unsaturated compounds, it will polymerize to
form an essentially linear polymer capable of forming fibres.
Polyvinylidene chloride was examined as a possible source of
useful synthetic fibres during the early 1930s. The lack of a suitable solvent caused spinning difficulties, however, and the polymer
was sensitive to heat.

The introduction of saran by Dow Chemical Co. in 1940
followed an intensive research project, the copolymer containing a small proportion of vinyl chloride being selected as most
satisfactory for fibre production. The polymer made by Dow was
spun by several firms, including Pierce Plastics and Firestone
Industrial Products Co., who marketed their saran fibre under
the names 'Permalon' and 'Velon' respectively.

Production of saran fibres was subsequently taken up in other
countries, and it is now established as a speciality textile fibre.

### TYPES OF POLYVINYLIDENE CHLORIDE FIBRE

The polyvinylidene chloride fibres produced today are copolymers
containing a small (less than 15 per cent) proportion of other

monomers, commonly vinyl chloride. Other monomers may be included in minor proportions, such as acrylonitrile.

The American F.T.C. regulations restrict the use of the term *saran* to those fibres spun from polymers containing at least 80 per cent of vinylidene chloride (see below), and this definition has come into general use for such polymers. It covers all the important polyvinylidene chloride type polymers now used for spinning fibres, and the properties of saran fibres produced by different manufacturers are sufficiently alike in general properties for the term saran to be of practical value.

Saran was produced originally in the form of heavy denier monofilaments, and it is still widely used in this form today. Monofilaments are spun in round cross-section, and also in a variety of flat and elliptical cross-sections.

Saran is commonly spun-dyed, and a wide range of coloured fibres is available.

## NOMENCLATURE

### Federal Trade Commission Definition

The generic term *saran* was adopted by the U.S. Federal Trade Commission for fibres of the polyvinylidene chloride type, the official definition being as follows:

*Saran.* A manufactured fibre in which the fibre-forming substance is any long-chain synthetic polymer composed of at least 80 per cent by weight of vinylidene chloride units

$$(-CH_2-C.Cl_2-).$$

## PRODUCTION

### Monomer Synthesis

#### (a) Vinylidene Chloride

There are a number of routes to the preparation of vinylidene dichloride, commonly via trichloroethane.

1,2-Dichloroethane may be chlorinated to trichloroethane (1), which is also formed by the chlorination of vinyl chloride (2) or ethylene (3).

Trichloroethane is converted to vinylidene chloride, either by pyrolysis at 400°C., or by treatment with lime (4). In either case, hydrochloric acid is removed.

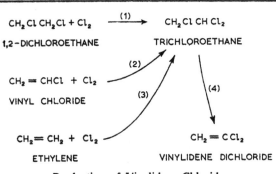

Production of Vinylidene Chloride

(b) *Vinyl Chloride.* See page 447.

### Polymerization

Vinylidene chloride and the vinyl chloride or other co-monomer are polymerized in aqueous emulsion in the presence of a catalyst. Commercial polymers are commonly of molecular weight in the region of 20,000–22,000.

### Spinning

The copolymer is melt spun through spinnerets at about 180°C., the filaments being quenched rapidly before being drawn to develop satisfactory tenacity.

Pigments may be incorporated in the molten polymer before spinning, titanium dioxide being used to provide a dull filament.

### Note

By suitable adjustment of the proportions of vinylidene chloride and vinyl chloride in the polymer, fibres can be made with softening points in the range 70–180°C. A typical commercial saran melts at about 160–170°C.

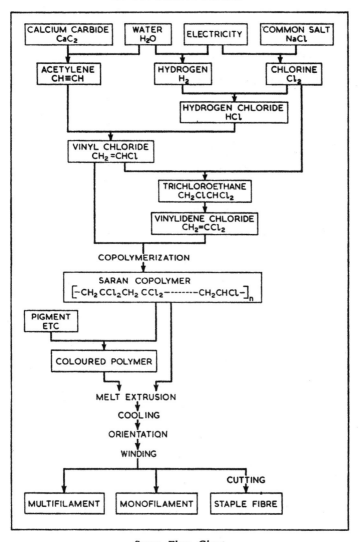

*Saran Flow Chart*

## PROCESSING

### Dyeing

Saran may be dyed by using techniques and dyestuffs similar to those used for vinyon and acetate fibres. Disperse dyes, for example, are used, but the fastness properties are poor.

Commercial saran fibres are commonly produced in a wide range of spun-dyed colours, and these are used in preference to dyeing.

## STRUCTURE AND PROPERTIES

### Fine Structure and Appearance

The filaments are smooth-surfaced and regular. They may be round, oval or flat in cross-section.

Saran is a faint golden-yellow or straw colour, and is translucent in non-pigmented form.

### Tenacity

Up to 20.3 cN/tex (2.3 g/den), wet or dry.
Std. Loop: 6.2–9.7 cN/tex (0.7–1.1 g/den).
Std. Knot: 8.8–15.0 cN/tex (1.0–1.7 g/den).

### Tensile Strength

1,050–3,150 kg/cm$^2$ (15,000–45,000 lb/in$^2$).

### Elongation

15 to 30 per cent, wet or dry.

### Elastic Recovery

98.5 per cent at 3 per cent elongation; 95 per cent at 10 per cent elongation.

### Average Stiffness

44.1–88.3 cN/tex (5–10 g/den).

### Average Toughness

0.16–0.26

### Specific Gravity

1.1–1.7

**Effect of Moisture**

Regain: 0.1–1.0 per cent.
Absorbency at 70°F. and 95 per cent r.h.: 0.1–1.0 per cent.
Moisture does not swell the fibre, and it has a negligible effect on the mechanical properties.

**Thermal Properties**

Softening point: 115–160°C.
Sticking point: 99–104°C.
Melting point: 171°C.
Saran fibres may be weakened at temperatures below the boiling point of water. At 100°C., saran loses about one-third of its strength.

*Flammability*

Saran is almost non-flammable, and will not support combustion.

**Effect of Age**

Negligible.

**Effect of Sunlight**

Good resistance, but discolours slightly on prolonged exposure.

**Chemical Properties**

*General*

Good resistance to bleaches and to most common chemicals. Not corroded by salt spray.

*Acids*

Excellent resistance to most acids in all strengths.

*Alkalis*

Excellent resistance to most alkalis, except ammonium hydroxide, which causes discolouration.

**Effect of Organic Solvents**

Not affected by alcohols or aliphatic hydrocarbons. Aromatic hydrocarbons, halogenated hydrocarbons, ketones, esters and ethers may be detrimental in varying degrees. (Temperature is an important factor in the effects of any of these materials.) Soluble in cyclohexanone, dioxan and tetrahydrofuran.

### Insects

Saran is not attacked by moth grubs or beetles.

### Micro-organisms

Saran is not attacked by mildew or bacteria.

### Electrical Properties

|  | Dielectric constant. | Power Factor (%) |
|---|---|---|
| 100 cycles: | 4.7 | 6 |
| 1,000 cycles: | 3.9 | 9 |
| 1,000,000 cycles: | 2.9 | 3 |

### Refractive Index

1.60

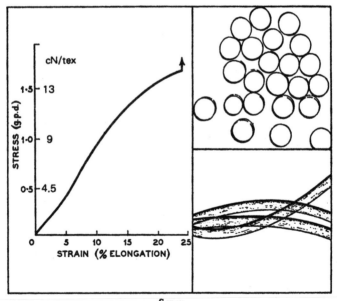

*Saran*

## SARAN IN USE

### General Characteristics

Saran is a flexible fibre, with a soft warm handle. The smooth rounded surface of the fibre contributes to the resistance to soiling and the easy removal of dirt.

### *Mechanical Properties*

Saran is unusually tough and durable, with excellent flex resistance. Woven fabrics have excellent abrasion resistance and a remarkable resistance to hard wear.

### *Specific Gravity*

The specific gravity of saran is fairly high. This would count against saran if the fibre had potentially important apparel end-uses.

### *Moisture*

The moisture absorption of saran is very low, and the mechanical properties of saran fabrics are unaffected by moisture. They are dimensionally stable, and wash and dry easily and quickly. The low moisture absorption contributes to the excellent stain resistance of saran fabrics; ink, food and drink, etc., may be removed with soap and water.

As saran is not normally used in making underwear and other garments worn next to the skin, the low moisture absorption does not detract from the value of the fibre as it does in the case of apparel fibres.

### *Thermal Properties*

Saran tends to soften at temperatures somewhat lower than those favoured for general textile use. Its low resistance to heat, coupled with its negligible moisture absorption, have restricted its use in garment fabrics.

The non-flammability of saran is an important asset, as in the case of polyvinyl chloride fibres, and saran fabrics are commonly used in draperies, etc., where a fire hazard is present.

### *Environmental Conditions*

Saran has excellent resistance to sunlight, ageing and general weathering conditions. It is completely resistant to attack by insects and micro-organisms.

*Chemical Resistance*

The resistance of saran to most common chemicals is excellent, and it will withstand most of the bleaches and other chemicals encountered in normal processing. It is attacked by certain solvents, but is not affected by those normally used in dry cleaning. Perchloroethylene should be avoided.

**Washing**

Saran will wash quickly and easily in soap and lukewarm water. It should be washed by hand, and great care is necessary to ensure that the temperature of the water is kept as low as possible.

**Drying**

Fabrics should be drip dried at room temperature. Tumble drying should not be used.

**Ironing**

Saran fabrics do not generally need ironing, but if ironing is necessary a wet press cloth must be used. The temperature must be as low as possible.

**Dry Cleaning**

Stoddard solvent is recommended for saran. Perchloroethylene should be avoided.

**End-Uses**

Despite the restrictions imposed on end-uses by the low softening temperature of saran, the special characteristics of the fibre have ensured it a market of considerable importance in the textile trade.

*Examples of Applications*

Car seat covers; luggage; filter fabric; handbags; fender cloths; drop cloths; rope; car and public vehicle upholstery; outdoor furniture tape and broad fabric; insect screening; beach umbrellas; doll hair; mannikin wigs; fishing lures; scouring pads; vacuum cleaner hose covering; grille fabrics; shade cloth.

## POLYVINYL ALCOHOL FIBRES

Fibres spun from polymers or copolymers of vinyl alcohol:

$$CH_2{=}CHOH \rightarrow \quad -CH_2-CH-CH_2-CH-$$
$$\qquad\qquad\qquad\qquad\qquad \underset{OH}{|} \qquad \underset{OH}{|}$$

Vinyl Alcohol   Polyvinyl Alcohol

## INTRODUCTION

Polyvinyl alcohol was first synthesized in Germany in 1924, and fibres were subsequently produced in 1931 by Wacker-Chemie G.m.b.H. under the name 'Synthofil'.

The polyvinyl alcohol molecule contains a great number of hydroxyl groups, and polyvinyl alcohol itself is soluble in water. The uses for 'Synthofil', therefore, were limited to specialized applications for which a water-soluble fibre is of value (see Alginate Fibres, page 148).

In the late 1930s, polyvinyl alcohol fibres attracted a great deal of attention in Japan. In 1939, I. Sakurada, S. Lee and co-workers at the Kyoto University discovered a process for producing a water-resistant polyvinyl alcohol fibre by dry heat treatment and acetalization. In the same year, M. Yazawa and his collaborators at the Kanegafuchi Spinning Co. Ltd. independently developed a wet heat treatment (heat treatment in a salt solution under pressure).

Efforts were made to manufacture water-resistant polyvinyl alcohol fibre by the Japan Synthetic Textile Research Association, Kanegafuchi Spinning Co. Ltd., Kurashiki Rayon Co. Ltd., and others. The work was interrupted by World War II, however, and it was not until 1950 that the Kurashiki Rayon Co. Ltd and the Nichibo Co. Ltd. began producing the first polyvinyl alcohol fibres under the trade names 'Kuralon' and 'Mewlon' respectively.

The output of polyvinyl alcohol fibre has increased rapidly in Japan. Fibre is also produced in other countries of the Far East and elsewhere.

## TYPES OF POLYVINYL ALCOHOL FIBRE

Polyvinyl alcohol fibres are commonly insolubilized after spinning by heat treatment and treatment with formaldehyde, and this type of fibre represents the bulk of the output. It is manufactured as continuous filament yarns, staple and tow. Staple is available in sizes and deniers suitable for processing on the cotton, woollen and worsted systems. Tow is processed on the usual types of tow-to-top systems.

Polyvinyl alcohol fibres are also acetalized with aldehydes other than formaldehyde, providing fibres of modified characteristics. Benzaldehyde is used, for example, to produce fibres of high resilience which are used in uniforms, ladies' and children's clothing.

### Water-Soluble Fibre

Polyvinyl alcohol fibre is produced without the heat and aldehyde treatments which bring about water-insolubility. These water-soluble fibres are used for special purposes, such as surgical threads. Mixed with other fibres or yarns, soluble polyvinyl alcohol fibres serve as scaffolding fibres and yarns; when the fabric has been made, the polyvinyl alcohol fibres are washed out with hot water.

## NOMENCLATURE

In Japan, polyvinyl alcohol fibres are known by the generic name *vinylon*. In the U.S.A., U.K., and other countries which use the term *vinyon* for polyvinyl chloride type fibres, there is danger of confusion between the two closely-similar terms, and the name *vinal* has come into use for describing polyvinyl alcohol fibres. This is the term adopted officially by the U.S. Federal Trade Commission.

## Federal Trade Commission Definition

The generic term *vinal* was adopted by the U.S. Federal Trade Commission for fibres of the polyvinyl alcohol type, the official definition being as follows:

*Vinal.* A manufactured fibre in which the fibre-forming substance is any long-chain synthetic polymer composed of at least 50 per cent by weight of vinyl alcohol units $(-CH_2CH.OH-)$ and in which the total of the vinyl alcohol units and any one or more of the various acetal units is at least 85 per cent by weight of the fibre.

## PRODUCTION

Vinyl alcohol is an unstable material, and polyvinyl alcohol is made indirectly by the hydrolysis of polyvinyl acetate.

### Monomer Synthesis

#### Vinyl Acetate

Vinyl acetate is made by the reaction of acetylene (or ethylene) with acetic acid in the presence of a catalyst:

$$CH \equiv CH + CH_3COOH \rightarrow CH_2 = CHOCOCH_3$$

Acetylene　　Acetic Acid　　　　Vinyl Acetate

### Polymerization; Saponification

The vinyl acetate is dissolved in methanol, and is polymerized with the help of a catalyst (e.g. peroxide or azo-compound), forming polyvinyl acetate (1). Caustic soda is added to the methanol solution, bringing about saponification of the polyvinyl acetate to polyvinyl alcohol (2). This is precipitated from the methanol solution, pressed and dried.

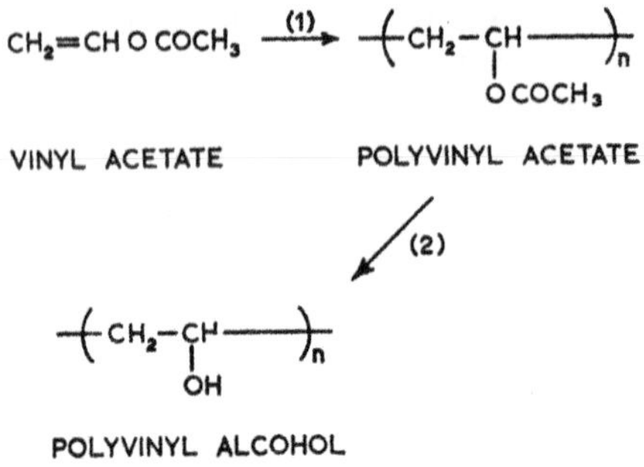

$$CH_2\!=\!CH\ O\ COCH_3 \xrightarrow{(1)} \left(CH_2\!-\!\underset{\underset{O\ COCH_3}{|}}{CH}\!\!-\!\!-\!\!-\right)_n$$

VINYL ACETATE                    POLYVINYL ACETATE

(2)

$$\left(CH_2\!-\!\underset{\underset{OH}{|}}{CH}\!\!-\!\!-\!\!-\right)_n$$

POLYVINYL ALCOHOL

Production of polyvinyl alcohol

### Spinning

#### Wet Spinning

Polyvinyl alcohol fibres are commonly produced by wet spinning. The polymer is dissolved in water to form a 14–16 per cent solution, which is filtered and pumped through spinnerets. The jets emerge into an aqueous coagulating bath containing sodium sulphate solution.

#### Dry/Melt Spinning

Polyvinyl alcohol fibres may also be spun by a process which combines features of dry and melt spinning. The polymer is dissolved in water under pressure and made into a highly concentrated solution (30–50 per cent). The hot molten mass is forced through spinnerets, and the jets emerge into a hot air stream which evaporates the solvent to leave solid filaments of polyvinyl alcohol. These are hot drawn.

### Insolubilization

Polyvinyl alcohol fibres produced by the regular wet spinning process are heat treated, e.g. at about 240°C. This produces a more compact fibre in which hydrogen bonding between the hydroxyl groups of polymer molecules is greatly intensified. This is confirmed by the increase in specific gravity which takes place, and by X-ray examination of the heat treated fibre.

Heat treatment is followed by acetalization, usually with formaldehyde. Acetal groups are formed, which may link adjacent hydroxyl groups on the same molecule (2), or create cross-links between hydroxyl groups on two adjoining molecules (3). The degree of acetalization achieved industrially is between 30 and 40 mol per cent.

$$
\begin{array}{ccc}
-CH_2-\underset{\underset{OH}{|}}{CH}- & & -CH_2-\underset{\underset{O}{|}}{CH}- \\
 & \xrightarrow{(1)} & \\
-CH_2-\underset{\overset{|}{OH}}{CH}- & & -CH_2-\underset{\overset{|}{}}{CH}-
\end{array}
$$

$$
-CH_2-\underset{\underset{OH}{|}}{CH}-CH_2-\underset{\underset{OH}{|}}{CH}- \; + \; CH_2O \xrightarrow{(2)} \; -CH_2-\underset{\underset{O}{|}}{CH}-CH_2-\underset{\underset{O}{|}}{CH} \quad O-CH_2-O
$$

$$
\begin{array}{ccc}
-CH_2-\underset{\underset{OH}{|}}{CH}- & & -CH_2-\underset{\underset{O}{|}}{CH}- \\
 & \xrightarrow[\;]{(3)} & \underset{\underset{O}{|}}{CH_2} \\
-CH_2-\underset{\overset{|}{OH}}{CH}- & + \; CH_2O & -CH_2-\underset{\overset{|}{}}{CH}-
\end{array}
$$

Insolubilization of Polyvinyl Alcohol

*Note*

Aldehydes other than formaldehyde may be used in the acetalization. Benzaldehyde, for example, provides a fibre of high resilience; aldehydes containing active groupings such as amino groups may be used to confer special dye affinity on the fibre.

## PROCESSING

### Desizing

The usual techniques are used, depending on the nature of the size that has been used. Polyvinyl alcohol and other water-soluble

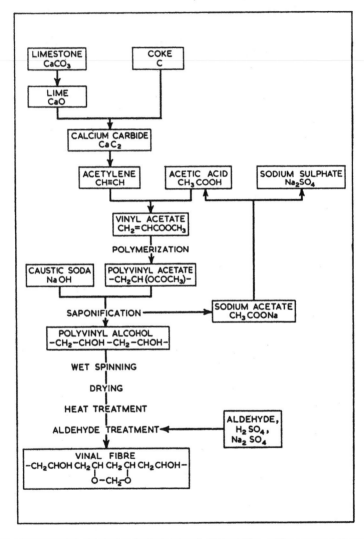

*Polyvinyl Alcohol (Vinal) Fibre Flow Chart*

sizes may be removed with warm water at 80–90°C. Starch is removed with an enzymatic desizing agent, e.g. diastase.

### Scouring

Alkaline scouring treatments, such as kier boiling, tend to shrink polyvinyl alcohol goods, and cause yellowing. They are best avoided.

A soap rinsing with neutral detergent is preferred, e.g. for 30 minutes at 80–90°C., using a 0.5 per cent solution.

### Bleaching

Bleaching by hydrogen peroxide, sulphur dioxide and similar agents is not effective. Chlorine bleaches are preferred.

Hypochlorite and chlorite bleaches are used effectively for the bleaching of polyvinyl alcohol fibres.

Optical bleaching agents may be used after the normal bleach.

### Dyeing

Polyvinyl alcohol fibres have a good affinity for dyestuffs; the fibres resemble cotton and other cellulosic fibres in having hydroxyl groups along the molecule. The following types of dyestuff may be used: direct, acid, basic, metal complex, sulphur, vat, naphthol, acetate.

(a) 100 *per cent Polyvinyl Alcohol Fibre*

#### Direct Dyestuffs

These are generally of low fastness to sunlight and washing, and are used for light shades where fastness is not an important factor.

#### Acid Dyestuffs

In general, acid dyestuffs too are of inadequate light and wash fastness when used with polyvinyl alcohol fibres.

#### Basic Dyestuffs

Basic dyestuffs are generally poor in light fastness when used on polyvinyl alcohol fibres. The fastness is improved by use of a mordant, but these dyes are not recommended where sunlight resistant is important.

### Metal Complex Dyestuffs

1 : 2 type metal complex dyestuffs are used very effectively for dyeing polyvinyl alcohol fibres; they are distinguished by excellent light fastness.

### Sulphur Dyestuffs

Some sulphur colours and sulphur vat colours have good dyeing affinity for polyvinyl alcohol fibres. The colours are not brilliant.

### Vat Dyestuffs

These are the most effective dyes for use with polyvinyl alcohol, providing a range of bright colours of excellent fastness. A wide range of shades is available.

### Naphthol Dyestuffs

Naphthol dyes are used for dyeing polyvinyl alcohol fibres, providing a range of bright, fast colours.

### Acetate Dyestuffs

Acetate dyestuffs are suitable generally for the production of light shades.

### (b) Polyvinyl Alcohol/Cellulosic Fibre Blends

### Sulphur and Sulphur Vat Dyestuffs

After-dyeing with metal complex dyes may be necessary to obtain good results.

### Vat Dyestuffs

After-dyeing with metal complex dyes may be necessary to obtain good results.

### Naphthol Dyestuffs

These dyes are used preferably when blends contain a high proportion of cotton.

### (c) Polyvinyl Alcohol/Wool Blends

Direct acid, acid mordant and mordant dyestuffs may be used.

Acetate dyes may be used for after-dyeing the polyvinyl alcohol fibre.

Direct and acid dyestuffs are used mainly for light shades.

## Printing

Resin pigment printing is used effectively with polyvinyl alcohol goods. After being printed and dried, the fabric is heat-treated to complete polymerization of the resin and set the pigment deep in the fibre.

Fabrics printed in this way are fast to sunlight and washing and colours are bright.

## Singeing

Polyvinyl alcohol fibre does not burn readily, and it is not easily singed effectively. The lumps of burned fibre tend to stick to the surface of the fabric.

If polyvinyl alcohol size has been used on the fabric, this must be removed by desizing before the cloth is singed. If this is not done, the heat may insolubilize the polyvinyl alcohol size, making its removal more difficult.

## STRUCTURE AND PROPERTIES

### Fine Structure and Appearance

Fibres are smooth-surfaced. They are white, with a silk-like lustre.

The cross-section is generally U-shaped, like a flattened tube. There is a pronounced skin layer, which is more crystalline than the core. Mean value of crystallinity is about 50 per cent.

| Tenacity<br>cN/tex<br>(g/den) | Staple | High-Tenacity Filament | Water-Soluble Filament |
|---|---|---|---|
| Dry | 33.5–54.7<br>(3.8–6.2) | 53.0–75.1<br>(6.0–8.5) | 26.5–35.3<br>(3.0–4.0) |
| Wet | 28.3–44.1<br>(3.2–5.0) | 44.1–67.1<br>(5.0–7.6) | — |
| **Elongation** (per cent) | | | |
| Dry | 13–26 | 9–22 | 13–20 |
| Wet | 14–27 | 10–26 | — |

|  | Staple | High-Tenacity Filament | Water-Soluble Filament |
|---|---|---|---|
| **Elastic Recovery** (per cent) | | | |
| From 3 per cent strain 65–85 | | 70–90 | 85–95 |
| From 5 per cent strain 50–60 | | – | – |
| **Initial Modulus** cN/tex | 220–618 | 618–1,589 | 441–795 |
| (g/den) | (25–70) | (70–180) | (50–90) |
| **Young's Modulus** (kg./mm²) | | | |
| | 300–800 | 800–2,000 | 600–1,000 |
| **Average Stiffness** cN/tex | 150–459 | | |
| (g/den) | (17–52) | | |
| **Average Toughness** | | | |
| | 0.41–0.52 | | |
| **Specific Gravity** | | | |
| | 1.26–1.30 | 1.26–1.30 | 1.26–1.30 |
| **Effect of Moisture** | | | |
| Regain (per cent) | 4.5–5.0 | 3.0–5.0 | 9.0 |
| Absorbency at | | | |
| 100 per cent r.h. | 12.0 per cent | | |

*Note*

The following properties refer to the regular (water-insoluble) polyvinyl alcohol fibre.

**Thermal Properties**

Polyvinyl alcohol fibre undergoes a shrinkage of 10 per cent at 220–230°C. At 220°C. it begins to turn yellow, and it shrinks and softens at 230–250°C.

In boiling water (30 minutes immersion) shrinkage is 0.2 per cent.

*Flammability*

Polyvinyl alcohol fibre does not burn readily.

**Effect of Age**

None

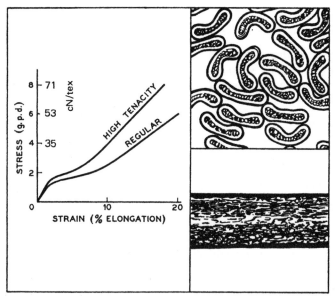

*Polyvinyl Alcohol Fibre ('Mewlon')*

### Effect of Sunlight

Slightly affected after 100 days exposure, but loses strength on more prolonged exposure. The colour remains good.

### Chemical Properties

*Acids*

Polyvinyl alcohol fibre has a good resistance to acids under normal conditions. Hot or concentrated mineral acids cause swelling and shrinkage.

*Alkalis*

Resistance is generally good. Strong alkalis cause yellowing, but tenacity is not affected.

*General*

Resistance is generally good.

**Effect of Organic Solvents**

Inert to animal, vegetable and mineral oils, and to most common organic solvents. Swelled or dissolved by phenol, cresol, formic acid and hot pyridine.

**Insects**

Completely resistant

**Micro-organisms**

Completely resistant

**Electrical Properties**

Surface resistivity (measured on yarns) $1.3 \times 10^{11}$ ohms/cm.

## POLYVINYL ALCOHOL FIBRES IN USE

Polyvinyl alcohol fibres have the flexibility that is associated with a flattened-tube type of cross-section, and the handle of fabrics made from these fibres is excellent. Polyvinyl alcohol fabrics are soft and warm, and feel very comfortable when worn next to the skin.

*Mechanical Properties*

Fabrics have high tensile and bursting strength, and excellent impact and abrasion resistance. They are extremely durable and hard wearing.

The elastic recovery of polyvinyl alcohol fibres is on the low side, and this would suggest that dimensional stability and wrinkle resistance may not be high.

*Specific Gravity*

The specific gravity is lower than that of cotton or rayon, and about the same as silk, wool and acetate.

*Moisture Relationships*

The regain is higher than that of most vinyl-type fibres due to the high proportion of hydroxyl groups which remain even in the acetalized fibre. Despite this, the mechanical properties are not unduly affected by water.

The ability to absorb moisture contributes to the comfort of garments worn next to the skin.

### Thermal Properties

The shrinkage/softening temperature range, beginning at about 220–230°C., is a little on the low side for some textile applications, but is adequate for most purposes.

Polyvinyl alcohol fibres do not heat-set as effectively as nylon, polyester and other thermoplastic fibres.

Fabrics made from polyvinyl alcohol fibres will burn only with difficulty, much depending, as always, on other factors such as finishes used, cloth construction, etc.

### Environmental Conditions

Polyvinyl alcohol fibres are resistant to insects, micro-organisms and other influences encountered in outdoor applications.

### Chemical Resistance

The high resistance to acids, alkalis and many other chemicals is an important factor in the industrial applications of polyvinyl alcohol fibres.

### Cost

Polyvinyl alcohol fibres are potentially very cheap, and their future would seem to lie in the field of hard-wearing, low-cost fabrics and garments.

### Washing

Polyvinyl alcohol fabrics may be washed without difficulty, no special precautions being necessary.

### Drying

Fabrics are quick-drying, and any of the normal methods may be used.

### Ironing

Garments may be ironed safely below 150°C. when dry, preferably at 100–130°C. (rayon setting). Wet fabrics should not be ironed, as there is a tendency for the material to harden.

**Dry Cleaning**

Polyvinyl alcohol fabrics are dry cleaned without difficulty, and are not affected by the solvents commonly used.

**End-Uses**

In Japan, where polyvinyl alcohol fibres have become of major importance, they are being used in virtually every textile field, ranging from the finest wearing apparel to the toughest of industrial applications.

*Apparel*

Staple fibre is used in 100 per cent form, or as blends with other fibres. Excellent materials are made by blending with cotton or rayon.

Apparel applications include materials such as denims, poplins, shirtings, serges, gabardines, suitings, linings, etc. These are made into all manner of garments, including uniforms, sportwear, suits, dresses, stockings, socks, gloves, hats, children's clothing and foundation garments.

In all these apparel applications, the hard-wearing qualities of polyvinyl alcohol are associated with warm, comfortable handle and easy-washability.

*Home Furnishing Fabrics, etc.*

Polyvinyl alcohol fabrics are used for curtains, upholstery, carpets, umbrellas, tablecloths, sheets and the like.

*Industrial Applications*

The durability, chemical resistance, water resistance, strength and resistance to outdoor exposure have enabled polyvinyl alcohol fibres to establish many important industrial end-uses. They are made into fishing nets, ropes, hoses, tarpaulins, conveyor belts, tyre cords, filter cloths, tents and sacks for grain storage.

Monofilaments are used for making synthetic bristles, and for nets used in cultivating seaweed.

## Water-Soluble Polyvinyl Alcohol Fibre

A small proportion of the output of polyvinyl alcohol fibre consists of fibre which has not been heat-treated or acetalized to render it insoluble in water. These water-soluble fibres are used for special applications in which their solubility is advantageous, e.g. surgical threads and scaffolding fibres. Fabrics of novel effects may be obtained by knitting or weaving yarns spun from blends containing polyvinyl alcohol fibres; lace and other openwork fabrics are made by incorporating polyvinyl alcohol yarns which are subsequently washed out.

Important uses include base cloth for Guipure embroidery, draw threads in knitting half-hose socks and sweaters, weft-less felts for paper making, support for low-strength yarns, e.g. low-twist wool and cotton yarns.

## POLYTETRAFLUOROETHYLENE FIBRES

Fibres spun from polymers of tetrafluoroethylene:

$$CF_2 = CF_2 \longrightarrow$$

TETRAFLUOROETHYLENE          POLYTETRAFLUOROETHYLENE
                                      (PTFE)

### INTRODUCTION

In 1938, a research chemist at E.I. du Pont de Nemours and Co.
Inc.. U.S.A., stored a quantity of tetrafluoroethylene gas under
pressure in a steel cylinder. When he tried subsequently to release
the gas from the cylinder, he found that part of it had turned
into an insoluble, non-melting, waxy white powder. The tetrafluoroethylene had polymerized, forming polytetrafluoroethylene.

The new polymer had unusual and potentially useful properties,
and efforts were made to develop it as a new type of plastic.
Many difficulties were encountered, however, and it was not until
1946 that polytetrafluoroethylene resins became available commercially in limited quantities. They were marketed by du Pont
under the trade name 'Teflon'.

'Teflon' is unusual in being extremely stable and inert. It has
an extraordinary resistance to chemicals, and is attacked only by
molten alkali metals, chlorine trifluoride and fluorine gas, hot
and at high pressure. There is no known solvent for the polymer
at normal temperatures, and it withstands elevated temperatures
much better than any other organic plastic. It begins to decompose at about 300°C. without melting.

During World War II, 'Teflon' was used for a number of
applications of extreme importance, including the uranium-235
production plant at Oak Ridge, U.S.A. Large-scale production
of the polymer began in 1950 at Parkersburg, West Virginia, and
'Teflon' has continued to serve as a special-purpose plastic since
that time.

In 1954, 'Teflon' fibre was introduced by du Pont, and the
unique properties of the polytetrafluoroethylene resin have proved
equally valuable in the textile field. 'Teflon' fibre is expensive

T                               509

and is produced in relatively small amounts, but it has found a number of highly specialized applications for which no other fibre is equally satisfactory.

## NOMENCLATURE

Polytetrafluoroethylene is commonly known by the initials P.T.F.E., which is a convenient way of avoiding the unwieldy chemical name. The fibres spun from the polymer are described as P.T.F.E. fibres, and they are commonly included in the more general term *fluorocarbons* or *fluoropolymers,* which include other materials with a high proportion of fluorine atoms as substituents on the carbon chain of the polymer molecule.

There is no U.S. Federal Trade Commission definition to cover P.T.F.E. fibres. The uses of the fibres are specialized and lie outside the general textile field; an official definition of the F.T.C. type would serve little useful purpose.

## PRODUCTION

### Monomer Synthesis

*Tetrafluoroethylene*

Calcium fluoride (fluorspar) is reacted with sulphuric acid to form hydrogen fluoride (1). This is reacted with chloroform, producing chlorodifluoromethane (2). Pyrolysis of the latter at 600–800°C. gives tetrafluoroethylene (3).

$$CaF_2 \;+\; H_2SO_4 \;\xrightarrow{(1)}\; 2HF \;+\; CaSO_4$$

$$2HF \;+\; CHCl_3 \;\xrightarrow{(2)}\; CHClF_2 \;+\; 2HCl$$

CHLOROFORM                CHLORODIFLUOROMETHANE

$$\xrightarrow{(3)}$$

$$CF_2CF_2 \;+\; 2HCl$$

TETRAFLUOROETHYLENE

## Polymerization

Tetrafluoroethylene is purified and polymerized under heat and pressure in stainless steel autoclaves in the presence of a peroxide-type catalyst. The reaction takes place rapidly, with the release of heat, and must be kept carefully under control. P.T.F.E. is formed as a white powder which is subsequently washed and dried.

## Spinning

P.T.F.E. is insoluble and it decomposes before melting. It cannot be spun, therefore, by the dry, wet or melt spinning techniques commonly used for producing fibres from synthetic polymers. It was necessary to devise special techniques for spinning the polymer.

In the production of P.T.F.E. fibre, the polymerization is carried out in such a way as to produce a dispersion containing about 15 per cent of the polymer. The fine particles in the dispersion are ribbon-like in shape.

When the polymerization is completed, the dispersion is extruded through a spinneret, the jets emerging into an aqueous coagulating bath consisting of a dilute solution of hydrochloric acid. The dispersion is coagulated, the particles of P.T.F.E. holding together as weak filaments in which the particles remain entirely separate.

The filaments are heated rapidly to about 385°C., where they are maintained for a few seconds. The polymer particles are sintered, and fuse into a coherent filament. This is quenched quickly, and drawn at room temperature to three or four times its original length.

P.T.F.E. fibres produced in this way are marketed as multi-filament yarns, staple fibre and tow.

## PROCESSING

### Bleaching

P.T.F.E. fibre has a mottled brown or tan appearance, but it can be bleached effectively.

Exposure of the fibre to the air for 3–6 days at 260°C. will bring about partial bleaching to a grey. Higher temperatures reduce the time necessary, but this may be offset by some degradation of the fibre.

Wet oxidation with hot mineral acid mixtures is the quickest and most effective way of bleaching P.T.F.E. fibre, producing a pure white.

### Dyeing

P.T.F.E. fibres are virtually undyeable.

### Sizing

In processing low-twist continuous-filament yarns through plying, twisting, spooling, warping and weaving, it is preferable to use a sizing agent. A 2 per cent coating of a polyvinyl alcohol type agent is recommended. It keeps the filaments together during quilting, warping or weaving, and may be removed easily by a subsequent scouring of the fabric.

Plied filament yarns or highly twisted single end yarns may be processed without sizing.

### Tying

A double weaver's knot is recommended for tying together two continuous filament yarns of P.T.F.E.

### Spinning Staple Fibre

The slick, waxy feel of P.T.F.E. continuous-filament yarn is undesirable in some applications, such as protective garments. In these cases, spun yarn made from P.T.F.E. fibre may be used.

The processing of a low friction fibre such as P.T.F.E. staple creates special difficulties. Card webs and sliver tend to fall apart under their own weight. The low cohesion of the filaments can be overcome to some degree by blending in 3 per cent of rayon staple. Powdering with rosin is also effective, and the addition of small quantities of asbestos may help with carding and spinning.

### Direct Spinning

Spun yarns of P.T.F.E. fibre are made satisfactorily by direct spinning. The filaments in a heavy continuous-filament, low-twist tow are broken in random fashion, and the fibres are twisted into a spun staple yarn. The troublesome carding step is thus eliminated entirely.

Excellent yarn uniformity has been achieved with this technique, and the resulting fabrics have a much less waxy feel than fabrics made from continuous filament P.T.F.E. yarns of comparable weight.

## Textured Yarns

Continuous filament P.T.F.E. yarns may be 'Taslan' textured to provide a more uniform, highly porous yet dimensionally stable fabric having a dry, spun-yarn-type handle.

## Heat Setting

For service at moderate temperatures (100°C.), heat setting may be achieved by conventional boil-off. Further exposures in boiling water will cause very little additional shrinkage. Considerable additional shrinkage of the boiled-off fabric may occur, however, if it is subjected to service temperatures well above the boil-off conditions.

## STRUCTURE AND PROPERTIES

### Fine Structure and Appearance

The carbon atoms forming the backbone of the P.T.F.E. molecule are completely surrounded by fluorine atoms, which act as a barrier that protects the carbon chain. P.T.F.E. fibres are extremely stable to heat and chemicals.

The molecules of P.T.F.E. are electrically neutral, and there are no strong polar forces binding the molecules together as in the case of polyamide, polyester, cellulosic and other fibre molecules. The molecules of P.T.F.E. have an extremely regular structure, however, and this makes possible a very close packing of the chains. P.T.F.E. fibres have a high degree of crystallinity, and the relatively weak but very numerous van der Waals forces combine to create a substantial intermolecular attraction within the crystallites.

The molecules of P.T.F.E. are easily deformed by gross mechanical forces, and the amorphous zones of the fibre are soft.

By contrast, the close packing of the large fluorine atoms round the carbon chain make the molecules relatively immobile to forces of the order that produce Brownian movement. The final transformation, therefore, from an interlaced fibrous structure to a freely-moving molten mass does not occur until quite high levels of thermal energy are reached.

P.T.F.E. fibre is thus soft and very flexible, and yet has a high melting point. The close packing of fluorine atoms around the carbon and the close fitting of molecules in the crystallites are also responsible for the high density of the fibre.

Filaments are smooth surfaced and of round cross-section. They are tan to brown in colour, but can be bleached white in strong oxidizing mineral acids.

## Tenacity

|  | *Multifilament* | *Monofilament* |
|---|---|---|
| cN/tex (g/den) | | |
| Std.: | 10.6–12.4 (1.2–1.4) | 4.4 (0.5) |
| Wet: | 10.6–12.4 (1.2–1.4) | 4.4 (0.5) |
| Std. Loop: | 9.7–11.5 (1.1–1.3) | – |
| Std. Knot: | 9.7–11.5 (1.1–1.3) | – |

## Tensile Strength

|  | *Multifilament* | *Monofilament* |
|---|---|---|
| $kg/cm^2$ | 2,205–2,625 | 980 |
| $(lb/in^2)$ | (31,500–37,500) | (14,000) |

## Elongation

|  | *Multifilament* | *Monofilament* |
|---|---|---|
| (per cent) Std.: | 15–32 | 52 |
| Wet: | 15–32 | 52 |

## Initial Modulus

123.6 cN/tex (14.0 g/den)

## Work of Rupture

0.12 g.cm./den.cm.

## Average Stiffness

| Staple: | 132.5 cN/tex (15 g/den) |
|---|---|
| Filament: | 19.4–62.7 cN/tex (2.2–7.1 g/den) |

## Average Toughness

0.12–0.15.

**Specific Gravity**

2.1.

**Effect of Moisture**

P.T.F.E. does not absorb moisture. Most non-wettable of all known fibres.

**Thermal Properties**

P.T.F.E. has the best thermal stability of the tough, flexible fibres. Some inorganic fibres, such as glass or asbestos, have better thermal stability, but are not as tough or chemical-resistant as P.T.F.E.

On being heated, P.T.F.E. shrinks to some extent. Fabrics may be pre-shrunk by a brief heat-treatment at a temperature above the proposed service temperature.

P.T.F.E. fibre loses its fibre properties and reverts to its massive form at 327°C. It retains a useful strength up to 205°C., and for certain applications can serve at temperatures as high as 288°C. At 290°C., decomposition products are lost at the rate of 0.0002 per cent per hour; at 430°C., at the rate of 1.5 per cent per hour.

P.T.F.E. retains a good set and abrasion resistance when heated, and can be used effectively for many applications at 205–275°C.

*Useful Environmental Temperature.* −73°C. to 275°C.

*Zero Strength Temperature.* 310°C.

*Gel Temperature.* 327°C.

*Specific Heat.* 0.25 B.T.U./lb./°F.

*Thermal Conductivity.* 1.7 B.T.U./hr./sq.ft./°F./in.

*Flammability.* Non-flammable; melts with decomposition.

**Effect of Sunlight**

Negligible.

**Chemical Properties**

*Acids*

P.T.F.E. fibre is completely inert, for example to boiling sulphuric acid, to fuming nitric acid or to aqua regia.

*Alkalis*

P.T.F.E. is completely inert, even to boiling saturated sodium hydroxide.

*General*

P.T.F.E. has an extraordinary resistance to chemical degradation. On the one hand, the carbon-to-carbon bonds are extremely strong, the only reagents that will break them being molten alkali metals. On the other hand, the fluorine atoms are packed so closely around the carbon chain that the carbon-to-carbon bonds in the chain are thoroughly protected from any reagent except fluorine gas at high temperature and pressure, or chlorine trifluoride.

**Effect of Organic Solvents**

The only known solvents for P.T.F.E. are certain perfluorinated organic liquids at temperatures above 299°C.

**Insects**

Not attacked.

**Micro-organisms**

Not attacked.

**Electrical Properties**

Excellent insulator.

**Adhesiveness**

Few materials will stick to P.T.F.E.

**Coefficient of Fibre-to-Fibre Friction**

About 0.2 – lowest of all known fibres.

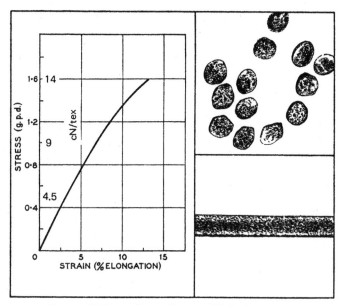

*Polytetrafluoroethylene* ('*Teflon*')

## P.T.F.E. FIBRES IN USE

P.T.F.E. fibres are unique in that they combine the resistance to chemicals, solvents and elevated temperatures associated with inorganic fibres, with the flexibility and toughness that are typical of many organic fibres.

From the outset, P.T.F.E. fibres have been extremely costly, and for this reason alone they could not be considered for general textile use. In addition, they have an unpleasant, greasy handle, no moisture absorption, high density and low modulus; even if cost was not an important factor, they would be of little interest for apparel applications.

The outlet for P.T.F.E. fibres must obviously lie in end-uses where performance is of greater significance than initial high cost, and such end-uses have created a relatively small but important market for P.T.F.E. fibres.

T* 517

*Toxicity*

Although P.T.F.E. itself is not poisonous in any way, it gives off toxic gases when heated above 204°C. The release of these gases increases as the temperature rises. Care should be taken, therefore, when P.T.F.E. is being used to ensure that small portions of P.T.F.E. fibre do not contaminate tobacco. A piece of P.T.F.E. lint picked up by a cigarette could evolve poisonous vapours which might cause discomfort if inhaled. Smoking should be prohibited whenever P.T.F.E. fibre is being processed.

**End-Uses**

The applications for P.T.F.E. fibres are almost entirely in the field of industrial specialties, where there are many small-volume uses for a fibre that combines toughness with unprecedented corrosion resistance.

*Braided Packing*

P.T.F.E. braids are used with great success in the packing of chemical pump shafts, where the material must withstand highly corrosive materials. In a pump handling 102 per cent fuming nitric acid, for example, a braided packing of P.T.F.E. fibre was still in good condition after 7 months, whereas the best previous packing lasted 2–3 weeks. The packing in a centrifugal pump handling molten urea mixture lasted 34 days, whereas the best previous packing lasted 2 to 5 days.

The low coefficient of friction of P.T.F.E. fibre contributes to its success in this application. Best results were obtained by impregnating the braid with P.T.F.E. dispersion.

*Filtration Fabrics*

The high resistance to corrosion of P.T.F.E. fabrics enables them to serve as filtration fabrics for special industrial applications. P.T.F.E. is used successfully for the filtration of hot, corrosive liquids and gases of many types.

*Gaskets, etc.*

The chemical resistance, softness and toughness of P.T.F.E. fibre are characteristics which serve it well in the production of gaskets for pipe flanges handling corrosive liquids. Other uses include laundry pads, laundry roll covers, special conveyor beltings, electrical tapes and wraps for corrosive services, protective clothing, corrosion-resistant cordage, anti-stick bandages, and hose and V-belt curing tapes.

## Bearings

The low coefficient of friction and high load bearing characteristics of this fibre have resulted in its wide acceptance in heavy duty bearings where low relative speeds are involved. Bearings may be fabricated from woven fabrics or composite materials containing P.T.F.E. fibre flock.

# FLUORINATED ETHYLENE-PROPYLENE COPOLYMER FIBRES

Fibres spun from fluorinated ethylene-propylene copolymer:

F.E.P. RESIN

## INTRODUCTION

The introduction of P.T.F.E. plastics and fibres has been followed by the development of many resins based upon copolymers of tetrafluoroethylene. One of the most important of these is a copolymer of tetrafluoroethylene and hexafluoropropylene, which is known as fluorinated ethylene propylene copolymer, or F.E.P.

Fibres spun from F.E.P. are very similar to P.T.F.E. fibres, except that the F.E.P. fibres are thermoplastic, melting at about 290°C.

## PRODUCTION

F.E.P. resins are produced by the copolymerization of tetra-fluoroethylene and hexafluoropropylene:

TETRAFLUORO–          HEXAFLUORO–          F.E.P. RESIN
ETHYLENE             PROPYLENE

## PROPERTIES

Similar to P.T.F.E., but the fibres are thermoplastic, melting at about 290°C.

## POLYVINYL FLUORIDE FIBRES

Fibres spun from polymers or copolymers of vinyl fluoride:

$$CH_2 = CHF \rightarrow -CH_2-CHF-CH_2-CHF-$$

## PRODUCTION

Polyvinyl fluoride monofilaments are produced by extrusion of polymer followed by orientation.

## STRUCTURE AND PROPERTIES

*Tenacity* 19.4–38.8 cN/tex (2.2–4.4 g/den)

*Tensile Strength* 3,500–7,000 kg/cm$^2$ (50,000–100,000 lb/in$^2$)

*Elongation* 15–30%

*Initial Modulus* 10,500–38,500 kg/cm$^2$ (150,000–550,000 lb/in$^2$)

*Elastic Properties* 100% recovery up to 10% extension

*Relative Stiffness* 176.6–397.3 cN/tex (20–45 g/den)

*Relative Toughness* 0.3–0.6 (est)

*Coefficient of Friction* 0.1

*Resistance to Abrasion* Very good

*Refractive Index* 1.42

*Specific Gravity* 1.76

*Effect of Moisture* Surface resists wetting and does not retain water. Adsorption 0.04% Shrinkage in water at 100°C. after 20 minutes, 4–20%

*Thermal Properties*

Melting point: 170°C
Usable temperature range: −62° to 150°C.
Loss of tensile strength at 100°C.: 28%
Life at 150°C.: unlimited
Flammability: self-extinguishing and non-dripping
Specific heat: 0.33 cal/g/°C.
Thermal conductivity: 2.9 x 10$^{-4}$ cal/sec/cm$^2$/°C/cm

*Effect of Sunlight* Excellent resistance

*Chemical Properties* Very good resistance to most common chemicals. Resistant to most common acids, oxidants and solvents, except fuming sulphuric acid, primary aliphatic amines and acetone

*Resistance to Ageing* Excellent

*Insects: Micro-organisms* Excellent resistance

*Electrical Properties*

|  | 60 cyc | 10$^3$ cyc | 10$^6$ cyc |
|---|---|---|---|
| Dielectric constant: | 8.4 | 8.0 | 6.6 |

| Dissipation factor: | 0.049 | 0.018 | 0.17 |
| --- | --- | --- | --- |

Dielectric strength
(volts/mil) 3.18 mm     260
     0.2  mm    1280

Volume resistivity
(ohm.cm.)     $2 \times 10^{14}$

## POLYVINYL FLUORIDE FIBRES IN USE

Monofilament has been woven into filter cloths and installed
in a number of pulp mills to take advantage of the excellent
resistance to chlorine and chlorine dioxide. Filter cloths are used
in other corrosive environments at temperatures ranging from
$-62^{\circ}$ to $150^{\circ}$C. Other applications include mist eliminators,
surgical sutures and electrical braid.

## POLYVINYLIDENE DINITRILE FIBRES

Fibres spun from polymers or copolymers of vinylidene dinitrile:

$$CH_2=C \overset{CN}{\underset{CN}{<}} \quad \rightarrow \quad -CH_2-\overset{CN}{\underset{CN}{C}}-CH_2-\overset{CN}{\underset{CN}{C}}-$$

Vinylidene Dinitrile      Polyvinylidene Dinitrile

### INTRODUCTION

Vinylidene dinitrile was synthesized in 1947 in the research laboratories of the B.F. Goodrich Co., U.S.A. Polymers and copolymers were made, which led eventually to the development of a new type of synthetic fibre. This was produced under the trade name of 'Darlan', which was changed later to 'Darvan'.

The manufacture of 'Darvan' was discontinued in the U.S. in mid-1961, the rights having been sold in 1960 to Celanese Fibres Co., U.S.A. Celanese Fibres Co. and Farbwerke Hoechst undertook further joint development of polyvinylidene dinitrile fibres in West Germany, through Bobina Faserwerke G.m.b.H., the name 'Travis' being adopted for the fibre sold in Europe.

### NOMENCLATURE

#### Federal Trade Commission Definition

The generic term *nytril* was adopted by the U.S. Federal Trade Commission for fibres of the polyvinylidene dinitrile type, the official definition being as follows:

*Nytril.* A manfactured fibre containing at least 85 per cent of a long-chain polymer of vinylidene dinitrile ($-CH_2-C(CN)_2-$) where the vinylidene dinitrile content is no less than every other unit in the polymer chain.

#### Note

The information on polyvinylidene dinitrile fibres which follows is based upon the fibre 'Darvan' as produced in the U.S.A. until 1961.

## PRODUCTION

'Darvan' is a copolymer of vinylidene dinitrile and vinyl acetate made by polymerization of a mixture of the two monomers. The polymer has the following empirical structure:

$$
\begin{array}{cccc}
H & H & H & CN \\
| & | & | & | \\
-C & -C & -C & -C- \\
| & | & | & | \\
H & O & H & CN \\
& | & & \\
& CO & & \\
& | & & \\
& CH_3 & &
\end{array}
$$

### Monomer Synthesis

*Vinyl Acetate.* See page 463.

### Vinylidene Dinitrile

There are several possible routes to vinylidene dinitrile, including the following:

(a) Malonitrile and formaldehyde are reacted to form tetracyanopropane (1). This is then heated to drive off malonitrile and form vinylidene dinitrile (2).

Production of vinylidene dinitrile from malonitrile

(b) Acetic anhydride and hydrogen cyanide are reacted to form 1—acetoxy—1, 1—dicyanoethane (1). Removal of acetic acid by pyrolysis results in vinylidene dinitrile (2).

CH₃CO\
       \O  +  HCN ──(1)──> CH₃CO CN|COCOCH₃|CN
CH₃CO /

ACETIC ANHYDRIDE          HYDROGEN          1-ACETOXY-1, 1-DICYANOETHANE
                          CYANIDE

(2)

CH₂=C with CN above and CN below  +  CH₃ COOH

VINYLIDENE                ACETIC
DINITRILE                 ACID

Production of vinylidene dinitrile from acetic anhydride

### Polymerization

A mixture of vinylidene dinitrile and vinyl acetate in 50:50 mol ratio is polymerized by heating a solution of the two monomers in benzene in the presence of a peroxide catalyst. The polymer is precipitated from the solution as it is formed, and is filtered off and washed.

### Spinning

The copolymer of vinylidene dinitrile and vinyl acetate is dissolved in dimethyl formamide, and the solution is pumped through a spinneret. The jets of solution emerge into a coagulating bath containing water; filaments of solid copolymer are precipitated and hot stretched before being crimped and cut into staple fibre.

## PROCESSING

### Bleaching

'Darvan' may be bleached effectively with acidic sodium or

calcium hypochlorite or sodium chlorite. Alkaline hydrogen peroxide should not be used.

### Dyeing

'Darvan' can be dyed with disperse, cationic or azoic dyes. It has little or no affinity for direct, acid, metallized, chrome or vat dyes.

Pastel shades are produced on 'Darvan' by using disperse dyestuffs or combinations of disperse and cationic dyes. Medium and deep shades are obtained with disperse or cationic dyes. Azoic dyes developed with $\beta$-oxynaphthoic acid are used for black, navy and some red shades.

Medium and deep shades require carrier or pressure dyeing. Vigorous after-scouring is needed to remove excess dyestuffs and carrier (if used).

Strong alkaline conditions, dry temperatures above 162°C., and wet temperatures above 120°C. should be avoided.

## STRUCTURE AND PROPERTIES

### Fine Structure and Appearance

*Molecular Structure*

Vinylidene dinitrile will copolymerize with many monomers, and it tends to form alternating copolymers rather than random copolymers. The structure of the copolymer with vinyl acetate, for example, is virtually the same no matter what proportions of the two monomers are used. The monomer units alternate to form a polymer containing the two units in 50:50 molar ratio. In practice, the polymer is made from a mixture of monomers in this ratio.

The alternation of monomers in the polymer structure of 'Darvan' is probably due to the strong electron-attracting forces of the two nitrile groups on a single carbon atom. Hydrogen bonding results in strong intermolecular attraction between polymer chains. This results in high second order transition temperatures in vinylidene dinitrile copolymers. The transition temperature for an equimolar copolymer of vinylidene dinitrile and vinyl acetate is 171°C. This is some 110°C. higher than for a comparable copolymer of acrylonitrile and vinyl acetate.

526

X-ray diffraction patterns show an almost complete absence of crystallinity in 'Darvan' fibre, with very slight evidence of orientation. The fibre thus has no first order transition point.

## Fibre Form

'Darvan' has a flat, curled cross-section. It is slightly off-white.

## Tenacity

*Dry*: 17.7 cN/tex (2.0 g/den)
Wet: 15.0 cN/tex (1.7 g/den)

## Tensile Strength

2,100 g/cm$^2$ (30,000 lb/in$^2$)

## Elongation

30 per cent, wet or dry.

## Elastic Recovery

From 3 per cent extension: 100 per cent.
From 5 per cent extension: 85 per cent.

## Initial Modulus

176.6–220.7 cN/tex (20–25 g/den)

## Yield Point

Stress: 6.6 cN/tex (0.75 g/den)
Strain: 2–3 per cent.

## Average Stiffness

53.0 cN/tex (6 g/den)

## Average Toughness

0.3.

## Specific Gravity

1.2.

## Effect of Moisture

Regain: 2–3 per cent.
After 3 minutes in water at 100°C., 'Darvan' fabrics show a

shrinkage of 1 per cent; after 30 minutes at 120°C., they shrink between 1 and 15 per cent.

### Thermal Properties

Softening/melting point: 170–176°C. Dimensionally stable at 150°C.

### Effect of High Temperature

After 8 days' dry heat at 165°C., test samples of 'Darvan' retained their original tensile strengths almost unchanged. After 4 days' dry heat at 180°C., the tensile strength was 70 per cent of the original.

### Flammability

'Darvan' compares in ease of ignition and rate of flame travel with untreated cotton, acetate and viscose rayon. 'Darvan' fabrics melt-burn. The apparent ignition temperature is 477°C.

### Effect of Age

Nil.

### Effect of Sunlight

'Darvan' has a high resistance to the effects of direct sunlight. Continuous filament yarns exposed in Florida for 24 months retained 88 per cent of their original strength.

Fabrics made from 'Darvan' staple fibre exposed to Arizona sun for 5 months suffered no measurable loss of strength, and retained over 85 per cent of their original strength in 36 months.

### Chemical Properties

### Acids

Resistance to acids is good. After 4 hours in 10 per cent sulphuric acid or nitric acid at 100°C., there is a loss of strength amounting to 6–30 per cent.

### Alkalis

'Darvan' has a good resistance to dilute sodium hydroxide at low temperatures. After 168 hours in 0.5 per cent sodium hydroxide at 44°C., 'Darvan' suffers a loss of strength amounting to 6–30 per cent. The fibre is degraded by heating for 4 hours in 5 per cent sodium hydroxide at 75°C.

## General

'Darvan' has a good general resistance to attack by chemicals in common use. The fibre shows no loss of strength after 4 hours in a 10 per cent zinc chloride solution at 100°C.

### Effect of Organic Solvents

'Darvan' is insoluble in acetone and in methylene chloride, and is not affected by the solvents used generally in dry cleaning. It dissolves at room temperature in dimethyl formamide.

### Insects

Not attacked.

### Micro-organisms

Not attacked.

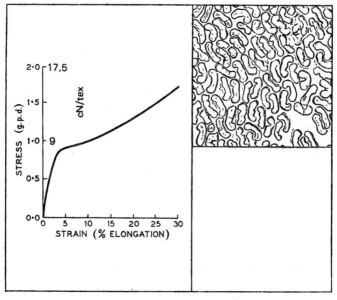

*Polyvinylidene Dinitrile ('Darvan')*

529

## POLYVINYLIDENE DINITRILE FIBRES IN USE

'Darvan' is unusual among synthetic fibres in the combination of softness and resilience which endows it with a most attractive handle. It is very like wool in this respect.

### Mechanical Properties

'Darvan' is a medium strength fibre, comparable with rayon, and it has good elastic recovery. Fabrics are durable and long-lasting, being comparable with acrylic fibres in their wearing properties. They have good crease-retention and wrinkle-resisting characteristics.

### Specific Gravity

The specific gravity is low, a factor which contributes to the good covering properties of 'Darvan' fabrics.

### Moisture Relationships

With a regain of 2–3 per cent, 'Darvan' is intermediate between the hydrophobic fibres, such as polyolefins, and the more absorbent fibres such as nylon. The accumulation of static electricity could prove troublesome.

### Thermal Properties

The softening temperature of 'Darvan' is on the low side, and care is needed in processes involving elevated temperatures. A safe ironing temperature of 160–175°C. is recommended.

Despite this low softening temperature, 'Darvan' retains its mechanical properties well at temperatures close to the softening point.

Flammability is comparable with that of cotton, rayon and acetate.

### Environmental Conditions

'Darvan' fabrics are rot-proof, insect-proof and resistant to sunlight. They will withstand exposure outdoors for long periods without deterioration.

### Chemical Resistance

Fabrics of 'Darvan' are resistant to most of the chemicals and solvents encountered in normal uśe, but their chemical resistance is not as high as that of fibres such as polyvinyl chloride or polyester types. 'Darvan' fabrics should not be subjected to Kier boiling.

## High Bulk Yarns

When fibre is crimp-set under steam pressure, the crimp becomes more permanent. Fibres in yarns spun from crimp-set 'Darvan' appear to have lost their crimp, but a few minutes in boiling water, or even a few weeks at room temperature, will permit the fibres to revert to their crimped state, giving a lofty high-bulk yarn.

'Darvan' tow may be converted to bulky yarns by the Turbo-Perlok system. The differential shrinkage attained is considerably less than that of the acrylics, but it is enough to give a somewhat lofty yarn.

## Heat Setting

Fabrics made from 'Darvan' may be heat-set.

### Washing

'Darvan' fibre absorbs only a small amount of water, and fabrics wash readily in water and detergent. Dimensional stability is good.

### Drying

'Darvan' fabrics dry quickly and easily, drip-drying being especially effective. They may be tumble dried at temperatures up to 60°C.

### Ironing

Fabrics have good wrinkle resistance, and do not generally need ironing. If this is considered necessary, 'synthetic' setting should be used. Maximum safe ironing temperature is 160–175°C.

### Dry Cleaning

There are no difficulties inherent in the dry cleaning of 'Darvan' fabrics, as the fibre is resistant to all common dry cleaning solvents.

**End-Uses**

'Darvan' combines excellent handle with resilience and excellent resistance to degradation. Its uses are essentially those in which these characteristics are of particular importance.

*Deep Pile Fabrics*

Pile fabrics containing 100 per cent 'Darvan' pile are soft and resilient, with little tendency to mat. They may be dry cleaned effectively.

*Suitings*

'Darlan'/wool worsted suitings have good wrinkle resistance, excellent crease retention, good durability and excellent pilling resistance. Woollen type fabrics benefit from the soft luxurious handle of the 'Darvan'.

*Knitted Goods*

100 per cent 'Darvan' crimp-set, high-bulk yarns are used in women's sweaters, which have a remarkably soft, wool-like handle. They are capable of withstanding repeated machine washings and dryings.

## POLYSTYRENE FIBRES

Fibres spun from polymers or copolymers of styrene:

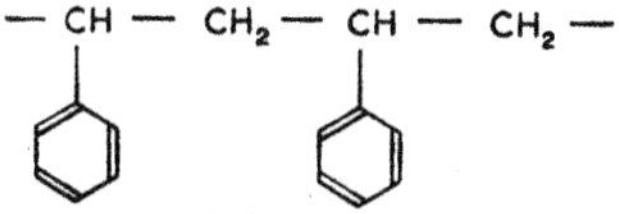

$$- CH - CH_2 - CH - CH_2 -$$

Polystyrene

### INTRODUCTION

Polystyrene is one of the most important synthetic plastic materials, and the polymer is available in large quantities at relatively low cost. The monomer, styrene, was discovered in 1831, and its polymerization was observed shortly afterwards. It was not until the late 1930s, however, that polystyrene became of major importance as a plastic.

Polystyrene may be extruded to form monofilaments, and these have been produced for some years for specialized uses, such as brush bristles.

### PRODUCTION

#### Monomer Synthesis

*Styrene*

Styrene is produced by reaction of benzene and ethylene, both of which may be obtained from coal or petroleum. Ethylene, in addition, is made from alcohol.

## Polymerization

Styrene polymerizes readily to a transparent, glass-like plastic. It will polymerize quickly on standing at room temperature, and the process can be accelerated with the help of heat and catalysts.

Polymerization is commonly carried out in solution, or simply by allowing the liquid itself to polymerize in the mass. Styrene is also polymerized as an emulsion, including the special form of emulsion technique called pearl polymerization. This consists of stirring the styrene with water and a dispersing agent in such a way that the styrene forms droplets about the size of a pin-head. These droplets polymerize to form little beads or pearls of polystyrene.

$$CH = CH_2 \longrightarrow ---CH-CH_2-CH-CH_2---$$

STYRENE
(VINYL BENZENE)

POLYSTYRENE

## Extrusion

Monofilaments are made by extrusion of polystyrene through heated dies, followed by several stages of drawing.

## STRUCTURE AND PROPERTIES

### Molecular Structure

Monofilaments are extruded usually from 100 per cent polystyrene. The phenyl groups forming the side chains on polystyrene molecules are so large and bulky as to interfere with the close-packing of the long molecules. It has long been assumed, therefore, that polystyrene could not be expected to provide strong fibres of the type spun from polymers capable of a high degree of crystallinity.

The development of polymerization techniques in recent years, however, has made possible the production of isotactic polystyrenes which are highly crystalline and melt sharply at 218–220°C. (cf. isotactic polypropylenes, page 567). These polymers could become of commercial importance in the fibre field.

## POLYSTYRENE FIBRES IN USE

Polystyrene monofilaments are used largely as brush bristles.

## 4. POLYOLEFIN FIBRES

Fibres spun from polymers or copolymers of olefin hydrocarbons, such as ethylene and propylene:

$$CH_2=CH_2 \rightarrow \quad -CH_2-CH_2-CH_2-CH_2-$$
Ethylene  Polyethylene

$$CH_3-CH=CH_2 \rightarrow \quad -CH_2-CH-CH_2-CH-$$
$$| \qquad |$$
$$CH_3 \qquad CH_3$$

Propylene  Polypropylene

### INTRODUCTION

In common with many other compounds containing a double bond, olefins are capable of undergoing addition polymerization. Ethylene and propylene, for example, polymerize as shown above.

Olefins are open-chain unsaturated hydrocarbons, and polyolefin molecules have the backbone composed of a succession of carbon atoms which is typical of all vinyl-type polymers.

The table on page 538 lists some of the simpler olefins and their polymer structures. These are all alpha-olefins, in which the double bond lies between the first and second carbon atoms. The remaining portion of the monomer molecule forms pendant groups attached to the side of the polymer chain.

### Polyisobutylene

The first olefin to be polymerized successfully was isobutylene, as long ago as 1873.

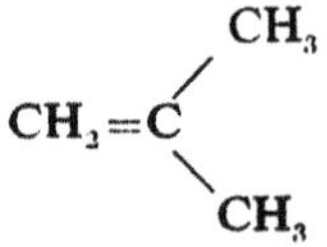

$$CH_3=C \overset{CH_3}{\underset{CH_3}{\diagup}}$$

Polyisobutylene was obtained as a viscous liquid, but it was not until the 1930s that useful linear polymers of isobutylene were made. A rubber-like polyisobutylene was marketed by I.G. Farbenindustrie A.G. in Germany, under the trade name 'Oppanol'.

It had excellent electrical properties and a high resistance to acids and alkalis.

### Polyethylene

During the 1920s, many attempts were made to polymerize other olefins, notably the simplest one, ethylene. The double bond in the ethylene molecule, however, is much less reactive than that in isobutylene, which is activated by the two methyl groups in the molecule. Polymerization of ethylene proved a difficult technical problem, and it was not until the early 1930s that chemists in Imperial Chemical Industries Ltd., England, polymerized ethylene successfully.

During World War II, polyethylene (polythene) developed into an important plastic, and many attempts were made to produce useful fibres from it. Monofilaments were spun and used for special applications, but the polymer was too low melting and the filament too weak to permit of its widespread use as a textile fibre.

In 1954, the development of the Ziegler Process for the polymerization of ethylene (see page 542) provided polyethylenes of higher melting point, which were capable of being spun into strong fibres. Since then, polyethylene filaments and monofilaments have been produced in relatively small amounts, and have established a number of small-volume applications where their special properties are of value. The melting point of these improved polyethylene fibres, however, is still too low (about 135°C.) to permit them to become of real importance in the general textile field.

Until the late 1950s, polyethylene remained the only olefin fibre to be manufactured on a commercial scale, and there seemed little prospect of olefin fibres generally becoming anything more than specialized fibres of relatively minor commercial importance.

### Polypropylene

During the early 1960s, the polyolefin fibre situation underwent a dramatic change. Polyolefins were no longer to be regarded merely as fibres of limited industrial application. Instead, they became of immense potential importance, and showed promise of ranking alongside the polyamides, polyesters and acrylics as general textile fibres.

This change in the status of polyolefin fibres was brought about by the successful development of polypropylene fibres. These fibres display a combination of properties which can serve them well in applications that extend throughout the textile field as a whole. In addition, they are made from a cheap raw material, propylene, which is available in almost unlimited quantities from the petroleum industry.

## ALPHA-OLEFINS AND THEIR POLYMERS

| MONOMER | | POLYMER |
|---|---|---|
| (1) ETHYLENE | $CH_2=CH_2$ | $-CH_2-CH_2-CH_2-CH_2-$ |
| (2) PROPYLENE | $CH_2=CHCH_3$ | $-CH_2-\underset{CH_3}{CH}-CH_2-\underset{CH_3}{CH}-$ |
| (3) BUTENE-1 | $CH_2=CHCH_2CH_3$ | $-CH_2-\underset{CH_2CH_3}{CH}-CH_2-\underset{CH_2CH_3}{CH}-$ |
| (4) PENTENE-1 | $CH_2=CH(CH_2)_2CH_3$ | $-CH_2-\underset{(CH_2)_2CH_3}{CH}-CH_2-\underset{(CH_2)_2CH_3}{CH}-$ |
| (5) 3-METHYL-BUTENE-1 | $CH_2=CHCH(CH_3)_2$ | $-CH_2-\underset{CH(CH_3)_2}{CH}-CH_2-\underset{CH(CH_3)_2}{CH}-$ |
| (6) 4-METHYL-PENTENE-1 | $CH_2=CHCH_2CH(CH_3)_2$ | $-CH_2-\underset{CH_2CH(CH_3)_2}{CH}-CH_2-\underset{CH_2CH(CH_3)_2}{CH}-$ |
| (7) STYRENE | $CH_2=CH-C_6H_5$ | $-CH_2-\underset{C_6H_5}{CH}-CH_2-\underset{C_6H_5}{CH}-$ |

### POLYOLEFINS

#### Specific Gravities and Melting Ranges

| Polymer | Specific Gravity | Melting Range °C |
|---|---|---|
| Polyethylene (branched) | 0.91-0.93 | 107-121 |
| Polyethylene (linear) | 0.95-0.97 | 130-138 |
| Polypropylene (isotactic) | 0.90-0.92 | 165-170 |
| Polybutene-1 (isotactic) | 0.92 | 125-130 |
| Polypentene-1 (isotactic) | 0.87 | 75-80 |
| Poly-3-methyl-butene-1 (isotactic) | 0.91 | 300-317 |
| Poly-4-methyl-pentene-1 (isotactic) | 0.83 | 240-250 |
| Poly-4-methyl-1-hexene (isotactic) | 0.86 | 188 |
| Polystyrene (atactic) | 1.04-1.05 | 99-110 |
| Polystyrene (isotactic) | 1.10 | 240 |

Irregularities, such as branching of chains and end-groups of different composition (which are more important the lower the mean molecular weight) may impair crystallinity in a polymer. Some of these irregularities result in a melting point which is lower than that of the 'ideal polymer', and which is therefore less sharply defined than that of pure substances of low molecular weight. This gives rise to a transition region, which begins when the smallest and more irregular crystals melt, and which attains its upper limit (which is often more sharply defined with the structurally more pure polymers) when the largest crystals melt.

### Other Polyolefin Fibres

The table above lists some of the crystalline polymers which may be made from simple olefins, with their melting points and densities. For most textile applications, it is necessary that a fibre should retain dimensional stability on heating to at least 100°C. If the fibre is to be used in making apparel fabrics, it should be capable of being ironed without softening, and this requires stability to much higher temperatures.

Several of the simple olefins melt at temperatures high enough to warrant consideration for general textile use. Polypropylene, poly-3-methyl butene, poly-4-methyl-1-pentene, and poly-4-methyl-1-hexene all melt above 165°C., and could be regarded as potentially useful textile fibres.

Below this temperature region, the range of application becomes limited, and fibres can be used only where they do not have to withstand more than a modest rise in temperature above normal. Polyethylene is the only comparatively low-melting polyolefin to be used commercially in fibre-production, and its range of application is temperature-restricted in this way.

Practical experience has shown that all the polyolefins melting below polypropylene are inadequate from the point of view of textile fibre production. And some of those melting above polypropylene may be disregarded because of the high cost of raw material; polymers made from them would be too high-priced to be competitive.

Polypropylene itself is economically the most attractive of these higher-melting polyolefins. Another polymer which has come under commercial scrutiny is poly-4-methyl-1-pentene. The raw material, 4-methyl-1-pentene, is potentially available at reasonable cost, and in adequate quantity.

Unfortunately, early studies on this polymer showed that its physical properties deteriorate more rapidly than anticipated at elevated temperatures. Where polypropylene fibres retain some 50–60 per cent of their room temperature strength at 100°C., for example, fibres spun from poly-4-methyl-1-pentene retain only 30 per cent.

It seems unlikely, therefore, that any of these simple olefins will offer a serious challenge to the cheap and readily available propylene as a raw material for polyolefin fibre production. Polypropylene is the one polymer in this class in which the outstanding advantages of polyolefins as fibre-forming materials may be realized.

The large-scale commercial production of polyolefin fibres in the foreseeable future will probably be centred on two polymers. On the one hand, we have polyethylene which is being produced in enormous quantity as a plastic; monofilaments spun from this polymer have become established in many specialized fields.

On the other hand, we have polypropylene, which has become an important man-made fibre with a wide range of textile applications.

## TYPES OF POLYOLEFIN FIBRE

Two types of fibre, polyethylene and polypropylene, dominate the polyolefin fibre field. As already indicated, polyethylene fibres are of relatively minor importance, serving in specialized applications; polypropylene fibres, on the other hand, are of much greater significance in the textile field.

## NOMENCLATURE

### *Federal Trade Commission Definition*

The generic term *olefin* was established by the U.S. Federal Trade Commission for fibres based on polyolefins. The official definition is as follows:

*Olefin.* A manufactured fibre in which the fibre-forming substance is any long-chain synthetic polymer composed of at least 85 per cent by weight of ethylene, propylene or other olefin units, except amorphous (noncrystalline) polyolefins qualifying under category (1) of Paragraph (j) of Rule 7.

### Note

In the section that follows, the two commercially-available types of polyolefin fibre are discussed individually:

1. Polyethylene Fibres.
2. Polypropylene Fibres.

## (1) POLYETHYLENE FIBRES

Fibres spun from polymers or copolymers of ethylene:

$$CH_2 = CH_2$$

## INTRODUCTION

### (a) *High-temperature Process*

The polymerization of ethylene was achieved during the early 1930s by chemists in Imperial Chemical Industries Ltd., England,

and a patent covering the process was applied for in 1936 (B.P. 471,590).

The conditions used in this polymerization process were unusually severe. Pressures of 1,000–2,000 atmospheres and temperatures of 150–200°C. were required, and the process was activated by traces of oxygen. The product was a solid resembling paraffin wax.

The industrial development of ethylene polymerization presented many difficulties, but by 1939 polyethylene was commercially available in the U.K. Licences were granted by I.C.I. Ltd. to du Pont and Bakelite (a division of Union Carbide) in the U.S.

Polyethylene's combination of excellent dielectric and mechanical properties made it invaluable as an insulating material. During World War II, the entire output was used in high-frequency radar equipment, submarine cables and other essential applications. Production was expanded continually to meet the growing wartime needs.

At the end of the war, new applications for polyethylene were found, not only in the electrical insulation field but in the production of films, sheet, tubing, extruded and moulded products. Polyethylene became one of the most important of the post-war plastics, and production has increased steadily in recent years.

Polymerization of ethylene by the high-pressure/high temperature process does not result in straightforward linear molecules of polyethylene. The molecules are branched, and polymer produced by this method may have as many as 30 branches for every 1,000 carbon atoms in the molecular chain.

Branching restricts the ability of polymer molecules to pack together, and prevents them aligning themselves into the orderly patterns that make for regions of crystallinity. Polyethylene made by high-temperature/high pressure polymerization is not highly-crystalline material, and this is reflected in its properties, especially with respect to the characteristics of filaments spun from it. The melting point of polyethylene made by this process, for example, is comparatively low—about 110–120°C.

## (b) *Low-temperature Processes*

In 1954, Professor Karl Ziegler of the Max Planck Institute in Germany discovered a new technique for the polymerization of ethylene, using organo-metallic catalysts (Belg. Pat. 533,362).

Similar processes were developed by the Phillips Petroleum Company (Belg. Pat. 530,617) and by Standard Oil Co. (U.S. Pat. 2,691,647) in the U.S.A.

These new processes bring about the polymerization of ethylene at much lower pressures, and at temperatures below 100°C. Under these conditions, the molecular chains are much less branched than those produced by the high-temperature/high-pressure processes. Ziegler-type polyethylene, for example, has some 4-5 branches per 1,000 carbon atoms in the molecular chain; Phillips-type polyethylene has fewer than 2.5 side branches per 1,000 carbon atoms. The high-temperature process, by contrast, produces a polyethylene with 25-30 side branches per 1,000 carbon atoms.

In addition to providing polymers in which the molecules have fewer branches, these low-temperature processes yield polymers of higher molecular weight.

The molecules of polymer from low-temperature polymerization processes, with fewer side branches, are able to pack together more effectively than those from the high-temperature process. The low-temperature polymers are more highly crystalline than those made by high-temperature polymerization, and this affects the physical properties of the polymer. The density of low-temperature polymer, for example, is higher than that of the high-temperature polymer; the long unbranched molecules of the former can pack closer together, so that the weight per unit volume is increased. Ziegler-type polyethylene has a density of 0.95, and Phillips-type polyethylene of 0.96. The density of the earlier type of polyethylene is 0.92.

This difference in density is commonly used in referring to the different forms of polymer. Polyethylene made by the high-temperature/high-pressure (earlier) process is called *Low-density Polyethylene*; polymer made by the low-temperature/low pressure (later) processes is called *High-density Polyethylene*.

By 1956, high-density polyethylene was commercially available. It was being produced not only by the Ziegler and Phillips processes, but also by a modification of the original I.C.I. process. The increased degree of crystallinity and the higher molecular weight affected the properties of the polymer in ways which extended its range of practical applications. The melting point, for example, was now higher: 130-138°C. compared with 110-120°C. for the low-density polyethylene.

### Fibres from Low-Density Polymer

When supplies of polyethylene became available for general use in the late 1940s, the successful development of nylon had already stimulated interest in the possibility of producing other types of synthetic fibre. Much had been learned about the spinning of fine filaments by extrusion of molten polymers at high temperatures. It was natural, therefore, that the newly-available synthetic polymer, polyethylene, should be considered as a source of synthetic fibres.

The 'clean' shape of linear polyethylene molecules suggests that they are capable of packing together into the orderly arrangements that result in regions of crystallinity. Extruded filaments of polyethylene might be expected to form strong fibres when stretched to orientate the molecules.

Unfortunately, polymerization of ethylene at high temperature and pressure does not produce straightforward linear molecules of polyethylene. The extent of branching is such as to prevent the molecules packing together into the ordered patterns that make for a high degree of crystallinity.

Despite the low degree of crystallinity of this type of polyethylene, and the comparatively low molecular weight, the polymer can be extruded and drawn to form filaments of moderate strength. The early post-war polyethylene was spun into comparatively thick monofilaments which found their way into a number of practical applications. But the tenacity of the material was too low to allow of the production of filaments fine enough for general textile use.

These early filaments of low-density polyethylene had many interesting properties. They were chemically inert, and were quite unaffected by water. They were flexible and resilient, and were not attacked by micro-organisms or insects. Their general characteristics were such as to encourage further development by pioneering firms. Heavy spun-dyed filaments were woven into fabrics which were tested experimentally in a number of applications, including car-seat covers and furnishing materials.

The early polyethylene fabrics suffered, however, from serious shortcomings. Dimensional stability was poor and abrasion resistance low; the filaments deteriorated rapidly in sunlight; the softening point was too low for normal textile use. It soon became clear that fabrics made from the early low-density polyethylenes were not going to be a commercial success.

Despite these setbacks, a few firms continued their development of polyethylene fibres, notably I.C.I. Ltd. and Courtaulds Ltd. in the U.K., and Reeves Bros. Inc. and National Plastics Products Company in the U.S.A. Their efforts brought a steady improvement in the quality of fibre produced from the polymer then available. But real progress was to come eventually from the developments in polymerization technique that led to the production of high-density polyethylenes.

### Fibres from High-Density Polymer

By 1956, high-density polyethylene was commercially available; it was being produced not only by the Ziegler and Phillips processes, but by a modification of the original I.C.I. process too.

This new type of polymer made possible the spinning of much finer filaments than those obtainable from low-density polyethylene. As anticipated, the reduced degree of molecular branching and the higher molecular weights of the new polymers brought remarkable improvements in the physical properties of the fibres. High-density polyethylene fibres were stronger than the low-density types; they reached tenacities comparable with that of nylon. The softening range was higher; 130–138°C., compared with 110–120°C. for the low-density fibres. The stiffness of the fibres had increased.

Despite these and related improvements in physical properties, high-density polyethylene retained deficiencies which have continued to restrict its development as a textile fibre. Filaments spun from it have low resilience, and are subject to relatively high deformation under stress (creep). The softening point is still too low to meet the requirements of normal textile use. The filaments tend to split lengthwise, causing practical difficulties in processing.

Added to these shortcomings are others inherent in the polyolefin structure. The lack of any affinity for water, for example, precludes dyeing by normal techniques.

## TYPES OF POLYETHYLENE FIBRE

Polyethylene is commonly produced today by one or other of the two polymerization processes outlined above, the polymers differing in density and other physical characteristics depending upon the process used.

The two forms of polyethylene made available in this way are described as

> (a) Low-density Polyethylene,
> (b) High-density Polyethylene.

Both types of polyethylene may be spun into monofilaments, and it is in this form that the bulk of polyethylene fibre is produced. Some heavy denier multifilament yarns are also available, and finer denier multifilament yarns are spun in small amounts.

Polyethylene monofilaments are available in a range of diameters and spun-dyed colours. They are produced normally in round cross-section, but may be extruded also in flat, oval and other cross-sections to meet special requirements.

## NOMENCLATURE

### Olefin

Polyethylene fibres are defined as *olefins* under the U.S. Federal Trade Commission definition (see page xxvi).

### Polyolefin

Ethylene is chemically a member of the olefin class of hydrocarbons, and polyethylene is a polyolefin. Polyethylene fibres are thus a type of polyolefin fibre.

### Polythene

The term polyethylene is often used in the shortened form, *polythene* and polyethylene fibres are sometimes called *polythene fibres*.

## PRODUCTION

### Monomer Synthesis

### Ethylene

Ethylene is obtained from petroleum processing.

### Polymerization

#### (a) *High Pressure/High Temperature Process*

Ethylene is polymerized by heating at temperatures in the region of 150–200°C., and pressures of 1,000–2,000 atmospheres. The reaction is promoted by traces of oxygen or other catalysts. Polyethylene is produced in the form of a molten material which solidifies to a waxy solid.

#### (b) *Low Pressure/Low Temperature Process*

Ethylene is polymerized at much lower pressures and at temperatures below 100°C. A variety of catalyst systems may be used. The process developed by Professor Karl Ziegler in 1953–54 made use of organometallic compounds, e.g. of lithium, sodium and aluminium, in conjunction with a small amount of transition metal compound, e.g. titanium tetrachloride.

### Spinning

#### (1) *Low-density Polyethylene Fibres*

Low-density polyethylene is extruded into monofilaments of round flat or other cross-sections, using extrusion techniques similar to those used in making monofilaments of other thermoplastic polymers, e.g. nylon, saran, etc.

Molten polyethylene is held, for example, at about 205°C., and extruded through dies of appropriate shape. Filaments emerging from the die are cooled to 15–60°C., and are then passed round a set of orienting rolls which draw them to between 4 and 10 times their original length. The draw ratio depends upon the type of polymer used.

The oriented monofilaments are collected on spools or tubes.

#### (2) *High-density Polyethylene Fibres*

High density polyethylene is extruded into monofilaments using techniques similar to those used for the low-density polymer. The extrusion temperature is preferably about 210°C.

Drawing of the extruded monofilaments is carried out at a higher temperature. The filaments are heated by hot water, steam or hot air to 100–125°C., and are then passed round heated rolls at 115–120°C. They are drawn to a higher degree than the low-density filaments, commonly in the ratio of 10:1. The temperature of the drawing is critical, and is held within close limits.

## PROCESSING

### Dyeing

Polyethylene fibres cannot be dyed effectively by normal dyeing techniques. Coloured filaments are produced by dispersing pigments in the molten polymer before extrusion, and a range of these spun-dyed filaments is available.

## STRUCTURE AND PROPERTIES

The molecules of polyethylene are commonly branched, the degree of branching depending upon the conditions under which the ethylene polymerization takes place. The polymerization technique also influences the average molecular weight and the molecular weight distribution of the polymer. Polymerization can be controlled to provide a 'tailor-made' polymer with specified branching and molecular weight characteristics.

The nature of polyethylene in these respects has an important influence on the mechanical properties of fibres spun from it. Decrease in the degree of branching, and increase in molecular weight, result in increased tensile strength and stiffness, and higher softening point.

The physical properties of the fibre are also influenced by the conditions under which it is spun and stretched. As the degree of orientation increases, for example, so does the tensile strength increase and the elongation at break decrease.

The subsequent treatment of the oriented fibre is important too. Heat treatment at temperatures below the softening point will influence the flexural strength, elastic recovery and shrinkage.

The properties of polyethylene fibres are thus subject to great variation, depending upon the average molecular weight, the size distribution of the molecules, the degree of branching, and the way in which the orientation of the molecules is controlled. The mechanical properties of a polyethylene fibre produced by one manufacturer may differ considerably from those of fibre produced by another manufacturer. But despite these differences, the fact that both fibres are polyethylenes will confine these variations within recognizable limits.

## Low-density Polyethylene

Polyethylene produced by the high-pressure/high-temperature process may have as many as 30 branches for every 1,000 carbon atoms in the molecular chain.

## High-Density Polyethylene

Polyethylene produced by the low-pressure/low temperature process has fewer branches in the molecule. Ziegler-type polyethylene for example, has some 4–5 branches per 1,000 carbon atoms in the molecular chain; Phillips-type polyethylene has fewer than 2.5 side branches per 1,000 carbon atoms.

The polymers produced by these processes are of higher molecular weight than those produced by the high-pressure process.

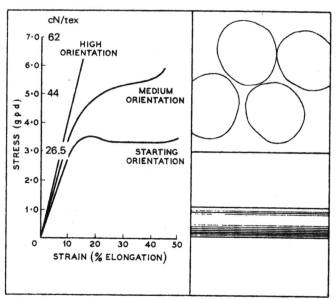

*Polyethylene (High Density Type)*

**Fine Structure and Appearance**

Polyethylene fibres are spun commonly in round cross-section, but may be produced in other cross-sections for special applications. The fibres are smooth-surfaced and of waxy appearance.

**Tensile Strength**

The branched molecules of low-density polyethylene do not permit of the high degree of crystallinity and orientation that is possible with linear molecules of high-density polymer. The tensile strength of low density polyethylene monofilaments is low; tenacity around 8.8–13.2 cN/tex (1.0–1.5 g/den). Linear polyethylene monofilaments, on the other hand, may be three or four times as strong, and are comparable in this respect with nylon. Tensile strength is 2,100–5,950 kg/cm$^2$ (30,000–85,000 lb/in$^2$); tenacities are 70 cN/tex (8 g/den) or more.

Examples of the tenacities of various grades of polyethylene monofilament are shown in the table on page 551.

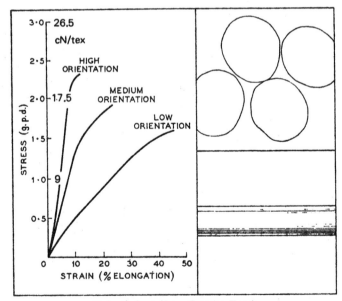

*Polyethylene (Low Density Type)*

PHYSICAL PROPERTIES OF POLYETHYLENE FILAMENTS
COMPARED WITH OTHER SYNTHETIC FILAMENTS

| Polymer | Specific Gravity | Tenacity (g./den.) | Tensile Strength (p.s.i.) | Elongation at Break (%) | Stiffness Modulus (p.s.i.) | Shrinkage 100°C.,20 min. (% orig.length) | Softening Range (without load) °C |
|---|---|---|---|---|---|---|---|
| **Polyethylene** | | | | | | | |
| (1) Branched, low mol. wt. (19,000–20,000)(Reevon 400) | 0.92 | 1.0–1.5 | 15,000 | 45–50 | 20,000–25,000 | 50–60 | 85–96 |
| (2) Branched, med. mol. wt. (23,000–25,000)(Reevon 600) | 0.92 | 2.0–2.3 | 23,000 | 25–30 | 25,000–30,000 | 40–50 | 105–110 |
| (3) Modified branched (Reevon 670) | 0.93 | 3.0–3.3 | 37,000 | 17–25 | 40,000–50,000 | 25–35 | 118–120 |
| (4) Modified linear. (Reevon 760) | 0.945 | 5.0–5.5 | 65,000 | 14–20 | 75,000–85,000 | 10–15 | 120–124 |
| (5) Straight linear (Reevon 700) | 0.95 | 5.5–7.0 | 80,000 | 10–20 | 90,000–110,000 | 5–10 | 124–130 |
| | 0.96 | 6.5–8.0 | 90,000 | 10–20 | 130,000–150,000 | 5–10 | 126–132 |
| **Polypropylene** | | | | | | | |
| Isotactic (Reevon 800) | 0.90 | 5.5–7.0 | 80,000 | 15–25 | 150,000–180,000 | 10–15 | 149–155 |
| **Nylon 66** | | | | | | | |
| (1) Regular (Dry) | 1.14 | 4.5–6.0 | 80,000 | 25–30 | 400,000–600,000 | 5–10 | 243–250 |
| (Wet) | | 4.0–5.2 | 70,000 | | | | |
| (2) High Tenacity (Dry) | 1.14 | 6.0–9.0 | 120,000 | 15–28 | 550,000–650,000 | 5–10 | 243–250. |
| Polyvinylidene Chloride (Saran) | 1.65 | 1.8–2.2 | 40,000 | 20–30 | 55,000–95,000 | 10–15 | 115–137 |

**Elongation**

The elongation at break of low-density polyethylene monofilaments may be as high as 50 per cent. The more highly orientated linear polyethylene monofilaments, on the other hand, may have elongations at break of only 10 per cent (see table, page 551).

**Elastic Properties**

Polyethylene filaments generally are flexible and resilient, the low-density material being more flexible than the high-density type. Increased orientation results in increased stiffness (see table, page 551).

*Creep Characteristics*

Polyethylene fibres tend to undergo creep when subjected to a persistent load over long periods of time. The degree of creep increases with decrease in chain-branching, the non-recoverable elongation being more pronounced in the high-density polymer than in the low-density polymer.

The Phillips-type polyethylene, with less than 2.5 side branches to every 1,000 carbon atoms in the chain, has the worst creep properties. Ziegler-type polyethylene, with 4–5 side branches per 1,000 carbon atoms is better. And low-density polyethylene, with perhaps 20–30 branches per 1,000 carbon atoms, shows the lowest degree of creep of all.

**Specific Gravity**

The specific gravity increases as molecules are able to pack more closely together. Low-density polyethylene monofilaments have specific gravities in the region of 0.92; highly-crystalline high-density (linear) polyethylene has a specific gravity of 0.95 to 0.96 (see table, page 551).

**Effect of Moisture**

The moisture absorption of polyethylene is virtually nil, and for most practical purposes can be regarded as such. Dry and wet strengths are identical, and moisture has no effect on the other mechanical properties of the fibre.

*Water Absorption* (ASTM Test Method D570-54T): 0.01 per cent.

## Thermal Properties

### Softening Point

The softening point of polyethylene rises as the degree of crystallinity increases. A low-density orientated filament of low molecular weight will soften in the range 85–96°C., whereas a high-density orientated filament will soften in the range 126–132°C. (see table, page 551).

### Effect of Low Temperature

Polyethylene retains its flexibility to very low temperatures; it is outstanding in this respect.

Brittleness temperature (ASTM Test Method D746–55T) is less than −114°C.

### Effect of High Temperature

Polyethylene does not degrade readily on heating. With the help of stabilizers, it may be heated for short periods up to 315°C. without decomposing or yellowing. This heat stability makes melt-extrusion of the polymer a practical proposition.

### Shrinkage Properties

Polyethylene fibres do not show any appreciable increase in crystallinity on heating below the softening point, and permanent heat-setting does not take place. Some degree of heat-setting is possible, however, due to the thermoplastic nature of the polymer; the tendency to shrink can be reduced by holding orientated filaments at 17–26°C. below the melting point for a short time. This relieves internal stresses.

In general, orientated filaments are dimensionally unstable, and tend to retract in length when heated. The extent of the shrinkage depends upon the type of polymer and the heat-treatment it has received after orientation.

A low-density polyethylene will commonly shrink, on heating, much more than a high-density monofilament (see table, page 551).

### Flammability

Polyethylene burns slowly in air, but fine filaments tend to melt and drop away before propagating a flame. Various inorganic

compounds and pigments are used to improve flame resistance.
Flammability (ASTM Test Method D635–44):
Slow, 1.0 in./min.

*Specific Heat.* 0.47–0.50 Cal./°C./gram.

### Effect of Sunlight

The polyethylene molecule is attacked by oxygen, the reaction
being stimulated by ultra-violet light. Low-density polyethylene
is more susceptible to oxidation than high-density polymer.

The effect of light is particularly serious when the polymer is
in the form of fine filaments, which have a high surface-volume
ratio. In the early days of polyethylene fibre, light-sensitivity was
a serious drawback, but the incorporation of stabilizers has
effected great improvements.

### Chemical Properties

*Acids*

Polyethylene fibres have a high resistance to acids at all concen-
trations, and up to comparatively high temperatures. They are
attacked by nitric acid (oxidation – see below).
*Resistance to Acids* (ASTM Test Method D543–56)
Weak acids: No effect
Strong acids: Oxidizing acids attack slowly.

*Alkalis*

Polyethylene fibres are highly resistant to alkalis at all concen-
trations and up to comparatively high temperatures.
*Resistance to Alkalis* (ASTM Test Method D543–56)
Weak alkalis: No effect
Strong alkalis: No effect.

*General*

Polyethylene, being a paraffin hydrocarbon, is inherently inert.
Polyethylene fibres are highly resistant to a wide range of chemicals at ordinary temperatures. They are susceptible to attack by
oxidizing agents.

### Effect of Organic Solvents

Polyethylene fibres are insoluble in most common organic solvents at room temperature. They swell and may ultimately dissolve in some chlorinated hydrocarbons and aromatic solvents, e.g. benzene, toluene and xylene. Solutions are obtained at 70–80°C.

Mineral and vegetable oils are absorbed and tend to swell the fibres, especially at high temperatures.

In general, resistance to solvents increases with increased crystallinity; high-density (linear) monofilaments are more resistant than low-density (branched) monofilaments.

### Insects

Polyethylene fibres are not digested by insects and other living creatures. They may be bitten through in an attempt to reach other materials that are used as food.

### Micro-organisms

Polyethylene fibres are completely resistant to bacteria, mildew and other micro-organisms.

### Electrical Properties

Polyethylene made rapid headway as a plastic, largely through its excellent electrical properties. It is an outstanding electrical insulator, especially to high-frequency currents.

The electrical characteristics of the polymer are as follows:

*Dielectric Strength* (Volts/Mil. Short time, $\frac{1}{8}$ inch thickness) (ASTM Test Method D149–59): 500–515.

*Dielectric Constant* (ASTM Test Method D150–54T):
    1 kc.     2.30–2.41
    1 mc.     2.30–2.41

*Volume Resistivity* (ohms/cm.) (ASTM Test Method D257–58): $4 \times 10^{15}$.

*Dissipation (power) Factor* (ASTM Test Method D150–54T):
    1 kc.     0.00023
    100 kc.     0.00032

### Other Properties

Polyethylene is odourless and non-toxic.

*Hardness of Polymer*
  Rockwell R (ASTM Test Method D785–51): 40–45
  Shore D (ASTM Test Method D676–449T): 65–70

*Abrasion Resistance of Polymer* (CS17 Wheel, mg. loss/1,000 cycles). (ASTM Test Method D 1044–55): 5.6–6.1

## POLYETHYLENE FIBRES IN USE

The manufacture of polyethylenes, of both the low- and high-density types, is now an important branch of the world's plastics industry. Polyethylene is a cheap and readily-available fibre-forming material. But, as yet, only a relatively small proportion of the world output of the polymer is used in fibre manufacture. And the prospect of polyethylene fibres being used on a large scale as general textile fibres seems remote.

Production of polyethylene fibres is restricted almost entirely to the extrusion of monofilaments or heavy-denier multifilament yarns, e.g. 1,100–3,300 dtex (1,000–3,000 den).

Some low-denier multifilament yarns are spun from high-density polyethylene, but these are of minor commercial importance. The development of polypropylene fibres, with important advantages including higher softening point, superior strength, resilience and processability, has left little incentive for intensive study of polyethylene fine-denier yarns.

The commercial development of polyethylene fibres in the general textile field has, from the very beginning, been hampered by low softening point, high shrinkage. low stiffness, poor creep characteristics, and an inability to take dyes. These and other shortcomings have prevented widespread acceptance of polyethylene fibres since they first became available, and they are still restricting its use to specialized non-apparel applications.

### Low-Density Polyethylene

The poor tensile properties and low melting point of low-density polyethylene fibres are deficiencies inherent in the polymer, and little improvement in these respects can be expected from developments in processing. The poor resistance to ultra-violet light shown by early fibres is no longer a significant problem; protection is given by incorporating special pigments and stabilizers in the polymer.

tection is given by incorporating special pigments and stabilizers in the polymer.

Low-density fibres have a particularly good low-temperature flexibility, and they are used in applications where this characteristic is important, for example, in aerial tow-targets. The low softening point is also turned to advantage by using low-density polyethylene fabrics as interlinings in shirt collars.

Low-density polyethylene fibres are used for many industrial applications, including ropes and cordage, filtration fabrics and protective clothing. As resilient shrinkable yarns they are used in the production of three-dimensional fabrics.

## High-Density Polyethylene

The introduction of high-density polyethylene has overcome some of the deficiences of the low-density polymer fibres. The increased crystallinity of high-density polyethylene filaments has brought improved tensile strength (up to four times that of low-density polyethylene) and stiffness (up to five times that of low-density polyethylene), and a higher softening point.

Fibres spun from high-density polyethylene are dimensionally-stable in boiling water, and will withstand temperatures up to about 110°C.

Despite these improvements in fibre properties, there are still inherent deficiencies which restrict the use of high-density polyethylene fibres to specialized applications. The softening point is still too low for general textile use, especially with respect to apparel fabrics; dyeing remains a problem, resilience is poor, and creases persist in fabrics after folding.

The applications found for high-density polyethylene fibres have followed the pattern established by low-density polyethylene fibres. They make use of the special characteristics of polyethylene in fields where the shortcomings are of little or no importance. Polyethylene fibres have the lowest density of all commercial fibres, other than polypropylene; they have excellent dielectric properties, they are resistant to a wide range of chemicals, including acids and alkalis; they are tough and strong, and they retain their strength and flexibility at temperatures well below zero. Even at −100°C., polyethylene fibres have not become brittle.

High-density polyethylene fibres are used in aerial tow-targets, in high-altitude balloons and in equipment to be used in Arctic conditions, where the retention of strength and flexibility at low temperatures are invaluable characteristics.

The strength, lightness, water- and rot-resistance of polyethylene fibres have enabled them to become established in the marine cordage field. There are great advantages in ropes and nets that float, do not rot and do not absorb water.

Furniture fabrics, car upholstery fabrics, curtains, protective clothing, tarpaulins and filter fabrics are other applications in which the special properties of polyethylene fibres enable them to compete effectively with other fibres, despite their inherent shortcomings which restrict their use in general textile fields.

## Irradiation

When polyethylene is irradiated with gamma rays or with high-speed electrons, cross-links are formed between the polymer molecules. The movement of the molecules relative to one another is restricted, and the polymer becomes more resistant to softening when heated. If cross-linking is sufficiently extensive, the polyethylene will no longer melt.

## End-Uses

### Twines and Nets

High density polyethylene monofilament yarns are widely used for the manufacture of twines and netting for the fishing industry. The main features of polyethylene twines and nets are as follows:

*Rot-resistance.* The inherent resistance of polyethylene to micro-organisms and to chemical attack make rot-proofing treatments unnecessary.

The strength and other mechanical properties of polyethylene nets and twines are unaffected by immersion in the sea or by burial in the ground for long periods. Nets and twines made from natural fibres will rot under comparable conditions.

A trawl made from high-density polyethylene yarn ('Courlene X3') was lost at sea, and recovered almost a year later. The twines showed no deterioration, and the trawl was put into immediate service again.

*Hard Wear.* The high initial strength, coupled with excellent wet abrasion resistance, is maintained throughout the long life of a net. Experience has shown that nets made from high-density polyethylene require a minimum of mending.

*Ease of Handling.* Polyethylene nets are lighter than those made from natural fibres. This makes the net easier to handle, and saves time in shooting and hauling in. There is less drag, resulting in lower fuel costs. Polyethylene does not absorb water, and a soaked net will commonly weigh only about one-third as much as a wet manila net.

*Cleanliness.* Polyethylene filaments are smooth-surfaced, and do not cling to sand particles, marine growth and other unwanted materials. This simplifies the work of the crew.

*Resistance to Extreme Cold.* Polyethylene retains its flexibility at very low temperatures, and does not become rigid under freezing conditions. The wet knot strength of twines increases as the temperature decreases. These factors are very important when fishing takes place in Arctic waters, where the air temperature may fall to $-10°C$. or lower.

*Buoyancy.* The specific gravity of polyethylene is less than that of water, and nets float naturally. This reduces the chance of a net fouling the propellers, and cuts down the number of floats that are required on nets. The mouths of the trawls remain wide open.

*Stability.* Polyethylene nets do not shrink on immersion in water. They retain their shape and mesh size, minimizing the danger of an infringement of regulations.
Single knots only are necessary in the manufacture of nets and trawls, so long as the knots are first pulled tight. There is no need for heat-treatment to stabilize the knots.

### Ropes

Ropes made from high-density polyethylene monofilament yarns are now being used for a wide range of applications, from painters on dinghies to mooring ropes for tankers.

Experience with mooring and gig ropes on tugs has proved that polyethylene is eminently suitable for this type of heavy work. Three-strand mooring ropes of 16.5 cm (6½ in) diameter are still in excellent condition after two years of continuous use.

Heavy ropes of 20 and 23 cm (8 and 9 in) circumference are used in a variety of constructions, including 4 x 2 plaited. Finer ropes of 6.3–9 cm (2½ – 3½ in) circumference are now accepted in the fishing industry as quarter ropes, while cod-lines, life-lines, mooring ropes, etc., are also used in increasing quantities.

## Main Characteristics

The most important features of polyethylene ropes are as follows:

*Lightness.* Polyethylene ropes are only two-thirds the weight of manila or sisal ropes of equal circumference. A polyethylene rope is only about half the weight of a manila Grade 1 special quality rope of equal strength, and 41 per cent the weight of a comparable sisal rope.

When the ropes are wet, the difference in weight is even more striking, as polyethylene does not absorb water.

*Strength.* Polyethylene ropes are about 33 per cent stronger than manila ropes (Grade 1 special quality) and 50 per cent stronger than sisal ropes of equal circumference. Polyethylene ropes of the same strength as manila or sisal ropes are thus of 13 and 19 per cent less circumference respectively.

*Rot-resistance.* Sea-water, acids, alkalis and other materials commonly encountered in use have no deleterious effects on polyethylene ropes. They need never be dried out, and are ready for use at a moment's notice.

*Resiliency.* Polyethylene ropes are very resilient and easy to use. They do not harden when wet, nor do they freeze or harden even in the most severe weather conditions.

*Abrasion Resistance.* Abrasion resistance of polyethylene ropes is good, and they are very durable. The surface of the rope may become fluffy after prolonged wear and tear, but there is no appreciable loss of tensile strength by the rope itself.

*Buoyancy*. Like nets and twines, polyethylene ropes will float on water, reducing the risk of fouling propellers.

*Colouration*. Ropes are commonly made from mass-coloured fibre. The colours are very fast to light and washing.

### Filtration Fabrics

One of the earliest applications for polyethylene yarns was in the production of industrial filtration fabrics. The range of properties offered by polyethylene is particularly suitable for this application, and the fibre is being used increasingly for this purpose.

### Main Characteristics

The main features of polyethylene fabrics with respect to industrial filtration fabrics are as follows:

*Strength*. Polyethylene yarns retain their high strength in the presence of water, and fabrics are not weakened during the filtration of water solutions.

*Chemical Resistance*. The excellent resistance of polyethylene to a wide range of chemicals and solvents is of obvious value in a fabric used for filtration of industrial liquids.

*Abrasion Resistance*. Filtration fabrics may be subjected to considerable abrasion during use, and the high abrasion resistance of polyethylene enables the fabrics to withstand such treatment.

*Cake Release*. In the filtration of sewage liquids, solid particles build up into cakes between the layers of filters. These cakes are removed perhaps several times a day by opening the filter presses, and it is important that the solid material should come away cleanly and easily from the filter fabric. Polyethylene is particularly good in this respect.

*Rot-resistance*. The complete resistance of polyethylene to attack by micro-organisms is of great value in many filtration applications, such as the filtration of sewage. Cotton and other susceptible fibres are attacked rapidly under such conditions.

*Dimensional Stability*. Filtration fabrics must be able to withstand pressure used in forcing the liquid through the press. Polyethylene fabrics are heat-set to provide excellent dimensional stability.

*Cost*. In sewage filtration, which is one of the most important industrial filtrations processes, cotton fabrics have long been used in the filter presses. Polyethylene fabrics are initially more expensive than cotton, but their extra useful life outweighs this extra cost. Experiments carried out by Spenborough Corporation, England, showed that in sewage filtration, cotton cloths cost 0.81 pence per pressing, compared with 0.492 pence per pressing for a polyethylene ('Courlene X3') fabric. Cotton disintegrated after 76 pressings, whereas the polyethylene fabric was satisfactory up to more than 350 pressings.

Other synthetic fibre fabrics were used in this year-long practical experiment, but they had the disadvantage of blinding after some 200 pressings; their costs were in the region of 0.621 pence per pressing.

Blinding refers to the blocking of the filtration fabric to the point at which washing does not free the fabric of filtered particles.

## Blinds, Awnings and other Outdoor Fabrics

Polyethylene offers an attractive range of properties to the manufacturer of blinds, awnings, deck-chair and other fabrics for use outdoors.

### Main Characteristics
The main characteristics of interest in this respect are as follows:

*Colour and Transparency*. Pigmented polyethylene is available in a wide range of attractive colours. The yarns are commonly translucent, but opaque materials may be obtained by using darker coloured warp and weft yarns. Tests show that 95 per cent of the ultra-violet radiation in sunlight is dispersed on the surface of the polyethylene fabric.

*Rot-resistance*. Polyethylene fabrics are completely resistant to micro-organisms and insects, and are ideal in this respect for use outdoors. Polyethylene fabrics need not be dried before storage.

*Light-resistance.* Polyethylene fabrics are stabilized to resist degradation stimulated by light, and have a long outdoor life. The pigments locked in the filaments before spinning are fast to light and water, and colours remain bright.

*Chemical Resistance.* Outdoor fabrics are subjected to attack by the polluted air of towns and cities, which contains a variety of acids and other corrosive chemicals. Polyethylene is unaffected by these pollutants, which bring about the rapid decay of 'more sensitive fabrics.

*Water Resistance.* Polyethylene retains its strength and tear resistance when wet. It does not absorb moisture, and is quick-drying. Fabrics are stain resisting, and are easily cleaned with detergent and water. Polyethylene fabrics may be coated to make them waterproof if desired.

*Abrasion Resistance.* Outdoor fabrics must withstand rough handling, and are often subjected to great wear and tear. The abrasion resistance of polyethylene is excellent and outdoor fabrics are tough and hard-wearing.

*Low-temperature Resistance.* Under the severest winter conditions, polyethylene remains strong and flexible.

*Density.* The low density of polyethylene in comparison with other fibres is a great advantage in deck-chair and similar fabrics. The lightness of polyethylene fabrics simplifies handling problems in outdoor furniture which is constantly moved about.

*Signwriting.* Awnings and the like are often required to carry advertisements and signs, and polyethylene fabrics offer no difficulties in this respect. Special inks are available for this purpose.

*Heat-setting.* Dimensional stability is excellent in heat-set polyethylene fabrics, and a variety of attractive embossed effects may be obtained by heat-treatments.

## (2) POLYPROPYLENE FIBRES

Fibres spun from polymers or copolymers of propylene:

$$CH_2=CHCH_3$$

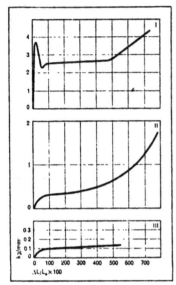 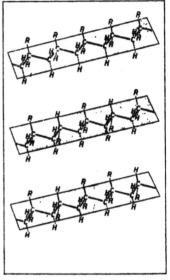

*Polypropylene: Steric Structure*

**Right:** The main chain of carbon atoms forming the 'backbone' of a molecule of vinylic ($CH_2=CHR$) polymer may be considered as a zig-zag lying on a plane.

(1) *Isotactic Polymer* (top). The R groups are all on one side of the plane.

(2) *Syndiotactic Polymer* (middle). The R groups lie alternately above and below the plane.

(3) *Atactic Polymer* (bottom). The R groups appear in any order above or below the plane.

**Left:** Stress-strain curves of (I) isotactic polypropylene, (II) stereo-block polypropylene, and (III) atactic polypropylene.

The differences between the curves are not due to the different molecular weights of the three types of polypropylene but to differences in their steric structure – *Ciba Review.*

## INTRODUCTION

The successful polymerization of ethylene, using organometallic catalysts, was followed by attempts to use this type of catalyst for polymerizing other olefins, notably propylene. Polypropylene of high molecular weight had been made in 1952 by Fontana, but the polymers obtained were oils and greases. In 1954, Professor Giulio Natta of Milan Polytechnic, Italy, discovered that certain Ziegler-type catalysts could bring about polymerization of propylene to linear polypropylenes of high molecular weight.

Natta was able to separate his polypropylene into a number of polymers, using differences in their solubility. He found that some polymers had a density as high as 0.91, whereas others were as low as 0.85. They varied also in having different melting points, and some crystallized where others remained amorphous. These differences were found even in polypropylenes of similar molecular weights.

X-ray and infra-red investigations showed that the differences between the polypropylenes were due to different steric structures of the polymers. Certain steric structures permitted an ordered arrangement of the molecules into crystalline regions, providing the solid, high-melting, high-density polypropylene; other molecular structures were incapable of packing together in this orderly fashion, and these molecules formed the viscous, low-melting, low-density polypropylene.

### Stereoregularity

The molecule of polypropylene consists of a long chain of carbon atoms, with methyl groups forming appendages which stand out from the sides of the chain. This three-dimensional structure of the molecule permits it to exist in a variety of forms which differ in their spatial arrangements.

Examples of different spatial arrangements in a polymer of an alpha-olefin such as propylene are shown on page 566. The backbone of the polymer molecule consists of a zig-zag chain of carbon atoms which may be considered as lying in the plane of the paper. Every side-grouping (R) may then lie in one of two positions; it may be above the plane of the paper, or below it.

Professor Natta recognized that certain basic types of polymer structure were made possible by these alternative positions of the

## (1) ISOTACTIC POLYPROPYLENE

$$-CH_2 \; \underset{\overset{|}{CH_3}}{CH} CH_2 \; \underset{\overset{|}{CH_3}}{CH} CH_2 \; \underset{\overset{|}{CH_3}}{CH} CH_2 \; \underset{\overset{|}{CH_3}}{CH} CH_2 \; \underset{\overset{|}{CH_3}}{CH} CH_2 -$$

## (2) SYNDIOTACTIC POLYPROPYLENE

$$-CH_2 \; \underset{\overset{|}{CH_3}}{CH} CH_2 \; \underset{\underset{CH_3}{|}}{CH} CH_2 \; \underset{\overset{|}{CH_3}}{CH} CH_2 \; \underset{\underset{CH_3}{|}}{CH} CH_2 \; \underset{\overset{|}{CH_3}}{CH} CH_2 -$$

## (3) ATACTIC POLYPROPYLENE

$$-CH_2 \; \underset{\overset{|}{CH_3}}{CH} CH_2 \; \underset{\underset{CH_3}{|}}{CH} CH_2 \; \underset{\overset{|}{CH_3}}{CH} CH_2 \; \underset{\underset{CH_3}{|}}{CH} CH_2 \; CH CH_2 -$$

*Polypropylene*

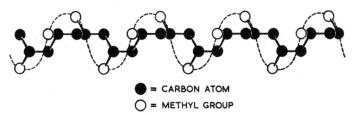

= CARBON ATOM
= METHYL GROUP

*Isotactic polypropylene.* The bulky methyl groups take up angles of 120° with respect to neighbouring methyl groups. This makes possible a regular helical or spiral structure as shown.

---

side groups, and he gave them names by which they are now generally known.

An *isotactic structure* is one in which there is a regular repetition of units of the same configuration. The side groups are all located on the same side of the backbone plane, as in the upper diagram of figure on page 564.

A *syndiotactic structure* is one in which there is a regular sequence of alternating units of opposed configurations. The side groups are located in sequence of alternate sides of the backbone plane, as in the middle diagram.

An *atactic structure* is one in which the units are distributed irregularly along the molecular chain. The side groups are located on either side of the backbone chain in random fashion, as in the lower diagram.

In the case of polypropylene, methyl groups are attached to the backbone chain of carbon atoms, and the three corresponding steric structures are shown on page 566.

(1) *Isotactic Polypropylene* has all the methyl groups on the same side of the backbone plane.
(2) *Syndiotactic Polypropylene* has methyl groups on alternate sides of the plane.
(3) *Atactic Polypropylene* has methyl groups distributed in random fashion on both sides of the plane.

The steric arrangement of the polypropylene molecule has an important influence on the properties of the polymer. Isotactic and syndiotactic polypropylenes (i.e. the 'tactic' polymers) are regular structures, and their regularity enables the molecules

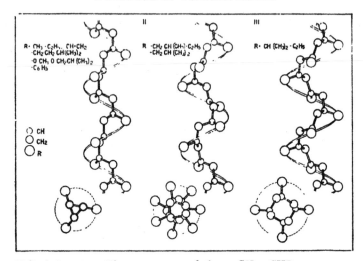

*Helical Structure.* The asymmetry of the $-CH_2-CHR-$ monomer unit of the isotactic poly-α-olefin makes the formation of symmetrical crystals impossible save in the case of helical chains—the only case where all the monomer units of a poly-α-olefin can attain structural equivalence. All the main chains of isotactic polymers are therefore helical in their crystalline state. Shortly after its discovery in 1954, the first crystalline polymer of propylene was found to be isotactic and to have a helical chain. Indeed, crystallinity of the vinylic polymers can result from two simple structures only, namely the isotactic and the syndiotactic.

The figure above shows the different types of helices present in isotactic polymers of known structure. Three-dimensional crystals require not only a regular structure of the individual chains, but also a regular packing of the chains in a direction at right angles to the axis of the chain. If only the first condition is met, and not the second, as in orientated and quenched materials, smectic and incompletely crystallized structures of lower density and greater transparency result – *Ciba Review.*

to pack together in an ordered fashion. These polymers are crystalline.

Atactic polypropylene, on the other hand, is an irregular structure, and the molecules cannot assume a crystalline structure. This type of polymer is amorphous.

The first crystalline polymer of propylene was examined shortly after it was made in 1954, and found to be isotactic. The side groups in this polymer were arranged in helical fashion, the concept of the flat plane of the backbone being only a convention which enables us to represent the steric structure on paper. The figure on page 568 shows the different types of helices present in isotactic polymers of known structure.

*Stereoblock Polymers.* In addition to the isotactic, syndiotactic and atactic types of polymer, there are intermediate types having isotactic or syndiotactic sequences so short that crystallizable and non-crystallizable parts coexist in the same long molecule. These are 'stereoblock' polymers. They are less crystalline than ideal isotactic or syndiotactic polymers, and their mechanical properties are intermediate between those of the tactic and the atactic polymers. Stereoblock polymers thus have lower tensile strength but greater elastic elongation than isotactic polymers (see figure on page 570).

*Stereospecific Polymerization Processes*
The polymerization processes developed by Karl Ziegler, Giulio Natta and their colleagues were important not only in providing

---

Polymerization. Head and Tail Linkages. Polymers with a chemically regular structure are formed by asymmetrical monomers ($CH_2$=CHR) only when the monomer units are linked in a particular order, i.e head-to-tail or head-to-head-tail-to-tail sequence.

Head-to-tail sequence is frequent, but in some polymerizations it may be accompanied by irregularities due to random head-to-head or tail-to-tail linkages. If these irregularities exceed certain limits, they may lead to a reduction or complete disappearance of crystallinity – *Ciba Review*.

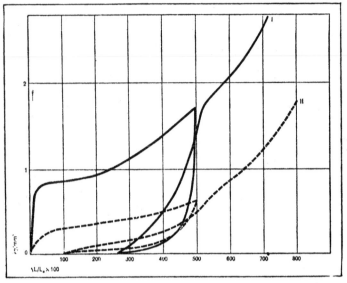

*Stereoblock Polypropylenes. Stress-Strain Diagrams.* Stereoblock polymers have lower tensile strength and greater elastic elongation than isotactic polymers. (I) refers to a polymer with a higher molecular weight than (II) – *Ciba Review.*

---

linear polymers of propylene and other alpha-olefins, but in establishing such control over the polymerization that polymers of desired steric structure could be made. The process could be controlled to provide an isotactic polypropylene, or a syndiotactic polypropylene, or an atactic polypropylene. The technique has become known as stereospecific polymerization.

This remarkable development in polymer chemistry was achieved by using very special types of catalyst and by controlling the reaction conditions with great care. The production of an isotactic polymer, for example, requires not only regular 'head-to-tail' linking of the monomer molecules (see figure on page 569), but also a constant opening of the double bond (always *cis* or always *trans*) and constant positioning of the monomer with respect to the plane in which the double bond lies.

The techniques used previously in polymerizing alpha-olefins could not bring about a predetermined positioning of the monomer in relation to the growing chain. Regularity of positioning is achieved only by accurate orientation of the monomer towards the catalysts at the stage immediately preceding its addition to the growing chain.

The catalysts used by Ziegler in the low-temperature polymerization of ethylene made possible the first stereospecific techniques. Ethylene itself is a symmetrical molecule, and the linear polyethylene molecule does not display steric differences of the isotactic, syndiotactic and atactic type. Also, ethylene is a more reactive substance than propylene and the higher olefins, and the Ziegler catalysts used for polymerizing ethylene do not necessarily polymerize propylene at all, much less polymerize it to a stereoregular polymer in high yield.

Before the discovery of stereospecific polymerization, the only linear polyolefins of high molecular weight which could be made were those derived from olefins with a symmetrical structure (ethylene, isobutylene). But the new sterically-controlled polymerization processes make possible the polymerization of olefins higher that ethylene, and the copolymerization of these with ethylene. High molecular weight products are obtained which, depending upon their structure and composition, may be of commercial value.

**Commercial Development**

The first patent applications on polypropylene were filed by Professor Giulio Natta in conjunction with Professor Karl Ziegler in 1954. They covered the method and composition of matter, with specific claims for the production of fine filaments.

Production began in Italy by the firm of Montecatini Societa Generale, first in one and then in two plants. By 1957, commercial quantities of isotactic polypropylene became available from Montecatini in Italy. Meanwhile plans for large-scale production of the polymer had been made in the U.S., U.K. and other countries, and by 1962 several manufacturers in the U.S., and I.C.I. Ltd and Shell Chemical Co. in the U.K. had begun producing polypropylene.

Much of the initial impetus behind this development of polypropylene came from the potential it offered as a plastics material. Polypropylene is similar in many respects to polyethylene, but

571

the increased temperature-resistance, greater stiffness and strength, better surface finish and other properties give it advantages over polyethylene in many applications. Like polyethylene, it is made from a raw material which is available in almost unlimited quantity at a low cost.

Plans for the production of polypropylene were pushed ahead with great enthusiasm, the U.S. production capacity for 1964 being set at more than 250 million kg a year, rising to about 450 million kg a year by 1967. In other countries, notably Italy, the U.K. and Japan, polypropylene production went ahead with equal vigour, the planned total capacity of countries outside the U.S.A. being greater than 450 million kg a year by 1967.

The *actual* production of polypropylene, however, was very much slower than anticipated, and the development of polypropylene fibres did not proceed as rapidly as the more optimistic forecasts had predicted. It was not until the early 1960s that small quantities of Italian-made staple fibre and multifilament yarns began to appear on the market. Soon, this was followed by supplies of fibre from U.S., British, Japanese and other producers, but even by 1963 the U.S. consumption of fine textile grade olefin fibres was still below 5 million kg per annum.

The reasons for this comparatively slow development of polypropylene fibre were partly economic and partly technical.

*Economic Factors*

In the early stages of polypropylene development, too much emphasis was placed upon the prospective low cost of the isotactic polymer itself. The raw material, propylene, is available in quantity at a low cost, and many prophets assumed that the polypropylene polymer and fibre would necessarily have an impressive and immediate cost advantage over competitive products.

Early forecasts of polypropylene as cheap as, or cheaper than, polyethylene proved to be unrealistic. They did not take into account the very heavy development costs, which added considerably to the cost per kg of polypropylene produced.

In the U.S., where the development of polypropylene was pushed ahead with great enthusiasm, the cost of polymer was maintained initially at around 20 cents per kg. Between 1961 and

1963, it fell somewhat, but still remained too high for fibre to be produced at a really competitive price. By 1964, the price of polyethylene dropped, bringing polypropylene down too, reaching a level of 11 to 14 cents per kg, depending on grade. The price of polypropylene staple fibre in late 1964 was 29 cents per kg.

Meantime, the prices of established synthetic fibres were also moving down, partly to meet the threat offered by polypropylene in certain markets. In late 1964, nylon staple was selling in the U.S. at $0.45 to $0.40 per kg, polyester fibre at $0.44 and acrylics at $0.49 to $0.36 per kg, depending on grade and denier. By this time therefore polypropylene had achieved a substantial cost advantage, which was increased still further by its low density (i.e. more bulk per kg).

## Technical Factors

In the development of polypropylene fibres, a number of technical problems have had to be faced.

### 1. Heat Sensitivity

Polyolefins are inherently prone to degradation by oxidation, which becomes progressively more serious as the temperature rises. The first commercial polypropylenes tended to undergo considerable molecular breakdown at the extrusion temperatures (110–190°C. higher than the melting point). As better-grade polymers became available, the heat-stability improved.

This problem of heat stability is important, too, in its influence on the behaviour of the fibres themselves, and it has been necessary to devise stabilizing systems which will prevent or reduce the degradation caused by elevated temperatures. Individual polypropylene manufacturers have developed their own systems for this purpose, and a range of fibre-grade polymers has become available, each with its own heat-stability characteristics.

### 2. Light Stability

The oxidative breakdown of polypropylene is accelerated by light, and the unprotected fibre is sensitive to ultra-violet radiation. As in the case of heat-sensitivity, the answer has been to develop stabilizer systems. These may include ultra-violet absorbers and combinations of several antioxidants acting synergistically. Non-extractable systems are preferred for general use, and are essential for some applications.

## 3 *Melting Point*

The softening and melting points of polypropylene are low by comparison with most (but not all) established synthetic fibres. They are sufficiently high to enable polypropylene fibres to be used in general textile applications, including apparel uses, and do not raise any real difficulties in this respect. Polypropylene fabrics can be ironed, but a greater degree of caution is required than is necessary with most other established fibres. This limitation has tended to be a promotional disadvantage which weighs more heavily than it should.

The comparatively low softening point has proved a more serious disadvantage by placing limitations on the temperatures which may be used in finishing operations. In blends with cotton or rayon, for example, polypropylene may be subjected to finishing procedures which would normally involve heating to temperatures higher than it can stand without softening. It is necessary, therefore, to use finishing techniques which do not involve temperatures greater than about 125°C.

## 4. *Spinning and Processing*

The extrusion of polypropylene, and the processing of the fibre, do not present any basic difficulties. The techniques used are essentially similar to those already developed for melt spinning other synthetic polymers, notably polyamides and polyesters.

The stages of spinning and processing had to be adapted to suit the characteristics of the new polymer, however, and controlled to provide for requirements of specific end uses. These factors raised some problems.

Variations in the characteristics of different polypropylenes were already causing difficulties, as outlined above. But the fibre-manufacturer found also that he was required to produce more grades of polypropylene fibre than was usual with other types of synthetic fibre.

Initially, polypropylene was introduced as a high-strength fibre which compared in tenacity with nylon. Once the heat-stability and other polymer problems had been solved, there was no difficulty in producing fibres with tenacities in the region of 53 cN/tex (6 g/den).

It was then found, however, that for certain applications it was preferable to use a more resilient crimped staple fibre with a lower tenacity, e.g. in the region of 26.5 cN/tex (3 g/den). For some applications, on the other hand, such as ropes and fishing nets a high tenacity continuous filament polypropylene was required with tenacity of 71–80 cN/tex (8–9 g/den).

The ability to influence the physical properties of polypropylene by varying the conditions of spinning and processing has made polypropylene into one of the most versatile of all synthetic fibres. But it has, at the same time, created unusual problems for the fibre-manufacturer who must spin and process polypropylene under precisely controlled conditions.

Control of elongation and shrinkage, crimping and crimp retention became necessary to provide fibres with special characteristics, such as high bulk performance, or for blending with other fibres. Special consideration had to be given to the production of polypropylene filaments which would be subjected to the various bulking and texturing processes now in widespread use.

## 5. Dyeing

Unmodified polypropylene fibre is virtually undyeable by means of the standard products and techniques used in the textile trade. This problem has been a major factor in slowing up the acceptance of polypropylene as a general textile fibre.

The use of pigmented fibre has proved satisfactory for many applications. Modification of the polymer or of the fibre structure to promote dyeability has not proven to be a promising line of attack on this problem. There are now several modified polypropylene fibres on the market which are dyeable but they represent a very small part of total production.

### Future Prospects

When the dyeing and other problems have been solved, polypropylene fibres will be available for unrestricted use in blankets, knit goods, upholstery fabrics, sweaters and other apparel fabrics. It will be in widespread use in the women's hosiery market.

It is safe to predict that the next decade or two will see a tremendous demand for an inexpensive, serviceable, available textile fibre that will be needed in very large quantities. At the present time, polypropylene is the only fibre developed or theoretically feasible that can combine serviceability with economy and availability. It seems certain, therefore, that polypropylene fibres must become the 'cotton' of synthetic fibres before the end of the twentieth century has been reached.

## TYPES OF POLYPROPYLENE FIBRE

By varying the conditions of polymerization, spinning and processing, it is possible for the manufacturer to influence the physical properties of polypropylene fibres to a remarkable degree. In consequence, there is a very wide choice of polypropylene fibres on the market, which differ in physical characteristics over a considerable range. Each manufacturer markets the types of fibre which he believes will satisfy the needs of his particular customers.

In addition to this variation in the properties of polypropylene fibres *per se*, variations are introduced in attempts to modify the dyeability and other characteristics of the fibre. Thus, fibres are now on the market in which affiinity for particular types of dye has been increased by physical or chemical modification of the polymer.

Polypropylene fibres are produced in the form of multifilament yarns, monofilaments, staple fibre and tow. There is a wide range of pigmented fibres and there is a limited number of dyeable polypropylene fibres.

Many manufacturers are now producing polypropylene fibrillating film or slit-film fibres.

### NOMENCLATURE

*Olefin*

Polypropylene fibres are defined as *olefins* under the U.S. Federal Trade Commission definition (see page xxvi).

## Polyolefin

Propylene is chemically a member of the olefin class of hydro-carbons, and polypropylene is a polyolefin. Polypropylene fibres are thus a type of polyolefin fibre.

## PRODUCTION

### Monomer Synthesis

Propylene is a constituent of the mixtures obtained from thermal and catalytic cracking processes in the petroleum industry. It is available in virtually unlimited quantity and at potentially very low cost.

### Polymerization

The actual conditions under which stereospecific polymerization of propylene is carried out are not disclosed in detail by the manufacturers. Ziegler-type catalysts are used (see page 542), such as an organo-metallic compound of aluminium in the presence of titanium trichloride, and the reaction is carried out under 10 atmospheres pressure at less than 80°C.

The catalyst in this reaction controls the way in which the polymer is built up, feeding each monomer molecule to the end of the polymer chain in such a way that it adds on in the desired position. Thus, an isotactic polypropylene molecule is built up, possessing the characteristics which are inherent in this shape of molecule, as distinct from the atactic or syndiotactic molecules (see page 566).

### Spinning

Polypropylene fibres are made by extrusion of molten polymer, followed by drawing to orientate the molecules and crystals in the filaments. Polymer of high viscosity is spun, in order to obtain optimum fibre properties. The filaments are cooled in air as they emerge from the spinneret, and are collected on bobbins.

Bundles of filaments are hot-drawn, twisted and packaged to provide multifilament yarns containing, for example, 10 to 500 filaments of 2.2 to 16.5 dtex (2−20 den).

In the production of staple fibre and tow, bundles of fine filaments are combined to form a tow containing hundreds of
thousands of individual filaments. The tow is drawn and crimped,
and then cut to an appropriate staple length.

### Spinning and Processing Conditions

Polypropylene crystallizes so rapidly that the undrawn filaments
are highly crystalline. In this respect, polypropylene differs from
other synthetic polymers which are melt-spun, such as polyamides
and polyesters.

The unusually high crystallinity renders fibre-production very
sensitive to conditions of spinning and subsequent processing, and
permits a fine degree of control over fibre properties.

The temperature at which extrusion is carried out, commonly
80°C. or more above the polymer melting point, and the conditions under which the filaments are cooled, affect the nature
and extent of crystallinity. Rapid cooling or quenching results in
small crystals, whereas slow cooling allows larger crystals to form.

The degree of orientation achieved by drawing the filaments
influences the mechanical properties; the greater the degree of

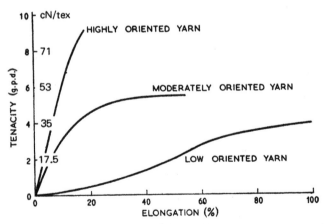

*Stress-Strain Curves for Polypropylene Fibres – Courtesy Hercules
Powder Co.*

578

stretch, for example, the higher the tensile strength and lower the elongation.

The stretched fibre may be further conditioned by heat-treatment at temperatures below the softening point. Heat-setting, for example, affects the elastic recovery, shrinkage characteristics and flex resistance of the fibre.

The effects of spinning conditions and subsequent treatment of the fibre are due largely to the way in which they influence the amorphous orientation and crystalline structure of the polypropylene. The ability to control morphology during spinning in this way is used to advantage, permitting the production of fibres with a wider range of properties than would otherwise be possible.

### Stabilization

Degradation of polypropylene takes place primarily through oxidation, which is encouraged by heat and light. Stabilization of polypropylene is essential to confer light and heat stability, and substances are commonly added for this purpose before the molten polymer is spun.

Stabilizers are mainly compounds capable of deactivating free radicals which promote the oxidative chain reaction. Compounds which are primarily U.V. absorbers are of little value in the stabilization of fibres (too high surface/volume ratio).

*Heat Stability.* Heat stabilization can be provided for polypropylene fibres which enable the fibres to retain useful mechanical properties after exposure to air for extended periods at temperatures of, for example, 120°C.

A particular problem arises when polypropylene fibres are exposed to solvents or aqueous solutions for long periods, as the stabilizers tend to be removed by extraction. Repeated laundering, for example, may result in the gradual loss of some stabilizers. Special formulations have been developed, however, to meet this difficulty.

*Light Stability.* The use of polypropylene fibres in applications such as car upholstery requires that the fibre should be able to withstand long exposure to light at elevated temperatures. Finding stabilizers for such applications has been a challenging problem.

Pigmented formulations have been devised, and are used commercially in carpets, upholstery and window channel fabrics.

For reasons of economy, it is preferable to stabilize polypropylene fibres to meet the requirements of specific end-uses, rather than to use fibre which has been stabilized to give adequate protection in all applications. Manufacturers commonly market a range of polypropylene fibres of differing stability characteristics.

## PROCESSING

### Scouring

The size may be removed form the warp yarns of a polypropylene or polypropylene-viscose or polypropylene-cotton blend fabric by an appropriate scouring process.

Scouring may be carried out in winch or jigger, using, for example, a solution containing 2–4 grams per litre of sodium carbonate and 2–3 grams per litre of non-ionic synthetic detergent (e.g. Anionico SCL) for 45 minutes at 80–95°C. Caustic soda (2 ml./litre conc. 36 Bé) may be used instead of sodium carbonate.

After thorough rinsing, the fabric is dried in the tenter frame, bearing in mind that the dried fabric can withstand 125–130°C. without suffering damage.

### Whitening

Polypropylene fibre itself does not require whitening, but it is often necessary to whiten blends containing other fibres. In general, the processes commonly used for the individual fibres of the blend may be used. Fluorescent brighteners may also be added.

### Dyeing

Polypropylene fibre, in its unmodified form, is virtually undyeable by any of the familiar dyestuffs and techniques. This factor has, more than any other, restricted the fibre's progress in the apparel and general textile fields.

The reasons for polypropylene's lack of dyeability are twofold. On the one hand, the paraffinic chemical structure of the polymer is almost completely non-polar; there are no centres of chemical affinity by which dyestuffs can be held in the fibre. On the other hand polypropylene fibres are hydrophobic and almost entirely impermeable to water; water-borne dyestuffs cannot penetrate readily into the interior of the fibre.

Since polypropylene fibres became available, every effort has been made to devise satisfactory dyeing techniques. A number of processes have now been developed, with varying degrees of success. Some techniques are used on the unmodified polypropylene, and do not involve changing the chemical constitution of the polymer. Other techniques require chemical modification of the polymer structure in order to achieve their effects.

## 1. Techniques based on Unmodified Polypropylene

### Conventional Dyeing Processes

Every class of dyestuff has been used in attempts to dye polypropylene fibres. Some have achieved a certain degree of colouration, but none has provided a commercially useful colour range.

Flat, pale shades with poor fastness properties, for example, may be obtained by dyeing with selected vat dyes using the acid process in preference to the normal alkaline process. Acid must be added slowly to the reduced liquor, as precipitation may otherwise occur, the liquors becoming unstable and the dyeings unlevel.

Swelling agents, dyestuff solvents and dyeing aids have some slight effect, but the colourations are, in most cases, either not resistant to washing or dry cleaning or to light.

· More satisfactory results were obtained when hydrocarbonsoluble dyes were applied at temperatures above the boiling point of water. A dye bath of molten urea, for example, was used to carry out the dyeing at 135°C. Techniques of this sort, however, deviate from normal dyehouse practice and are not readily acceptable.

### Mass Colouration

Polypropylene, like polyethylene, may be coloured effectively by adding suitable pigments to the molten polymer before spinning. The light, wash and dry-cleaning fastness of pigmented fibres is outstanding, but the addition of pigments to the melt inevitably increases the production costs. The extra cost has not, however, proved too exorbitant in practice.

The main disadvantage of pigmented fibres lies in the need to store and market a wide range of colours of every grade and type of fibre. In recent years, as experience in the production and use of pigmented fibres has grown, the technique has become economically acceptable, and the pigmented fibre continues to make steady progress.

### Addition of Dye Receptors

The dyeability of polypropylene fibres may be improved by mixing dye-receptive substances with the polymer before it is spun. Additives of this sort include (a) metallic compounds, (b) polymers, and (c) miscellaneous low molecular weight compounds.

A number of fibres are now on the market, in which dyeability has been increased by introducing additives of this type. The techniques used in dyeing fibres of this type are specific to each fibre, and depend upon the nature of the additive.

### 2. *Techniques based on Chemically-Modified Polypropylene*

As the non-polar nature of the polypropylene molecule is partly responsible for the dyeing difficulties, an obvious approach to the problem is to modify the polymer to improve its dye-receptivity. Many developments of this sort have been made, and modified polypropylene fibres are now on the market. They include copolymers made by incorporating a dye-receptive monomer into the polymerization process, or by grafting dye-receptive segments on to the main polymer chain.

Chemical treatment of the fibre, such as halogenation, has also been used to improve the dyeability of polypropylene by altering its chemical structure.

These techniques, and others like them, bring about fundamental changes by altering the chemical nature of the fibre, and they inevitably change the characteristics of the fibre too. These changes may be small, the fibre retaining its polypropylene characteristics with only minor changes. They may, on the other hand, be so severe as to create a fibre in which the properties differ significantly from those of a typical 100 per cent polypropylene fibre.

#### Singeing

Polypropylene fibre fabrics, in common with most fabrics made from synthetic fibres, have a tendency towards pilling. This may be avoided by light singeing treatments.

Owing to the comparatively low softening point of polypropylene fibre, particular care is necessary in carrying out the singeing operation. This is especially so in the case of fabrics of 100 per cent polypropylene.

Singeing is carried out on the usual machines, and one or both sides of the fabric may be treated. The distance and inclination of the flame are set in the normal way, but singeing is usually carried out at high speed e.g. 140–150 meters per minute. This is about twice the speed commonly used for cotton and viscose fabrics.

Alternatively, the speed may be set at 70–80 metres per minute, and the number of burners reduced to half that used for cotton.

When light and loose-weave fabrics are treated, it is advisable to wet them before singeing.

Singeing usually brings about some superficial softening of polypropylene fabrics, causing some stiffening. This can be eliminated by passing the fabrics through a calender.

### Synthetic Resin Treatment to Improve Hand

The hand of a fabric may be improved and its tendency towards pilling reduced by treatment with special finishes consisting of synthetic resins, usually in emulsion form. These finishes are especially useful in the case of loose-woven fabrics and fabrics made from ply yarns, such as loose-woven or twill-weave upholstery fabrics.

Modern resin-based finishes of this type are usually able to resist removal by laundering and dry-cleaning, and they may be considered as permanent finishes.

### Creaseproof Treatments

Fabrics of 100 per cent polypropylene fibre cannot be treated effectively with creaseproof resin, as the fibres are so inert that the resins are not fixed permanently to them.

Fabrics made from polypropylene fibre blended with cellulosic fibres, however, may be creaseproofed by means of resins which act upon the cellulosic components.

Owing to the comparatively low softening point of the polypropylene fibre, it is necessary to select finishes that do not require heating to temperatures above 130°C. This restricts the selection considerably, and many of the most effective of the crease-proofing finishes cannot be used.

**Waterproofing**

*Acrylic Resin Finishes*

Fabrics of 100 per cent polypropylene may be waterproofed by coating with acrylic resins dissolved in organic solvents, or in the form of aqueous emulsions. This results in an impermeable fabric, and is not recommended for apparel applications.

*Silicone Finishes*

Polypropylene fabrics may be given a high degree of water repellency by treatment with silicone finishes. These have the advantage of leaving the fabric air-permeable.

*Chromium Complexes of Stearic Acid*

Finishes based on the chromium complexes of stearic acid may be used to impart good water-repellency to polypropylene fabrics.

**Permanent Antistatic Finishing**

For certain applications, such as laboatory coats, overalls, etc., it is necessary to provide an antistatic finish that remains effective throughout the life of the garment. A number of products are available for this purpose.

**Calendering**

Calendering must be carried out very carefully, bearing in mind the relatively low softening point of. polypropylene fibres. It is recommended that the temperature should be no more than 90°C., and the cylinder load 15–20 tons. It is possible to use 30–40 tons on the silking calender.

**Foam Backing**

Foam-backed fabrics may be made satisfactorily from 100 per cent polypropylene fabrics, and also from blended polypropylene/cotton and other fabrics.

The procedures used for applying polyurethane foams are similar to those used with fabrics made from other fibres. Acrylic resins may be used as adhesives, or the foam may be applied by fusion techniques.

### Moulding

Polypropylene fibre is thermoplastic, softening at about 150°C. and melting at about 160–170°C. Fabrics consisting wholly or in part of polypropylene fibre may be moulded under proper conditions of time and temperature into almost any desired shape.

This mouldability may be used in innumerable applications where a fabric is required in a three-dimensional shape, such as brassières, hats, handbags, inner and outer linings for luggage, panels for furniture and car upholstery.

### *Fibre Content and Fabric Type*

A minimum of 70–80 per cent polypropylene fibre in the fabric is recommended in order to ensure good mouldability. The fabrics may be made from staple or continuous filament yarns, and may be in almost any type of weave. Very dense fabrics require higher temperatures than thinner fabrics.

Non-woven fabrics must have high and evenly-distributed strength properties in all directions. The main problem presented by non-woven fabrics is the readiness with which they tear during the heating cycle, due to shinkage. Chemical bonding agents can often overcome this difficulty.

### Heat-Setting

In common with other thermoplastic fibres, polypropylene fibres and fabrics can be heat-set against subsequent shrinkage and deformation.

A temperature of 130°C. is the maximum practicably-usable temperature for heat-setting. Fabrics set at this temperature will resist shrinkage in boiling water, and are stable up to 120°C.

Blended fabrics containing more than 30 per cent polypropylene can be heat-set, the effectiveness increasing with the proportion of polypropylene in the blend.

Pleats and creases may be heat-set in polypropylene fabrics or blends, and these fabrics may also be moulded into three-dimensional shapes (see above). They may be calendered and embossed to give permanent effects.

Heat-setting imparts resistance to wrinkling during washing.

### Bonding

A variety of adhesives may be used in bonding polyolefin fabrics and yarns, including (1) contact adhesives containing volatile solvents, (2) hot melt resins, and (3) epoxy compounds.

## STRUCTURE AND PROPERTIES

### Molecular Structure

Unlike the natural fibres and most of the important synthetic fibres, polyolefins have no polar groups. Prior to the development of isotactic polypropylene, it was considered that polar groups were essential to the production of textile fibres of high strength. They provided the forces which held the linear molecules together as they lay alongside one another in the orientated fibre.

The presence of polar groups in a linear molecule *does* make the manufacture of high-strength fibres possible from polymers of relatively low molecular weight. Polyamides yield strong fibres, for example, when the molecular weight lies between 10,000 and 30,000.

In polypropylene, it is crystallinity alone that provides tensile strength by preventing slippage of the linear molecules. And higher molecular weights are necessary to obtain the necessary strength. Polymers of molecular weight greater than 50,000 are needed, for example, to eliminate intracrystalline flow, corresponding to tenacities of at least 44–53 cN/tex (5–6 g/den).

Monofilaments with tenacities above 88.3 cN/tex (10 g/den) require polymers of molecular weight greater than 100,000.

The average molecular weights of polypropylene used in fibre manufacture are commonly higher than 200,000. High-density polyethylene fibres, by contrast, have molecular weights in the region 50,000 to 150,000, and low-density polyethylene 20,000 to 25,000.

The linear molecule of isotactic polypropylene is thus, on average, twice as long as a linear molecule of high-density polyethylene, and more than 10 times as long as the molecule of low-density, branched polyethylene.

### Degree of Crystallinity

The crystallinity of a polypropylene fibre, which has such an important influence on tensile strength and other mechanical properties, is controlled primarily by the nature of the polymer itself. The degree of crystallinity attainable depends upon the average molecular weight and molecular weight distribution, and on the content·of isotactic polymer. It may vary between two extremes, represented by a completely amorphous atactic polypropylene on the one hand, to the 100 per cent crystalline isotactic polypropylene on the other. The normal commercial polymer has a degree of crystallinity of about 60 to 70 per cent.

The degree of crystallinity of a polypropylene fibre is influenced also by the treatment to which the fibre is subjected during processing. Stretching and heat-treatment may affect the way in which the linear molecules are packed into their ordered crystalline patterns.

The high degree of crystallinity achieved in isotactic polypropylene fibres is reflected in the high strength and stiffness, and the low elongation of the fibres.

## Isotacticity Index

The content of isotactic polypropylene is measured by estimating the amount of residue remaining after extraction of the polymer with boiling n. heptane; it is expressed as the *isotacicity index*. Commercial polypropylene has an isotacticity index of over 90 per cent.

### Fine Structure and Appearance

Polypropylene fibres are produced as colourless filaments, and in a range of pigmented shades. They are smooth-surfaced and are somewhat waxy in appearance. The filaments are often of round cross-section, but they are also produced in a variety of other cross sections for special applications. Diameters lie commonly between 0.02–0.5 mm (0.0008–0.02 in), but heavier monofils are produced for use as bristles.

### Tensile Strength

Polypropylene fibres are produced in a variety of types of differing tenacities designed to suit varying market requirements.

Most applications are adequately served by fibre of medium tenacity, and commercial staple and continuous filament yarns produced for general textile uses have tenacity 26.5–44.1 cN/tex (3–5 g/den); tensile strength 2,450–4,200 kg/cm$^2$ (35,000–60,000 lb/in$^2$).

yarns reaching 80 cN/tex (9 g/den) are made. For special

purposes, yarns of tenacity up to 115 cN/tex (13 g/den) are produced.

the other hand, does not need to be of very high tensile strength. High resilience is a more desirable characteristic, and fibres of lower tenacity are produced for this type of application.

Knot and loop strengths are usually some 10–15 per cent lower than the strength of the straight fibre.

**Elongation**

Commercial polypropylene monofilaments have an elongation at break in the region of 15 to 25 per cent. Multifilament yarns are in the range of 20 to 30 per cent, and staple fibre 20 to 35 per cent.

**Elastic Properties**

The elastic properties of polypropylene fibres, in common with other mechanical properties, may be varied over a wide range by choice of polymer and processing conditions. Fibres can be produced to meet the requirements of specific applications with regard to elastic properties.

*High Tenacity Fibre*

The elastic recovery properties of commercial high-tenacity fibres are excellent, and similar to those of nylon. Immediate recovery after 10 per cent elongation is about 90 per cent with virtually no permanent set.

*Medium Tenacity Fibre*

*Initial Modulus.* The modulus of elasticity on 10 per cent extension is in the range 263–795 cN/tex (30–90 g/den) for multifilaments, and 221–353 cN/tex (25–40 g/den) for staple.

*Elastic Recovery.* Elastic recovery at 5 per cent elongation is 90–98 per cent for multifilaments and 90–95 per cent for staple.

*Low Tenacity Fibre*

Yarns of tenacity 17.7 cN/tex (2 g/den) are made under special conditions of heat-treatment at high temperature. These yarns show a 95 per cent immediate recovery from 50 per cent extension, and complete recovery after 5 minutes.

## Creep Characteristics

Cold flow or creep characteristics of polypropylene are satisfactory, and a great improvement on polyethylene, but do not reach the standard set by polyamide or polyester fibres. Polypropylene fibres will undergo cold flow of up to 0.5 per cent of the original length when subjected to a load of 13.2 cN/tex (1.5 g/den) for 16 hours at room temperature.

## Flex Resistance

Excellent.

## Specific Gravity

As in the case of polyethylene fibres, the specific gravity of polypropylene varies with the degree of crystallinity. Amorphous polypropylene has a specific gravity of 0.85; commercial fibres are in the range 0.90–0.91, and highly crystalline fibres reach 0.92–0.94.

Polypropylene fibres are thus the lightest of all commercial textile fibres, even the highly crystalline fibres being lighter than all but polyethylene.

## Effect of Moisture

Polypropylene is a paraffinic hydrocarbon, and it does not absorb water. The moisture regain of polypropylene fibres is so small as to be insignificant, and water has no effect on tensile strength and other mechanical properties.

Water does not cause any noticeable degradation in polypropylene fibres. Fibres subjected to boiling water or steam for long periods show no loss of strength.

## Thermal Properties

### Softening Point; Melting Point

The softening point of polypropylene fibres is in the region of 150°C., and the fibres melt at 160–170°C. The softening and melting points of specific polypropylenes are determined by the nature of the polymer and by the way in which crystallinity has been influenced during treatment of the fibre after spinning.

### Effect of Low Temperature

Polypropylene fibre retains its flexibility to temperatures of

589

−70°C. or lower. It does not reach the remarkable standard set by polyethylene in this respect, but its low temperature flexibility is excellent for most practical purposes.

*Effect of High Temperature*

The mechanical properties of the fibre deteriorate with increasing temperature below the softening point, but polypropylene performs better than polyethylene in this respect.

*Shrinkage Properties*

Shrinkage of polypropylene fibres depends greatly upon the treatment the fibre receives during processing. In boiling water, monofilaments may shrink as much as 15 per cent after 20 minutes; multifilament and staple yarns may shrink between 0 and 10 per cent.

*Flammability*

Polypropylene is a hydrocarbon, and it will burn. On being exposed to a flame, however, the fibre melts and draws away from the flame, extinguishing itself.

Polypropylene fabrics exceed the requirements of Class 1 of the ASTM Standard for textile fabrics. Tested according to B.S.2963, they are self-extinguishing and therefore of low flammability (B.S.3121).

Construction, additives, finishes and the presence of other fibres have a considerable influence on the burning characteristics of any particular fabric or structure.

For the purposes of fire insurance, polypropylene fibre is included in the same class as wool.

*Thermal Conductivity*

The following table lists the thermal conductivities of polypropylene and other important textile fibres:

| *Fibre* | *Thermal Conductivity* (relative to air 1.0) |
|---|---|
| Polypropylene | 6.0 |
| PVC fibre | 6.4 |
| Wool | 7.3 |
| Cellulose acetate | 8.6 |
| Viscose | 11.0 |
| Cotton | 17.5 |

Polypropylene has the lowest thermal conductivity of all commercial fibres, and in this respect is the 'warmest' fibre of all.

### Effect of Sunlight

Like polyethylene, polypropylene is attacked by atmospheric oxygen, and the reaction is stimulated by sunlight. Polypropylene fibre will deteriorate on exposure to light, but it may be protected effectively by means of stabilizers.

### Chemical Properties

*Acids.* Excellent resistance, similar to polyethylene.

*Alkalis.* Excellent resistance, similar to polyethylene.

### General

Polypropylene is inert to a wide range of chemicals. Its resistance and susceptibilities are similar to those of polyethylene (see page 554), but its high crystallinity tends to make it more resistant than polyethylene to those chemicals which degrade olefin fibres.

### Effect of Organic Solvents

Excellent resistance, generally similar to polyethylene. There is no known solvent for polypropylene at room temperature.

### Insects

Polypropylene cannot be digested by insect and related pests, such as white ants, dermestid beetles, silverfish and moth larvae. Polypropylene fibre is not liable to attack unless it becomes a barrier beyond which the insect must pass to reach an objective. In this case, the insect may cut through the fibre without digesting it.

### Micro-organisms

Polypropylene fibre will not support the growth of mildew or fungi. Some micro-organisms, however, may grow even on the very small amounts of contaminants which may be present on the surface of fibres or yarns in use. Such growth has no effect on the strength of any materials made from polypropylene fibre.

### Electrical Properties

Polypropylene is an excellent insulating material, and since the absorption of moisture is so extremely small there is little or no change in the electrical properties at high humidities – an important point in electrical applications.

### Coefficient of Friction

Polypropylene has a relatively high coefficient of friction against smooth surfaces, particularly metal or porcelain. The coefficient of friction of yarn against, for example, a yarn guide decreases with increasing tension.

The coefficient of friction against a matt guide, i.e. a guide with a discontinuous surface, is much less than against a polished guide, owing to the smaller actual area of contact between the yarn and matt surface.

*Coefficient of Filament to Filament Friction*

| | |
|---|---|
| Static friction (average speed 2.5 cm./min.) | 0.32–0.42 |
| Dynamic friction (average speed 95 cm./min.) | 0.29–0.40 |

### Other Properties

*Handle*

Polypropylene has a much less waxy feel than polyethylene, and its fabrics have a pleasanter handle.

*Environmental Stress Cracking*

Polypropylene does not show any tendency to 'craze' or develop surface cracks when subjected to stresses in the presence of detergents or other substances.

### Identification of Polypropylene Fibre

The following tests will help in the identification of polypropylene fibres:

(a) *Appearance*

Polypropylene seen through the microscope is smooth and featureless in appearance. Circular, trilobal and delta cross-section filaments are commonly encountered.

(b) *Burning Test*

When a flame is brought up to a polypropylene yarn, the fibres melt and retract. A bead of molten polymer is formed, which on

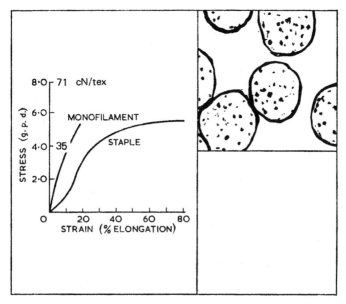

*Polypropylene*

continued contact with the source of heat will burn with a blue-and-yellow flame similar to that of a candle. A characteristic odour is given off when the burning material is extinguished.

(c) *Density*

Polypropylene and polyethylene are the only textile fibres which are lighter than water, and this is the basis of one of the most useful methods of identification.

A test sample is cut into lengths of ¼ inch (6 mm.) or less, teased into individual fibres and stirred into water containing a little wetting agent (1 g./litre Lissapol N). If the small pieces of fibre float, they are almost certainly polyethylene or polypropylene. These may be distinguished by their melting points; polypropylene lies in the region 160–170°C., and polyethylene in the region 110–140°C., depending on the type of polymer.

## POLYPROPYLENE FIBRES IN USE

### General Characteristics of Polypropylene Fibre Goods

Initially, polypropylene monofilaments made their way into applications which had already been pioneered by polyethylene. They offered higher strength, increased toughness, resilience, abrasion resistance and creep resistance, and a higher melting point; these properties were allied to the water-resistance, chemical inertness and other properties that had enabled polyethylene to compete in a range of applications.

Polypropylene monofilaments on this basis, were soon entrenched in a number of fields where they competed very effectively with other fibres; these fields included ropes and cordage, fishing nets and twines, filter fabrics, protective clothing and the like.

With the introduction of fine-denier multifilament yarns and staple fibre, the range of potential applications for polypropylene fibre was extended into the general textile and apparel fields. This brought them into competition with established natural and synthetic fibres in applications where the shortcomings of polypropylene fibres often place them at a disadvantage. Despite the many attractive characteristics of polypropylene fibres, less desirable properties – such as dyeing difficulties and relatively low melting point – have tended to limit progress.

### Processing Behaviour

Polypropylene fibres have a filament-to-filament coefficient of friction higher than that of any other textile fibre. This, in combination with crimp stability and low static charge accumulation, makes for excellent processing characteristics. Polypropylene fibres blend easily and effectively with other textile fibres.

### Covering Power

The low specific gravity of polypropylene fibre gives it a covering power which is, weight for weight, greater than that of any other textile fibre. This has been particularly helpful in the development of polypropylene blankets, upholstery, carpet and apparel fabrics.

### Thermal Insulation

The thermal insulation characteristics of a fabric are determined largely by the amount of air entrapped in the fabric. The thermal conductivity of polypropylene fibre is, however, lower than that of other fibres, and in this respect polypropylene is the 'warmest' fibre in commercial use.

*Crease Resistance*

The ability of a textile fabric to resist crease formation during use is influenced by the physical nature of the fibre itself. In the case of polypropylene fibre, this characteristic varies with the molecular weight of the polymer, and with the conditions of spinning and drawing. Polypropylene yarns are produced in which the crease resisting properties are at optimum value for applications where this is important.

In general, the crease-resistance of polypropylene fibres is of the same order as that of wool. Unlike wool, however, polypropylene fibre does not lose its high crease-resistance when wet.

*Shrinkage*

By controlling the conditions of processing, the shrinkage characteristics of polypropylene yarns may be varied over a considerable range.

Shrinkage of polypropylene yarns is quite low under the usual conditions of fabric scouring and dyeing. Special care must be taken, however, when high-temperature finishing treatments such as resin curing are used. Shrinkage may be high at elevated temperatures.

This shrinkage caused by high temperatures is not progressive, and polypropylene fabrics may be heat-set. In common with other synthetic fibres, heat-treated polypropylene yarns which have been allowed to shrink and become stabilized at a given temperature are essentially stable to subsequent treatment up to this temperature. This principle is used in providing dimensional stability to fabrics made from polypropylene fibres.

Fabric shrinkage under commercial dry-cleaning conditions with perchlorethylene tends to be excessive. This limits the use of polypropylene in some textile applications.

High-shrinkage polypropylene fibres may be prepared by the use of appropriate processing conditions. Fibres of this type are useful in the production of certain types of textured and bulked yarns. In blends, the high shrinkage of a polypropylene component of this type may be used to create bulked or puckered effects in yarns and finished fabrics.

## Creep

Polypropylene yarns generally exhibit more creep than polyester or nylon yarns, but the amount is so small that it is rarely of practical significance. It is usually appreciably less than the normal extension of the yarn under the load.

Where a rope or other construction has to withstand a continuous heavy load, it is usually designed so that the average loading is quite low, e.g. 4.4–8.8 cN/tex (0.5–1.0 g/den). This is necessitated by safety considerations. Under these conditions, the effect of creep is very small.

## Dimensional Stability

Dimensional stability of polypropylene fabrics is excellent, and the water-repellency of the fibre enables it to retain its high stability through repeated launderings. Polypropylene fabrics retain their shape in changing conditions of moisture and humidity, and on the strength of this stability these fabrics found a ready market in upholstery and industrial fabrics, laundry bags, seat belts and protective clothing. In such applications, polypropylene fibre provides high strength, low elongation, toughness, resilience and abrasion resistance which do not change under normal conditions of use.

## Abrasion Resistance

Polypropylene fibres have a high resistance to abrasion when dry, and even greater resistance when wet. The abrasion resistance of a blend containing polypropylene fibre increases in proportion to the amount of polypropylene fibre in the blend.

Abrasion resistance is of particular importance in applications such as carpets, where the ability to withstand wear is essential. In common with other mechanical properties, abrasion resistance may be influenced by the molecular weight of the polypropylene, cross-section, and the conditions of spinning and processing. Fibres may be produced with optimum abrasion resistance characteristics for special applications.

## Felting Behaviour

Polypropylene fibres, like most other synthetic fibres, do not felt. In blends with wool, polypropylene reduces the tendency to felt in proportion to the amount of polypropylene fibre in the blend.

## Wash and Wear Characteristics

Polypropylene fibre is unusually resistant to soiling. This is influenced in the main by two factors:

### (a) Electrostatic Attraction

Polypropylene fibres show little tendency to accumulate charges of static electricity through friction during use. They do not attract dust and dirt to the extent that most other synthetic fibres do.

### (b) Chemical Inertness

Polypropylene does not react chemically with the substances encountered in general use, nor is it attacked by common solvents, greases, oils, etc. It is not readily stained, therefore, and such staining as does take place is commonly superficial. The stain is held in the interstices of the fabric by capillary attraction, and is readily removed by laundering or dry cleaning.

## Hand and Draping Characteristics

Hand and draping characteristics depend greatly on the weaving and finishing of fabrics.

## Flex Resistance

Fibres have good flex resistance and moderately good recovery from bending, and these properties serve them well in carpets and floor coverings. Textured filament and bulked staple yarns have made good headway in these and other pile fabrics.

High flex resistance coupled with excellent loop- and knot-strength are important in the production of knitted goods, and polypropylene fibre has found many applications such as sweaters cardigans, swimwear and underwear.

## Pleat Permanence

Pleats may be heat-set in polypropylene fabrics, and the pleats show excellent retention. A blend of polypropylene fibre (60 per cent) and wool (40 per cent) will take pleats well; they may be ironed into the fabric at 140°C. In most cases, pleats may be pressed into polypropylene fabrics at 130°C., which gives a better margin of safety, especially in fabrics of 100 per cent polypropylene fibre.

## *Cheapness*

Perhaps the most important of all the features of polypropylene to be taken into account when assessing its use for a particular application is its cost. Polypropylene is potentially the cheapest of all synthetic fibres, and this must give it a telling advantage in competition with other fibres where technical differences are marginal.

### Washing

The resistance of polypropylene fibres to a wide range of chemicals including acids, alkalis and normal textile bleaches, enables fabrics to withstand all the chemical conditions commonly encountered in washing and laundering. In addition, the fibre does not absorb moisture, and it does not felt.

Washing may be carried out without difficulty, using temperatures up to 100°C. It is preferable to use warm water at about 40°C., with soap, soap powder or detergent. The goods should be rinsed well, and a small amount of detergent or softening agent may be added to the final rinse to increase softness of handle and antistatic properties.

### Drying

Polypropylene fibres do not absorb moisture, and fabrics will dry quickly at room temperature. Spin drying or drip drying are preferred; tumble drying should be used with great care to ensure that the goods tumble freely.

### Ironing

The softening point of polypropylene fibre is lower than that of most other commercial textile apparel fibres, and ironing must be carried out with care. A steam iron may be used, or an ordinary iron in conjunction with a damp cloth. If ironing is done without either steam or cloth, a low setting should be used, bearing in mind that the fibre softens at about 140°C.

### Dry Cleaning

Polypropylene goods may be dry cleaned satisfactorily with trichloroethylene or white spirit at temperatures below 50°C.; polypropylene behaves in this respect like wool. Perchloroethylene is not recommended as a dry cleaning agent, as it may affect the dimensional stability of the fabric.

**End-Uses**

*Ropes*

The principal requirements of a fibre to be used in making ropes for maritime, agricultural and industrial use include the following:

(1) high strength, wet or dry
(2) resistance to repeated loading and flexing
(3) resistance to abrasion
(4) minimum water absorption
(5) lightness
(6) resistance to weathering, light, seawater, chemicals, solvents, micro-organisms and other potentially degradative agents encountered in normal use.

This is a formidable combination of properties, and few commercial fibres are able to satisfy all requirements to a high degree. Polypropylene fibres compete favourably with other fibres, natural and synthetic, in this respect, and they have found an important outlet in the field of rope manufacture.

Polypropylene ropes will float in water, and this can be a significant advantage in certain circumstances. It reduces the losses when ropes fall overboard, for example, during fishing and other marine operations.

*Fishing Gear*

Amongst the most important properties of yarns to be used in fishnets, trawls and lines are:

(a) high strength, wet and dry
(b) great toughness
(c) low weight
(d) low moisture absorption
(e) dimensional stability
(f) resistance to degradation by weathering, light, seawater, chemicals, solvents, micro-organisms and other potentially degradative agents encountered in normal use.

This combination of requirements is essentially similar to the requirements listed in the section on ropes (above). In the past, cotton and manila and a few other natural fibres have, as in the case of ropes, been used for making fishing gear. Their shortcomings were such, however, that synthetic fibres made rapid progress in this field, and nylon in particular has been able to take over a large part of the production of fishing gear since the end of World War II. More recently, nylon has been joined by polyester, polyethylene and polyvinyl alcohol fibres which have all secured a place in the world market for fishing gear.

In 1961, polypropylene was introduced into this field, and it has made rapid progress in direct competition with its established synthetic fibre rivals. By 1964, for example, the production of polypropylene nets in Great Britain exceeded the overall total of nets produced from all other synthetic fibres.

The reasons for the ready acceptance of polypropylene in the production of fishing gear are to be found in its unique combination of properties, and in its low cost.

Many polypropylene fibre producers are now marketing high-tenacity polypropylene multifilament yarns for use in the production of fishing gear. These yarns commonly provide a tenacity in the region of 71 cN/tex (8 g/den), and extension in the region of 20 per cent.

Monofilament yarns may also be used with advantage for certain applications. Fishnets made from monofilament have properties similar to those of the equivalent multifilament nets, with, however, a higher resistance to abrasion and usually a stiffer handle.

## Chain Warps in Carpets

The yarns, e.g. of spun cotton, which are normally used for chain warps in carpets may be replaced by high tenacity polypropylene yarn. A 120 filament polypropylene yarn of 570 denier (633 decitex) with 5 turns/inch (197 turns/metre) 'S' twist, for example, is substantially stronger than 3/9s c.c. (15/3 metric) cotton yarn. In addition it has some three times the runnage and, depending upon the price of the cotton yarn, it effects a price saving of about 40–50 per cent.

Carpets made with polypropylene chain warps are similar in appearance and flexibility to carpets made with conventional cotton chain warps. The abrasion resistance of the pile is identical for conventional carpets and those containing polypropylene filament chain warps.

Dimensional stability and tuft anchorage of the polypropylene chain warp carpets are at least equal to those of cotton chain warp carpets.

## Tufted Carpets

Polypropylene multifilament yarns have gained wide and rapidly-growing acceptance for the production of tufted carpets.

Bulked yarns, in particular, offer many advantages in this application, among which the following are important:

*Hand.* The finished carpet has a firm, lofty feel, without harshness.

*Cover.* The natural bulk or cover of polypropylene, with its low specific gravity, is greater than that of any other fibre. This, coupled with a 'blooming' action of the yarn during finishing, produces a carpet with more cover per kg, than can be obtained with any other fibre.

*Soiling and Stain Resistance.* Dirt and stains are removed easily from polypropylene carpets with warm water and detergent.

*Resilience.* The larger ratio of fibre diameter to denier, the natural resilience of the fibre, and the larger fibre to area relationship in carpets made from polypropylene produce a fabric with a good resistance to foot traffic and furniture marks.

*Wear.* Polypropylene fibres are tough and have a high abrasion resistance. Carpets made from them have outstanding wearing qualities, which are particularly noticeable on stair-nosings and in high-traffic areas.

*Colour Fastness.* The colours of pigmented polypropylene fibres, as used in the production of tufted carpets, are fast and permanent, and will last the life of the carpet.

*Static.* Carpets of polypropylene are almost static-free. They show little tendency to deliver shocks to people who have walked across them, and then touch metal objects. This absence of static charge also contributes to the soil-resistance of polypropylene carpets.

## Carpet Production

Polypropylene bulked continuous filament yarn may be used efficiently for high-speed tufting on a variety of different types of equipment. A few simple adjustments and precautions are all that is necessary to ensure excellent performance and a minimum of off-quality carpet.

## Sewing Thread

High strength, excellent chemical resistance, versatility and low cost have contributed greatly to the acceptance of polypropylene fibre for sewing thread used in multiwall bag and other industrial sewing applications.

Polypropylene yarns give approximately 65 per cent greater coverage than yarns of the same weight per length or denier in cotton or rayon. Polypropylene of 1,155 dtex (1,050 den), for example, provides the same bulk or coverage as a theoretical 1,897 dtex (1,725 den) rayon filament. Coupled with the inherently higher strength of polypropylene, this bulk-yield advantage permits a greater yardage per kg, per similar thread cross-section.

## Laundry Nets

Polypropylene continuous multifilament yarns are used with advantage in the production of laundry nets. The following comparisons have been made with nylon in this application:

(a) Polypropylene nets shrink only half as much as nylon nets.

(b) Polypropylene nets last twice as long as those made from nylon.

(c) After 150 washings, polypropylene nets were still in use, having suffered little loss in strength. Nylon nets were of no further use after 70 washings.

(d) Polypropylene nets have a better resistance to bleach than nylon nets.

## Blankets

Compared with wool, viscose rayon, cotton or acrylic fibres, poly-propylene is the strongest, lightest and most extensible fibre of these commonly-used blanket fibres. It also has the lowest mois-ture regain, and shows negligible shrinkage or felting during washing.

Polypropylene fibre has proved particularly suited to the pro-duction of blankets, and the manufacture of blankets of different types, weights and sizes is increasing rapidly.

Polypropylene blankets have low flammability (comparable with wool), and they are light in weight compared with blankets of similar cover and thickness made from other fibres.

Because of the high strength and extensibility of polypropylene, more severe raising treatments may be used than are possible with wool or other weaker fibres. This also results in bulkier, warmer blankets, as loftier naps can be produced with deterioration in blanket strength or increase in fibre shedding during use.

Polypropylene blankets may be laundered quickly and effect-ively, but they should not be ironed or hot pressed. They should not be dry-cleaned, as this has a deleterious effect on the handle.

## Carpets, Moquettes and Rugs from Staple Fibre

Polypropylene staple fibre is now being widely used in the pro-duction of carpets, moquettes and rugs, in which applications it offers a combination of properties that enable it to compete effec-tively with other fibres.

Polypropylene fibre is light and strong, and has a very good resistance to abrasion. It does not felt, even after prolonged wear. It does not soil or stain, and is of very low flammability. The types of polypropylene fibre used in carpet production are commonly pigmented, and the colours are very fast to light and rubbing

Manufacturers produce grades of polypropylene fibre which are specially designed for use in carpets, with moderate tensile strength about 26.5 cN/tex (3 g/den) and high elongation (about 80%). Fibres of this type have excellent elastic properties which render them particularly suited to use in carpets. They have a high resilience of the same order as wool.

Polypropylene carpets, rugs and moquettes have excellent dimensional stability, are easily cleaned and dried, do not pick up

603

dirt readily, and resist staining. The colour fastness of pigmented fibre is good, and abrasion resistance is high.

Carpets and rugs are light in weight, do not felt with washing or wear, and are completely resistant to insects and micro-organisms. The fibres melt when subjected to heat, for example, from a cigarette stub, and can be regarded as self-extinguishing. Marks left on a carpet are no larger than the stub which made them.

Polypropylene is affected by light to a degree comparable with polyamide fibres.

### Upholstery

The production of upholstery fabrics from polypropylene staple fibre is an application of rapidly-growing importance. In this field, the properties of polypropylene fibre offer many advantages over competitive fibres. The low specific gravity results in lightness and excellent covering power. High abrasion resistance, resistance to felting, stain resistance, low flammability, negligible water-absorption, resistance to micro-organisms and insects; all these are useful characteristics in a fibre that is to be used for upholstery fabrics. The pigmented polypropylene fibres are also fast to light, to rubbing and to washing.

Upholstery fabrics are commonly made from 100 per cent polypropylene fibre, and many manufacturers produce staple fibre which is suitable for this purpose. Fabrics are generally backed with latex compounds.

Upholstery fabrics made from polypropylene are easily cleaned using lukewarm water and synthetic detergents. Dry cleaning should be avoided.

In common with other types of polypropylene fabric, they show excellent stability to wear. The high abrasion resistance is an important characteristic in this application, as are polypropylene's resistance to micro-organisms and insects, and its low flammability.

### Knitwear

Polypropylene fibre is making steady progress in the knitwear field. It offers many attractive properties in this application, including an excellent hand, loft, dimensional stability, colour fastness, easy care, and all-round wearing qualities.

Polypropylene yarns are also strong, tough and highly resistant to flexing abrasion and chemical attack. The low specific gravity and high covering power are also advantageous in the knitwear field.

All the standard texturing processes may be used on polypropylene yarns, including false twisting, stuffer box crimping, edge crimping and other systems.

Polypropylene yarns have low shrinkage in hot water, good resiliency, and high wrinkle-resistance.

### Blends with Rayon

The blending of polypropylene with cellulosic fibres, such as rayon, is an excellent way of controlling shrinkage caused by relaxation crimping. Tensile strength and abrasion resistance are also improved by incorporation of the polypropylene fibre.

Fabric shrinkage may be reduced by as much as 60 to 70 per cent in this way at very low cost.

### Paper

The addition of polypropylene staple fibre to wood pulp used in paper-making affects the physical properties of latex-saturated paper in a number of ways. In particular, it brings about a marked increase in tear strength. Polypropylene-strengthened papers of this type are valuable for many speciality applications.

### Conveyor Belting

Polypropylene multifilament yarn combines many of the desirable properties of a conveyor belting yarn, including high strength for low cost, high modulus, high extension at break and high work of rupture. Also, polypropylene yarn is completely insensitive to moisture.

There are two main problems in the use of polypropylene yarn for conveyor belting. Firstly, it is difficult to bond to rubber or P.V.C. Secondly, it melts at about 165°C. and shrinks below this temperature if held in an unrestrained condition.

The problem of adhesion has tended to restrict the use of polypropylene fibre in belting applications to blend fabrics with cellulosic staple fibre. Blends with cotton and 'Durafil' provide reinforcement fabrics which are competitive with other types of belting fabric on a strength/cost basis.

## Apparel Fabrics

### Ladies' Hosiery

Pigmented continuous filament polypropylene yarns provide comfortable, snag-resistant stockings.

### Knit Pile Fabric Backing Yarns

The strength and light weight of polypropylene staple fibre serve it well in this application.

### Men's Stockings

Excellent stockings are made from blends of polypropylene staple with acrylic fibre.

### Dress Knitwear

Blended with rayon, polypropylene staple fibre provides knitwear of excellent hand. Blended with wool, it is used for sweaters and other garments of this type.

Continuous filament textured yarns are also used.

### Cellulosic Blends

Blends of polypropylene staple fibre with cotton and other cellulosic fibres show great promise as apparel fabrics. Blends with cotton containing up to 20 per cent polypropylene may be ironed without difficulty. The development of low-temperature curing resins has increased the versatility and scope of blended materials of this sort.

## Filtration Fabrics

Staple fibre and continuous multifilament yarns are woven into fabrics that have excellent chemical resistance, high strength and long life.

## Tyre Cord

Tyre cords are made from staple, multifilament and monofilament yarns. Polypropylene can be made in very high strengths for this application, but the low melting point places some restriction on use.

## Miscellaneous Uses

### Fiberfil

Polypropylene fiberfil is used in sleeping bags, mattresses and quilted fabrics. It is light, washes easily and is completely resistant to insects and micro-organisms.

### Car Upholstery

Polypropylene monofilament yarns provide upholstery fabrics with long life, soil resistance and low static characteristics.

### Knit Pile Boot Linings

Pigmented polypropylene staple is used for this application, providing excellent thermal properties, improved pile retention and resistance to insects, mildew, etc.

## POLYPROPYLENE SPLIT FILM FIBRES
### (FIBRILLATING FILM)

### INTRODUCTION

It has long been known that fibrous materials could be produced by splitting synthetic polymer films. A patent covering this technique was filed during the 1930s, but little practical use of the process has been made until comparatively recent times. The commercial development of polyolefins has stimulated renewed interest in the production of split film fibres.

Many polyolefin manufacturers are now producing split film fibres, notably from polypropylene, which offers the optimum price/properties combination for the type of uses foreseen for these fibres. Polypropylene split film fibres may be as much as 25 per cent cheaper to produce than the corresponding multifilament yarn, and for this reason they are making rapid headway in markets that were previously closed to polypropylene fibres.

Outlets for polypropylene split film fibre include mooring ropes, fishnet twine, ropes and cords, packaging twines, fishing nets, baler twine, hose reinforcement and electrical cords. Special types of polypropylene split film fibre suitable for weaving have been developed; they are used, for example, in the production of carpet backings and packing fabrics.

## PRODUCTION AND PROCESSING

Many methods have been investigated for the conversion of synthetic polymer films into fibrous material. These include scribing, cutting, abrading, air blowing and other techniques aimed at splitting the film along weak spots which would develop into cracks.

Initially, the production of fibre from film was regarded as a responsibility of the manufacturer, but this attitude has now changed with the development of fibrillating film which is converted to fibrous yarn by spinning under standard conditions of tension, speed and twist level. The manufacturer supplies fibrillating tape or film, which is used as raw material by the spinner.

This technique of using fibrillating film avoids the complete breakdown of the film into fibrils, which was a feature of many early techniques. There is no sacrifice of the advantages of cohesion, knot stability and strength which products based on partly-fibrillated yarns possess.

## STRUCTURE AND PROPERTIES

The properties of yarns made from split film are basically similar to those of conventional polypropylene multifilament yarns. The split film yarns tend to have a firmer handle, however, than the conventional type of yarn.

## POLYPROPYLENE SPLIT FILM FIBRE IN USE

The initial development of polypropylene split film has been in the fields of coarser industrial products, rather than in the finer industrial or apparel yarns. Use of the film offers great economies in raw material and processing; several stages of processing are eliminated by using a fibrillating tape in place of conventional fibres, for example, in the production of twines and ropes.

The use of split film has made fastest progress in the highly-competitive markets where cheap, strong rotproof materials are required, e.g. ropes and trawl twines, packaging twine and baler twine.

## Packaging Twine

Single-twist twines made from polypropylene split film have the properties desirable in packaging twines, e.g. of the heavier type used in tying bundles of newspapers, parcels, letters, etc. They are also kind to the hands and are trouble-free when used in automatic tying machines.

## Ropes

Polypropylene split film ropes compare favourably in strength with other synthetics of comparable price, show extremely low creep, and have an excellent resistance to kinking or hockling. Because of their high resistance to compression, these ropes do not readily melt when running over winch barrels, fair-leads, pulley-blocks or other bearing surfaces.

Because of their high stretch resistance, low compressibility and low energy absorption, polypropylene split film ropes are safer to handle than some other types of rope, and they do not tend to exhibit 'whiplash' when suddenly released from strain. They are especially suitable for applications where a maximum resistance to stretch is required, rather than for outlets where a maximum energy absorption is needed. This makes them suitable particularly for end-uses such as fishnet mounting ropes, mooring and hauling ropes, and cargo slings.

## Baler Twine

This is one of the largest potential markets for polypropylene split film. Attempts to replace the commonly-used sisal with other types of fibre have generally failed on economic grounds, and on poor knot stability. Baler twine made from polypropylene split film fibre has good knot stability, low weight, good regularity, and high strength.

## Fishnet Twines

For certain types of bottom trawl and Danish seine net, it is sometimes desirable to have a twine with a greater stiffness than that obtainable with polypropylene multifilament yarns. Twines based on polypropylene split film have the required stiffness, with a higher straight and knot strength than polyethylene twines commonly used, a higher resistance to stretch and creep, and a knottability comparable with that of the natural fibre twines.

## 5. POLYURETHANE FIBRES

Fibres spun from polymers made by a reaction taking place between small molecules, in which the linkage of the molecules occurs through the formation of urethane groups ($-NHCOO-$).

Linear polyurethanes may be made by reaction of a glycol with a di-isocyanate. The reaction of butane diol, for example, with toluene di-isocyanate results in the formation of a polyurethane in the following way:

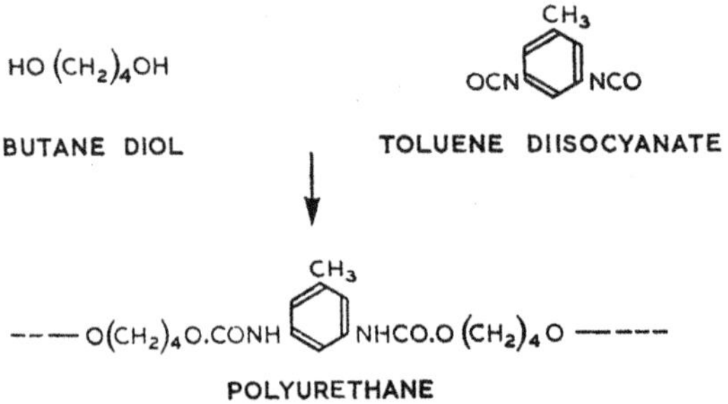

HO $(CH_2)_4$ OH

**BUTANE DIOL**

$CH_3$

OCN     NCO

**TOLUENE DIISOCYANATE**

$CH_3$

$----O(CH_2)_4O.CONH$     $NHCO.O(CH_2)_4O$ $------$

**POLYURETHANE**

### INTRODUCTION

More than a century ago, it was discovered that a reaction took place between the isocyanate group, $-N=C=O$, and the hydroxyl group, $-OH$, resulting in the formation of a urethane group, $-NHCOO-$. During the reaction, the hydrogen atom of the hydroxyl group migrates to the nitrogen atom of the isocyanate; the residue of the alcohol is transferred to the carbon atom of the isocyanate group:

$$R-NCO+R'-OH \rightarrow \quad R-NHCOOR'$$

During the late 1930s, this reaction was used for making polymers, by using it to link together two types of small molecule, one of which contains two isocyanate groups (i.e. a di-isocyanate) and the other two hydroxyl groups (i.e. a glycol). The polymerization which occurs in this way does not involve the loss of water or other small molecules, and it is therefore an addition rather than a condensation polymerization.

## 'Perlon' U

During World War II, a fibre was spun in Germany from a polyurethane formed by the reaction of 1,4-butanediol with hexamethylene di-isocyanate:

$$HO(CH_2)_4OH + OCN(CH_2)_6NCO \rightarrow$$
$$[-O(CH_2)_4OCONH(CH_2)_6NHCO-]_n$$

This first polyurethane fibre was marketed as 'Perlon' U. It bore a general resemblance in many respects to nylon, but it was inferior to nylon as a textile fibre. It melted at 180°C., for example, which is lower than is desirable in an apparel fibre. It was stiff, with a harsh wiry handle; it had a moisture absorption lower than nylon.

'Perlon' U achieved some limited success during World War II, finding markets in rather specialized applications such as brush bristles, filtration fabrics and other industrial uses. It did not compete effectively with the rapidly developing nylon, however, and polyurethane fibres of this type have made little headway since the war.

### Elastomeric Fibres

In recent years, polyurethanes have achieved a new and increasing importance in the textile world. They have become the basis of a novel type of elastomeric fibre which is known generically as *spandex* (see definition below).

Elastomeric fibres are those which display elasticity characteristics associated with natural rubber; they will stretch to several times their original length, and on release will snap back quickly to recover their original length almost completely.

Natural rubber filaments have long been used in the textile industry to provide stretch properties in fabrics and garments. Rubber filaments will 'give' under the action of a force, and the stretched filaments will then exert a recovery force as they try to return to their original shape. 'Power stretch' fabrics used in elastic webbings, support and foundation garments, for example, derive their properties from filaments of rubber or similar materials incorporated in the fabric.

Natural rubber filaments have been supplying the power to stretch fabrics for more than a hundred years, and they have established a position of some importance in the textile industry. Unfortunately, natural rubber has several shortcomings with respect to its use as a textile material (see page 153), and fibre chemists have long been seeking to develop elastomeric materials which would be superior to rubber in recovery force, resistance to abrasion, chemical stability, dyeability and other properties.

The elasticity of natural rubber derives from its long, folded polymer molecules, which are linked together at intervals by the chemical bonds introduced during vulcanization. When a filament of vulcanized rubber is pulled, the long molecules unfold and the rubber stretches. The extent of the deformation is restricted by the links between the molecules, and when the tension is released the long molecules tend to revert to their relaxed, folded state. So the filament springs back to its original form.

Elastic fibres produced from natural rubber have excellent elasticity, but the tensile strength and force of recovery from stretch are less than adequate for the production of lightweight garments. In addition, the double bonds which remain in the rubber molecules after vulcanization impart chemical reactivity, especially with respect to oxidising agents. Finally, the hydrocarbon nature of the rubber molecule results in a low acceptance of dyes.

In the search for new types of elastomeric fibre, chemists have sought to develop molecular structures which would provide the fundamental requirements of rubber-like elasticity, but did not have the disadvantages inherent in the hydrocarbon structure of rubber itself.

### Elastic Polyamides

In rubber, the tie-points holding the long molecules together at intervals consist of covalent chemical bonds introduced during vulcanization. It was realized, however, that linear polymer molecules could be linked effectively by tie-points resulting from hydrogen bonding developed by polar groups in the molecules. If a polymer molecule could be made, for example, in which long segments of 'amorphous-type' molecule were linked together by polar groups, the requirements of rubber-like elasticity might be met. The polar groups would establish strong forces between the molecules, providing tie-points separated by segments on non-polar material.

A great deal of research was carried out in attempts to make elastomeric fibres from polyamides in this way. Polyamides were made by condensing dicarboxylic acids with a mixture of two diamines, one of which had substituted (secondary) amine groups. The polymer from this condensation included two elements necessary for the establishment of high elasticity; (1) a pliable constituent of low melting point with poor interchain bonding force (the N-substituted polyamide segment, and (2) an interchain bonding constituent (the unsubstituted polyamide with its hydrogen bonding capability).

Elastic polyamides of this type were made, for example, from hexamethylene diamine and sebacic acid (i.e. nylon 6.10 type) in which a proportion of the hexamethylene diamine was replaced by a diamine carrying bulky butyl groups. Those regions of the molecules formed from the hexamethylene diamine and sebacic acid were able to align themselves closely, and develop strong hydrogen bonds between the molecules; those regions of the molecules formed from the substituted diamine, on the other hand, were unable to align themselves, forming regions of amorphous polyamide.

When these fibres were stretched, the molecules in the amorphous regions were able to unfold, the degree of stretching of the fibres as a whole being restricted by the powerful bonding of the molecules in the crystalline regions. When the tension was released, the molecules in the amorphous regions tended to revert to their original positions as the fibre recovered from its stretch.

### Segmented Polyurethanes

In the random copolymers produced from mixed diamines, the average length of the two constituents was rather small. The sequences of N-substituted polyamide units were too short to permit of the development of the chain flexibility necessary for high rubber elastic force, while the sequences of unsubstituted polyamide units were too short to prevent rupture during stretching or heating (rupture during stretching causes flow; rupture during heating results in low softening point). What was needed was a polymer molecule in which longer segments of amorphous-type structure were separated by segments of molecule capable of developing powerful hydrogen bonds.

The solution to these problems was found in the development of polyurethane polymers in which segments of the molecule are deliberately tailored to perform the desired functions. The lack of flexibility of the pliable constituent (with low melting point) was overcome by using preformed segments of molecule of considerable size (the 'soft' segments); the cohesion of the interchain bonding constituents ('hard segments') was assured through the use of urethane or urea groups.

The development of these segmented polyurethane polymers has provided the textile trade with an entirely new type of elastomeric fibre—the spandex fibre. Derived from a polymer that differs fundamentally from the hydrocarbon polymer of natural rubber, the spandex fibre is stronger than rubber, and has a greater 'recovery power'. It is a white or clear and near-transparent fibre, capable of being dyed to match other fibres in a fabric. Globe's 'Sheerspan' produces sheer-looking fabrics in white, and its high degree of dye receptivity results in a fabric with sharp, deep and distinct shades.

Spandex fibres have a high resistance to chemicals, sunlight and other degradative influences, and may be washed repeatedly without ill-effect.

The first spandex fibre, du Pont's 'Lycra', was introduced on to the market in 1958, and commercial production began in 1960–61.

Since then, many firms have followed suit, and there are a number of spandex fibres now available.

## TYPES OF POLYURETHANE FIBRE

The process used in producing segmented polyurethane fibres (see 'Production' below) is such as to allow of almost infinite variation in the chemical structure of the polymer that is formed.

Modern segmented polyurethane fibres may be considered as falling into one or other of two general types which differ in the chemical structure of the 'preformed' segments of the molecule. In this respect, segmented polyurethane fibres are either

      (a) polyether types, or
      (b) polyester types.

Both types of segmented polyurethane fibre are produced by fibre manufacturers in considerable variety, providing a range of elastomeric fibres which meet many different end-use requirements.

The polymers formed by linking preformed segments of polyether or polyester molecules, via the urethane group, may be essentially linear molecules, or they may be branched and/or cross-linked into three dimensional structures. The linear types are generally capable of melting, and will dissolve in appropriate solvents. The three-dimensional types may be completely insoluble and non-melting.

Fibres spun from these segmented polyurethanes may be in the form of monofilaments, or they may be multifilament yarns in which a number of fine filaments have coalesced after spinning. Monofilaments may also be produced by extruding sheets of the polyurethane and then cutting this into filaments.

## NOMENCLATURE

### Elastomeric Fibre

The segmented polyurethane fibres now on the market are all characterized by the high extension and snap-back recovery associated with rubber-like elasticity. They are therefore properly described as elastomeric fibres in that they are fibres which behave in a rubber-like way.

The term *elastomeric fibre* does not, of course, relate to the chemical structure of a fibre, and it should be realized that it is not synonymous with 'segmented polyurethane' or 'spandex'. Both of these latter terms are based upon the chemical structure of the fibre (see below).

### Federal Trade Commission Definition

The generic name *spandex* was adopted by the U.S. Federal Trade Commission for fibres of the segmented polyurethane type, the official definition being as follows:

*Spandex.* A manufactured fibre in which the fibre-forming substance is a long-chain synthetic polymer composed of at least 85 per cent of a segmented polyurethane.

## PRODUCTION

Spandex fibres are spun from segmented polyurethanes made by a series of chemical stages, as follows:

(1) Production of low molecular weight polymer (pre-polymer)
(2) Reaction of pre-polymer with di-isocyanate
(3) Coupling of isocyanate-terminated pre-polymer to form segmented polyurethane.

### (1) Production of Low Molecular Weight Polymer (Pre-polymer)

The first step in the synthesis of a spandex fibre is to create the soft segment of the molecule, i.e. that segment that provides the amorphous region in which the unfolding of the molecules permits extension of the fibre to take place.

Two classes of compounds, polyesters and polyethers, are commonly used for this rubbery soft segment, the materials produced being polymers of low molecular weight (500 to 4,000) with reactive hydroxyl groups at each end of the molecule. They are macroglycols, which are made by normal polymerization techniques.

*Polyesters* are made by condensation of dicarboxylic acids with a slight excess of glycol; the condensation takes place until there are glycol units at each end of the polymer molecule, which thus has hydroxyl end groups (A).

*Polyethers* are made by the ring-opening polymerization of epoxides or cyclic ethers (B).

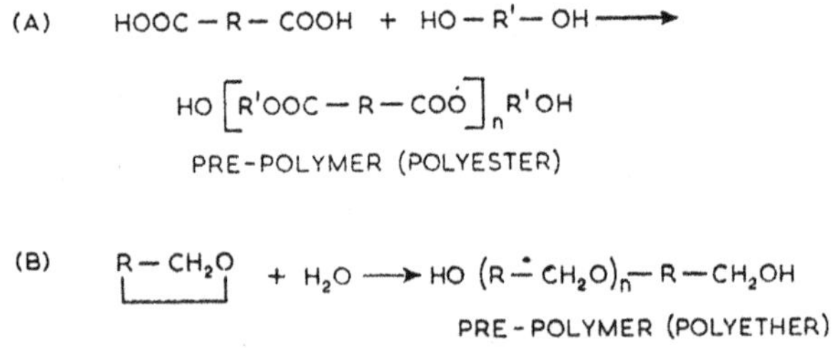

(A)    $HOOC - R - COOH + HO - R' - OH \longrightarrow$

$$HO \left[ R'OOC - R - CO\overset{..}{O} \right]_n R'OH$$

PRE-POLYMER (POLYESTER)

(B)    $R - CH_2O + H_2O \longrightarrow HO \left(R \overset{..}{-} CH_2O\right)_n - R - CH_2OH$

PRE-POLYMER (POLYETHER)

Production of Pre-polymer

The structure of the soft segment influences the properties of the resultant polymer, e.g. melting point, flexibility and chemical stability, and the selection of the type of soft segment to be used depends upon the type of fibre required, and on availability and cost of raw materials.

## (2) Reaction of Pre-polymer with Di-isocyanate

The next step in the production of the segmented polyurethane is to convert the soft segment macroglycol into a pre-polymer which has isocyanate end groups. The macroglycol is reacted with an excess of di-isocyanate, the hydroxyl groups on the ends of the macroglycol molecules reacting with isocyanate groups to form urethane groups. If two moles of di-isocyanate are used per mole of macroglycol, for example, a pre-polymer is formed in which each molecule now has isocyanate end-groups.

---

$$HO-Pol-OH \ + \ 2OCN-R-NCO \longrightarrow$$

PRE-POLYMER

$$OCN-R-NHCOO-Pol-OCONH-R-NCO$$

Isocyanate treatment of pre-polymer

---

The reactivity of the di-isocyanate depends upon its structure. In general, the aromatic di-isocyanates are more reactive than the non-aromatic. Aromatic di-isocyanates are also commercially available, and they are used predominantly in the reaction used in building spandex polymers.

## (3) Coupling of Isocyanate-terminated Pre-polymer to form Segmented Polyurethane

The final step in making the segmented polyurethane consists in the creation of the hard segment by 'chain extension', or coupling of the isocyanate-terminated pre-polymer by reaction with low molecular weight bifunctional compounds, such as glycol or diamine. The reaction product is a polymer having hydrogen bonding sites in the form of urethane or urea groups, at least two of which will occur in the resulting 'hard segment'.

(1) GLYCOL

$OCN - Pol - NCO + HO - R - OH \longrightarrow$

$- OCONH - Pol - NHCOOROCONH - Pol - NHCOO -$

(2) DIAMINE

$OCN - Pol - NCO + H_2N - R^1 - NH_2 \longrightarrow$

$- NHCONH - Pol - NHCONH - R^1 - NHCONH - Pol - NHCONH -$

Chain extension with glycol or diamine.

---

This final chain extension stage may be carried out also by addition of water to the isocyanate-terminated prepolymer, instead of a glycol or a diamine. Water may be added, for example, in quantity sufficient to react with a proportion of the terminal isocyanate groups, forming pre-polymer molecules with an isocyanate group on one end and an amine group on the other end (1). When this polymer is heated, the amine and the isocyanate groups react to bring about further polymerization and cross-linking of the molecules (2).

---

$OCN - Pol - NCO + H_2O \xrightarrow{(1)} OCN - Pol - NH_2 + CO_2$

$OCN - Pol - NH_2 + OCN - Pol - NH_2 \xrightarrow{(2)} OCN - Pol - NHCONH - Pol - NH_2$

Chain extension with water

---

This reaction results in only one urea group for each two isocyanate groups, carbon dioxide being liberated as a by-product. The evolution of gas during chain extension may be regarded as desirable in the production of polyurethane foams, but it is usually to be avoided in the production of polyurethane fibres.

Spandex polymers produced in this way may be essentially linear molecules, in which end-to-end linking of bifunctional molecules is the predominant reaction. Such materials are commonly soluble in appropriate solvents.

If branching occurs during the polymerization process, however, e.g. by reaction of the isocyanate end-group with an active hydrogen in the molecular chain, the polymer may build up into a branched or cross-linked three-dimensional structure. Such polymers may reach the stage of being insoluble in any solvent, and incapable of melting.

**Spinning**

The technique used in spinning spandex fibres depends upon the type of polymer that is spun. Some segmented polyurethanes, for example, are essentially linear molecules, and are soluble in solvents. Other segmented polyurethanes may be branched or cross-linked structures which are insoluble.

### (a) *Linear (Soluble) Polyurethanes*

Soluble polyurethanes are dissolved in an appropriate solvent, and the solutions may be extruded through spinnerets into a coagulating bath (wet spinning) or into an atmosphere which removes the solvent (dry spinning). The techniques are essentially the same as those used for spinning 'hard' synthetic fibres, due allowance being made for the fact that spandex fibres are elastic.

### (b) *Branched or Cross-linked (Insoluble) Polyurethanes*

When the molecule of polyurethane is allowed to grow into a three-dimensional structure, it is insoluble and cannot be spun by the above techniques. In this case, a 'chemical spinning' process may be used. The isocyanate-terminated pre-polymer is spun at a stage when it forms a viscous dope, the jets emerging into a gaseous or liquid environment containing a chain extender which diffuses into the fibre and reacts. The pre-polymer molecules are linked into their final form, producing the branched or cross-linked polyurethane in fibrous form.

Spandex fibres may be spun as monofilaments, or as multi-filament yarns in which a number of fine filaments have coalesced after spinning. Square section monofilaments may be produced by extruding sheets and then cutting these into filaments.

An important feature of spandex fibres is that they may be spun in very fine filaments. The finest rubber yarns are commonly of about 167 dtex (150 den), but spandex fibres may be produced as 44 dtex (40 den) or finer.

## PROCESSING

**Scouring**

Fabrics containing spandex fibre may be scoured with an emulsified solvent such as perchloroethylene at about 82°C. The fabric is after-scoured in a fresh scour bath containing detergent but no solvent, to remove the last traces of solvent and finish.

### Bleaching or Whitening

Bleaches may be used on fabrics containing spandex fibres, but care should be taken in the selection of the bleach. Individual spandex fibres differ in their reactions to different bleaching conditions. Hypochlorite and sodium chlorite bleaches generally cause discolouration, and peroxide or perborate bleaches are preferred.

Optical whiteners may be used on spandex fibres, but they should be chosen carefully to provide optimum light fastness. Optical whiteners should be selected which give a colourless residue after the whitener has been broken by light.

### Dyeing

Spandex fibres generally have a marked affinity for a wide range of different types of dye, but colour fastness is achieved only with dyes substantive to the fibre. Basic dye sites on the fibre provide for bonding of the dye molecules through the acid groups in the acid dye. Greater fastness may be achieved by the use of top-chrome acid dyes, but the colour obtained with these dyes is less brilliant. Disperse dyes are also used, especially where colour-fastness is not critical.

### Finishing

After whitening or dyeing, a resin finish may be applied to spandex-containing fabrics to improve hand and body. Certain special finishes which provide good whiteness retention on exposure to atmospheric fume contaminants are recommended for white or lightly-dyed fabrics. They provide good protection in severely contaminated atmospheres.

### Heat-Setting

Spandex fibres are thermoplastic, and may be heat-set like other thermoplastic fibres. Heat-setting is used to ensure dimensional stability of fabrics and garments.

The reaction to heat shown by spandex fibres is comparable to that of most heat-settable man-made fibres and no special handling short of variation of conditions is required. Fabrics requiring heat-setting should be handled with the following principles as basis:

(1) Temperatures should be kept as low as possible, consistent with effective setting.

(2) Tension should be kept to a minimum, consistent with control of fabric dimensions.

(3) Exposure time should be balanced with temperature to avoid over-setting with a consequent decrease in power.

(4) Heat-setting should not in principle be used to compensate for shortcomings in fabric construction and manufacture.

## STRUCTURE AND PROPERTIES

### Fine Structure and Appearance

*Molecular Structure*

Spandex fibres may be regarded as 'block' copolymers in which long flexible sections of the molecule are joined by urethane links to shorter stiffer sections. The chemical structure of the polymers may be varied through an infinite range, to provide fibres of the desired characteristics. Modern spandex fibres are derived usually from low molecular weight polyethers or polyesters (macroglycols).

When the polymer is spun into fibres, the molecules are established in a state of random disorder. If a stretching force is applied to the fibre, the folded or coiled sections of the molecules tend to straighten out and become aligned. The short, stiff sections, however, are bonded to one another by intermolecular links derived from hydrogen bonds or van der Waals' forces. These bonded regions act as 'anchor points' which prevent the molecules sliding past each other to take up new permanent positions relative to one another.

The distortion of the fibre under the action of the stretching force is thus limited to the extension permitted by the straightening of the folded molecules. When the stretching force is released, the molecules revert to their folded state, and the fibre returns (ideally) to its original length.

The segmented polyurethanes have two characteristic features which explain their superior physical properties relative to conventional rubber structures.

621

First, the long-chain polyurethane molecules are synthesized from preformed, soft-segment polymer blocks, and the hard segments forming the tie-points are spaced more regularly along the chains than those in randomly vulcanized rubber. In rubber, the occurrence of tie-points close together may limit the flexibility and hence the elastic effectiveness of the in-between soft-segments; this is avoided in the case of the segmented polyurethanes. Also, the more regular network of the polyurethanes should result in greater elongation before breaking, by minimizing the number of chains which are stretched prematurely to the breaking point.

The second characteristic of polyurethanes is the occurrence of tie-points which may be broken and re-formed during stretching. This behaviour minimizes the concentration of points of stress, leading to an even more regular network structure with the advantages outlined above.

Finally, the hard-segment bonding is not necessarily limited to the tying together of molecules two-by-two, as is the case with covalent cross-links. The effect of multiple hard-segment 'packages' provides additional reinforcement similar to that obtained in conventional rubbers by the use of active fillers, such as carbon black.

*Fibre Form*

Spandex fibres are produced as monofilaments, e.g. of round cross-section, or as partly-fused multifilaments. Monofilaments made by cutting thin sheets of polymer may also be produced.

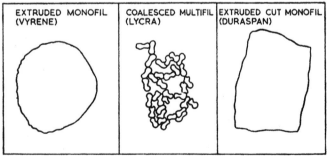

*Spandex Fibres: Cross-sections*

Spandex fibres are commonly clear and near-transparent, or white. The colour may deteriorate somewhat with age.

### Tenacity; Tensile Strength

Because of their segmented structure, spandex fibres may be made stronger than natural rubber filaments. The breaking tenacities of spandex fibres are 4.9–8.8 cN/tex (0.55–1.0 g/den), compared with 2.2 cN/tex (0.25 g/den) for natural rubber.

Tensile strengths of the spandex fibres are in the range 616–994 kg/cm$^2$ (8,800–14,200 lb/in$^2$).

Tensile properties are affected only slightly by water.

### Elongation

It is a characteristic of spandex fibres that they are capable of being stretched to several times their original length. The breaking elongations range from 450 to 700 per cent and may vary according to denier with the same type of spandex fibre.

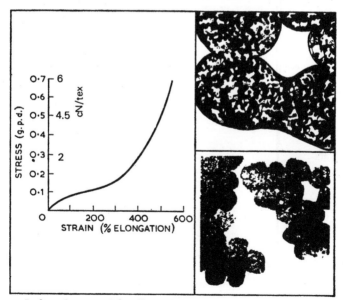

*Left and upper right: 'Lycra'. Bottom right: Glospan S–5*

**Elastic Recovery**

Spandex fibres have a snap-back rubber-like elasticity, but recovery from stretching to a given extent is not usually as complete as in the case of natural rubber. In some fibres of this type, there is a small residual extension which is not recovered after stretching, i.e. a degree of permanent set. This is not usually a progressive effect, reaching a constant value after a few repeated cycles of loading and unloading. The 'permanence' of the set is also a matter of degree; the set gradually diminishes with time when the fibre is allowed to relax. Recovery is speeded up by increased temperature.

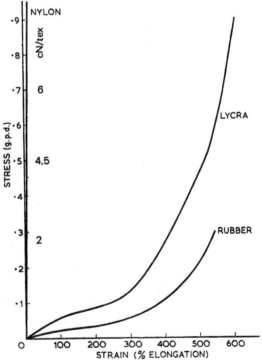

Load-elongation curves for spandex and rubber fibres.

This time-recoverable set is a result of temporary deformation of the molecules as a result of the viscosity of the system, rather than being due to molecular flow. Genuine 'permanent set' is induced by stretching the fibre so far that rupture of some of the interchain tie-points occurs. These bonds re-form after molecular flow has taken place, and cause an unrecovered strain.

As seen in the table below, set increases with the degree of stretch in the case of 'Lycra', particularly above a stretch of 300 per cent.

*Set Induced by Various Degrees of Stretch
Glospan S7 and S5; 467 dtex (420 den)*

|  | Set (per cent) | |
|---|---|---|
| *Stretch (per cent)* | *S5* | *S7* |
| 200 | 4 | 2 |
| 300 | 10 | 5 |
| 500 | 29 | 13 |
| 600 | 40 | 16 |

*Note.* The fibre was cycled five times. on 'Instron' tensile tester to designated stretch. Set expressed as increase in original length of fibre.

**Modulus of Elasticity**

Spandex fibres have a low modulus of elasticity, e.g. about 1/1,000 that of a conventional 'hard' fibre such as nylon or cotton. They have a higher modulus of elasticity than natural rubber, however. To achieve a given stretch, spandex fibres require a force twice as high, over the entire range of elongation, as the force required by rubber.

**Specific Gravity**

1.2 to 1.25.

**Effect of Moisture**

Most spandex fibres have a moisture regain of the order of 1.0 to 1.3 per cent ('Vyrene' has a lower than average regain of about 0.3 per cent).

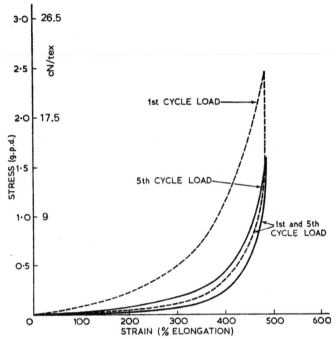

Typical stress-strain curves for spandex fibre ('Effective tex' is the tex at the point of measurement).

---

**Thermal Properties**

Spandex fibres are commonly thermoplastic, sticking becoming noticeable in some cases as low as 150°C., but in others as high as 280°C. Melting points are in the range 230–290°C.

**Effect of Sunlight**

Resistance generally is excellent, with discolouration occurring in some cases on prolonged exposure.

**Effect of Age**

Spandex fibres show little deterioration on ageing.

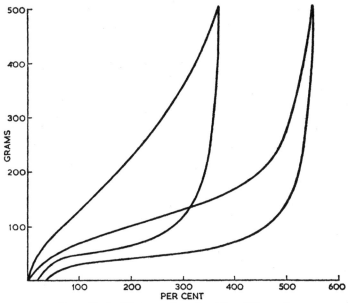

*Stress-Strain Diagram, Spandex Fibre ('Vyrene')*

This diagram shows the stress-strain relationships of a 'Vyrene' spandex fibre, 75's count, first and sixth cycle of extension. The modulus is higher than that of a rubber thread.—*Lastex Yarn and Lactron Thread Ltd.*

---

**Chemical Properties**

*Acids*

Resistance varies according to the type of spandex fibre. Most fibres have a good resistance to cold dilute acids, but may be attacked under more severe conditions.

*Alkalis*

Resistance to alkalis is generally good.

*General*

Good resistance to most common chemicals, including cosmetic oils and lotions. Good resistance to oxygen.

The resistance to bleaches is generally good, but some care is necessary in the selection of bleaches for use with specific fibres. Discolouration generally occurs, for example, when hypochlorites or sodium chlorite bleaches are used, and peroxide and perborate bleaches are preferred.

## Effect of Organic Solvents

Spandex fibres have a good resistance to most common solvents, including dry-cleaning solvents. They may be affected by prolonged exposure to unsaturated hydrocarbons, but are not generally affected by saturated hydrocarbons.

## Insects

Completely resistant.

## Micro-organisms

Completely resistant.

## SEGMENTED POLYURETHANE (SPANDEX) FIBRES IN USE

### General Characteristics

An elastic fibre is characterized by a high breaking elongation (in excess of 100 per cent, and usually 450 to 700 per cent), a low modulus of elasticity (about 1/1,000 that of a conventional 'hard' fibre such as nylon or cotton), and both a high degree and a high rate of recovery from stretching. The following table lists typical properties of a spandex fibre, rubber and nylon.

|  | Tenacity (cN/tex) | Elongation (per cent) | Modulus of Elasticity (cN/tex) | Recovery from 100 per cent stretch (per cent) |
|---|---|---|---|---|
| Spandex ('Lycra') | 8.0 | 550 | 0.4 | 95 |
| Natural rubber | 2.7 | 540 | 0.2 | 97 |
| Textile nylon | 37.1 | 26 | 220.8 | — |

Because of their segmented structure, spandex fibres may be made stronger than rubber fibres. To achieve a given stretch, spandex fibres require a force twice as high, over the entire range of elongation, as the force required by rubber. In contrast to a hard fibre such as nylon, however, spandex fibres have a very low modulus of elasticity.

In many of the applications for elastic fibres, one of the most important properties is recovery force after stretching and partial relaxation. Attention must be given, therefore, to both the 'load' and the 'unload' cycles in the stress-strain curves. There is a time-dependent difference in stress between load and unload cycles resulting in a loss of energy which is dissipated by heat.

This hysteresis results from a stress decay which reaches a constant value after repeated cycling, no change being noted beyond the fifth cycle. The hysteresis affects not only recovery force but also degree of recovery.

The lack of recovery, usually referred to as 'set' is only partially permanent. If the fibre is allowed to relax completely, the set diminishes gradually with time.

Because of stress decay and set as factors in recovery force, it is necessary to assess yarns under conditions which approximate the degree of stretch and cycling the yarn will undergo during manufacture into fabrics, and during wear. The extent of stretch required in elastic yarns varies according to the type and required elasticity of the fabric into which it is to be manufactured. Figure-controlling garments, such as women's girdles, are designed to have stretch 50 to 100 per cent greater than the elongation during wear. The restraining force of such garments is related directly to the recovery force of the elastic fibre at its elongation when it is in use.

The term 'effective power' has been used for the recovery force per unit of linear density of a yarn at a given elongation. The figure on page 632 shows the effective power of the du Pont spandex fibre 'Lycra' and of rubber over a range of available stretch. The spandex fibre has about twice the effective power of rubber at the important elongations at which additional stretch of 50 to 100 per cent is available.

Analytical procedures, based on the principles given above, have been devised for predicting the performance of a yarn incorporated in a given fabric from measurement of the stress strain properties of the yarn. For these measurements, test conditions were established which simulated the stretch conditions to which yarn would be subjected during manufacture of the fabric. The following table shows a comparison of 'power' for the yarn and for the fabric.

*Correlation of Effective Power of Yarn and Woven Fabric*

| | Effective Power ($cN$/effective tex )[1] | | | |
| | 'Lycra' Spandex | | Natural Rubber | |
| | 50% | 70% | 50% | 70%[2] |
|---|---|---|---|---|
| Uncovered yarn | 0.78 | 1.15 | 0.36 | 0.51 |
| Covered yarn | 0.76 | 1.22 | 0.34 | 0.47 |
| Covered yarn removed from fabric[3] | 0.75 | 1.13 | 0.37 | 0.50 |
| Fabric[3] | 0.66 | 1.10 | 0.42 | 0.54 |

*Notes*: [1] 'Effective tex' is the tex at the point of measurement.

[2] Percentages indicate actual fabric stretch, or, in the case of yarns, simulation of 'in-fabric' stretch, of 50 or 70 per cent.

[3] Stretchable woven fabric, 100 per cent total stretch available.

The data in this table show good agreement of values for the original yarn, the yarn in covered form, the covered yarn after weaving and subsequent removal from the fabric, and the fabric itself.

The durability of spandex fibre in garments, where flex, abrasion, and needle-cutting are very important, is high. This durability results from the regularity of polymer structure and the strength of the interchain hydrogen bonding. The 'power' of two commercial power-net fabrics — 467 dtex (420 den) 'Lycra' and 56 dtex (50 den) is virtually unchanged after 500,000 flex cycles at elongation of 100 per cent.

The following table shows the durability and dimensional stability of the spandex-fibre fabric when it is subjected to cyclic flexing at elongation approaching the limit of stretch of the fabric. Spandex-fibre fabrics of two different weights show high resistance to flex and low growth compared to their rubber-fibre counterparts. Moreover, because of the durability of spandex fibres, very fine fibres (e.g. 22 dtex; 20 den) can be produced. The availability of these fine yarns has opened up new applications for elastic fibres.

*Flex Resistance of Elastic Yarns in Hosiery*

|  | Cycles to Rupture[1] | Washings | Growth[2] (per cent) |
|---|---|---|---|
| 'Lycra' spandex, 308 dtex | 50,000 | 18 | 17 |
| 'Lycra' spandex, 154 dtex | 50,000 | 18 | 10 |
| Rubber, 550 dtex | 5,000 | 10 | 45 |

*Notes*: [1] A piece of sock top 2.5 cm. wide was subjected to cyclic flexing at elongation of 100 per cent.

[2] 'Growth' means increase in length of the sample subjected to cyclic flexing relative to original length.

### Chemical Properties

Spandex fibres were the first elastomeric fibres to accept dyestuffs readily. This dyeability results from the chemical nature of the polyurethane molecule, which contains active groups capable of holding appropriate dyes.

Dyeability has opened up many new applications to spandex fibres, which were served inadequately or not at all by natural rubber yarns. Bare spandex fibres may now be incorporated in fabrics, and dyed to match other fibres in the fabric.

As would be anticipated from the chemical structure of the segmented polyurethane, spandex fibres are resistant to hydrolysis. 'Lycra', for example, retains 100 per cent power after boiling in water for 1 hour at pH 3 to 11. In addition, spandex fibres have a good resistance to ultra-violet radiation, oxygen, heat, perspiration, body oils and lotions. They are physically and chemically stable to a range of heat, pH and reagent conditions such as are used in processing operations on other fibres.

#### Washing

Spandex fibres are not affected by the conditions used in washing fabrics made from other fibres, and no special precautions are necessary. Soap and detergent may be used, and garments may be washed effectively by machine or by hand.

#### Drying

No special precautions are needed in drying fabrics containing spandex fibres, other than the avoidance of unnecessarily high

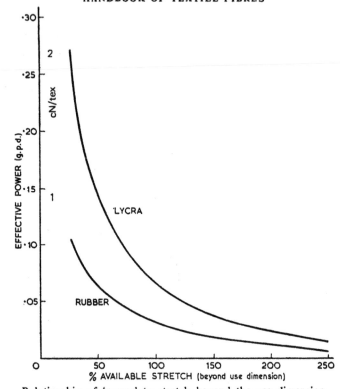

Relationship of 'power' to stretch beyond the use dimension.

temperatures. In general, the drying conditions may be selected to suit the base fabric.

### Ironing

It is generally unnecessary to iron the types of fabric in which spandex fibres are used, but ironing may be carried out effectively if necessary. A low temperature setting should be used.

### Dry Cleaning

Spandex fibres are not affected by the usual dry-cleaning solvents, and garments may be dry cleaned without difficulty.

**End-Uses**

Spandex filaments are used in three forms:

    (1) Bare Filaments
    (2) Covered Yarns
    (3) Core-spun Yarns
    (4) Core-twisted Yarns.

*Bare Filaments*

Natural rubber filaments are commonly used as covered yarns, in which the rubber is protected and hidden from view by a sheath of 'hard' fibre such as nylon, cotton, etc. Because of their high resistance to abrasion, whiteness, high 'power' and inherent dyeability, it is possible to use spandex fibres in the uncovered or 'bare' form, and this has now become common practice in many sections of the stretch fabric field.

The major applications of bare spandex fibres are in the production of foundation garments, swimwear and hosiery. The fabrics used include power-nets, tricot, lace, and circular knits. The rapid penetration of spandex fibre into these markets followed the reduction in cost resulting from elimination of the expensive covering operation. All ranges of fabrics, from heavy to lightweight garments are now produced.

In the power-net field, spandex fibre has not only taken over a major portion of the market, but it has also enabled the market to expand by increasing the versatility of the fabrics. This has been particularly noticeable in the lightweight fabrics aimed at areas such as the teenage market. Dyeability of spandex has been an important factor in this respect, allowing of the production of a wide range of colours and shades when spandex is dyed, for example, in combination with nylon.

The introduction of fine count spandex fibres and of beamed spandex fibres has permittted tricot manufacturers to produce a wide range of nylon and spandex tricot stretch fabrics. Spandex can be knitted in combination with nylon in the form of half-gauge or full-gauge warp combinations to provide a variety of fabric textures and constructions. The fact that tricot fabrics are knit, coupled with the stretch of the spandex fibre, allows the production of tricot fabrics with considerably more stretch in both directions than was hitherto possible.

The use of finer counts allows of the production of fabrics of a high degree of 'femininity' for the foundation trade. A variety of tricot fabrics may be created by using fine count spandex fibres under various heat-setting conditions.

### Covered Yarns

Rubber filaments are covered by winding yarns of 'hard' fibre round them in spiral fashion. Either continuous filament or spun staple yarns may be used, and two layers wound in opposite directions are commonly used to provide a balanced structure.

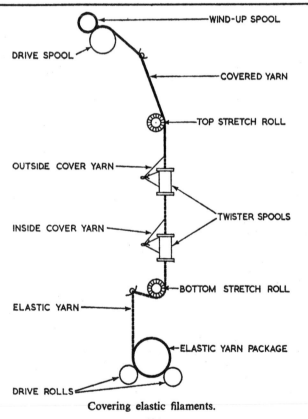

Covering elastic filaments.

Spandex yarns may be covered in this fashion, using standard types of covering machine. The spandex yarn is strong, uniform and highly resistant to abrasion, permitting the use of very fine covering yarns without risk of damage to the spandex filaments.

In the covering machine, the elastic filament of rubber or spandex passes through the centre of a hollow spindle which rotates at high speed. As it rotates, the spindle wraps the covering yarn spirally on to the elastic filament, which is held under a controlled degree of stretch during the covering operation.

Covered spandex yarns are used primarily for the foundation garment trade. Covering provides maximum fabric power with controlled stretch, the cover setting a limit to the amount of stretch that can take place. This type of yarn is made into power-nets, woven lenos, taffetas, satins, narrow fabrics and braids.

The purpose of covering spandex yarns is rather different from that of rubber. Spandex yarns are covered primarily to prevent the slippage of extended thread within the unstretched fabric. Where the fabric construction (e.g. power-net, tricot) provides a sufficiently firm anchorage, then marked threads can be used, but in woven constructions, for example, it is preferable to use covered, core-spun or core-twisted yarns.

### Core-Spun Yarns

A core-spun yarn is one in which a non-elastic fibre sheath is spun around a core of spandex or other elastomeric yarn. This type of yarn is made by introducing the elastomeric filament, stretched to a carefully-controlled degree, into the spinning frame where staple fibre is being spun, in such a way that the staple fibre is twisted into a yarn with the elastomeric filament forming a core. Any type of staple fibre, such as cotton, wool, acrylic, polyester, etc., may be used in forming the spun sheath, and all types of spinning systems may be used.

The power and stretch of the resulting yarn depend upon the denier of the elastomeric yarn, and the degree of stretch under which it enters the spinning zone. Only a comparatively small proportion (e.g. 5–7 per cent) of elastomeric yarn is used, as a rule, the amount depending upon the end-use for which the yarn is intended.

It is obvious that the structure of a core-spun yarn is different from that of a covered yarn made by applying spirally-wound yarns to an elastomeric core. The core-spun yarn may be made

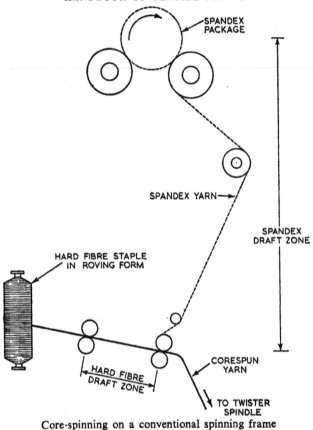

SPANDEX PACKAGE

SPANDEX YARN→

SPANDEX DRAFT ZONE

HARD FIBRE STAPLE IN ROVING FORM

CORESPUN YARN

HARD FIBRE DRAFT ZONE

TO TWISTER SPINDLE

Core-spinning on a conventional spinning frame

in a wide variety of different forms by varying the spinning conditions, type of staple fibre and the subsequent treatment of the yarn.

As they emerge from the spinning frame, core-spun spandex yarns have a tendency to relax as the elastic core recovers from the stretch imposed during spinning. By using adequate tension, however, the core-spun yarn may be held in the stretched state while it is woven or knit into fabrics.

The versatility of core-spun spandex yarns is increased by the fact that they can be heat-set. This may be carried out effectively at temperatures as low as 110°C., with exposure times of about 30 minutes, or at higher temperatures for shorter times.

Heat-setting of yarns or fabrics provides a broad range of effective power, from zero to the maximum. The process is important because it provides a means of adjusting the elastic properties of fabrics to the requirements of the application.

Core-spun spandex yarns are used for making woven and knit goods of many types.

## Woven Fabrics

Fabrics of a wide range of weights, from lawns and batistes to

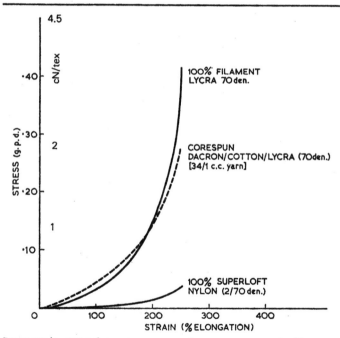

Stress-strain curve for core-spun spandex yarn compared with curves for spandex and stretch nylon.

heavy ducks, have been made from core-spun spandex yarns with polyesters, acrylics, polyamides and their blends with natural fibres. Spandex of 44–78 dtex (40–70 den), for example, may be used for lightweight fabrics, and 156–312 dtex (140–280 den) for poplin or duck weights.

## Knit Fabrics

Two general classes of knit garments are made from core-spun spandex yarns, (1) low-power or easy-stretch fabrics, and (2) high-power or restraining fabrics.

Knitted fabrics are inherently stretchable by reason of the fabric structure, but the recovery characteristics are often poor. Core-spun spandex yarns introduced into these fabrics will supplement the recovery properties of the knitted fabric, adding to the dimensional stability of the knitted garment.

Spandex yarns are widely used in the warp knitting industry, where the combination of high power and low weight is advantageous.

The production of Raschel power-nets for foundation garments is a major outlet for spandex fibres.

## 6. MISCELLANEOUS SYNTHETIC FIBRES

## GLASS FIBRES

Fibres spun from sodium calcium silicate and related substances forming the materials known commercially as glass.

### INTRODUCTION

The knowledge that fibres could be made from glass is probably as old as glass itself. Molten glass is viscous like treacle, and on being touched with anything, it will 'string out' to form a filament when it is drawn away. As glass is in a molten condition during its manufacture, these filaments must have been discovered at an early date. Nature herself produces glass fibres of this type from molten volcanic glass that is spun into fibres by the wind.

The ancient Egyptians were skilled in the art of drawing coarse filaments from rods of glass made by rolling molten glass with a metal bar. The rods, about as thick as a pencil, were reheated at one end, filaments being drawn away from the molten material. These coarse filaments were used to ornament glass vessels in the New Kingdom from about 1600 B.C. They can be regarded as the first synthetic fibres made by man.

The Romans perfected the art of using glass filaments for decorating their glassware, and developed a technique in which a glass vessel was spun on a potter's wheel, drawing off glass filaments from a molten rod so that the filaments were wound continuously on to the rotating vessel.

In the sixteenth and seventeenth centuries, the great Venetian glassmakers mastered the art of glass filament manufacture which they, too, used for decorating their glassware. Even at this time, however, there was no real attempt being made to use glass filaments in textiles, despite the fact that the manufacturing technique of producing the filaments had reached a high degree of perfection.

### Robert Hooke; René-Antoine Ferchault de Réaumur

In his publication *Micrographia* (1665), Robert Hooke, the famous English physicist mentioned the drawing of fine filaments from a heated glass rod, and speculated about the possibility of producing synthetic fibres.

In 1713, the French physicist René-Antoine Ferchault de Réaumur described the making of glass filaments for the manufacture of imitation heron feathers: Two men work together, the one heating a piece of glass over a flame and the other plucking a filament from the softened glass with a glass hook, which he attaches to the rim of a wheel like that used for spinning, and then reels up the filament by revolving this wheel rapidly.

Réaumur speculated as follows on the possibility of using glass filaments in textiles: 'If we knew how to manufacture glass threads as pliable as those in which spiders envelop their eggs, these threads could be interwoven. Even if glass is not pliable, it cannot be said—to use the expression—that it is not "textile".' Réaumur experimented in the production of fine glass filaments, and succeeded in making filaments finer than those of silk. They were, however, very short.

## Incombustible Glass Fibre Lamp Wicks

The comparatively coarse glass fibres which were produced at that time were so brittle as to be of little value in yarns or fabrics, and their applications lay outside the field of textiles. In 1822, for example, the use of glass filaments in the production of incombustible lamp wicks was patented by two British inventors, Alexander and David Gordon.

## Early Spinneret Process

One of the earliest references to the use of glass filaments in fabrics is found in a report issued in 1842. It describes glass filaments and fabrics made by Louis Schwabe, a prominent Manchester silk weaver and supplier to Queen Victoria and the French Court. Schwabe developed his own technique of spinning glass filaments, and his machine was demonstrated at a Manchester meeting of the British Association. It produced glass filaments by drawing molten glass through small orifices, and may thus have been the forerunner of the modern technique of spinning manmade fibres by extrusion of liquids through spinnerets.

## Old Method of Spinning Glass

From the end of the eighteenth century onwards, glass filaments were commonly produced by the old technique of drawing them from the end of heated glass rods. The filaments were wound on to a narrow-rimmed reel (the 'wheel') 102 cm (40 in) in diameter, which was kept turning at about 650 r.p.m. by a crank-driven rope belt. After a time, a skein on the reel was cut through at one point to make many filaments each 3 m (10 ft) long. In 1850, this technique was used at the von Brunfautsche Glasspinnmanufaktur in Vienna for making glass filaments as small as 6 microns diameter.

## Nineteenth-Century Bicomponent Glass Fibre

The glass filament spinning process described above is still used today in scarcely modified form in the cottage industries of the Thuringian and Bohemian forests, to make the medium-fine filaments for 'Angel's Hair' and other Christmas-tree decorations.

'Angel's Hair' is interesting as an early example of the bicomponent fibre which is becoming of great interest today (see page 413). It has a permanent crimp which results from the drawing of a single filament from a rod of lead crystal to which has been welded a strip of plate glass. The durability of the crimp is due to the different coefficients of expansion of the two components of the filament.

During the mid-nineteenth century, the use of glass filaments was restricted to lacework, ladies' hats, wall-coverings, lampshades and the like, and it was not until 1893 that they were first used in clothing fabrics. In that year, Edward Drummond Libbey of Toledo, Ohio, U.S.A., made a dress with a natural silk warp and a glass thread weft (filling) for the American stage star Georgia Cayvan. A similar dress was made for Princess Eulalia of Spain.

One of these dresses, which was shown at the Columbian Exposition in Chicago in 1893, is still on display today in the Toledo Museum of Arts. The glass filaments used in the manufacture of these early dresses and fabrics were of 34 to 51 microns thickness.

A description of the method of manufacturing glass filaments and fabrics at the end of the nineteenth century appears in an 1897 issue of the German periodical *Seide*: 'Some time ago now, a workshop was opened in the Passage Jouffroy in Paris in which (glass filaments) are spun and woven by male and female operatives in full view of the public. Heated in melting ovens to a temperature of 1000 to 1200°C., rods of glass are drawn out into extraordinarily thin filaments and wound on to a wooden cylinder

4 metres (13 feet) in diameter and revolving at 400 r.p.m. When the cylinder is full of spun glass the filament is transferred to small spools which can be inserted in shuttles. The glass filament is combined with one of silk to make the weft in a fabric with a silk or cotton warp. The looms used for this weaving are hand operated and have Jacquard heads. Furniture and clothing materials, umbrella and necktie fabrics are woven in the public sight. But a metre of curtain material, for instance, costs 100 francs!'

### Modern Glass Fibres

The production of glass filaments suitable for textile use requires that they should be flexible enough to stand up to normal wear and tear. This is achieved not by changing the composition of the glass itself, but by making the filaments so fine that they can bend without breaking.

Thick rods of glass will bend only very slightly, but the minimum radius to which a glass rod or fibre may be bent without breaking decreases with decreasing rod or fibre diameter. Yarns consisting of very fine fibres may be bent quite sharply without breaking, as it is unlikely that any individual fibre will be bent to breaking point when the yarn is flexed.

The early methods of producing glass filaments were capable of spinning very fine filaments suitable for textile use. But the processes were not able to produce uniform fine filaments at a cost low enough for commercial textile applications.

From 1912 onwards, methods of spinning glass filaments cheaply were developed, notably by centrifugal processes in which the molten glass is thrown from holes in a metal spinner rotating at high speed. The early high-speed production techniques provided tangled, coarse filaments, however, of diameter greater than 20 microns, and the fibres were not suitable for textile use.

The production of textile-type fibres presupposes that they can be made in diameters smaller than about 12 microns, in adequate quantities and at an economic cost, in the form of a continuous strand, or of a sliver of staple fibres of adequate length, both being suitable for spinning into a yarn.

During the 1930s, this was made possible by the development of glass fibre production processes, notably by Owens-Corning Fiberglass, an American company formed by the Owens Illinois and Corning Glass Companies. This firm developed three processes before 1939:

(1) *Continuous Filament Process.* In this process, strands of glass fibre of very great length (e.g. several thousands of yards) were drawn from platinum bushings and used for twisting and doubling into yarns.

(2) *Long Staple Fibre Process.* In this process, streams of glass were allowed to flow from a bushing baseplate into the slot of a steam blower. The streams were drawn into long (76 cm; 30 in) single fibres which were collected on a rotating perforated drum which was kept under suction. The web so formed was drawn off after a quarter-turn of the drum and drafted into a sliver of more or less parallel fibres. The sliver was spun into coarse wool-like yarns and fabrics.

(3) *Short Staple Fibre Process.* A process similar to (2) was developed for the manufacture of large quantities of short (13–100 mm; ½–4 in) fibres. Streams of molten glass were attenuated by the blowing of steam or air, but this time through the use of higher pressure the fibres were drawn and ripped into short lengths due to the strong turbulence of the blowing medium.

In processes (1) and (2), individual electrically-heated melting units or bushings were used, but in process (3) metal bushings were fixed to the forehearths of big glass-melting tanks. It is only in comparatively recent times (1950s) that textile bushings have been fixed to glass-tanks for the production of continuous filament glass yarns.

From 1938 onwards, the traditional process of drawing glass filaments from rods was also perfected by Glas-Wolle K.G.W. Schuller and Co. (Germany) and other firms. Special processes for the production of superfine filaments with diameters of 1 to 3 microns have since been developed, e.g. by Owens-Corning Fiberglass and by the S.A. des Manufactures des Glaces de Saint-Gobain, Société du Verre Textile.

These modern methods of spinning glass fibres have opened the way to the manufacture of glass yarns and fabrics suitable for a wide variety of industrial, furnishing and apparel uses. The development of new types of glass has increased the versatility of glass fibre, and extended still further its potential market.

Between 1937 and 1967, the glass fibre industry expanded three-fold, and it continues to increase. Initially, glass fibres were used largely in insulation and filtration applications. The introduction of glass-fibre-reinforced plastics brought a new market which is still expanding rapidly, and the acceptance of glass fibre into genuine textile applications is making steady headway.

## TYPES OF GLASS FIBRE

Glass is made in a wide variety of different compositions, and fibre may be spun from virtually any glass to provide material suited to particular applications.

In general, there are two main types of textile glass fibre in large-scale commercial production; 'E' glass and 'C' glass. Both types are similar, but each is designed to serve to advantage in specific end-uses.

'*E*' *Glass* is a boro-silicate glass of low alkali content. It has a very high resistance to attack by moisture, and has superior electrical characteristics and high heat resistance.

'*C*' *Glass* has superior resistance to corrosion by a wide range of chemicals, including acids and alkalis. It is widely used for applications where such resistance is required, e.g. in chemical filtration.

Glass wool fibre for non-textile applications is also spun from glasses of other compositions. Continuous filament is spun from 'A' glass (alkali glass; window glass).

## FORMS OF GLASS FIBRE AVAILABLE

Glass fibre is produced in two basic forms; continuous filament and staple fibre.

### Continuous Filament

Continuous filament glass fibres are made usually from 'E' glass. They are produced in a range of filament diameters, with an upper limit in the region of 12 microns (for textile applications). The increasing brittleness of filaments of diameter greater than this renders them of little value as textile fibres. Textile glass filaments with a diameter as small as 2.5 microns are commercially available ('Beta' fibres) and filaments of diameter 1 micron and less may be made for special applications.

Continuous filaments are produced in the form of strands containing many individual filaments—e.g. from 51 to 4,000— depending on specific requirements.

## Staple Fibre

Glass staple fibres are made usually from 'C' glass. They are produced in a range of filament counts and lengths, e.g. from 20–38 cm (8–15 in).

### Commercial Products

Glass continuous filament strands and staple fibre are commonly marketed by the manufacturers in a variety of made-up forms.

## Continuous Filament Yarn

This is made by twisting and/or plying a number of continuous filament strands. The number which are twisted or plied together affect the yarn's strength, diameter and flexibility.

Continuous filament yarns are commonly sized (e.g. 2 per cent starch-oil) to facilitate subsequent handling and fabrication operations. They are also furnished in a variety of treated forms, e.g. dyed, waxed, pre-saturated, vinyl-coated; they may also be combined with other textile fibres.

Continuous filament yarns are fabricated into cords and sewing threads. They are widely used, for example, as reinforcement in electrical insulation materials, wire and cable, plastics, etc., as filtration materials, and in decorative fabrics.

## Staple Fibre Yarn

Yarns spun from staple fibre are woven into fabrics used for wet and dry filtration operations. They are used also as reinforcement in conveyor belts handling hot materials, and as braids in electrical insulation applications.

Staple fibre yarn is commonly supplied in combination with flame-proof waxes, lacquers and commercial insulation varnishes.

## Staple Sliver

This is a low-cost wadding material. It is used, for example, in aquarium filtration applications. Woven into fabric, it provides electrical turbine generator thermal blankets.

## Bulk Staple Fibre

A fluffy, bulky fibre that is used for air and liquid filtration, pharmaceutical wadding and dunnage.

## Fine Fibres—Unbonded

A mass of soft, fluffy fibre, ranging in diameter from $\frac{1}{2}$ to 3 microns. They are used for 'all glass' papers and high efficiency filtration applications.

## Bonded Staple Sliver

A ribbon of parallel fibres bonded together with an alkyd resin. It is used as a filler, and also as an outer braid for many electrical cable applications.

## Cordage

Cordage is made by twisting, plying and cabling continuous filament yarns. It is commonly available in a variety of diameters ranging, for example, from 0.4–4 mm (1/64–10/64 in). It may be untreated, or treated with various coatings.

Cords are used in cable wrapping seals, reinforcement of high pressure steam hose, etc.

## Sewing Thread

Sewing threads are made from very fine continuous filament yarns. They have the highest tensile strength, flexibility and resistance to high temperatures of any textile sewing thread.

## Scrim

This is a low cost, non-woven reinforcement fabric made from continuous filament yarn in an open mesh construction. It is made by coating length-wise yarns with a hot-melt adhesive and applying cross-yarns while the adhesive is still molten. Asphalt base and polyethylene adhesives are used, for example, in making scrim.

Glass fibre scrim is moisture-resistant and rotproof. The glass yarns provide exceptional tear and tensile strength in two directions, and materials reinforced with glass fibre scrim have a very high dimensional stability.

Glass fibre scrim is available in a wide variety of sizes. It can be combined—in the laminating and extruding processes—with paper, film and foil. The resulting products are used extensively in the packaging industry.

## Mat

A non-woven material used primarily in plastics reinforcement. It is distributed in a random pattern to ensure maximum uniformity in the finished laminate. Mats are treated with various bonding resins to provide optimum compatibility with the laminating resin, and the desired handling and fabrication characteristics.

There are several types of mat available:

*Chopped Strand Mat* consists of chopped strands bonded together by a resin. It is used in most applications of reinforced plastics usually when large and complicated shapes are to be made. Examples are corrugated sheets, boats, motor car bodies, building components, containers and the like.

*Continuous Strand Mat* consists of uncut continuous strands held together in sheet form by a bonding resin. It is used mainly in matched metal die moulds, where relatively deep and complex contours require maximum 'draw' characteristics.

*Needled Mat* consists of cut strands which are needled to a carrier tissue. It is used where a particularly bulky reinforcement mat is wanted.

*Bonded Mat or Staple Tissue* is a thin highly porous mat made from monofilaments of type 'C' glass arranged in a veil-like pattern. This mat is used mostly in the reinforcement of bitumen which is applied to buried pipes as protection against corrosion. It is also used as a surfacing mat on top of other reinforcement materials to produce a smooth resin-rich surface.

## Roving

This is a low-cost, high-strength reinforcement material made by gathering a number of continuous filament strands and winding them into a cylindrical package. It is available in two forms: continuous strand roving and spun strand roving.

647

*Continuous Strand Roving* consists of parallel strands which provide high unidirectional strength.

*Spun Strand Roving.* This roving consists of one or more continuous strands looped back and forth upon themselves. It is held together by twisting it slightly and applying a cohesive sizing. The roving has less unidirectional strength than continuous strand roving, but it possesses greater bulking characteristics.

Continuous strand rovings are used in processes where they are chopped into short lengths. These chopped strands give a non-directional reinforcement to plastics similar to chopped mats.

Rovings can also be woven into fabrics, continuous filament rovings giving higher strengths and spun rovings better interlaminar adhesion.

Rods and other profiles can be produced by drawing resin-impregnated continuous and spun strand rovings through dies and a curing oven. Rovings may also be wound on to formers which can be rotated to provide cylindrical vessels and similar bodies which possess great strength. (Filament Winding).

### Chopped Strands

These are available in a wide variety of fibre lengths. They are used as a reinforcement for resins, and for reinforcing putties, caulking compounds and foam rubber. They are also used for gypsum wallboard reinforcement.

### Milled Fibres

These are made from hammer-milled continuous filament strands. They are used where shorter fibre lengths are required for reinforcement applications.

### Fabrics and Tapes

These are made from continuous filament yarns, rovings and staple yarns, by weaving on conventional looms.

'Broad fabrics' refer to woven glass fabrics 46 cm (18 in) or wider. 'Narrow fabrics' or tapes refer to woven glass that ranges in width from 6.4–200 mm (¼ – 8 in).

Fabrics and tapes may be processed through moulding, laminating and coating techniques. They are used in applications requiring the most exact control over thickness, weight and strength, including industrial filtration, electrical insulation and decorative fabrics.

## Dough Moulding Compound

This is a mixture of short glass fibre strands with resin and fillers, which is dried but not cured. The compound is finally cured by the laminator in matched metal dies with the application of heat.

## NOMENCLATURE

Glasses used for the manufacture of fibres are either soda-lime-silicates or boro-silicates (such as 'E' or 'C' glass). Other materials can be added to the silica network to modify the properties of the resulting glass.

For this reason, it is not possible to give a general chemical formula for glass fibres, but each glass composition from which they are drawn may be accurately defined in chemical terms by the ingredients which go into the melt.

### Federal Trade Commission Definition

The generic term *glass* was adopted by the U.S. Federal Trade Commission for fibres of this type, the official definition being as follows:

*Glass.* A manufactured fibre in which the fibre-forming substance is glass.

## PRODUCTION

### Glass Manufacture

Glasses of many different compositions are made by the glass industry, the type produced being selected to suit the end-uses for which it is required. Silica sand (silica) and limestone (calcium carbonate) may be regarded as basic ingredients, to which are added varying amounts of other materials such as soda ash (sodium carbonate), potash (potassium carbonate), aluminium hydroxide or alumina (aluminium oxide), magnesia (magnesium oxide) or boric oxide.

The glasses commonly used in making textile fibres—'E' glasses and 'C' glasses—are made from compositions of the following type:

'E' Glass

| Ingredient | Amount (% approx.) |
| --- | --- |
| Silica | 52.5–53.5 |
| Lime | 16.5–17.5 |
| Magnesia | 4.5–5.5 |
| Alumina | 14.5 |
| Soda; potash | Less than 1.0 |
| Boric oxide | 10.0–10.6 |

'C' Glass

| Ingredient | Amount (% approx.) |
| --- | --- |
| Silica | 62.0–65.0 |
| Lime | 6.0 |
| Potash | 1.0–3.0 |
| Alumina | 1.0 |
| Soda | 11.0–15.0 |
| Boric oxide | 3.0–4.0 |

The ingredients are charged into a furnace, where they are fused at high temperature, forming molten glass. Filaments may be spun direct from this melt, or the glass may be formed into marbles of 16 mm (5/8 in) diameter. The marbles are inspected, and any that contain impurities are discarded. The others are then passed to the spinning machines, fiberizing units or 'bushings' as they are called in the glass fibre industry.

### Fiberising

(A) *Continuous Filament Process*

Continuous filament strands are produced by allowing molten glass to flow through perforated tips on the baseplate of a platinum melter ('bushing'), drawing away a stream of glass from each tip to form a fibre, and attaching the fibres to a high-speed winding collet. The bushings may be mounted on glass-melting tanks, or they may form the basis of independent fiberizing units, being fed with glass by remelting glass marbles.

As the 50 or more filaments are drawn from the orifices, they are brought together into a strand. A lubricating size is applied to facilitate subsequent processing.

If the strand is to be used for reinforcement of plastics, the lubricant or size will be chosen to be compatible with polyester or epoxide laminating resins, and the strand may be converted into various forms of chopped strand mats, rovings or chopped strands.

If the strands are intended for use in textile weaving, they may be processed through conventional textile processes into suitable yarns. These yarns are available in standard basic counts e.g. of 5.5, 11, 22, 33 and 66 Tex.

The diameter of the filaments produced in this way depends upon the rate at which the glass is drawn from the orifices, the size of the orifices, and the viscosity of the melt. Production rates are extremely high, reaching 3 km (1.9 miles) a minute and more. One marble will provide as much as 160 km (100 miles) of filament.

## (B) *Staple Fibre*

There are a number of methods of producing glass staple fibre, the most important of which may be considered as the Centrifugal, Jet and Rod Drawing Processes.

### *Centrifugal Process*

In this process, molten glass is thrown out of holes in the base of a metal spinner rotating at high speed. Fibres of this type are bonded into a web, and are used in heat and sound insulation. This technique is not generally used for producing textile grade glass fibres.

### *Jet Process* (Steam Blowing, Batwool or Staplefibre Process)

Staple fibre is produced by a process in which molten glass flows under gravity through holes in the base of platinum bushings, as in the production of continuous filament. The streams of glass are then drawn into fibres by the action of high-speed streams of turbulent gas or steam. The solidified fibres are broken by the turbulence into staples of up to 38 cm (15 in) and collected into a web on a revolving vacuum drum. The web may then be guided from the drum and drafted into the form of glass staple sliver. This is then processed on conventional textile machines for weaving into glass fibre staple fabrics.

## Rod Drawing Process

This is a modern development of the traditional process for making glass filaments. In a typical modern machine, 125 glass rods, usually about 4 mm. (0.1575 inch) in diameter, are mounted vertically and adjacent to one another in a so-called 'spinning frame'. The rods are kept moving slowly downwards, and are simultaneously melted at a temperature of about 1200°C., either by individual, movable and adjustable gas burners or by electric heating coils in a fireclay chamber. Drops of glass fall away from the ends of the rods, drawing glass filaments after them. These are led via an inclined plane on to a rapidly revolving cylinder (usually 101.6 cm; 40 in wide, and rotating at 800 r.p.m., so taking up 2,475 m; 2,750 yd per minute!) on to which they are wound next to, but independent of, one another. Lateral movement of the spinning frame next forms a glass filament web on the cylinder, from which it is at intervals cut and after doubling formed by horizontal doffing and repeated slight stretching into glass fibre slubbing.

Alternatively, the filaments are merely drawn off and out by the cylinder, not being taken up but being doffed, after about three-quarters of a rotation, by a combined lifting and cutting apparatus. Their flow direction is then deflected and they are tossed on to a stationary sieve and collected and doffed as a glass fibre sliver composed of fibres of unequal length.

## PROCESSING

### Sizing

The application of a suitable size is important when glass fibres are to be subjected to textile processing. The size must lubricate the fibres to minimize the effect of fibre-fibre friction, and hold the individual filaments together in the strand. At the same time, the

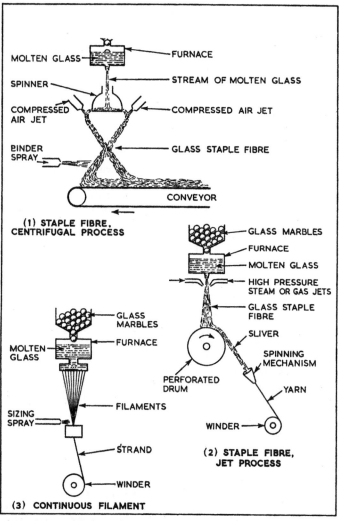

*Glass Fibre Production.* Three important methods of making glass fibre are shown above.

(1) Staple Fibre. Centrifugal Process.
(2) Staple Fibre. Jet Process.
(3) Continuous Filament.

653

size must not make the strands adhere in the package.

Dextrinized starch gum, gelatine, polyvinyl alcohol, hydrogenated vegetable oils and non-ionic detergents are commonly used.

## Heat-Cleaning

Glass fabrics and tapes are usually combined with other materials such as coatings and resins. In order to ensure that the union of glass with these materials is effective, it is usually necessary to remove the sizing that was applied to the yarn during the manufacture of the goods. This may be done by heat-cleaning at high temperatures.

There are various techniques and conditions under which heat-cleaning may be carried out. Continuous heat-cleaning, for example, may be achieved by passing fabric through an oven at about 650°C. The organic material is burned away to leave the white fabric.

For some applications, such as the use of glass fabrics in the production of melamine laminates, a partial heat-cleaning may be carried out, Some of the organic matter is caramelized, the starch being converted to carbon which remains on the fabric.

## Heat-Setting

Glass is a thermoplastic material, and the strains introduced during processing and production of yarns and fabrics may be relieved by treatment at an adequate temperature. In the Coronizing treatment, for example (see page 655), dimensional stability is conferred on glass fabrics by a heat-treatment at 650°C. This is a form of heat-setting but it is not a true heat-setting such as that which occurs in the heat-treatment of a partially crystalline fibre.

## Dyeing

Glass fibres absorb only a negligible amount of water, and their affinity for dyestuffs is virtually nil. It is not possible to colour glass by using the normal dyeing techniques. Many specialized processes have been developed, however, for the colouration of glass fibre fabrics. An early technique was to mix pigment into the molten glass before spinning, but this has not attained real commercial importance, Most modern processes rely upon the application of some sort of coating to the fibre, the coating being either pigmentèd before application or coloured subsequently by dyeing.

### (1) *Pigmentation of Molten Glass*

The technique of adding finely-dispersed pigments to the melt before spinning, as used in the production of many types of spun-dyed synthetic fibres, may be used in the colouration of glass fibre.

In view of the high temperature at which glass fibre is spun—around 1,200°C.—it is necessary to use inorganic pigments. A range of pastel shades may be obtained satisfactorily in this way.

Production of spun-dyed glass fibre suffers from the drawbacks inherent in this process. The manufacturer must allocate spinning machines to the production of specific colours, or be prepared for tedious and expensive cleaning operations in changing from one colour to another. Also, each additional colour increases the storage problems, and the range may be restricted on this account.

### (2) *Coronization*

Coronization is a system of treatment of glass fabric which brings about significant changes in the characteristics of the cloth. It makes possible, too, the application of colour in an effective way.

The colouration of glass fabric by Coronization takes place in the following stages.

(a) Glass fabric is passed through a dispersion of colloidal silica.

(b) The silica-impregnated fabric is passed through an oven at 650°C. for 5–15 seconds. Sizes and other organic materials are burned away, leaving only the fine particles of silica adhering to the surface of the glass fibres.

During this heat-treatment, the fibres are relaxed and the weave is set, thus establishing the dimensional stability of the fabric.

(c) The heat-treated fabric is passed through a bath containing a resin dispersion, commonly a butadiene-acrylonitrile copolymer, and coloured pigment. The fabric then passes through a curing oven at 160°C. The resin is cured on the cloth, binding the pigment to the fibres.

(d) The pigmented cloth now passes through another padding process, in which a solution of stearatochromic chloride is applied. This is followed by drying in another oven at 160°C.

Stearatochromic chloride has a high affinity for glass, and it establishes a powerful bond between the glass and the resin coating.

Coronized fabrics display increased abrasion resistance, crease resistance and water repellency. The fabric acquires a softer handle, and becomes more flexible, Ageing and sunlight resistance are improved; fabrics can be washed, and dried by rolling in a towel. They do not need ironing.

### Printing

The Coronizing process may be modified to permit the screen printing of glass fibre fabrics. The stages in the process are as follows:

(a) Silica treatment.

(b) Heat treatment at 650°C.

(c) Application of cationic softening agent to lubricate and protect the fibres.

(d) Screen printing, using a printing paste containing pigment and resin latex which has been thickened with alginate.

(e) Curing at 160°C. for 5 minutes.

(f) Padding in stearatochromic chloride, followed by drying at 120°C. for 15 minutes.

Fabrics printed in this way have good washfastness and resistance to crocking. They are crease resistant and water repellent, and have good drape and handle. They may be washed effectively, and need no ironing.

### Finishing

A wide variety of finishing treatments is available for use with glass fibre goods. Most of these are intended to increase the efficiency of glass goods when used as reinforcement in plastics, and are not therefore of immediate interest in the consideration of glass as a textile fibre.

There are, however, some finishing treatments which are of great importance in the effect they have on glass in textile applications.

## Coronizing

Coronizing is a finishing treatment which has a significant effect on glass fibre and fabrics used for textile applications. It increases the resistance to abrasion, water repellency, crease resistance and flexibility of glass fabrics, and improves the handle to an impressive degree. It also makes possible the effective colouration of glass material.

## Corrosion Resistant Finish

Dispersions of polytetrafluoroethylene are applied to glass fabrics followed by drying and the application of pressure at temperatures in the region of the P.T.F.E. sintering temperature. Cloth produced in this way is used for filtration fabrics that must withstand corrosive conditions, pump diaphragms and the like.

## Antistatic Finish

Glass fabrics are heat-cleaned and then dipped in compositions containing potassium isobutyl polysiloxanolate. The treated material is heated at 150°C. for up to 30 minutes, dried and rinsed with dilute hydrochloric acid. It is then dried and washed in water.

## Strengthening Finishes

Glass fabrics are coated with a resinophobic material that prevents adhesion of resinous materials to the surface of the glass fibre. The treated cloth is then coated with a resin finish, which penetrates into the interstices of the fabric without adhering to the individual strands of fibre. The freedom of movement which is left to the fibres improves the tearing and bursting strengths and the flex resistance of fabrics to be used in heavy-duty fabrics such as awnings, tarpaulins and tent materials.

## STRUCTURE AND PROPERTIES

### Fine Structure and Appearance

### Molecular Structure

The term 'glass' describes a range of materials made by fusing together one or more of the oxides of silicon, boron or phosphorus, with certain basic oxides, e.g. sodium, potassium, magnesium, calcium, and cooling the product rapidly to prevent crystallization taking place.

Glass is thus a mixture of silicates, some water-soluble, like sodium and potassium silicates, and others water-insoluble, like calcium and magnesium silicates. It is an amorphous material, in which the atoms do not take up the ordered positions associated with regions of crystallinity, and it differs in this respect from those synthetic organic polymers which produce fibres of partially crystalline structure. It is a supercooled liquid with such a high viscosity that no perceptible flow takes place.

*Fibre Form*

Glass fibres are smooth-surfaced and commonly of circular cross-section. They are transparent.

**Tenacity**

Dry: 53–64.5 cN/tex (6.0–7.3 g/den); wet: 34.4–41.5 cN/tex (3.9–4.7 g/den)
Std. loop: 8.0–9.7 cN/tex (0.9–1.1 g/den)
Std. Knot: 15.9–19.4 cN/tex (1.8–2.2 g/den)

**Tensile Strength**

14,000–15,400 km/cm$^2$ (200,000–220,000 lb/in$^2$).

**Elongation** (per cent)

Dry: 3.0–4.0 ('C' type: 4.5)
Wet: 2.5–3.5

**Elastic Recovery**

100 per cent

**Poisson's Ratio**

0.22

**Hysteresis**

None

**Creep**

None

**Average Stiffness**

2,843 cN/tex (322 g/den)

**Specific Gravity**

2.54 ('C' type: 2.49)

**Hardness**

(Moh Scale) 6.5

**Effect of Moisture**

Absorbency: up to 0.3 per cent (surface)
Regain: Nil

**Thermal Properties**

|                          | 'E' Type | 'C' Type |
|--------------------------|----------|----------|
| Softening point (°C.)    | 846      | 752      |
| Strain point (°C.)       | 507      | 435      |
| Annealing point (°C.)    | 657      | 585      |
| Specific heat            | 0.19     |          |

Flammability: Glass fibre does not burn.

**Effect of Age**

None.

**Effect of Sunlight**

None.

**Chemical Properties**

*Acids*

Glass fibres are resistant to acids of normal strength and under ordinary conditions. They are attacked by hydrofluoric, concentrated sulphuric or hydrochloric, and hot phosphoric acids

*Alkalis*

Hot solutions of weak alkalis, and cold solutions of strong alkalis will attack glass, causing deterioration and disintegration.

*General*

Highly resistant to all chemicals in common use.

**Effect of Organic Solvents**

Glass fibre is not attacked by organic solvents, but the size on the yarn may be attacked.

**Insects**

Not attacked.

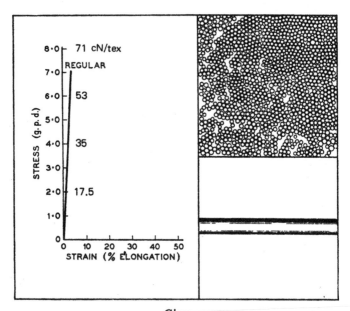

*Glass*

### Micro-organisms

Not attacked.

### Electrical Properties

The following values were obtained from measurements on bulk glass:

|  | 'E' Glass | 'C' Glass |
|---|---|---|
| Dielectric strength (volts/mil) | Up to 2,800 | — |
| Dielectric constant (22°C.) |  |  |
| $10^2$ cycles | 6.43 | — |
| $10^4$ cycles | — | 7.14 |
| $10^{10}$ cycles | 6.11 | 6.79 |
| Power factor (22°C.) |  |  |
| $10^2$ cycles | 0.0042 | — |
| $10^4$ cycles | — | 0.009 |
| $10^{10}$ cycles | 0.0060 | 0.013 |
| Volume resistivity – solid glass (ohms/c.c.) |  |  |
| 22°C. | $2-5 \times 10^{12}$ | — |
| 715°C. | $10^7$ | — |
| 788°C. | $10^6$ | — |
| 871°C. | $10^5$ | — |
| 982°C. | $10^4$ | — |
| 1093°C. | $10^3$ | — |
| 1260°C. | $10^2$ | — |

### Coefficient of Friction

With glass: 1.0 for clean glass.

### Optical Properties

Index of refraction (at 550 millimicrons; 0°C.): 1.549 ('C' Type; 1.541).
Clarity: transparent.
Ultra-violet transmission: opaque.

### Acoustical Properties

Velocity of sound: 5,786.4 m/sec (18,000 ft/sec)
Acoustic impedance (g./cm²/sec.): $1.4 \times 10^6$
Velocity of crack propagation: 1,603.25 m/sec (5,260 ft/sec)

## GLASS FIBRE IN USE

### General Characteristics

*Physical Properties*

Glass is a heavy fibre, with a specific gravity of the order of aluminium. It has good transparency.

The moisture absorbency of glass is negligible, and the fibre shows no swelling or shrinkage. This is a useful characteristic in applications where dimensional stability in the presence of water is desirable, and in electrical applications. It is detrimental in apparel applications, however, where absorption of moisture is a useful attribute.

Water affects the tensile properties of glass fibre, bringing about a reduction in tenacity after a very short time. This is due to the leaching out of some of the soluble materials from the glass, including alkali silicates. 'E' glass, with its low alkali content, is better than 'C' glass in this respect.

The high coefficient of friction of glass against glass, taken in conjunction with the high flexibility of fine glass filaments, makes for poor abrasion resistance. The movement of filament against filament during use brings about fracture of the filaments, and the creation of a hairy fabric. The use of suitable lubricants and finishes minimizes the effects; metal coatings, for example, may be applied to glass fibres to reduce surface friction.

*Mechanical Properties*

Glass fibre has the highest strength-to-weight ratio of any fibre, and one of the lowest elongations. These characteristics are useful in applications requiring high dimensional stability, such as the use of glass fibre as reinforcement in plastics. The low elongation is not a useful attribute in apparel fabrics, however, where resiliency is an important factor.

Glass exhibits almost perfect elasticity, returning instantly and completely to its original dimensions on release from strain. This elasticity is exhibited over a very small range, as indicated by the low elongation of the fibre.

*Thermal Properties*

Glass has an excellent resistance to the effects of heat over a wide temperature range. Fabrics show an increase in strength up to about 205°C., after which the strength and flexibility begin to fall. At 370°C., glass filaments retain 50 per cent of their original strength; at 538°C. they retain about 25 per cent. The effect of temperature on glass depends greatly upon the composition of the glass.

Glass is completely non-flammable, and this is one of the most important factors in its textile applications. Glass fabrics are used where resistance to the spread of flame is of overriding importance.

The high heat conductivity of glass is a useful attribute in electrical insulation applications, where close-woven glass fabrics dissipate heat rapidly. Despite this high heat conductivity, however, glass fibre finds an important outlet as heat insulation material. In this case, it is used in the form of mats and waddings, where insulation is provided by the entrapped air.

### Electrical Properties

Glass is an excellent electrical insulator, and glass fibre finds important outlets in the electrical field. 'E' glass is designed specially for this application.

### Chemical and Biological Properties

Glass has a high resistance to most chemicals. It is, however, attacked by alkalis, which disintegrate it. Glass fibres may also be attacked by some strong mineral acids and by phosphoric acid.

The resistance of glass to biological degradation is complete.

### Washing

Glass fabrics wash easily, but they should be subjected to a minimum of mechanical action. They are best hand-washed, without rubbing, squeezing or drying. Mechanical washing may be carried out carefully.

### Drying

Glass fabrics are best drip dried. They should not be tumble dried.

### Ironing

In general, ironing of glass fibre fabrics is unnecessary. Ironing may be carried out, however, using a 'cotton' setting.

### Dry Cleaning

Glass fabrics may be dry cleaned, but great care must be taken to avoid mechanical action.

### End-Uses

*Insulation*

Glass fibres are widely used for electrical, thermal and acoustical insulation purposes. They are less bulky and more efficient in many respects than other insulators.

In electrical insulation, glass fibre offers a combination of high strength, non-flammability, excellent corrosion resistance, good dielectric strength, high heat conductivity and excellent moisture resistance. End-uses include rotating equipment, transformers, switchgear, wire and cable insulation, and other applications of this type.

For thermal insulation, glass fibre is used in the form of mats and waddings, where the high heat conductivity of the glass itself is a factor of minor importance compared with the insulation provided by the trapped air. Glass wool is used, for example, in the insulation of houses, and for lagging hot water and steam pipes.

*Reinforcement in Plastics*

The use of glass fibre as reinforcement in plastics has become the largest single end-use for textile-type fibres. Glass fibres reinforce plastics in the same way as steel reinforces concrete. The high strength of the fibres, coupled with their resistance to stretch, ensure a product of great impact strength and high dimensional stability. The non-flammability of the fibres, and their resistance to corrosion and biological attack, have added to their efficiency in this application.

Glass-fibre-reinforced plastics are used in building boat hulls, in car bodies, aircraft, radio and television cabinets, hot air ducts, flexible tubing, petrol (gasoline) storage tanks, building panelling, translucent sheet, furniture and innumerable other end-uses where high strength, durability, light weight and resistance to deterioration are important factors.

*Industrial Filtration*

Glass fibre and fabrics are used for filtering gases and liquids in many industrial operations.

## Tyre Cords

Glass yarns used as reinforcement in radial ply tyres result in better miles-per-gallon, greater tread life, improved cornering and cooler running than tyres reinforced with other types of yarn. This is a major market for glass fibres, especially in the U.S.A.

## Belting

Glass fibre is used as reinforcement in industrial belting, including conveyor belts for handling hot materials and driving belts, particularly toothed timing belts for car engines and industrial machinery.

## Textiles

Glass fibres have much to recommend them as textile fibres. They are strong and stable to moisture, heat and other influences; they are not easily soiled or stained, and they can be cleaned readily in soap and water; they are non-flammable. But the deficiencies of glass as an apparel textile fibre greatly outweigh its advantages, and there is little prospect at the present time of glass fibre becoming a really important fibre for general textile use.

Glass fabrics have a poor resistance to abrasion, the filaments breaking as they rub against each other during use. Glass fibres have only a very small elongation, and do not have the 'give' that is so desirable a characteristic of a textile fibre. They do not absorb moisture, and the fabrics are uncomfortable against the skin. They cannot be dyed by normal techniques.

Despite these drawbacks to the use of glass fibres for general textile applications, the fibres have established important end-uses. They are made into fireproof fabrics for curtains and draperies in cinemas, theatres and other public buildings, and the newer types of fibre (e.g. Beta fibre) are now being used increasingly in domestic applications of this type. Tablecloths made from glass fibre, for example, are not damaged by cigarettes left burning on them.

Glass fibres are used for apparel fabrics in special applications such as car racing suits and suits for astronauts.

## ALUMINIUM SILICATE FIBRES

Fibres spun from aluminium silicate, with or without the addition of minor amounts of other materials:

$$Al_2O_3.SiO_2$$

### INTRODUCTION

Glass, asbestos and mineral wool fibres provide excellent service in specialized applications for fibres up to temperatures of about 540°C. At higher temperatures, they tend to become ineffective, and there is a need for special-purpose fibres which can withstand temperatures above this level. A number of ceramic-type materials have been spun into fibres for this purpose, among them aluminium silicate.

Aluminium silicate fibres were developed originally for use in jet engines, providing a high-temperature-resistant (540–1,260°C.), light-weight, strong, fibrous material of low thermal conductivity. These fibres have since become of great value in many applications outside the jet engine field, including high temperature filters, packings, gaskets, insulation (thermal, electrical and acoustical) and the like.

The manufacturing process for producing aluminium silicate fibres is relatively simple, and the raw materials – alumina and silica – are cheap and readily available. These fibres can be produced at lower cost than other types of high-temperature-resisting fibres, such as the silica fibres.

### TYPES OF ALUMINIUM SILICATE FIBRE

Aluminium silicate fibres are produced typically as short staple fibre, or as long staple, textile grade fibres. The long staple type is produced as fine or medium fibres to meet particular requirements.

*Short staple fibres* are made from an aluminium oxide-silica melt which contains small amounts of sodium and boron (sodium borate) to help in controlling the melt. Staple length is 6.35–25.4 mm (¼–1 in); fibre diameter 0.5–10μ.

*Long staple fibres* of controlled diameters are produced from an aluminium oxide-silica melt which contains a small proportion of zirconia. Staple length 12.7–254 mm (½–10 in); fibre diameters: fine, 1–8 (mean 5) $\mu$; medium coarse, 3–25 (mean 10) $\mu$.

## FORMS AVAILABLE

Aluminium silicate fibre is available in a number of forms, including short staple bulk fibre; paper; long staple bulk; both short and long staple blanket; rope; yarn; roving; tape; coating cement; chopped fibre; washed fibre; block; board; tubes; milled fibre; felt; cloth; tubular shapes; cord; hollow braid; wicking; square braid; tamping mix; laminates; preforms; hydraulic setting compositions. Non-rigid forms are capable of being made into rigid or resilient forms by addition of a liquid, refractory hardening agent termed Rigidiser.

## NOMENCLATURE

### Ceramic Fibres

Aluminium silicate fibres are related chemically to the silicate mixtures that form the basis of ceramic (pottery) materials, and they are often included in a general classification called *ceramic fibres*. This is an imprecise and unsatisfactory term which is, however, widely used.

### Inorganic Fibres

Aluminium silicate is an inorganic material, and fibres spun from it are *inorganic fibres*.

### High Temperature Fibres

The growing importance of fibres capable of retaining useful properties at elevated temperatures has led to the use of a general descriptive term *high temperature fibres*. This term is commonly applied to fibres which can be used above about 400°C., and aluminium silicate fibres qualify for inclusion in the group.

667

## PRODUCTION

### Aluminium Silicate

A 1:1 mixture of alumina and silica, containing a small proportion of sodium borate (for short staple) or zirconia (for long staple) is fused in an electric furnace.

### Fibre Production

Molten aluminium silicate is poured from the furnace in a small stream. As it falls, the stream of liquid meets a blast of compressed air or steam. The stream of aluminium silicate is disintegrated into fine droplets or particles, and these are attenuated into fine fibres as they are blown through the air. The fibres are collected on a mesh screen, forming a mat of fibre which is collected for processing.

## STRUCTURE AND PROPERTIES

### Fine Structure and Appearance

#### *Molecular Structure*

Aluminium silicate fibres are amorphous in structure, similar in this respect to glass. If they are held at temperatures above 1,000°C. and below the melting point for long periods of time, the material undergoes devitrification; the amorphous material acquires an ordered crystalline pattern. This change makes the fibre more brittle, lower in tensile strength and more easily abraded. There is also some evidence of shrinkage.

#### *Fibre Form*

Aluminium silicate fibres are white. They are smooth-surfaced and of round cross-section.

### Tensile Strength

Long staple, fine: 12,600 kg/cm$^2$ (180,000 lb/in$^2$)
medium coarse: 8,106 kg/cm$^2$ (115,800 lb/in$^2$)
coarse: 3,500 kg/cm$^2$ (50,000 lb/in$^2$) (est.)

**Specific Gravity**

2.73.

**Thermal Properties**

*Melting point*: above 1,760°C.

*Maximum Use Temperature*: about 1,260°C.
Aluminium silicate fibre shows no tendency to melt or sinter even at temperatures up to 1,370°C., but there is a gradual change from the amorphous glassy structure to the crystalline structure at temperatures above 1,000°C., the change proceeding faster at higher temperatures.

*Specific Heat*: Bulk fibre, measured at 60°C., per ASTM designation C–351: 0.20.

*Thermal Conductivity*: the diagram on page 670 shows curves of thermal conductivity against mean temperatures for four fabricated forms.

**Chemical Properties**

*Acids*

Good resistance to most common acids under normal conditions, but attacked by hydrofluoric acid and phosphoric acid.

*Alkalis*

Good resistance to dilute alkalis at normal temperature, but attacked by strong alkalis.

*Metals*

The fibres exhibit non-wetting characteristics to molten aluminium, zinc, etc.; they are frequently used for the containment of these molten metals.

*General*

Good general resistance to most common chemicals. Resists both oxidizing and reducing atmospheres. No significant change on exposure to hydrogen at 1,300°C.

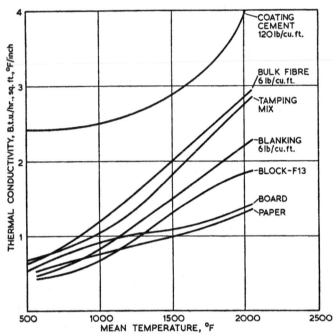

*Aluminium Silicate Fibre; Thermal Conductivity.* This graph shows the thermal conductivity of different forms of aluminium silicate fibre ('Fiberfrax') plotted against temperature.

### Effect of Organic Solvents

The fibres are wetted by organic solvents, but resist attack by most of them.

## ALUMINIUM SILICATE FIBRES IN USE

### General Characteristics

Aluminium silicate fibres are used largely for insulation purposes, filling a gap that exists from about 400°C., where glass, mineral wool and asbestos are often inadequate, to about 1,260°C. It can be used for short exposures at higher temperatures, but there is some loss of resilience and some shrinkage. The devitrification which takes place at these temperatures does not, however, affect the insulating properties of the fibre.

Extremely low thermal conductivity is one of the outstanding features of aluminium silicate fibres, values ranging from less than $\frac{1}{4}$ to 3 B.T.U./hr./sq.ft./°F./inch of thickness, depending upon temperatures and the form of the material.

Aluminium silicate provides a lightweight, high temperature heat barrier with excellent thermal shock resistance, flame resistance, resilience, chemical stability and electrical properties.

The fibres of aluminium silicate interlock. There is no brittle structure to develop stresses during sudden heating or chilling.

**End Uses**

Aluminium silicate fibres are used for a variety of high temperature applications, including thermal, acoustical and electrical insulation, filters, packings and gaskets, containment and conveyance of molten/non-ferrous metals.

## *Bulk*

One of the major uses in bulk form is in high temperature insulation. End-uses include furnace hot-topping, heating element cushions and jet engine blankets. The bulk fibre is also used as an expansion-joint packing in kilns and furnaces to reduce heat losses and help maintain uniform furnace temperatures.

*Brazing Furnaces.* Short staple fibre is used in place of asbestos in continuous brazing furnaces to cushion aluminium and copper parts.

*Furnace Rolls.* Many of the insulation applications for which bulk fibre is used also require that the insulation should be vibration resistant, i.e. fibres must not be shaken loose during service. Long staple fibres provide better vibration resistance than short staple; this is an important factor, for example, when bulk fibres are used for insulating rolls that convey steel and alloy sheet metal through annealing furnaces.

*Gas Filters.* Long staple aluminium silicate fibre is used for filtration of hot gases in the region of 760°C.

## Batts

Batts are formed by bulking the fibre into a blanket-like shape, which is held together mechanically by wire or a similar fastening device.

*Air Conveyors.* When extremely hot materials are being conveyed, the fibre in batt form is placed in the bottom of an air chamber. Hot solids from roasting or calcining operations are fed into the upper part of the chamber, and air passing through the fibrous batt keeps the solids suspended for transfer to other processing areas. The air may also serve to cool the solids. The fibre batt is suitable for such uses because of its high temperature resistance, its air permeability, and the fact that it does not deteriorate when hot suspended solids are deposited on the batt during shutdown.

*Engine Silencers.* The good acoustical absorption characteristics of aluminium silicate fibres, coupled with resistance to high temperatures and vibration, makes them suitable material for acoustical insulation for jet engines.

## Paper

Aluminium silicate paper is made in a range of thicknesses, e.g. 20, 40 and 80 mils. Applications for paper include high temperature linings, filters, gaskets, fabrication aids and electrical applications.

*High Temperature Linings.* The thin gauge and low thermal conductivity of aluminium silicate papers make them useful for high temperature lining materials. Many non-ferrous metals, such as aluminium, magnesium, and some brasses and bronzes do not readily wet or attack the fibre. Paper made from it may therefore be used to line metal ducts, ladles, crucibles, spouts, launder systems, ingot moulds and other metal components used to handle these molten metals.

Aluminium silicate paper is also used as a lining for combustion chambers in oil-fired domestic heating units. The paper protects the steel wall of the chamber from direct flame impingement.

*Gaskets.* The paper is used for gaskets on pressure vessels, flanges, orifice blocks and other equipment that encounter high temperatures and make use of moderately low pressures. Because of the relatively high porosity of the paper, it can be impregnated with silicones, fluorocarbons and similar heat-resistant impregnants to adapt the paper to higher pressure gasket applications.

*Processing Aids.* Aluminium silicate paper can also be used as process and fabrication aids. It is used successfully, for example, as a braking cushion in making brazed stainless steel honeycomb sandwich structures. Dimensional stability and inertness are primary requirements.

## Blocks or Boards

Preformed blocks are made of short staple fibres and suitable fillers and binders which do not reduce the 1,260°C. maximum service temperature of the material. Aluminium silicate blocks or boards reduce hot wall temperatures in industrial furnaces, kilns and combustion chambers. The block provides lower heat losses and lighter walls as compared with many conventional refractory fire bricks.

Devitrification may take place in combustion applications, owing to the high temperatures involved. The temperature gradient through the block is such, however, that devitrification is generally confined to the hot face of the block. The block often appears to be stronger after exposure to devitrifying conditions than before. Exposure to high heat conditions does not appear to have a serious effect on the insulating properties of the blocks, and such blocks may be re-used, even though the hot faces may be more easily abraded and show some surface shrinkage.

*Oil-fired Furnaces.* Blocks or boards line the combustion chambers of oil-fired furnaces. The use of aluminium silicate blocks reduces slag build-up on the chamber wall, and little if any corrosion from direct flame impingement occurs. The lightweight block also eliminates the need for a forced air 'after purge' of the furnace. Purging is now done entirely by bleed-down of the primary air pressure chamber and by natural draught through the secondary air control damper and blower.

*Tunnel Kilns.* The good resistance of aluminium silicate blocks to thermal shock is of particular advantage when the blocks are used as roof tiles in tunnel kilns. Tunnel kiln operations occasionally involve jam-ups because of movement of skidrails or ware-supporting setter plates. Substantial production time may be lost in waiting for the kiln to cool slowly in order to avoid fracture or spalling of the insulating brick. Kilns insulated with aluminium silicate block can be cooled more rapidly because of the material's superior resistance to thermal shock. Also, the block may be re-used after the tunnel has been dismantled and the jam-up straightened out.

*Electric Furnaces.* Aluminium silicate blocks are used for lining electric furnaces used to form graphite tubes. The blocks provide effective insulation while maintaining a furnace temperature of 900°C. for several days. The blocks have proved satisfactory over several years of operation, without deteriorating in either oxidizing or reducing atmospheres or after contact with pitch vapours and sulphur gases.

## Textiles

Several types of aluminium silicate textile products are available, made from long staple fibres. These include blankets, roving, yarn, rope, tape and broad woven textile goods. Yarns reinforced with glass filaments, alloy wire and other materials are also made.

*Blankets.* Aluminium silicate fibre blankets are used as high temperature insulation in furnace tops, sidewalls and ducts; turbine exhaust lines, and superheated piping. In piping applications, the blanket is particularly well suited for insulating fabricated section such as flange and valve covers and elbows. These insulations are usually assembled by encasing the blanket in a flexible alloy screen or heat-resistant cloth and stitching the assembly together. Removable coverings may be made up in this way.

The degree of insulation obtainable with these blankets is indicated by an installation in an incinerator used to burn exhaust fumes generated in a paint spray operation. The blanket reduced a combustion chamber temperature of about 900°C. to a safe level of about 65°C. on the outside shell.

Aluminium silicate blankets are also used as a packing between firebrick linings of a petroleum refining furnace. The firebrick expands and contracts with fluctuating temperatures in the furnace, and the blanket acts as an expansion joint filler as well as a seal against gas leakage.

Combinations of aluminium silicate blankets and glass fibres or asbestos materials offer interesting possibilities. The blanket is used to reduce the temperatures to a safe range for the glass or asbestos. A flange cover used for superheated steam piping reduced a hot face temperature of 600°C. to about 50°C. on the cold face through 63.5 mm (2½ in) of aluminium silicate blanket and 38.1 mm (1½ in) of glass fibre blanket. The assembly is about one-half the weight of conventional insulation.

*Other Textiles.* A variety of applications for other ·textile products include gaskets and seals in heat-treating furnaces, rope packings for moulds (between the hot top and the ingot mould) in steel manufacture, bag filters for collecting hot carbon and chemicals, and electrical wire winding and wrapping.

## Castables

Castable aluminium silicate fibre compositions are available, which incorporate high temperature binding materials. Aluminium silicate offers extremely good thermal shock resistance, low thermal conductivity and much lower density than previously available castable refractories. In addition to drying lids and pressure vessel linings, castable aluminium silicate fibre has been used on boiler access doors, and as incinerator linings, duct insulation and induction furnace shielding.

*Drying Lids.* Castable aluminium silicate is used in lids for drying and settling ceramic linings in casting pots. The lid is exposed to a temperature cycle of room temperature to more than 1,093°C. Burner blocks are mounted on the lid, and an air-gas mixture is burned to provide a heat source. In operation, the castable surface is suspended over the casting pot and thus is in direct contact with the flame sweep during firing. Lids made from other types of castable refractories weigh 675–900 kg (1,500–2,000 lb). For an equivalent thickness, the aluminium silicate lid weighs only 257.5 kg (350 lb).

*Pressure Vessels*. Castable aluminium silicate compositions are used as linings for pressure vessels. The lining is installed on the inner wall of the vessel, conforming to the contours of the top and bottom, as well as the inlet and outlet ports. In one application, the vessel is designed to withstand 21 kg/cm$^2$ (300 lb/in$^2$) gas pressure at 815$^o$C. The lining, applied 51 mm (2 in) thick, permits the original 51 mm (2 in)–thick stainless steel wall to be replaced with a 13 mm (½ in) carbon steel wall.

## Moulded Shapes

Rigid, structural moulded shapes of aluminium silicate fibres and high temperature binders are used for many applications. These include re-usable risers for non-ferrous casting, which have the advantage that the riser is not wetted by molten aluminium or magnesium. Other applications are combustion chamber liners, baffles and flame detectors, small furnace and oven liners, and can setter plates and shields. Corrugated sheeting serves as spacers in high temperature absolute filters and may be adaptable to base-plates for heating element wires.

*Impregnated Mouldings*. Rigid moulded products made from aluminium silicate fibre, because of their relatively high porosity, may be impregnated by vacuum and immersion techniques with commonly available binders, including inorganic binders, rubber, phenolic, epoxy and silicone solvent or latex-base materials.

## Other Applications

Aluminium silicate fibres are being used commercially and experimentally in a great number of applications which show particular promise for the future.

*Smog Control*. Incineration of waste material at temperatures higher than those commonly employed can reduce contamination of the atmosphere. Burned at 760°C., raw sewage produces no objectionable odours or air pollution. Aluminium silicate fibre insulation may be used effectively in the design. of plants for burning waste at these increased temperatures.

*Filtration of Radioactive Particles.* Aluminium silicate fibre is used in filtering radioactive particles from hot gas streams in. atomic energy plants.

*Missiles and Rockets.* The resistance of lightweight aluminium silicate fibre to erosion and thermal shock is advantageous in missile and rocket insulation.

*Fire-resistant Products.* The non-flammability of aluminium silicate fibre is a useful property in the manufacture of fireproof safes, files, desks, panels and partitions and the like.

## METALLIC FIBRES

Fibres produced from metals, which may be alone or in conjunction with other substances.

### INTRODUCTION

Filaments of metal have been used as decorative yarns since the very earliest times. Metal threads, for example, were used by the Persians for producing the intricate and attractive patterns in their carpets, the ancient Egyptians wove threads of gold into the fabrics of their ceremonial robes.

These metal filaments were made by beating soft metals and alloys, such as gold, silver, copper and bronze, into thin sheets, and then cutting the sheets into narrow ribbon-like filaments. The filaments were used entirely for decorative purposes, providing a glitter and sparkle that could not be achieved by other means.

As textile fibres, these metal filaments had inherent shortcomings which restricted their use. They were expensive to produce; they tended to be inflexible and stiff, and the ribbon-like cross-section provided cutting edges that made for a harsh, rough handle; they were troublesome to knit or weave, and they had only a limited resistance to abrasion. Apart from gold, the metals would tend to tarnish, the sparkle being dimmed with the passage of time.

Despite these shortcomings, the metallic ribbon-filament has remained in use for decorative purposes right up to the present day. The development of modern techniques of surface-protection has brought cheaper metals into use; aluminium foil, for example, may be anodized and dyed before being slit into filaments which are colourful and corrosion-resistant.

Ribbon-filaments are now manufactured in considerable quantity, e.g. as tinsel, but they remain an essentially decorative material. The filaments are weak and inextensible, and are easily broken during wear; they lack the flexibility that is essential in a genuine textile fibre.

#### Multicomponent Metallic Filaments

In recent years, the ribbon filament of metal has undergone a transformation which has changed the commercial outlook for this ancient product. The metal of the filament is now sandwiched between layers of plastic, which protect it from the atmosphere and from other corrosive influences.

The multicomponent filaments produced by slitting sandwich materials of this type are stronger and more robust than the filaments cut from metal foil alone. They retain the glitter of the metal during prolonged periods of use, and have a soft, pleasant handle. Coloured pigments may be added to the adhesive used in sticking the plastic films to the metal foil or metallized film.

Metallic fibres of this type are now widely used in the textile industry, and are produced in a range of colours and forms by many manufacturers. They remain, however, essentially decorative materials, and their applications are restricted to this type of use.

### Single-component Metallic Fibres

With the rapid development of the missile and space-vehicle, the single-component metallic fibre has assumed a new importance. Metals are being made into filaments so fine that they achieve a degree of flexibility which enables them to serve as practical textile fibres. These fine-filament metallic fibres may be spun and woven and knitted on normal textile machinery, and they have physical and chemical characteristics that render them of great importance in many textile applications.

Many types of metallic filament are now being produced commercially; some, like stainless steel, are making good progress in a variety of textile end-uses.

## TYPES OF METALLIC FIBRE

Metallic fibres used in the modern textile industry are of two main types:

(1) Metallic (Single-component) Fibres, or Metallic (S.C.) Fibres.
(2) Metallic (Multi-component) Fibres, or Metallic (M.C.) Fibres.

In the section which follows, the two types of metallic fibre are considered separately under these headings.

## NOMENCLATURE

The term *metallic fibre,* in its general sense, means simply a fibre that is made from metal. Fibres of this type also come into the category of *inorganic fibres.*

### Federal Trade Commission Definition

The generic term *metallic* was adopted by the U.S. Federal Trade Commission for fibres of this type, the official definition being as follows:

*Metallic.* A manufactured fibre composed of metal, plastic-coated metal, metal-coated plastic, or a core completely covered by metal.

## (1) METALLIC (SINGLE-COMPONENT) FIBRES

### INTRODUCTION

The ductility of metals makes possible their conversion into wire, by drawing the metal through dies that become successively smaller. This process resembles, in some degree, the production of a synthetic fibre such as nylon by extruding molten polymer through the orifice of a spinneret. In each case, the material is being forced into a continuous length of narrow-diameter rod by forcing it through a hole of appropriate diameter.

The production of wire is a traditional human craft that has its origins in the earliest days of the metal-working civilizations. As the technique of wire-drawing became more refined, it became possible to produce wires of smaller and smaller diameter, but the nature of the process set a limit on the fineness of the wire that could be produced economically in this way.

Fine wire is, in effect, a monofilament of metal, and it has long been woven into fabric-like structures such as window screens, wire support structures in industry, filter screens, and the like. The metal wires used for these applications, however, do not have the flexibility that is a characteristic of a genuine textile fibre; they retain much of the inherent stiffness of the bulk metal from which they are formed. They are quite rigid materials, and this rigidity imposes mechanical difficulties in the weaving or knitting operations; it limits handling and production rates, and restricts the end-use applications of woven wire fabrics.

On the other hand, metals have inherent characteristics that could be of very great value in many modern textile applications; they could serve in end-uses for which organic, glass and other 'normal' textile fibres are inadequate.

During the late 1950s, the development of high-speed flight and space travel created demands for flexible materials that would be capable of withstanding unique environmental conditions. These materials would be used, for example, in the manufacture of protective garments, antenna membranes, parachutes and other structures used in the space programme. It became apparent that the requirements for such applications could be met by the production of fibres of adequate flexibility from high-temperature metals or alloys, e.g. of the stainless steel type.

The flexibility of materials is a function of cross-sectional area. As the diameter of a rod is reduced by one-half, its flexibility increases by a factor of four. The problem of producing metal fibres of adequate flexibility resolves itself, therefore, into that of producing filaments of the necessary fineness. In the case of a metal like stainless steel, with a modulus of 29,000,000, it is necessary to reduce the diameter of a rod to form a filament of approximately one-half-thousandth of an inch (12 microns, approx.) to attain a flexibility equivalent to that of a 3.3 dtex nylon filament.

During the 1960s, methods of producing metal filaments of the required fineness were developed. Initially, the process was so expensive that the filaments were restricted in their use to essential military and aerospace applications, where low volume and high cost could be tolerated. Subsequently, development of the process has led to increased production and a substantial lowering of costs. Metal fibres have become available in quantities and at a cost which entitles them to consideration as genuine textile fibres.

## TYPES OF METALLIC (S.C.) FIBRE

Continuous filament yarns and staple fibre are produced from a number of different metals and alloys, the stainless steels (300 series 18/8 alloys) being particularly important. Nickel base super alloys, such as Chromel R and Karma, and several of the refractory metals such as niobium and tantalum have also been formed into filaments of 12 microns and finer.

These metal fibres are available as twisted multifilament yarns, tow and staple fibre.

Multifilament yarns are heat-set and without torque. They are typically of 90 and 100, 12 micron diameter filaments, equivalent dtex 110–330 (den 100–300) approx.; 25 micron diameter filament yarns are also available for industrial applications such as high temperature conveyor belting or cordage, where high tensile strength is particularly important. Filaments of 8 $\mu$ diameter and finer are available.

Core-spun and wrapped yarns may be made with organic and glass fibres, the metal being either core or wrap.

Filament yarns may be bulked by the conventional processes to provide an unusually elastic textured yarn which lends high cover to fabrics.

Staple fibre is available usually in the form of sliver.

## PRODUCTION

Metallic (S.C.) filaments are produced by drawing either single filaments or multifilament strands.

Single filament drawing produces excellent quality metallic fibres, but multifilament drawing offers the best possibility of reducing cost and increasing availability. One method of drawing multifilament yarns is to enclose a bundle of 2 mil. (0.002 inch) wire in a sheath of a dissimilar alloy. The sheathed bundle is drawn to the required extent, and the sheath is removed with nitric acid to leave the multifilament yarn ready for sizing, warping and weaving.

## PROCESSING

### Staple Fibre

Staple fibre may be produced from continuous filament tow by processing through the Perlok System for fibre breaking. Tension control, special guides, especially-hardened breaker bars, etc., are required, but the process is essentially the same as that used for breaking organic filament tow.

The sliver formed in this way of 9.8–15.2 cm (4–6 in) staple, may then be processed by conventional gilling techniques (e.g. Warner-Swasey pin-drafter). Slivers as light as 16.7 g/m (15 g/yd) may be made with very few fibres in the cross-section.

Metal fibres may be made more amenable to processing by roughening their surface to improve drag or drafting characteristics. This property is reflected in excellent fibre-to-fibre interaction during drawing, and high translation of fibre strength to yarn strength at appropriate twist multipliers.

Roving and spinning processes may be applied in a conventional manner and yarns of 100 per cent stainless steel, for example, have been spun successfully.

### Blends

The primary application for stainless steel fibre has been in blends with organic fibres. Many important end-uses require blends containing very small proportions of stainless steel fibres.

On the American Worsted System, the blend is generally achieved at the pin-drafter. Sliver of steel fibre is introduced to the pin-drafter with an appropriate number of organic fibre tops. Three or four passes through the pin-drafters achieve a good blend with as little as $\frac{1}{6}$ per cent by weight of stainless steel fibre in the final yarn. Worsted yarns of 60s count and higher, containing $\frac{1}{6}$ per cent steel, may be made in this way.

When a tow is being used, the blend may be started at the filament-breaking machine by introducing stainless steel and organic fibre tow simultaneously.

Worsted-yarn blends containing small proportions of metal fibre may also be made satisfactorily by introducing Hood or Turbo-broken metallic fibre sliver along with Pacific Converted organic fibre or top to the pin-drafters for gilling and blending. Yarns made in this way are more applicable to fine worsted woven fabrics than the all-Turbo yarns described above, which have been used primarily in knitted structures.

Conventional opening, carding and spinning systems are suitable for processing metal fibre in blends with wool or man-made fibres for use in woollen spun carpet yarns. The various settings on the machines are determined by the predominant fibre in the blend.

Metal fibre is fed to the card web after the first breaker section.

The tow end is fed from an overhead or side-feed position into the centre of the card web before delivery to the centre or side draw feed. Care should be taken to unwind the tow end from the side of the package and not over the end of the package to prevent unnecessary twisting. The metal fibre should be added to the blend subsequent to Peralta or burr crushing rolls.

The advantages of using this technique are as follows:

1. No alterations in machine settings or production outputs are necessary to manufacture carpet yarns containing metal fibre.

2. Uniform blending of the small percentages of the metal fibre is obtained. Depending on the width of the card web, weight percentages of metal fibre in the spun yarn will vary from approximately $\frac{1}{8}$ to $\frac{1}{2}$ per cent.

3. The low percentage additions by weight of metal fibre are obtained by direct feed.

4. Time-consuming blending techniques which are necessary with staple fibres are avoided.

5. The additional advantage of significant reduction of in-process static in operations subsequent to the addition of metal fibre is obtained.

Cotton-length stainless steel fibres may be made by roller-drawing a Hood-broken sliver on super long draft equipment. Blending of organic fibre and steel fibre may then be accomplished by appropriate doublings.

## Non-wovens

Staple steel fibre may be processed into 100 per cent carded and garnetted webs to form non-woven structures. It may also be fed to air-lay web-forming systems as sliver or card web to form uniform fibre density webs. These webs may be compacted by various means to form highly efficient, high capacity filter media, vibration isolators, etc.,

### Dyeing

Stainless steel (304 type) fibre may be dyed with certain commercial textile dyes to produce dark shades.

## STRUCTURE AND PROPERTIES

The properties of a metal fibre are essentially those of the metal itself. The general characteristics of metal fibres as they relate to their use in textiles are discussed in the next section.

Details of the Structure and Properties of stainless steel fibres are given on page 690.

## METALLIC (S.C.) FIBRES IN USE

### General Characteristics

Metals have a number of general characteristics which can be of great practical use in textile applications.

### *Mechanical Properties*

The potential high yield and fracture strengths of most metals make possible the production of fabrics which exceed in strength those that can be produced from existing organic or glass fibres. The molecular architecture of metals imposes a characteristic behaviour under stress which differs from that of a typical organic textile fibre. The stress-strain curve of a metal fibre, for example, shows a completely elastic behaviour up to a yield point. It is devoid of a viscous component such as we find in organic high polymers, and which we identify as primary creep; this enables textile structures to be made with design loads far in excess of those achievable with organic materials.

The absence of a significant viscous behaviour characteristic in metal fibres ensures dimensional stability in certain structures which, at best, would be difficult to achieve with conventional textile materials.

The most significant difference between organic fibres and metal fibres lies in the Young's Modulus of Elasticity, which is most readily recognized as the rigidity or stiffness of a material. Most organic textiles have a Young's Modulus of around 500,000, whereas most metals exceed 10,000,000. Stainless steel, for example, has a modulus of 29,000,000.

It was this high modulus of elasticity, and consequent stiffness, which previously excluded metals from use as textile materials; it placed severe limitations on the processability and end-uses of the fibres. It is only with the advent of modern ultra-fine metal fibres that adequate flexibility has been achieved.

*Specific Gravity*

Most metals have a high density in comparison to those of organic materials, and this can be a serious disadvantage in the use of metal fibres for many applications. Stainless steel, for example, has a specific gravity of 7.88, compared with nylon at 1.14 and polypropylene at 0.9. It is difficult, therefore, to use the denier or tex designations for fineness, or the cN/tex or g/den values for tenacity, without compensating for this wide difference in density.

The high theoretical strengths of some metals, however, will result in high strength/weight ratios, as well as the high modulus-to-weight ratios now realized.

*Chemical Properties*

The resistance of metals to chemicals at room and elevated temperatures covers ranges which differ generally from those of organic fibres. Metal fibres can be exploited, on this account, in many military and industrial end-uses where specific chemical resistance is a requirement. There are, however, applications in which metals may suffer corrosion due to chemical action.

*Thermal Properties*

Many metals are able to tolerate and endure temperatures higher than those which organic fibres will generally withstand. Most organic fibres may be used in practice up to 120–260°C., whereas metals can be expected to perform useful structural functions in temperature ranges well in excess of 260°C. Metal textile fibres can operate usefully, for example, at temperatures well over 1,000°C.

Most metals are good conductors of heat, and metal fibres may function as heat-sinks when used in adequate proportion in blends. Thus, metal fibres may protect organic fibres in the blend from the effects of elevated temperature, by conducting away heat before the temperature of the fabric is raised to a level where damage to the organic fibre may occur.

This property is of significance in applications such as protective clothing, the metal fibre acting as a safeguard against the effects of brief exposure to high temperatures. It is also useful in industrial applications such as dryer felts in paper-making, where the thermal conductivity of the blended yarns contributes to drying efficiency.

## Electrical Properties

*Resistance Heating.* Metals, as a rule, are excellent electrical conductors. Within the range of high conductivity, however, there is a broad spectrum which includes metallic materials whose resistance to the flow of electric current enables them to be used as resistance heating elements.

*Elimination of Static.* The high electrical conductivity of most metal fibres provides an effective method of eliminating static electric charges. Blends containing low percentages of metal fibre provide fabrics that are essentially static-free. If a single metal fibre is incorporated in the cross-section of a yarn, and if the metal fibres along the yarn are within a critical distance from each other (without necessarily touching), sufficient electrical continuity is provided to obviate the accumulation of high concentrations of static electricity.

In such a structure, the elimination of static charges may be of much greater economic significance than at first appears. Apart from the nuisance created by clinging of garments, static electricity encourages soiling of fabrics by attracting particles of airborne dirt. And without the electric charge to hold it to the fibre, such dirt as does accumulate on the fabric is easily removed. This means reduced maintenance costs in applications such as carpets, and ease-of-care properties in apparel fabrics.

In some environments, the possibility of a spark arising from a static charge may be a potential hazard. When inflammable fluids are being handled, e.g. in fuelling stations or hospital operating rooms, the danger of explosion may be materially reduced if clothing is static-free.

The use of metal fibres in static control provides an antistatic treatment that is permanent, and lasts the life of the garment. Wear, laundering and dry cleaning, heat, light or common chemicals will not influence the retention or effectiveness of the electrical conductivity of the yarns or fabrics. Furthermore, by the very nature of this technique of static control, the effectiveness of the treatment is completely independent of the moisture content of the fabric, its finishes, or the environment. The antistatic properties are retained even after fabrics are dried in an oven to constant weight.

Most organic antistatic agents depend for their effectiveness on the absorption of moisture from the atmosphere, and their effectiveness will commonly be reduced as the relative humidity falls. This is unfortunate, as under these conditions the accumulation of static is most evident.

Organic antistatic agents are often wax-like substances in which soil may become embedded, and they tend to act as soil-traps. The incorporation of metal fibres as antistatic agents does not in any way contribute to soiling of the fabric.

*Pilling.* In structures composed of bulked, lofty yarns, the elimination of static reduces the tendency to pilling. Static electricity encourages the accumulation of lint, dander and fly, which serve as nuclei around which unlike charged fibres aggregate on the face of the fabric to form an incipient pill.

*Magnetic Properties.* The magnetic properties of some metals make possible the development of novel textile applications.

*Micro-wave Reflectivity.* Fabrics made from blends containing a low percentage of metal fibre have a high micro-wave reflectivity. This makes possible the production of micro-wave reflective garments, reflective tow targets and other military applications where radar reflectivity is important.

## END-USES

Continuous filament metal yarns may be woven into a range of fabric constructions, including the plain weaves, twills and satins. They may be knitted on both warp and weft knitting systems, and braided, flat and tubular; they may be used in filament wound composite systems with various resin matrices.

The properties of these textile structures confirm the expectations of high temperature tolerance, high flex-life, excellent abrasion resistance, corrosion resistance and dimensional stability. The structures are, in fact, true textiles, manifesting drape and hand as a textile technologist judges these characteristics.

The fabrication of metal fibre textiles is achieved on conventional textile processing equipment at commercial production rates, and without inordinate modifications. In some instances, in order to achieve good running characteristics, special finishes are applied to the yarns to improve internal filament-on-filament abrasion resistance and to add lubricity to the yarn surface to reduce wear on guides, needles and other machine surfaces. Certain of these finishes, when left on the filament, contribute to the end-use performance of the fabric by enhancing wear life, corrosion resistance, colouring and bonding to other materials.

Staple fibre yarns spun from 100 per cent metal fibre, or from blends containing metal fibre in all proportions, may be knitted or woven into a wide range of fabrics.

Very short fibre, with aspect ratios of 1 to 100 to 1,000 are used as reinforcement in resin, metal or ceramic matrices, lending structural, electrical, thermal and magnetic properties to the composite structure.

## Carpets

Blends of metal fibre with carpet fibres provide carpets that are essentially static-free. As little as $\frac{1}{6}$ to $\frac{1}{3}$ per cent of a stainless steel fibre, for example, will reduce static to below the level which is noticeable to an individual.

## Industrial Applications

Metal fibres, such as stainless steel fibres, may be used in many industrial applications. They provide reinforcement in mechanical rubber goods, including tyres, high-temperature-tolerant flexible conveyor belts, Fourdrinier wire and paper machine wet and dryer felts, filtration fabrics, both woven and non-woven, cordage, braided hose and webbing where high flex-life, corrosion resistance, dimensional stability and strength are demanding performance characteristics.

Raschel knitted commercial fish netting and cargo handling systems also hold attractive prospects for metal fibre textiles. Sewing threads are made effectively from metal fibres.

## Apparel Fabrics; Home Furnishings, etc.

Low percentage blends of metal fibres with bulked acrylics, nylon, polyester and wool are made into knitted fabrics such as men's hose, ladies' sweaters, double-knit fabrics, woven worsted fabrics and specialized work clothing fabrics.

The incorporation of higher percentages of metal fibres in these applications makes possible the resistance heating of the materials, e.g. in various items of clothing, and in home furnishings such as draperies, upholstery fabrics, bedding, etc. Static-free and electrically heated transportation fabrics hold considerable promise for an expanding market.

## STAINLESS STEEL FIBRES

Fibres spun from stainless steel.

### INTRODUCTION

Stainless steel was one of the first metallic (s.c.) fibres to be developed commercially for textile applications. A number of firms have taken part in the development, and several stainless steel fibres are now available. The information which follows relates primarily to a fibre of 304 type austenitic stainless steel which is essentially a high grade 18/8 chrome/nickel alloy steel.

### PROCESSING

#### Dyeing

Metallized dyes give deep shades.

### STRUCTURE AND PROPERTIES

#### Fine Structure and Appearance

Grey filament of near-round cross-section. Filament diameter 4 to 50 microns. Coefficient of variation of filament diameter for nominal 12 microns is approximately 4 per cent. This figure applies to the variation from filament to filament and also along the length of a particular filament.

### Tenacity

Standard: 22.1–28.3 cN/tex (2.5–3.2 g/den)
Wet: 22.1–28.3 cN/tex (2.5–3.2 g/den)
Std. loop: 15.9–21.2 cN/tex (1.8–2.4 g/den)
Std. knot: 16.8–22.1 cN/tex (1.9–2.5 g/den)

### Tensile Strength

Annealed: 7,000 kg/cm$^2$ (100,000 lb/in$^2$)
  Yield strength: 3,500 kg/cm$^2$; 50,000 lb/in$^2$
Hard: 17,500–26,950 kg/cm$^2$ (250,000–385,000 lb/in$^2$)
  Yield strength: 15,400 kg/cm$^2$; 220,000 lb/in$^2$

### Elongation (per cent)

Annealed: 11
Hard: 1.5

### Elastic Recovery (per cent)

100 at 1 per cent; 66 at 1.5 per cent

### Modulus of Elasticity

Annealed: 2,030,000 kg/cm$^2$ (29,000,000 lb/in$^2$)
Hard:

### Average Stiffness

1475–1881 cN/tex (167–213 g/den)

### Average Toughness

0.019–0.024

### Abrasion Resistance

High

### Specific Gravity

7.9

### Effect of Moisture

Absorbency: Nil.

**Thermal Properties**

*Melting point*: 1,426°C. approx.

*Effect of High Temperature*

Less than 10 per cent strength loss at 426°C. Up to 90 per cent loss at 980°C.

*Thermal Conductivity*

Stainless steel is a poor conductor but a good resistor compared with copper or nickel.
Stainless steel is a good conductor of heat.

*Flammability*: Non-flammable.

**Effect of Sunlight**

Nil.

**Effect of Age**

Nil.

**Chemical Properties**

*Acids*

Resistant to nitric and phosphoric acids. Attacked by sulphuric and halogen acids.

*Alkalis*

Not affected by common alkalis.

*General*

The chemical properties of stainless steel fibre are similar to those of the bulk metal. Not attacked by common bleaches unless they are halogen derivatives.

**Effect of Organic Solvents**

Not attacked by common solvents.

**Insects**

Not attacked.

**Micro-organisms**

Not attacked.

**Electrical Properties**

Compared with natural and other man-made fibres, metal fibres generally are good conductors of electricity. Compared with copper, nickel and some other metal fibres, stainless steel is a poor conductor; stainless steel fibres may be used for resistance heating.

STAINLESS STEEL FIBRE IN USE

**General Characteristics**

Stainless steel fibres have the basic metallic fibre characteristics already described, providing textile goods of high modulus, high tensile strength, flex-life, tear-resistance, abrasion resistance and compressional resilience. The fibres have a high resistance to many types of severe chemical and physical environments.

**End-Uses**

The end-uses for stainless steel fibres include those described for metallic (s.c.) fibres generally.

In comparison with some other metals such as copper or nickel, stainless steel is a poor conductor of electricity. Stainless steel fibres are particularly suitable for the production of heated structures such as pads, draperies and blankets.

The thermal conductivity of stainless steel is put to good use in protective fabrics, the steel fibres acting as a heat-sink that protects other fibres with which it is blended. Small concentrations of steel fibre (about 5 per cent) blended with 'Nomex' high temperature nylon, for example, have a remarkable effect on the high temperature resistance of fabrics. The stainless steel fibre conducts away heat and prevents the temperature rising to a point at which the 'Nomex' becomes inadequate.

The relatively high cost of stainless steel fibre has tended to encourage the development of applications in which small proportions of steel fibre may be blended with other fibres, to produce valuable characteristics such as those described. As little as 1/5 per cent steel fibre blended into a carpet yarn, for example, reduces static almost to negligible amounts.

## Consumer Textiles

Stainless steel fibre is used in carpets, upholstery, worsted suitings, blankets, uniforms and work clothing.

## Industrial Textiles

*Filters.* Because of fine fibre diameters, it is possible to design structures with high internal surface and high fatigue life. In consequence, filtering efficiency is increased. Using fine metal fibres in non-woven structures, with specially prepared fibre surfaces, battery plaques, fuel cell electrodes and capacitors can be made.

The high temperature tolerance of stainless steel fibre is an important advantage in many industrial applications. For example, it is possible to carry out high-temperature filtration of hydraulic fluids, fuels and hot gases. This has resulted in higher production rates in many industrial processes.

Using stainless steel fibre, high efficiency operation may be achieved at temperatures up to 700°C.

Stainless steel fibres are now being used in a variety of industrial applications, in addition to those described. They include sewing thread, conveyor systems, cordage, fishing lines and nets, cargo restraining devices, tarpaulins, vibration isolators, heat transfer systems and paper-makers' belts.

## Fibre Reinforcement

Stainless steel fibres used as internal support in metals, plastics, ceramics, rubber and filament wound structures have many advantages. Systems of this type can meet the rigorous demands of many modern end-uses, providing high flex-life, tensile strength, dimensional stability and heat tolerance. The metal fibre conducts heat away from the matrix, which may therefore perform better.

Applications in which stainless steel fibres provide reinforcement in this way include power belting, hosing, conveyor belts and inflatable structures, including tyres.

## Medical Applications

Stainless steel is physiologically inert, and this is a valuable characteristic in many medical applications. These include the development of a cardiac pacemaker electrode system, where good electrical conductivity and long-term fatigue life are essential. Multifilament stainless steel yarns of fine diameters are used for surgical sutures.

The ease of textile fabrication of stainless steel fibres is advantageous in many aspects of medical engineering, including vascular prostheses, heart components, and orthopaedic devices for tendon and bone repair.

## (2) METALLIC (MULTI-COMPONENT) FIBRES

Fibres made from metal in association with other materials, commonly plastics.

### INTRODUCTION

Modern metallic fibres of the multi-component type are based largely on aluminium, which provides sparkle and glitter at a fraction of the cost of the early types of decorative fibre based for example, on gold.

The aluminium in these fibres is in the form of a narrow ribbon-filament of either (a) metal foil, or (b) a plastic film which has been vacuum-plated with vaporized aluminium. This is coated with a layer or layers of plastic film.

In these composite structures, the metal is protected from corrosive influences of its environment, and from mechanical damage. Multicomponent metallic fibres have achieved great popularity as decorative fibres, and are an important facet of the modern textile industry.

## TYPES OF METALLIC (M.C.) FIBRE

Metallic (m.c.) fibres may be made in almost infinite variety by using different metals and plastics in their manufacture. Aluminium is, however, the metal most commonly selected, and it is sandwiched between cellulose acetate butyrate, cellophane (cellulose) or polyester films.

The following are the types of yarn commonly produced:

(1) *Acetate Butyrate, Aluminium Foil.* A continuous flat monofilament composed of aluminium foil laminated on both reflective surfaces with cellulose acetate butyrate film.

(2) *Cellophane, Aluminium Foil.* A continuous flat monofilament composed of aluminium foil laminated on both reflective surfaces with cellophane film.

(3) *Polyester, Aluminium Foil.* A continuous flat monofilament composed of aluminium foil laminated on both reflective surfaces with polyester film.

(4) *Polyester, Aluminium Metallized Polyester.* A continuous flat monofilament composed of aluminium metallized polyester laminated on its metallized surface or surfaces with polyester film.

(5) *Polyester, Aluminium Metallized, Non-Laminated.* A continuous, flat monofilament composed of a single layer of aluminium metallized polyester protected on its metallized surface.

The acetate butyrate types of metallic fibre are best used for applications which are not subjected to wet processing of other than very mild forms. Polyester types will withstand wet treatments or dry-heat operations as commonly used with most manmade fibres, but reference should be made to manufacturers' recommendations regarding time, pH and temperature conditions.

## NOMENCLATURE AND TERMINOLOGY

In the U.S., the former Metallic Yarns Institute established minimum quality standards for metallic (m.c.) yarns for textile purposes (see table on page 698), and prescribed a standard system of designation and terms of reference for these yarns.

The following definition of a *metallic yarn* was established by the Institute, and in general it is still in common use:

*Metallic Yarn.* A continuous flat monofilament produced by a combination of plastic film and metallic component so that the metallic component is protected.

This definition differs somewhat from that established by the U.S. Federal Trade Commission (see page xxvii).

### Terminology

Metallic yarns are designated by a group of three symbols, each separated by a hyphen, setting forth the two dimensions of width, and gauge or thickness, and generic type.

1. *Width.* The width of the yarn is expressed as the fraction of an inch to which the yarn has been cut, viz., 1/32, 1/64, etc.

2. *Gauge (or Thickness).* The thickness or gauge of the yarn is expressed as the sum of the thickness of the plastic film and metallic component in hundred-thousandths of an inch, as a whole number, viz., 35, 50, 150, 200, etc.

3. *Generic Type.* The type of the yarn is expressed on the basis of two components of the laminate – the generic name of the plastic film and the metal.

The components are separated by a comma, viz., Polyester, Foil.

*Example:* A Polyester, Aluminium Foil Yarn, 1/64 inch wide and 150/100,000 inch thick, is expressed in the industry as:

$$1/64 - 150 - \text{Polyester, Foil}$$

A manufacturer's trade name or mark may accompany, but where utilized, either alone or in combination, the above must be separately stated or referred to.

## METALLIC YARNS – STANDARDS

The following minimal standards were established for metallic yarns by the U.S. Metallic Yarns Institute.

In determining standards for basic yarns, a 1/64-inch width

is used as a determinant. To calculate yield for yarns of other widths in the same category, it is necessary to apply a direct arithmetic proportion.

| Type | Yield (yd./lb.) | Breaking Strength (g.) | Elonga-tion (%) | Yield Pt. (g.) | Colour Tolerance to Light (hrs.) (Gold and silver) |
|---|---|---|---|---|---|
| 1A 270 Acetate Butyrate, foil | 10.500 | 160 | 30 | 70 | 100 |
| 1B 310 Acetate Butyrate, foil | 10,000 | 170 | 30 | 100 | 100 |
| 2 230 Cellophane, foil | 10,000 | 200 | 20 | n.a. | n.a. |
| 3 150 Polyester, Alum. foil | 16,000 | 175 | 80 | 125 | 100 |
| 4A 100 Polyester, Alum. Metallized Poly-ester | 28,000 | 175 | 80 | 90 | 100 |
| 4B 150 Polyester, Alum. Metallized Poly-ester | 18,000 | 250 | 80 | 125 | 100 |
| 5 50 Polyester, Alum. Metallized, Non-Laminated | 47,000 | 90 | 80 | 40 | 100 |

n.a.=not available.

## PRODUCTION

Modern metallic filaments are made by sandwiching a thin layer of metal, usually aluminium, between thin sheets of the appropriate types of plastic film. There are several ways of carrying out the lamination.

In the production of the Acetate Butyrate, Aluminium Foil (Type 1) metallic fibre, for example, a thermoplastic adhesive is applied to both sides of a sheet of aluminium 0.01 mm (0.00045 in) thick. The adhesive may or may not be coloured. The coated foil is then heated to about 90°C., and it is passed through a set of rollers together with two sheets of cellulose acetate butyrate, in such a way that the aluminium foil becomes the centre of the sandwich.

The laminated material is then slit into narrow ribbon-like filaments of the desired width, e.g. 3.2–0.2 mm (1/8–1/128 in).

In some types of metallic filament (e.g. Types 4 and 5), the centre of the sandwich consists of a plastic (e.g. polyester) film which has been vacuum-coated with aluminium. This may then be protected with layers of lacquer or plastic film as above.

Many types of coloured metallic filament are now produced, the metallic glitter of the aluminium shining through the coloured adhesive or the coloured outer plastic films. Gold-coloured filaments are made, for example, by using an orange-yellow dyestuff or pigment in the adhesive; silver is obtained by relying only on the natural glitter of the aluminium.

## PROCESSING

### Desizing

Conventional techniques and equipment may be used for polyester type metallic yarns.

### Scouring

Polyester type yarns are best scoured on rope or open-width machines, the squeeze rollers being adjusted to give as light a nip as possible in order to avoid unnecessary damage.

### Bleaching

Polyester type metallic yarns may be subjected to normal bleaching operations. Care should be taken to minimize exposure to alkaline conditions, temperatures in excess of 90°C. and prolonged cycle times.

*Note.* Coated polyester yarns are not recommended for high temperature conditions, i.e. in excess of 70°C.

### Dyeing

Some metallic (m.c.) fibres may be dyed effectively, the techniques and dyes being selected according to the type of plastic in the fibre. The acetate butyrate types should be treated generally as acetate, and dyeing carried out at temperatures not exceeding 70°C. Polyester yarns may be dyed at the boil, with the exception of most coated types.

It is generally more satisfactory to make use of coloured metallic yarns where possible, rather than to dye the metallic yarns in the fabric. Dyeing problems are commonly concerned with the avoidance of colouration of the metallic yarns during the dyeing of the base material of the fabric. This is usually a matter of careful choice of dyes, avoiding those which dye the plastic in the metallic fibre.

## STRUCTURE AND PROPERTIES

The properties of a metallic (m.c.) fibre depend upon the nature of the plastic film used in its production, and of the metal used as the centre of the sandwich.

In general, the fibres behave in a manner similar to man-made fibres spun from polymer on which the plastic film is based. Acetate butyrate metallic filaments, for example, have a resemblance to acetate fibres; polyester type metallic filaments are similar to polyester fibres in their general characteristics.

The nature of the aluminium layer inside the sandwich affects the properties of the metallic filament to a significant extent. In Types 1, 2 and 3, the aluminium is a continuous layer of foil; in Types 4 and 5, on the other hand, it is in the form of discrete particles which have been deposited on a layer of plastic film. The discontinuous layer of the latter types results in a finer, softer and more pliable filament, with properties which differ in many respects from those of the foil-type metallic fibres, as indicated below. The figures quoted refer to specific metallic fibres of the various basic types, but there is considerable variation in properties between fibres of the same type.

### Fine Structure and Appearance

Metallic (m.c.) fibres are flat, ribbon-like filaments, commonly 3.2–0.2 mm (1/8–1/128 in) width. They are smooth-surfaced, and may be coloured or uncoloured.

700

## Tensile Strength

The standards originally set up by the Metallic Yarns Institute, U.S.A., are shown in the table on page 698.

## Tenacity

Acetate Butyrate, foil: 2.6 cN/tex (0.3 g/den).
Polyester, foil: 6.2 cN/tex (0.79 g/den).
Polyester, metallized: 11.0 cN/tex (1.25 g/den).

## Elongation

See table on page 698.
Acetate Butyrate, foil: 30 per cent.
Polyester, foil: 140 per cent.
Polyester, metallized: 140 per cent.

## Yield Point

See table, page 698.

## Elastic Recovery

Acetate Butyrate, foil: 75 per cent at 5 per cent elongation.
Polyester, foil: 50 per cent at 5 per cent elongation.
Polyester, metallized: 100 per cent at 5 per cent elongation.

## Flex Resistance

Relative flex resistances of the main types are in the following ratios:
Acetate Butyrate, foil: 1
Polyester, foil: 18
Polyester, metallized: 70

## Abrasion Resistance

Acetate Butyrate, foil: fair.
Polyester, foil: good.
Polyester, metallized: excellent.

## Effect of Moisture

Regain: Acetate Butyrate, foil: 0.1 per cent.
Polyester, foil: 0.5 per cent.
Polyester, metallized: 0.25 per cent.

**Thermal Properties**

Softening point: Acetate Butyrate, foil: 205°C.
                Polyester: 232°C.

**Effect of Age**

Nil.

**Effect of Sunlight**

Some loss of strength on prolonged exposure.

**Chemical Properties**

*Acids*

Generally good resistance.

*Alkalis*

Acetate Butyrate: good resistance to weak alkalis; degraded by strong alkalis.
Polyester: ditto. Metal foil types are more resistant.

*General*

Acetate Butyrate: Similar to acetate yarn. Not affected by sea-water, chlorinated water, or perspiration. Generally resistant to bleaches, but sensitive to caustic soda used in peroxide bleaching. Also sensitive to copper sulphate and sodium carbonate at high temperatures.
Polyester: Generally good resistance.

**Effect of Organic Solvents**

*Acetate Butyrate.* Attacked by acetone, ether, chloroform, methyl alcohol, tetrachloroethane. Not attacked by benzene, carbon tetrachloride, ethyl alcohol, perchloroethylene, trichloroethylene.

*Polyester.* Attacked by acetone, benzene, chloroform, tetrachloroethane, trichloroethylene. Not attacked by carbon tetrachloride, ethyl alcohol, methyl alcohol, perchloroethylene, white spirit.

**Insects**

Not attacked.

**Micro-organisms**

Not attacked.

**Allergenic Properties**

Non-allergenic.

**Electrical Properties**

Metallic (m.c.) fibres conduct electricity, the metallized types having a lower conductivity than the foil types.

## METALLIZED (M.C.) FIBRES IN USE

### General Characteristics

*Appearance*

Metallic (m.c.) yarns are used in the textile industry almost entirely as decorative materials. They provide a metallic glitter and sparkle that cannot be obtained in other ways. The aluminium foil that provides the glitter in a modern metallic yarn is protected from corrosive materials of its environment by the plastic film in which it is enclosed. It remains untarnished through long periods of wear, and polyester types will withstand repeated launderings without losing their sparkle. Metallic yarns are not affected by seawater or by the chlorinated water of swimming pools, and are widely used in modern swimwear.

The dyestuffs used in colouring metallic fibres are usually fast to light, and the colour remains bright to match the sparkle from the aluminium foil.

*Mechanical Properties*

As metallic (m.c.) yarns are used primarily for decorative purposes, they do not as a rule contribute significantly to the strength of fabrics or garments. Nevertheless, they may be used as weft or warp yarns, and are strong enough to withstand the weaving and knitting operations. If necessary, the metallic yarns are combined with support yarns, such as nylon.

The plastic film of the metallic yarn is flexible, and the yarns are extensible to a degree that depends upon the type.

*Chemical Properties*

Aluminium will corrode and tarnish in air, and in contact with seawater, but in metallic fibres it is protected so effectively that it retains its glitter for long periods. The chemical resistance of a metallic filament is, in general, the chemical resistance of the plastic film. In the case of polyester films, this is outstanding.

If metallic fibres are held in contact with strong alkaline solutions for prolonged periods, the aluminium may be attacked at the unprotected edges of the ribbon. Metallic fibres should not, therefore, be subjected to alkaline reagents of significant strength.

Organic solvents, too, may attack the laminate adhesive or lacquer coating; great care should be taken in dry cleaning to ensure that an appropriate type of solvent is used.

*Thermal Properties*

The plastic films in metallic fibres are thermoplastic, and will soften at elevated temperatures. Delamination may occur if the fibres are heated, and acetate types in particular should be processed only at low temperatures.

The plastic films may be permanently embossed by heat and pressure, and special effects may be introduced into the fibres in this way.

**Washing**

Acetate butyrate types may be hand washed in lukewarm water with a mild soap. If processed as silks or woollens, they may be safely washed in home or commercial laundry equipment.

Polyester types may be washed at temperatures up to 70°C. Dimensional stability is good and crease resistance fair.

Most coated polyester yarns will not withstand treatments other than those used for silks or woollens.

**Drying**

Acetate butyrate types must be dried at as low a temperature as possible. Polyester types may be dried at higher temperatures as used for polyester fibres, with the exception of most coated types.

**Ironing**

Acetate butyrate types should be ironed at temperatures no higher than 105°C. Polyester types may be ironed at temperatures up to 130°C. 'Rayon' setting is preferable for both types.

**Dry Cleaning**

Metallic fibres may be dry cleaned without difficulty, provided care is taken in the selection of solvent to suit the type of fibre.

**End-Uses**

Metallic (m.c.) yarns are used for decorative purposes in almost every field of textile application. Important end-uses include women's dress goods, upholstery, curtains, table linens, swimwear, packaging, footwear, car upholstery, suits and hats.

# POLYUREA FIBRES

Fibres spun from synthetic linear polymers containing the urea grouping, —NH—CO—NH—, as part of the repeating unit.

$$n\,NH_2 - R - NH_2 \; + \; NH_2\,CONH_2 \longrightarrow \; \left[-R-NHCONH-\right]_n \; + \; 2n\,NH_3$$

DIAMINE          UREA                    POLYUREA

*Production of Polyurea*

## INTRODUCTION

Following the introduction of nylon, many types of condensation polymer were examined as possible sources of synthetic fibres. Among those considered as worthy of detailed study were the polyureas, formed by condensation of diamines with urea, and characterized by the recurring urea group, —NH—CO—NH—, in the polymer molecule.

Polyurea fibres were spun in America, Germany and the U.K. during the 1940s, but despite a considerable research effort they did not yield a fibre that was selected for commercial development. It was not until 1950 that polyureas again came under intensive investigation, this time in Japan, where the firm Toyo Koatsu Industries Ltd. began seeking new outlets for urea. This chemical, produced in very large quantities, is a cheap and readily available raw material for the production of synthetic products. There is an obvious attraction in using it as the starting point for the manufacture of synthetic fibres.

Within 5 years, Toyo Koatsu had developed a laboratory-scale process for the production of a polyurea fibre made by condensing urea with nonamethylene diamine.

$$n\,NH_2\,(CH_2)_9\,NH_2 \qquad + \qquad n\,NH_2\,CONH_2$$

NONAMETHYLENE DIAMINE              UREA

$$\longrightarrow \; \left[-(CH_2)_9\,NH\,CONH-\right]_n \; + \; 2n\,NH_3$$

POLYUREA

*Production of 'Urylon'*

707

By 1958, a pilot plant was in operation, producing 1 ton per day of the polyurea fibre – now known as 'Urylon' – and it seemed likely that large-scale development would be undertaken in a matter of a year or two. Since then, however, 'Urylon' has made little apparent progress as a commercial fibre.

The properties of 'Urylon' are comparable in general with those of nylon 6.6 and PET polyester fibres. The fibre melts at 240°C. and has a tenacity of 39.7–48.6 cN/tex (4.5–5.5 g/den), with an elongation of 15–20 per cent. It has a low specific gravity – 1.07 – which makes it the lightest synthetic fibre other than polyethylene or polypropylene. 'Urylon' has good chemical resistance, moisture regain of 1.7, and in other respects seems eminently satisfactory as a textile fibre.

The problem with 'Urylon' and other polyurea fibres lies not in the characteristics of the fibre itself, but in the economics of production. Urea is cheap and plentiful, but nonamethylene diamine is expensive; it has been made until recently almost entirely from rice bran oil.

In the section which follows, information on polyurea fibres relates primarily to 'Urylon', and was provided by Toyo Koatsu Industries Ltd.

## PRODUCTION

Polyurea used in 'Urylon' production is made by the condensation of urea with nonamethylene diamine (see page 707).

### Monomer Synthesis

(a) *Urea*

Urea is manufactured on a very large scale by the reaction of carbon dioxide with ammonia at high temperature and pressure. Ammonium carbamate is formed (1), and this decomposes to yield urea and water (2).

$$2NH_3 + CO_2 \xrightarrow{(1)} NH_2 COONH_4 \xrightarrow{(2)} NH_2 CONH_2 + H_2O$$

AMMONIUM CARBAMATE    UREA

*Production of Urea*

## (b) *Nonamethylene Diamine*

Rice bran oil contains oleic and linoleic acids. When the oil is reacted with ozone, the oleic acid forms pelargonic acid and azelaic acid (1), and the linoleic acid forms caproic acid and azelaic acid (2).

The azelaic acid is reacted with ammonia to form ammonium azelate (3), which is dehydrated to azelaic dinitrile (4) and then hydrogenated to nonamethylene diamine (5).

*Note.* The production steps from azelaic acid are similar to those used in making hexamethylene diamine from adipic acid in the production of nylon 6.6

(A)   $CH_3 (CH_2)_7 CH = CH (CH_2)_7 COOH + 2O_2 \xrightarrow{(1)}$

OLEIC ACID

$CH_3 (CH_2)_7 COOH$  +  $HOOC (CH_2)_7 COOH$

PELARGONIC ACID    AZELAIC ACID

(B)   $CH_3 (CH_2)_5 CH = CH - CH = CH (CH_2)_7 COOH + 2O_2 \xrightarrow{(2)}$

LINOLEIC ACID

$CH_3 (CH_2)_5 COOH$  +  $HOOC (CH_2)_7 COOH$

CAPROIC ACID    AZELAIC ACID

(C)   $HOOC (CH_2)_7 COOH + 2NH_3 \xrightarrow{(3)} (CH_2)_7 \begin{matrix} COONH_4 \\ COONH_4 \end{matrix}$

AZELAIC ACID    AMMONIUM AZELATE

$\xrightarrow{(4)} (CH_2)_7 \begin{matrix} CN \\ CN \end{matrix} \xrightarrow{(5)} (CH_2)_7 \begin{matrix} CH_2 NH_2 \\ CH_2 NH_2 \end{matrix}$

AZELAIC DINITRILE    NONAMETHYLENE DIAMINE

*Production of Nonamethylene Diamine*

**Polymerization**

Nonamethylene diamine‾ and urea are dissolved in water, using a slight molar excess of diamine. The solution is heated at atmospheric pressure, the temperature being raised from 100°C. to about 250°C. over a period of 15–18 hours. Vacuum is applied in the final stages.

The polymer is milky-white, with a melting point of 225–231°C. It contains only a very small amount of low-molecular-weight material, and may be spun directly.

**Spinning**

The polyurea is melt spun, and is hot drawn, e.g. in three stages at 100°C. (to provide fibre of tenacity 48.6 cN/tex; 5.5 g/den). It is wound on to bobbins, and may be cut into staple.

## PROCESSING

**Dyeing**

Polyurea fibre may be dyed with acid, disperse, direct and vat dyes. The polymer molecule provides excellent anchor-points for acid dyes at the twin imino groups of the urea linkage, and at the terminal amino groups.

Acid dyes provide colours of good light and wash. fastness. Tannic acid treatment is recommended to increase fastness.

Sulphur and basic dyes are not satisfactory.

## STRUCTURE AND PROPERTIES

**Fine Structure and Appearance**

Smooth surfaced filaments of round cross-section.

**Tenacity**

39.7–48.6 cN/tex (4.5–5.5 g/den).

**Elongation**

15–20 per cent.

**Elastic Recovery**

70 per cent recovery from 8 per cent extension after 30 seconds; 92 per cent recovery after 5 minutes.

**Specific Gravity**

1.07.

**Effect of Moisture**

Regain: 1.8 per cent.
Hydrolysis occurs on exposure to steam at 130°C.

**Thermal Properties**

*Softening point*: 205°C.

*Melting point*: 240°C.

*Effect of High Temperature*

Some loss of strength on continued exposure to air at 150°C.

**Effect of Sunlight**

Similar to nylon.

**Chemical Properties**

*Acids*

Good resistance. Retains more than 90 per cent strength after 10 hours at room temperature in 40 per cent sulphuric acid.

*Alkalis*

Good resistance. Retains more than 90 per cent strength after 10 hours at room temperature in 40 per cent caustic soda.

*General*

Good resistance to most common chemicals.

**Effect of Organic Solvents**

Similar to nylon.

**Insects**

Not attacked.

**Micro-organisms**

Not attacked.

## POLYUREA FIBRES IN USE

**General Characteristics**

*Handle, etc.*

Polyurea fibres have a soft, attractive handle, and fabrics are comfortable and warm when worn next to the skin.

*Mechanical Properties*

In general, the mechanical properties of polyurea fibres are comparable with nylon 6.6 and PET polyester fibres, with a closer resemblance to nylon.

*Thermal Properties*

The melting point of 'Urylon' lies slightly below that of nylon 6.6, and above that of nylon 6. Thermal stability is good up to 140°C., but the fibre degrades on exposure to temperatures above this.

In common with other thermoplastic fibres, polyurea fibres may be heat set, e.g. by heating for $1\frac{1}{2}$ minutes at 160°C. The fibre shrinks 13 per cent at 180°C.

*Chemical Properties*

Polyurea fibres have excellent chemical stability. They resist acids better than nylon, and alkalis better than PET polyester fibres.

*Moisture*

With a regain of 1.8 polyurea fibre has a lower moisture absorption under standard conditions than either nylon or PET polyester fibres. This could be a disadvantage in apparel uses.

*Specific Gravity*

Polyurea fibre is lighter than any synthetic fibres other than polyethylene and polypropylene. This could mean greater covering power than nylon or polyester fibres, with which it seems likely to compete.

## Dyeability

It is not possible to predict at this stage whether dyeability will be significantly better than that of nylon, but it seems probable that this will be so.

## End Uses

The end uses of polyurea fibres will lie in the same fields as those of nylon and PET polyester fibres.

## Industrial Uses

Polyurea fibres will probably find a useful outlet in fishing lines, nets and ropes, where its lightness will be advantageous. It could prove a useful reinforcement fibre for tyres, conveyor belts and the like.

## Apparel Uses

Polyurea fibres could find a variety of applications in the apparel field similar to those served by nylon. The lower moisture regain may prove disadvantageous in this respect.

## POLYCARBONATE FIBRES

Fibres spun from synthetic linear polymers containing the characteristic grouping —O—CO—O— as part of the repeating unit.

### INTRODUCTION

Polycarbonate resins have been used for some years as plastics, their applications being determined largely by their useful electrical and high-impact-resistance characteristics. Melting points are typically in the region 150–300°C., and some of the polycarbonates have good spinning properties. The low cost of the starting materials has sustained an interest in fibres spun from polycarbonate resins, and monofilaments are produced commercially on a limited scale.

### NOMENCLATURE

In the chemical sense, polycarbonates are polyesters derived from carbonic acid, $H_2CO_3$. They do not qualify for the description *polyester*, however, under the U.S. Federal Trade Commission definition.

### PRODUCTION

Polycarbonates may be made by condensing an aromatic dihydroxy compound with suitable carbonic acid derivatives, e.g. carbonyl chloride (phosgene) or esters of carbonic acid.

### (a) *Carbonyl Chloride (Phosgene)*

Carbonyl chloride may be condensed, for example, with p, p′–isopropylidene–diphenol ('Bis Phenol A'), which is readily obtainable from phenol and acetone. In the presence of alkali, the following reaction takes place:

714

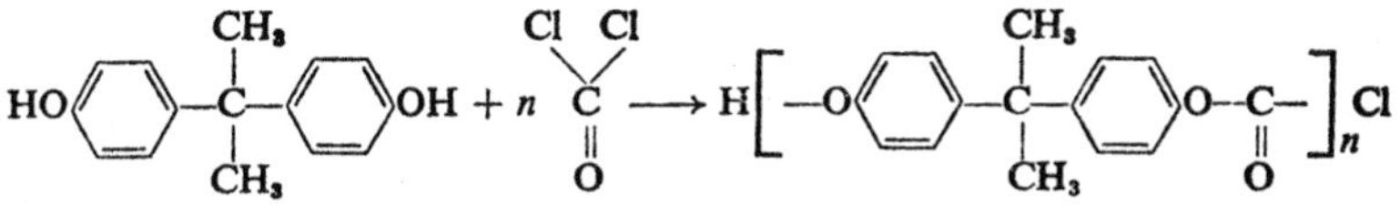

(b) *Carbonic Acid Esters*

An ester of carbonic acid, such as diphenyl carbonate, may be trans-esterified with bis phenols:

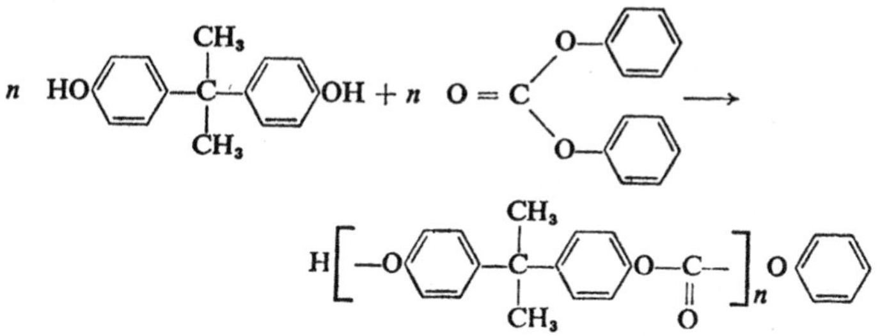

Filaments are melt spun.

## STRUCTURE AND PROPERTIES

| | |
|---|---|
| **Tenacity** | 16.8 cN/tex (1.9 g/den) |
| **Elongation** | 40–60% |
| **Initial Modulus** | 397.3 cN/tex (45 g/den) |

## POLYCARBONATE FIBRES IN USE

Monofilaments are used as temporary threads, e.g. basting threads in men's suits which can be removed subsequently by solvent treatment in a dry cleaning machine.

715

## CARBON FIBRES

### INTRODUCTION

In 1963 a team of British scientists, W. Watt, W. Johnson and L.N. Phillips, working at the Royal Aircraft Establishment, Farnborough, U.K., developed techniques for producing carbon fibres of high strength and outstanding rigidity. These fibres were in commercial production by 1968 and have since become of great importance, especially in the field of composites in which the fibres are embedded in resins or other materials.

Most of the important textile fibres in use today are derived from organic polymers, i.e., polymers in which the backbone of the molecular structure consists of carbon atoms to which are attached atoms of other elements, commonly hydrogen, oxygen and nitrogen.

It has long been known that pyrolysis of these fibres, such as rayon, could result in the removal of the non-carbon atoms to leave a filament consisting essentially of carbon. But the carbon atoms in these filaments are arranged in a more or less disordered form; the structure is amorphous rather than crystalline, and the filaments are weak and of little practical value. To achieve high strength and modulus, it was necessary to devise a process for producing carbon fibres which would orientate the carbon atoms and result in fibres of a high degree of crystallinity.

### PRODUCTION

The starting material for production of carbon fibres is commonly an acrylic fibre, such as 'Courtelle', in which the backbone of carbon atoms is attached to hydrogen atoms and CN groups. A three-stage heating process is used in converting the acrylic fibre to carbon.

The initial stage is to heat the acrylic fibres at 200–300°C under oxidising conditions. This is followed by a second stage when the oxidised fibre is heated in an inert atmosphere to temperatures around 1000°C. Hydrogen and nitrogen atoms are expelled, leaving the carbon atoms in the form of hexagonal rings which are arranged in oriented fibrils.

Finally, the carbonized filaments are heated to temperatures of up to 3000°C, again in an inert atmosphere. This increases the orderly arrangement of the carbon atoms which are organized into a crystalline structure similar to that of graphite. The atoms are in layers or planes which lie virtually parallel to each other. The planes are well oriented in the direction of the fibre axis, this being an important factor in producing high modulus fibres.

The mechanical properties of carbon fibres produced in this way are affected greatly by the conditions under which they are treated in the final stage, and by varying these conditions it is possible to produce fibres of different modulus and strength characteristics.

## STRUCTURE AND PROPERTIES

The properties of carbon fibres vary, depending on the conditions under which they are produced. The information which follows relates to a typical range of fibres.

*Fine Structure and Appearance.* Carbon fibres are black and smooth-surfaced, with a silky lustre. They are commonly of round cross-section, possibly with flattened sides.

*Ultimate Tensile Strength.* 1.80–2.40 kN/mm$^2$. (cf. steel 2.80–4.00).

*Breaking Extension.* 0.5%. (cf. steel 2.0%).

*Density.* 1.95 g/cm$^3$. (cf. steel 7.80).

*Stiffness.* 350–410 kN/mm$^3$ (cf. steel 207).

*Stiffness/Weight Ratio.* 180–210 (cf. steel 27).

*Elastic Properties.* Load/extension curve almost linear to break. Hookean behaviour. Perfectly elastic to break.

*Specific Gravity.* 1.75–1.85.

*Effect of Moisture.* Nil.

*Flammability*. Not flammable.

*Effect of Age, Sunlight*. Nil.

*Effect of Chemicals, Solvents*. Inert. Hot air oxidation and strong oxidising agents (e.g. sodium hypochlorite) cause some erosion.

*Effects of Insects, Micro-organisms*. Nil.

## CARBON FIBRE IN USE

Carbon fibres are characterised by high strength and great stiffness against bending and twisting forces. Steel fibres, which approach nearest to carbon in stiffness, are four times as dense as carbon, and carbon fibres have a very much superior stiffness to weight ratio.

The breaking extension of carbon fibres is low, and unsupported fibres are brittle. Applications lie very largely in the field of composites for specialised uses, where the high cost of carbon fibre relative to steel, glass and other reinforcing fibres is of minimal consequence. Carbon fibre composites are used, for example, in aircraft structural components, in brakes and engines. They have proved of immense value in space vehicles, where weight reduction is at a premium. As carbon fibres become cheaper with increased production, they are finding their way steadily into more mundane applications such as golf-club shafts, fishing rods, boats and submarines, pressure vessels in the chemical and allied industries.

Special grades of carbon fibre are used in protective clothing fabrics, where their inertness and heat resistance serve them well. Carbon fibre fabrics may be washed at 40°C (HLCC 6) and dried with a short spin tumble dry or calender. They may be ironed and dry cleaned.

719

Printed and bound by CPI Group (UK) Ltd, Croydon, CR0 4YY

03/10/2024

01040437-0001